Solid-State Electronics

Solid-State Electronics

GEORGE B. RUTKOWSKI
Head, Electronics Department
Electronic Technology Institute, Cleveland

HOWARD W. SAMS & CO., INC.
INDIANAPOLIS · KANSAS CITY · NEW YORK

FIRST EDITION

SECOND PRINTING—1975

Copyright © 1972 by Howard W. Sams & Co., Inc., Indianapolis, Indiana 46268. Printed in the United States of America.

All rights reserved. Reproduction or use, without express permission, of editorial or pictorial content, in any manner, is prohibited. No patent liability is assumed with respect to the use of the information contained herein.

International Standard Book Number: 0-672-20801-6

Library of Congress Catalog Card Number: 77-131132

Preface

This book was developed for use in electronics technology courses in the theory and applications of solid-state components. It can be used to follow typical electricity courses. A knowledge of dc circuits, through series-parallel circuits, and of ac circuits including reactive components, is a sufficient technical prerequisite. Since many technology students are limited to, or progressing from, a high-school-algebra background, the manipulations and derivations of equations are kept within straightforward algebraic techniques. The equations identified with asterisks have more complicated derivations which are provided in the Appendix.

The modern technician is expected to take over technical responsibilities that formerly were assumed by engineers. Therefore the valuable electronics technician, often called an Associate Engineer, must have more than a superficial understanding of the popular solid-state components in use today. To help technicians meet this challenge this book was written, not only to present the fundamentals, but also to develop the student's ability to select proper design components for solid-state electronic circuits.

The typical technology student obtains better understanding and retention if his studies are applied immediately. Consequently, a modified programmed style is used throughout the book. Each point discussed is followed by at least one worked example. The student is encouraged to work each example problem before referring to its solution. The answers to the odd-numbered end-of-chapter problems are provided in the back of the book. These problems, with the examples, make this book suitable for self-study as well as classroom use.

Copies of the original manuscript were used as the text in solid-state electronics courses at the Electronic Technology Institute (ETI), Cleveland, Ohio. Revisions were made from the results of these extensive field tests before publication of this book. A laboratory manual, with experiments specifically designed to demonstrate the circuits and principles discussed in this book, is available and has similarly been field tested.

This book uses the term *current* to mean electron flow. Thus "current" flows in the same direction as do electrons—from negative to positive. This convention is a standard policy in the military and most civilian technical schools in electronics.

I am grateful for the helpful suggestions from my colleagues—faculty members at ETI. I am especially indebted to Jerome Oleksy, Head of the ETI Physics Department, who contributed to the revisions of the manuscript and who wrote the laboratory manual, *Solid-State Electronics Laboratory Manual*.

GEORGE B. RUTKOWSKI

Contents

Chapter 1

SEMICONDUCTOR MATERIALS 11
1-1 Silicon and Germanium Atoms   1-2 Silicon and Germanium Crystals   1-3 Conduction in Pure Silicon and Germanium   1-4 Doping Silicon and Germanium

Chapter 2

GERMANIUM AND SILICON DIODES 21
2-1 The PN Junction   2-2 The Forward Biased Diode   2-3 The Reverse Biased Diode   2-4 V-I Characteristics of Diodes   2-5 Methods of Manufacturing PN Junctions

Chapter 3

DIODE CHARACTERISTICS SIMPLIFIED AND APPROXIMATED 31
3-1 The Ideal Diode   3-2 Simplification of Diode Characteristics   3-3 Diode Characteristics Approximated   3-4 Load Lines on Diode Characteristics   3-5 Reverse Bias Resistance of Diodes

Chapter 4

APPLICATIONS OF DIODES 47
4-1 Superposition of DC and AC   4-2 The Half-Wave Rectifier   4-3 The Full-Wave Rectifier   4-4 The Bridge Rectifier   4-5 Voltage Doublers   4-6 Rectifier Assemblies   4-7 Logic Circuits

Chapter 5

THE ZENER DIODE 94
5-1 V-I Characteristics of the Zener Diode   5-2 Power and Derating   5-3 The Zener as a Voltage Regulator   5-4 Practical Considerations   5-5 The Zener as a Voltage Limiter   5-6 Zeners and Rectifiers

Chapter 6

THE JUNCTION TRANSISTOR 119
    6-1 The PNP and NPN Junctions   6-2 Forward Bias on Base-Emitter Junctions and Its Effects   6-3 More on How I_E, I_C, and I_B Are Related

Chapter 7

THE COMMON-BASE CONNECTION 131
    7-1 Cutoff and Saturation Defined   7-2 Collector Characteristics and Load Lines   7-3 Superposition of DC and AC Signals in the Common-Base Amplifier   7-4 V-I Characteristics and Dynamic Resistance of the Emitter-Base Junction   7-5 Determining Voltage Gain, Current Gain, and Power Gain   7-6 High-Resistance Signal Source vs Ga'n   7-7 Signal and Load Lines   7-8 AC Load Lines   7-9 The Optimum Operating Point   7-10 Operating Point Affected by Leakage I_{CBO}

Chapter 8

JUNCTION TRANSISTORS IN THE COMMON-EMITTER CONNECTION.. 182
    Part I: DC Considerations
    8-1 I_C vs I_B (Cutoff, Saturation, and Leakage Defined)   8-2 Emitter Feedback for Stability   8-3 Collector Feedback for Temperature Stability   8-4 Emitter Feedback with Two Supplies for Temperature Stability   8-5 Emitter Feedback and Base Bias with a Voltage Divider   8-6 Collector Characteristics of the Common-Emitter Circuit   8-7 Load Lines of the Various Common-Emitter Circuits
    Part II: AC Considerations
    8-8 V-I Characteristics and the Dynamic Resistance of the Base-Emitter Junction   8-9 Determining A_e, A_i, and A_p of the Simple Base Biased Common-Emitter Amplifier   8-10 Gain With the Common-Emitter Amplifier Using Emitter Feedback   8-11 Gain With the Common-Emitter Amplifier Using Collector Feedback   8-12 Gain With Other Common-Emitter Amplifiers   8-13 AC Load Lines   8-14 The Optimum Operating Point   8-15 Useful "Rule of Thumb" Design Procedure

Chapter 9

THE JUNCTION TRANSISTOR IN THE COMMON-COLLECTOR CONNECTION 275
    9-1 Methods of Biasing the Common-Collector Amplifier   9-2 Determining Ga'ns A_e, A_i, and A_p for the Common-Collector Amplifier   9-3 AC Loads and Load Lines

Chapter 10

THE SILICON-CONTROLLED RECTIFIER 292
  10-1 Fundamental Characteristics of the SCR   10-2 Anode Characteristics of the SCR   10-3 Gate Trigger Characteristics of the SCR   10-4 Applications of the SCR   10-5 Glossary of Letter Symbols

Chapter 11

OTHER MEMBERS OF THE THYRISTOR FAMILY 344
  11-1 The Unijunction Transistor   11-2 Applications of the UJT   11-3 The Bidirectional Triode Thyristor (Triac)   11-4 The Bidirectional Diode Thyristor (Diac)   11-5 The Turnoff Thyristor (GCS) and the Four-Layer (Shockley) Diode

Chapter 12

GENERAL CHARACTERISTICS OF THE FIELD-EFFECT TRANSISTOR .. 393
  12-1 Concepts of the Junction Field-Effect Transistor   12-2 Fundamentals of the Insulated Gate Field-Effect Transistor   12-3 Modes of Operation   12-4 Parameters of FET's   12-5 Methods of Biasing FET's for Use as Amplifiers   12-6 Small-Signal Voltage Gain in FET Amplifiers   12-7 Large Signals and Load Lines   12-8 Selecting Components for FET Amplifiers   12-9 The Effects of Temperature Changes   12-10 A Practical Consideration

Chapter 13

INTEGRATED CIRCUITS 447
  13-1 General Construction of IC's   13-2 Diffused Components   13-3 Component Isolation   13-4 The NOR Gate With a Diode OR Gate and Inverter   13-5 The Transistor NOR GATE   13-6 The Dual Two-Input Gate IC   13-7 The Astable Multivibrator   13-8 The Operational Amplifier   13-9 Practical Considerations and Applications of the Operational Amplifier   13-10 The Operational Amplifier as a Multiplier   13-11 The Operational Amplifier as a Summing Amplifier   13-12 Miscellaneous Information about IC's

Chapter 14

MISCELLANEOUS SOLID-STATE COMPONENTS 515
  14-1 The Tunnel Diode   14-2 The Varactor   14-3 Photoconductive Cells   14-4 Photovotaic (Solar) Cells   14-5 Photodiodes and Phototransistors   14-6 The Light-Emitting Diode   14-7 The Light-Activated Thyristor

Appendix

DERIVATIONS OF CONSTANTS AND EQUATIONS 535
    A-1 The Half-Wave Rectifier   **A-2** The Bridge Rectifier   **A-3** The Dynamic Emitter-to-Base Resistance   **A-4** The End Points of an AC Load Line   **A-5** Derivations of Eqs. 8-10 and 8-12   **A-6** Derivations of Eqs. 8-16 and 8-18   **A-7** Derivations of Eqs. 8-23 and 8-26   **A-8** Derivations of Eqs. 8-34 and 8-36   **A-9** Derivation of Eq. 11-8   **A-10** Selecting Bypass and Coupling Capacitors

ANSWERS TO REVIEW QUESTIONS 551

ANSWERS TO PROBLEMS 559

INDEX 611

1

Semiconductor Materials

1-1 SILICON AND GERMANIUM ATOMS

All matter is composed of atoms. Each atom has a nucleus that is surrounded by revolving electrons. The nucleus is composed of neutrons and protons. The neutrons have no charge and are said to be neutral. Each proton, however, possesses a positive charge. The nucleus therefore has a positive charge because it contains neutral and positively charged components. The electrons are negatively charged. Each electron has a charge equal in magnitude but opposite in polarity to the charge of a proton. These negative electrons revolve around the positive nucleus in different *shells* as shown in Fig. 1-1. The maximum number of electrons in each shell follows the pattern

$$2, 8, 18, \ldots, 2n^2 \qquad (1\text{-}1)$$

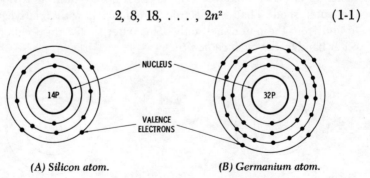

(A) Silicon atom. (B) Germanium atom.

Fig. 1-1. Electron shells of an atom.

where n is the number of the shell. The total number of electrons revolving about the nucleus of an atom is equal to the number of protons in the nucleus if the atom is neutral. For example, the silicon atom (Fig. 1-1A) has 14 protons in its nucleus, which means that when a silicon atom is neutral it has a total of 14 electrons in the shells around its nucleus. And according to the distribution given above, the first or innermost shell has $2n^2 = 2(1)^2 = 2$ electrons. The second shell has $2n^2 = 2(2)^2 = 8$ electrons. Since there are only four electrons more to account for, they revolve in the next, or third, shell. That is, the first, second, and outer shells of the silicon atom contain two, eight, and four electrons respectively. Similarily, it can be shown that since the germanium atom contains 32 protons in its nucleus, there are 32 electrons total in the shells. The distribution of these electrons from first to outer shell is 2, 8, 18, and 4.

It may be helpful to look at the neutral (uncharged) atom as one having a total number of electrons equal to the total number of protons, so that the charges of electrons and protons cancel each other. An atom is not neutral if it has *more* or *less* than the required complement of electrons. For example, a silicon atom with *five* electrons in the outer shell will have 15 total electrons. Since there are only 14 protons in the nucleus, the atom has a *negative* charge equal to the negative charge of the one extra electron. On the other hand, if the silicon atom should lose one of its outer electrons, leaving a total of 13, the atom has a positive charge equal to the charge of one proton. It should be pointed out here that when an electron is added to or taken from an atom, it is done in the outer shell. That is, electrons are moved, with relative ease, to and from the outer shell only. This outer shell is also called the outer band or the *valence band*.

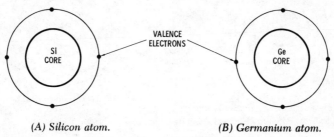

(A) *Silicon atom.* (B) *Germanium atom.*
Fig. 1-2. Valence band of an atom.

1-2 SILICON AND GERMANIUM CRYSTALS

In electrical and electronic theory we are interested primarily in the valence band because current in a material is supported by movement of electrons into and from the valence band, which will be shown later. From the electrical point of view, therefore, we can simplify the diagrams of the atoms we study as shown in Fig. 1-2. Note that only the valence band is shown. We may refer to the electrons in the band as *valence electrons*. The *core* represents the inner shells of electrons and the nucleus.

1-2 SILICON AND GERMANIUM CRYSTALS

If many silicon atoms are brought together under favorable conditions, a sharing of valence electrons takes place among the atoms as shown in Fig. 1-3. Note that the center atom shares its four valence electrons with four adjacent atoms. Likewise each of the four adjacent atoms shares its valence electrons with four atoms. These atoms are said to be in *covalent bonds*. The entirety of such atoms in covalent bonds is a crystal of silicon. Germanium atoms can similarly combine and form germanium crystal.

At low temperatures the pure silicon crystal is an insulator because there are few *free electrons* to support current. When silicon atoms are in covalent bonds, the valence electrons are held tightly to their *parent atoms*. A *free electron* is an electron that is not held by any atom and is free to migrate through the material. A *parent*

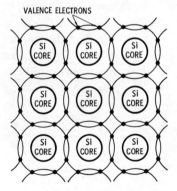
(A) *Silicon atoms in covalent bonds, sharing valence electrons.*

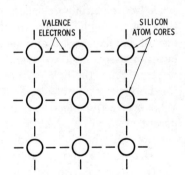
(B) *Simplified picture of silicon covalent bonds.*

Fig. 1-3. Covalent bonds of atoms.

atom is an atom to which a particular electron belongs in order to fill the atom valence band.

As the temperature of silicon crystal is increased, greater agitation of the electrons results. It may be helpful to visualize this if you assume that the electrons rotating around their parent atoms do so at a greater speed when the temperature is increased. This may cause electrons to occasionally break away from their parent atoms and they become free electrons. That is, they are free to move through the material to support current through it should a voltage be applied. In fact, as the temperature of silicon is increased, its conductivity increases because a greater number of valence electrons break away from their parent atoms and thus become free electrons. These electrons are called *thermally excited electrons.*

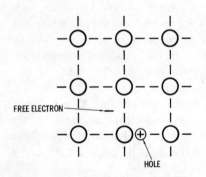

Fig. 1-4. Free electron and resulting hole (electron-hole pair) in silicon or germanium crystal.

A free electron broken away from its parent atom is shown in Fig. 1-4. Note that the atom that lost this electron is now missing one of its required four valence electrons. The place where this electron was is called a *hole*. That is, the hole represents a missing electron. Since the hole is an absence of an electron negative charge, the hole possesses characteristics of a positively charged particle. Whenever a free electron is produced by its breaking away from its parent atom in covalent bond, a hole appears also. This pair, free electron and hole, is called an *electron-hole pair*. Thus, we may say that as the temperature of silicon crystal is increased, more electron-hole pairs are produced.

Electron-hole pairs are also produced in ways other than temperature increases. If enough electromotive force (voltage) is applied to a crystal, electrons tend to be torn away from their

1-3 CONDUCTION IN PURE SILICON AND GERMANIUM

parent atoms. Also, light may produce more electron-hole pairs. It is on this latter principle that light-activated semiconductor devices work. As you progress through this text, you will see that most semiconductor devices, such as diodes, transistors, zeners, etc., are made basically from silicon or germanium crystals. The term *semiconductor* is used because it describes the "not-so-good" conductive properties of silicon and germanium crystals.

1-3 CONDUCTION IN PURE SILICON AND GERMANIUM

We may now see how current is supported by electron-hole pairs in silicon or germanium crystals. In Fig. 1-5, the negative signs

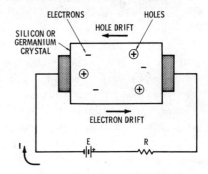

Fig. 1-5. Pure silicon or germanium crystals conduct current if temperature is high enough to produce electron-hole pairs.

represent the free electrons that have broken away from their parent atoms, and the holes that result from this are the circled positive signs. We can look at the voltage source E as if it is an electron pump. That is, source E tends to push electrons out of its negative terminal and draw them into its positive terminal, thus causing clockwise electron flow in the circuit. The source E injects electrons into the left side of the crystal and draws an equal number of electrons out of the right side of the crystal, resulting in a drift of free electrons from left to right within the crystal. The holes, on the other hand, behave like positively charged particles. They too are free to drift within the crystal, but with a voltage applied they drift in a direction opposite of the electron drift, as shown in Fig. 1-5. That is, holes being like positively charged particles are repelled by the positive terminal and attracted to the negative terminal of source E. This drift of holes within the crystal may be called *hole flow*. It may be further described as follows: As electrons are injected into the left side of the crystal shown in Fig. 1-5, some of these electrons will be captured by holes. An electron moving into a hole in this fashion is called *recombination*. As the hole and electron recombine in this way, the hole ceases to exist. Thus, as holes drift to the left each becomes filled with an electron and disappears. However, at the right side of the crystal, electrons are being drawn out of the crystal by the positive terminal of the source E. Some of the silicon or germanium atoms on the right side of the crystal are, in this way, forced to give up valence electrons. Thus, as holes disappear at the left, an equal number appear on the right. The holes on the right then drift to the left, eventually recombining with electrons.

Actually, the number of electron-hole pairs produced at normal room temperatures is quite small, especially in silicon crystals, and the amount of electron and hole drift is small too. Therefore, with few *charge carriers* (electrons and holes) in the crystal to support current through it, the crystal has relatively high resistance. As the temperature of the crystal is raised above room temperature, more electron-hole pairs are produced, which means that more charge carriers are available to support current, and the resistance of the crystal decreases. Of course, if the resistance of the crystal in the series circuit shown in Fig. 1-5 decreases, the current I in the leads increases. We may therefore conclude that increases or decreases in temperature of a crystal in a circuit can be detected by increases or decreases in the circuit current, which in the case of the circuit of Fig. 1-5 will cause increases or decreases in the voltage drop across the resistance R.

1-4 DOPING SILICON AND GERMANIUM

The number of free electrons and holes in pure semiconductor crystals is relatively small at typical room temperatures, and such crystals have little usefulness practically. The conductivity of a semiconductor can be increased by a process called *doping*. Silicon or germanium is doped by adding some other material to it during its fabrication. The atoms of this other material are called *impurity atoms*. These impurity atoms are usually of *pentavalent* or *trivalent* materials. A pentavalent material is one that has five valence electrons in each of its atoms. The trivalent material atoms have three valence electrons.

1-4 Doping Silicon and Germanium

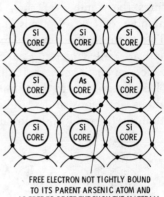

Fig. 1-6. N-type semiconductor with impurity donor atom shown in center.

FREE ELECTRON NOT TIGHTLY BOUND TO ITS PARENT ARSENIC ATOM AND IS FREE TO DRIFT THROUGH THE MATERIAL

Let us first consider silicon doped with a pentavalent impurity such as arsenic. Fig. 1-6 shows silicon crystal with an arsenic impurity atom. This arsenic atom shares four of its five valence electrons with adjacent atoms, but the remaining electron is free. That is, this extra or free electron is not tightly held to its parent atom and is *free* to support current through the material. Since this doped silicon has extra electrons, which are negative charge carriers, it is called *n-type* semiconductor. And since the arsenic atoms supply free electrons, they are called *donor atoms*. The presence of these extra free electrons significantly increases the ability of the semiconductor to conduct. We see in Fig. 1-7 that in the n-type semiconductor with a voltage applied, the free electrons drift toward the positive terminal and away from the negative terminal of the voltage source E. Though not shown, the semiconductor shown in Fig. 1-7 does have holes, too—produced when thermally excited electrons break away from their parent silicon atoms just as

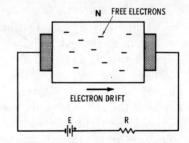

Fig. 1-7. Relatively large number of free electrons supplied by donor atoms in n-type semiconductor.

they did in pure silicon crystal. However, at normal room temperatures the number of such holes is very small compared to the number of free electrons that are provided by the donor atoms. Therefore, in the n-type material the electrons are in the majority and are called *majority carriers*. The holes, being in minority, are called *minority carriers*. It should be understood that while the free electrons are free to drift through the material, the donor and silicon atoms are more or less rigidly held in their places. That is, the atoms in the semiconductor are not free to drift.

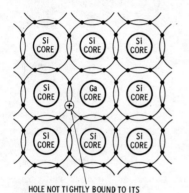

Fig. 1-8. P-type semiconductor with impurity acceptor atom shown in center of figure.

HOLE NOT TIGHTLY BOUND TO ITS PARENT GALLIUM ATOM AND IS FREE TO DRIFT THROUGH THE MATERIAL

In Fig. 1-8 we see a trivalent impurity, gallium in this case, in silicon crystal. Note that since the gallium atom has only three valence electrons, it cannot share one with each of the four adjacent atoms; there is an electron missing. Where an electron should be but is not, we have a hole. Thus, a trivalent impurity provides holes in the semiconductor.

These added holes allow electrons to move easily into them, and therefore the impurity atoms that provide these holes are called *acceptor atoms*. The holes behave like positively charged particles that are free to drift within the material. Their drift is toward the negative terminal and away from the positive terminal of the voltage source E, as shown in Fig. 1-9.

Holes support current through the semiconductor as was explained in Sec. 1-3. That is, for every electron that moves into a hole at the left side of the semiconductor of Fig. 1-9, an electron leaves at the right side and a hole appears in its place. Thus, the holes appearing at the right then drift to the left to be filled even-

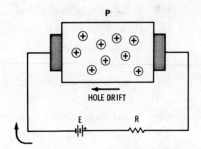

Fig. 1-9. P-type semiconductor has relatively large number of holes supplied by trivalent impurity.

tually by electrons. Though not shown in Fig. 1-9, the semiconductor has a few free electrons, too, which are thermally excited electrons that occasionally break away from their parent silicon atoms. At normal room temperatures, however, the number of such electrons is small compared to the number of holes present. Therefore, the holes are the *majority carriers,* and the electrons are the *minority carriers.* A semiconductor doped with trivalent impurity is called *p-type* because the majority carriers are holes which act like positively charged particles.

It should be understood that the n-type semiconductor *does not* possess a negative charge. Since it is composed of two types of atoms, silicon and pentavalent atoms, both types of which individually are neutral (uncharged), the n-type material likewise possesses no charge. For example, the arsenic atom is neutral with five valence electrons because it has a total number of electrons rotating the nucleus equal to the number of protons in the nucleus, while, similarly, the silicon atom with four valence electrons is neutral for the same reason. Therefore, neutral arsenic and silicon atoms together in n-type material have no charge. In the same way, a p-type semiconductor has no charge. Trivalent impurity atoms like gallium atoms are neutral with three valence electrons. These uncharged gallium atoms combined with uncharged silicon atoms, as in p-type material, cause the combination to be neutral.

REVIEW QUESTIONS

1-1. What is the *core* of a silicon atom?

1-2. Define a *neutral atom.*

1-3. Explain the difference between a *valence electron* and a *free electron.*

1-4. What is a *parent atom?*

1-5. How would you define a *hole?*

1-6. Define the *electron-hole* pair. How are electron-hole pairs produced?

1-7. What can be done to a semiconductor to provide it with more free electrons?

1-8. What does the term *recombination* mean?

1-9. What is a *doped* semiconductor?

1-10. What effect does doping silicon with a trivalent impurity have?

1-11. In what directions do free electrons and holes drift in a semiconductor when a voltage is applied?

1-12. Is pure silicon crystal a good conductor at room temperature?

1-13. In what ways can silicon crystal conductivity (ability to conduct current) be increased?

1-14. What are pentavalent and trivalent impurities from the point of view of their atomic structures?

1-15. Name one material that can be used as a pentavalent impurity, and name one that can be used as a trivalent impurity.

1-16. Compare *donor* and *acceptor* atoms.

1-17. What is an n-type semiconductor?

1-18. With no voltage applied to it, what charge does a piece of n-type semiconductor have?

1-19. What is a p-type semiconductor?

1-20. What charge does a piece of p-type semiconductor have in the absence of applied voltage?

1-21. What are the majority carriers and minority carriers in an n-type semiconductor?

1-22. What are the majority and minority carriers in a p-type semiconductor?

1-23. When a voltage is applied to a semiconductor, which of the following are free to drift: free electrons, holes, atoms, or parent atoms?

1-24. What word describes the merging of a free electron and a hole?

2

Germanium and Silicon Diodes

2-1 THE PN JUNCTION

The diode is among the simplest of semiconductor components. It is a junction of n-type and p-type semiconductors and may be referred to as a pn junction (see Fig. 2-1). If the junction is made of silicon p and n semiconductors, it is a silicon diode. If germanium p and n semiconductors are used, their junction forms germanium diode. The diode is often called a *unilateral conductor*. This means that it conducts well in one direction and not well in the other direction. Diodes are important components and are used

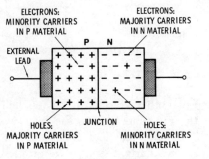

(A) *Physical construction.*   (B) *Schematic symbol of diode.*

Fig. 2-1. Pn junction diode with no bias voltage.

21

in almost all types of electronic equipment. Diodes are used as rectifiers in power supplies and as switches in logic circuits of computers. Home-entertainment equipment, industrial equipment, space and defense equipment, to name only a few, have diodes in their electronic circuits. The pn junction in Fig. 2-1A has no external voltage applied; therefore, no current flows in the external leads, and no current flows across the junction.

Note in Fig. 2-1 that the p material has many holes (majority carriers) and few free electrons (minority carriers). On the other hand, the n material has many free electrons (majority carriers) and few holes (minority carriers).

2-2 THE FORWARD BIASED DIODE

A forward biased diode is shown in Fig. 2-2. Note in Fig. 2-2B that the negative terminal of the voltage source is connected to the n material and the positive terminal is connected to the p material.

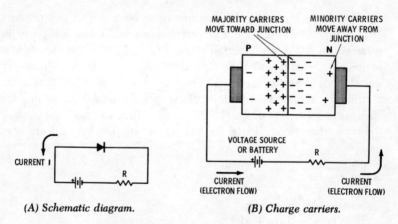

Fig. 2-2. Forward biased diode.

This causes the majority carriers of both materials to be repelled by the source terminals toward the junction; that is, the negative terminal of the source repels the majority carriers (electrons) in the n material to the left in this case. Similarly, the positive terminal of the source repels the majority carriers (holes) of the p material to the right. This causes the electrons of the n material to recombine at the junction with the holes of the p material. A continuous

2-3 THE REVERSE BIASED DIODE

recombination of these electrons and holes at the junction supports current across it, which causes current in the external leads.

A detailed account of this action may be given as follows: Electrons flow from the negative terminal of the voltage source into the right side of the n material. These electrons then drift through the n material toward the junction. At the same time, electrons are drawn from the left side of the n material by the source voltage as shown. As the electrons leave the p material, holes appear at the left side. These holes drift to the right toward the junction. At the junction the electrons and holes combine, and, as charge carriers, they disappear. As these charge carriers disappear at the junction, new ones, electrons and holes, appear at the right of the n material and at the left of the p material respectively. This action is continuous as long as forward bias is applied.

2-3 THE REVERSE BIASED DIODE

A reverse biased diode is shown in Fig. 2-3. In this case, the source terminals attract the majority carriers in both materials and thus draw them away from the junction. In this case, the majority carriers do not recombine and therefore do not support current across the junction, and nearly zero current flows in the external leads and resistance R.

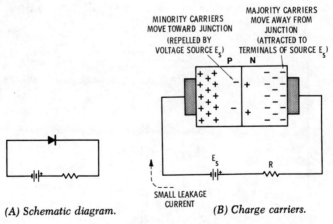

(A) Schematic diagram. (B) Charge carriers.

Fig. 2-3. Reverse biased diode.

Actually, some *leakage current* does flow across the reverse biased junction supported by recombinations of minority carriers at the junction. As mentioned in Sec. 1-4, minority carriers are holes in the n-type material and electrons in the p-type material. In the reverse biased junction, minority carriers are repelled by the voltage source toward the junction. At the junction they recombine and thus support a current across the junction, which results in a usually small leakage current in the external leads. Since the number of minority carriers at room temperature is relatively small, the leakage current is small, too. Increases in operating temperature, however, generate more electron-hole pairs, and this

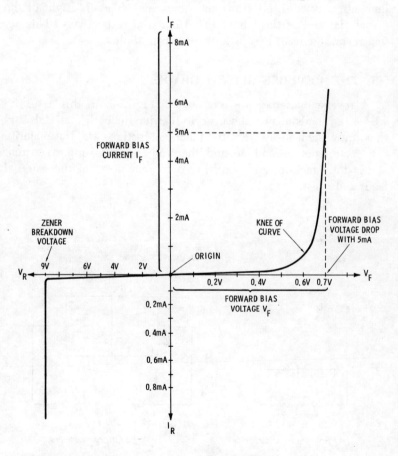

Fig. 2-4. Typical silicon diode *V-I* characteristic.

2-4 V-I CHARACTERISTICS OF DIODES

action will increase the number of minority carriers and the leakage current. Such increases in leakage current with increased temperature is especially noticeable in germanium diodes.

2-4 V-I CHARACTERISTICS OF DIODES

V-I characteristics are *volts-versus-amperes* characteristics as shown in Fig. 2-4. Note that this silicon diode does not conduct current well with less than a 0.5 V forward voltage V_F. Typically, a silicon diode has about a 0.7-V forward voltage V_F across it when conducting I_F; that is, when it is forward biased in a practical circuit. More will be said about this later.

When this silicon diode is reverse biased we see from the characteristic in Fig. 2-4 that with reverse bias voltages V_R up to about 8.5 V, the reverse current I_R increases in magnitude negligibly. However, at $V_R = 9$ V, the reverse current I_R suddenly increases. That is, this diode acts like a high resistance with reverse bias voltages V_R up to 9 V approximately. At 9 V the diode resis-

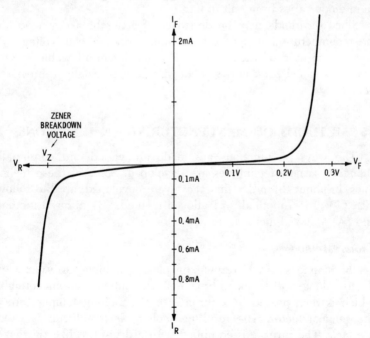

Fig. 2-5. Typical germanium *V-I* diode characteristic.

tance suddenly decreases, resulting in higher reverse current I_R. This sudden change is called the *zener effect*. The voltage at which this change occurs is called the *zener breakdown voltage* or simply *breakdown voltage* or *zener voltage*. This zener effect is caused by the reverse bias voltage V_R becoming strong enough to tear many valence electrons from their parent atoms and thus producing suddenly many electron-hole pairs in both p and n materials. This increases the number of minority carriers that support the relatively large reverse current at the zener breakdown voltage. The zener breakdown voltage value may be different with diodes of different types. That is, some diodes may have larger than a 9 V zener voltage, while others may have less. With ordinary diodes, if zener breadown occurs, it may destroy the diode. However, *zener diodes* are made to operate normally in the zener region of the *V-I* characteristics as will be shown later.

V-I characteristics of a germanium diode are shown in Fig. 2-5. Typically, its forward bias voltage drop is on the order of 0.2 V to 0.3 V. As with silicon, the zener breakdown voltage of the germanium diode varies from unit to unit.

Since the diode may be destroyed if zener breakdown occurs, the manufacturers specify the maximum reverse bias voltage V_R that can be used on their diodes. This maximum allowable reverse voltage is called *peak inverse voltage*, which is usually abbreviated *PIV*.

2-5 METHODS OF MANUFACTURING PN JUNCTIONS

The ways pn junctions are made are varied and continually changing as new techniques are developed. Some of these techniques, in about the order that they were developed, are the following: (1) grown method, (2) alloyed method, (3) diffused method, and (4) epitaxial method.

Grown Method

Pn junctions can be grown by placing a high-purity semiconductor along with some p-type impurity into a ceramic crucible which is then heated in a furnace to the melting temperature of the semiconductor. The resulting molten semiconductor is called the *melt*. The furnace is equipped with a drill-press-like rig that is capable of providing vertical and rotating motion on a vertical

2-5 METHODS OF MANUFACTURING PN JUNCTIONS

rod which can thus be dipped into and out of the melt. A small piece of crystal semiconductor, called a *seed*, is placed on the end of the rod and is then lowered into the melt. After touching the surface of the melt, the rod is drawn upward and rotated. Due to surface tension, the rod pulls along with it some of the melt, which subsequently cools and hardens. The crystal seed causes the hardening melt to crystallize also. That is, more crystal grows from the seed. This crystal is p type because initially p-type impurity was placed into the crucible. If during the pulling process n-type impurities are added to the melt, these n impurities counteract the p impurity already there, causing an n-type semiconductor crystal to be formed as the rod is raised. The rotating action tends to mix the impurities into the melt properly. In this way the semiconductor crystal column extracted from the melt can be made to have sandwiched p-type and n-type layers. Such a column can then be cut into many pn junctions.

Even with both p and n types of impurities mixed together in the melt, pn junctions can be made by a *rate grown method*. In this method the ceramic rod is pulled out of the melt at a varying rate. The way this works is based on the fact that p-type impurities stay molten at temperatures that harden the semiconducor material and the n-type impurity. Thus, if the rod is extracted slowly, the p-type impurity tends to run down and not get caught in the hardening process, and an n-type semiconductor forms. If the extraction rate is increased, the melt cools and hardens before the p-type impurities can run out. These p-type impurities dominate and make a p-type semiconductor. So if the rate of extraction is varied, pn junctions form in the column of the semiconductor being extracted from the melt.

Alloy Method

In this process, we can start with a piece of n-type semiconductor crystal and *fuse* or *alloy* a p-type region into it. This is done by first placing a small pellet of p-type impurity on an n-type semiconductor crystal as shown in Fig. 2-6A. This crystal and pellet are then heated so that the pellet melts into and combines with the crystal. The p-type impurities of the pellet dominate in the combined region, causing this region to become a p-type semiconductor as shown in Fig. 2-6B. This is called the *alloy method* of manufacturing pn junctions.

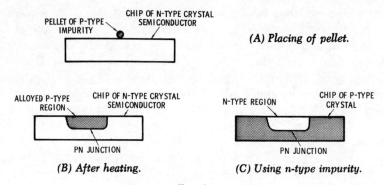

Fig. 2-6. Alloyed pn junctions.

We could have started instead with p-type semiconductor crystal and an n-type impurity pellet, which of course results in an n-type region alloyed into the p-type crystal as shown in Fig. 2-6C. In either case, pn junctions are formed.

Diffusion Method

This method starts with a thin piece of semiconductor, called a *wafer* or *chip*. Part of this wafer surface is exposed to vaporized (gaseous state) impurities and high temperatures. The impurity atoms diffuse through all exposed surface of the wafer. By suitable masking of the wafer and by controlling the temperature and time of exposure to the vaporized impurities, the location and amount of doping are controlled.

Thus, we can start with a wafer or chip of p-type crystal and lay a mask with a hole in it over the crystal. Both crystal and mask are then placed in a very hot environment containing vaporized n-type impurity. The hole in the mask exposes a portion of the crystal to the impurity vapor. In the region under this exposed area, n-type semiconductor material forms by diffusion. The result can be a pn junction as shown in Fig. 2-6C.

Epitaxial Method

This technique begins with a wafer or chip of very lightly doped semiconductor called a *substrate*. P-type and n-type regions can be diffused into the substrate as needed. For example, an n-type region can be diffused into the substrate, through a hole in a mask, as shown in Fig. 2-7A. Then a mask with a smaller hole is

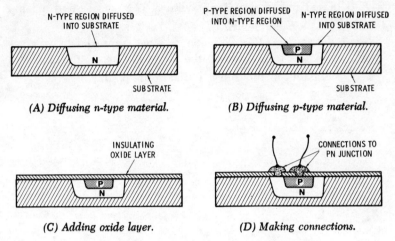

Fig. 2-7. Epitaxial method of forming pn junction.

placed over the wafer, and a p-type region is diffused into the n-type region as shown in Fig. 2-7B. Next, an oxide layer is formed over the entire surface as shown in Fig. 2-7C. Holes are then cut or etched into the oxide through which metal connections are made as shown in Fig. 2-7D. Obviously, the leads on the metal connection see a pn junction. More discussion of this method is given in a later chapter.

REVIEW QUESTIONS

2-1. In the reverse biased pn junction the majority carriers move (a) (toward), (away from) the junction while the minority carriers move (b) (toward), (away from) the junction.

2-2. In the forward biased pn junction recombinations of _____ carriers support current across the junction.

2-3. The silicon diode has a (larger), (smaller) forward voltage drop than the germanium diode.

2-4. A sudden increase in reverse bias current I_R at some reverse bias voltage V_R is called the _____ effect.

2-5. What factors may affect the reverse bias leakage current?

2-6. What is the meaning of the term *PIV*?

2-7. If a diode with the characteristic shown in Fig. 2-4 is forward biased in a series circuit with a 12 V source and a lamp rated 12 V, 6 W, will the lamp light? Why?

2-8. Referring to Ques. 2-7 above, what approximate voltages would you expect to find across the lamp and across the diode?

2-9. Name four methods of manufacturing pn junctions.

3

Diode Characteristics Simplified and Approximated

3-1 THE IDEAL DIODE

The ideal diode is a fictitious but useful component. Assuming that it exists helps one to analyze and easily build many types of electronic circuits. It has zero resistance (acts like a short) when forward biased and infinite resistance (acts like an open) when reverse biased. Its characteristic is shown in Fig. 3-1.

Note in Fig. 3-1 that forward current I_F flows, with forward bias, causing no forward voltage drop across the diode. The current I_F is limited by the circuit resistance R. Since the ideal diode acts like a short, the full source voltage E is across R, and therefore $I_F = E/R$ by Ohm's law.

On the other hand, with reverse bias, the ideal diode acts like an open as shown. Note in the ideal diode characteristic curve that the reverse current I_R is zero regardless of the reverse voltage V_R value and that there is no zener breakdown voltage.

To be more specific for example, note Fig. 3-2A. If we assume that the diode in this circuit is ideal, it is forward biased and acts like a short ($V_F = 0$). The full 12 V of the source appears across the resisance R. The forward bias current is, by Ohm's law,

$$I_F = E/R = 12/60 = 0.2 \text{ A}$$

If the ideal diode is reverse biased as shown in Fig. 3-2B, it acts like an open, and the full 12-V source voltage appears across the diode. The current $I_R = 0$, and the voltage drop across the resistance R is zero, too.

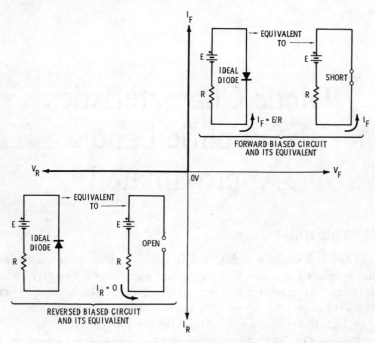

Fig. 3-1. *V-I* characteristic of ideal diode.

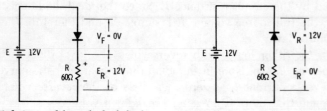

(A) *With forward biased ideal diode.*   (B) *With reverse biased ideal diode.*

Fig. 3-2. Basic ideal diode circuit.

While the real diodes actually used in circuits are not ideal, but instead typically have characteristics like those in Fig. 2-4 and Fig.

3-2 SIMPLIFICATION OF DIODE CHARACTERISTICS

2-5, it is convenient for many practical purposes to assume that they are ideal. As you will see, when the source voltage E is much greater than the forward voltage drop V_F of a real diode, and when the circuit resistance R is relatively high, it becomes quite practical to assume that real diodes are ideal.

3-2 SIMPLIFICATION OF DIODE CHARACTERISTICS

In Sec. 2-4 and Fig. 2-4, we learned that the conducting silicon diode has about a 0.7 V forward voltage drop for I_F values above the knee of the curve. For practical purposes in many cases, we may assume that the silicon diode drops a constant 0.7 V when

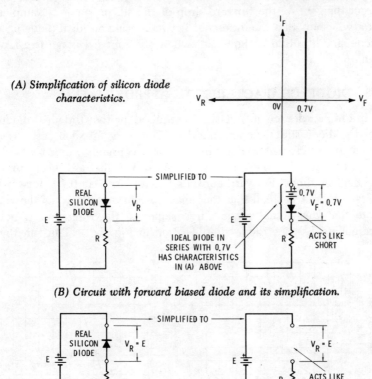

(A) *Simplification of silicon diode characteristics.*

(B) *Circuit with forward biased diode and its simplification.*

(C) *Circuit with reverse biased diode and its simplification.*

Fig. 3-3. Simplification of diode characteristics.

forward biased. Thus, the real diode characteristic (Fig. 2-4) may be *simplified* to the characteristic shown in Fig. 3-3A. This simplified characteristic shows that the silicon diode is assumed to be nonconducting with forward bias voltages up to 0.7 V, and that at 0.7 V the diode acts like a short circuit. Also, reverse bias voltages cause no reverse current I_R.

For example, returning to the circuit in Fig. 3-2A with the forward biased diode, we would show $V_F = 0.7$ V instead of zero if we use the simplified V-I curve, Fig. 3-3A. Likewise, then the resistor voltage $E_R = 11.3$ V instead of the full 12 V. Actually, the difference in these answers and the previous answers with the ideal diode are small, because in this case the source voltage 12 V is significantly larger than 0.7 V, the forward drop of the silicon diode. Examples later will show that the difference between using an ideal diode or a diode simplification is significant where the source voltage is a low value.

3-3 DIODE CHARACTERISTICS APPROXIMATED

If a higher degree of precision is required, neither the ideal diode nor the diode simplification may be satisfactory. In such cases, the actual diode characteristics can be more accurately *approximated*. For example, Fig. 3-4A shows a typical V-I forward bias curve for a silicon diode. Forward currents I_F with less than 0.4 V forward voltage V_F are so small that they may be considered zero as shown. Note that this actual diode curve slopes to the right and is not straight up like the diode simplification in Fig. 3-3A. Thus, in the

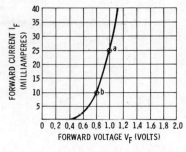 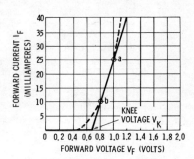

(A) *Actual characteristics of typical silicon diode.*   (B) *Approximation of typical silicon diode characteristics.*

Fig. 3-4. Silicon diode characteristics.

3-3 DIODE CHARACTERISTICS APPROXIMATED

real diode, the forward voltage V_F actually increases as the forward current I_F increases—much like in a resistive circuit where larger currents through a resistance causes more voltage across it. It is sometimes useful to approximate the actual characteristics of the diode with a straight line as shown in Fig. 3-4B. The exact position of the straight line is not extremely critical but should be drawn to have at least two points on it, not too close together, that are also on the actual characteristic curve, such as points a and b.

Note on the approximated characteristic (Fig. 3-4B) that the forward voltage V_F is about 0.7 V with very small forward currents I_F (less than 5 mA), while V_F increases to 1.2 V with $I_F = 40$ mA.

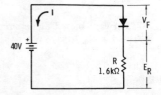
Fig. 3-5. Forward biased silicon diode circuit with $I = E_R/R = 25$ mA.

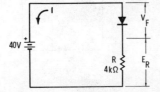
Fig. 3-6. Forward biased silicon diode circuit with $I = E_R/R = 10$ mA.

Thus, over the recommended range of forward currents, 0 to 40 mA in this case, the forward drop is never much more than about 1 V. This V_F is quite small compared to typical circuit operating voltages.

For example, suppose that we use a diode whose characteristic is shown in Fig. 3-4A, in the circuit of Fig. 3-5. Since the diode is forward biased and drops only a small voltage, most of the 40 V source is dropped across the resistor R. Thus, the current I in this circuit may be approximated by Ohm's law as follows:

$$I \cong \frac{E_R}{R} = \frac{40 \text{ V}}{1.6 \text{ k}\Omega} = 25 \text{ mA}$$

Going back to the characteristic in Fig. 3-4B, note that with about 25 mA forward current, the diode forward voltage drop is about 1 V. Thus, more accurately the drop across R is about 39 V, but this makes a negligible difference. That is, more accurately,

$$I = E_R/R = 39 \text{ V}/1.6 \text{ k}\Omega = 24.4 \cong 25 \text{ mA}$$

Similarly, if $R = 4$ kΩ is used in the circuit (Fig. 3-6), by Ohm's law and assuming that the full 40 V of the source is across R,

$$I = \frac{E_R}{R} \cong \frac{40 \text{ V}}{4 \text{ k}\Omega} = 10 \text{ mA}$$

From the characteristics of Fig. 3-4, $V_F = 0.8$ V when $I_F = 10$ mA. Therefore, more precisely, $E_R = 39.2$ V.

Example 3-1 The diode in the circuit of Fig. 3-7 has the characteristic shown in Fig. 3-9. Find the voltage across the resistor with each of the following conditions: (a) Assume that the diode is ideal. (b) Use simplification of diode characteristics. (c) Use approximation of diode characteristics.

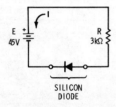

Fig. 3-7. Forward biased diode circuit with $I = E_R/R = 15$ mA.

Answers 3-1 (a) $E_R = 45$ V.—An ideal diode acts like a short circuit, thus placing the full 45 V of the source E across the resistor. (b) $E_R = 44.3$ V.—Using the simplified characteristics of the silicon diode we assume that the diode forward voltage drop V_F is a constant 0.7 V. The equivalent circuit in this case is shown in Fig. 3-9. (Note that a simplification of a real diode is an ideal diode in series with 0.7 V.) (c) $E_R = 44.2$ V.—Assuming that nearly the full 45 V applied to this circuit appears across the resistance R because the diode is forward biased, the circuit current $I \cong E_R/R = 45$ V/3 kΩ = 15 mA. Now referring to the characteristic of Fig. 3-8, we note that with $I_F = 15$ mA, voltage $V_F = 0.8$ V. Therefore $E_R = 45 - 0.8 = 44.2$ V.

As you can see, the differences between answers (a), (b), and (c) are very small. This is so because the source voltage E is much greater than the forward voltage drop V_F and the circuit resistance R is much greater than the diode forward biased dc resistance.

Example 3-2 Referring to Ex. 3-1, rework the problems with a reduced source voltage $E = 7.5$ V but with resistance R still equal to 3 kΩ.

Answer 3-2 (a) $E_R \cong 7.5$ V.—Again assuming that the diode is ideal, the full source voltage appears across the resistor R.

3-3 DIODE CHARACTERISTICS APPROXIMATED

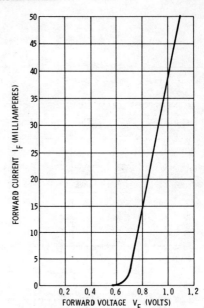

Fig. 3-8. Typical characteristics of a silicon diode as supplied by the manufacturer.

(b) $E_R \cong 6.8$ V.—Assuming that the diode characteristic can be simplified to Fig. 3-3A, we have a constant 0.7 V drop across the conducting diode and $E_R \cong 7.5 - 0.7 = 6.8$ V. (c) $E_R \cong 6.8$ V. —Initially assuming that the conducting diode drops a negligible voltage V_F, therefore $E_R \cong 7.5$ V. By Ohm's law the circuit current $I = E_R/R \cong 7.5$ V/3 kΩ = 2.5 mA. Now referring to Fig. 3-8, we see that this diode has $V_F = 0.7$ V when $I_F = 2.5$ mA. This means that, more accurately, $E_R \cong 7.5 - 0.7 = 6.8$ V.

Note now that the answer for (a) is different than answers (b) and (c) by a significant percentage. Since the methods used in solutions (b) and (c) are more accurate, the answer (a) might be unacceptable. You should keep in mind that with relatively small

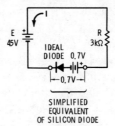

Fig. 3-9. Simplified equivalent of silicon diode has characteristics of Fig. 3-3A.

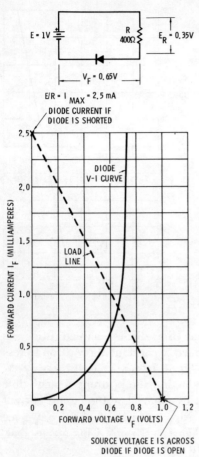

Fig. 3-10. Expanded characteristic and load line used when E and R are relatively low values.

values of source voltage E, a real diode in a circuit may not be assumed to be ideal unless significant error can be tolerated.

Example 3-3 Referring to Ex. 3-1, rework the problems with $E = 45$ V but with a reduced R. Let $R = 750$ Ω.

Answers 3-3 (a) $E_R = 45$ V.—Forward biased diode acts like a short circuit. (b) $E_R = 44.3$ V.—Diode simplified has a constant $V_F = 0.7$ V. (c) $E_R \cong 43.8$ V.—In this case the circuit current $I = E_R/R \cong 45/750 = 60$ mA. Referring to Fig. 3-8, we can only estimate that this diode has about 1.2 V forward voltage V_F when I_F is as high as 60 mA. In this case, therefore, $E_R = 45 - 1.2 = 43.8$ V.

The most accurate answer is (c) because we used the most accurate approximation of the real diode. If the resistance R were made still smaller, it could be shown that answer (c) would differ from (a) and (b) even more. However, if the resistance R of the circuit is too low, the current I_F may be too high; that is, it may exceed the maximum value allowed as specified by the manufacturer.

3-4 LOAD LINES ON DIODE CHARACTERISTICS

Suppose that both the applied voltage E and the resistance R are reduced significantly in the circuit of Fig. 3-7, say $E = 1$ V and $R = 400$ Ω, what is the resistor voltage drop E_R? In this case, since the applied voltage E is not much greater than the typical forward drop V_F of the diode, neither assuming that the diode is ideal nor using simplified diode characteristics will suffice if reasonable accuracy is to be expected. In this case we will use this diode characteristic expanded as shown in Fig. 3-10. Note that Fig. 3-10 is the lower portion of Fig. 3-8 with the vertical scale expanded (amplified). Graphs of characteristics like Figs. 3-8 and 3-10 are supplied by the manufacturers of diodes.

On the horizontal (voltage) scale of Fig. 3-10 an X is placed at 1 V, which is E, the source voltage value, and it is the maximum possible voltage that can appear across the diode if the diode were to open. On the milliampere scale an X is placed at 2.5 mA, which is the maximum possible current that can flow through the diode if it were to become shorted ($I_{max} = 1$ V/400 Ω $= 2.5$ mA). A straight line drawn through these two X's is called the *load line*. The point where this load line and diode curve cross is the *operating point*. The operating point graphically shows us the actual values of V_F and I_F of the circuit. In this case, directly below the operating point, we read $V_F \cong 0.65$ V, and to the left of the point we read $I_F \cong 0.875$ mA. Therefore, the drop across the resistor is $E_R \cong E - 0.65 = 0.35$ V. The application of load lines and operating points will come up again in transistor theory.

3-5 REVERSE BIAS RESISTANCE OF DIODES

Expanded reverse bias characteristics of the typical semiconductor diode are shown in Fig. 3-11. Note that with a small reverse

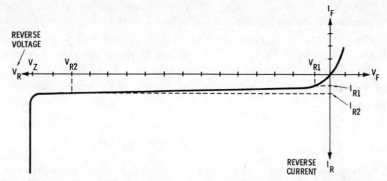

Fig. 3-11. Typical reverse bias characteristic of semiconductor diode.

voltage V_{R1}, a reverse current I_{R1} flows. This current I_{R1} may be considered thermally generated. This means that even at room temperature, about 25°C, electron-hole pairs are generated by heat. These electron-hole pairs provide minority carriers in both the p- and n-type materials that support leakage current I_{R1} across the junction. This thermally generated leakage I_{R1} is very small at room temperature. In silicon diodes it may be as low as 100 pA (100×10^{-12} A). Since I_{R1} is thermally generated, it varies with temperature, roughly doubling for every 6°C temperature rise in silicon diodes and roughly doubling for every 10°C rise in temperature in germanium diodes.

The reverse current I_R is also increased by increases in reverse voltage V_R. Note that as reverse voltage increases from V_{R1} to V_{R2}, the reverse current increases from I_{R1} to I_{R2}. Thus, the reverse biased diode acts like a resistance; that is, as voltage increases, current increases. However, since the reverse current typically increases only a few microamperes with several volts increase of V_R, the reverse resistance of the diode is on the order of many kilohms or megohms. The reverse resistance of silicon diodes may be 10 to 100 times greater than that in germanium diodes. This resistance is so large, especially in silicon diodes, that it is assumed to be infinite in most practical applications. That is, it is usually reasonable to assume that the reverse biased diode acts like an open. In high-resistance circuits, however, the reverse resistance of diodes may become a significant factor. For example, suppose that you have a 50 V source, a diode, and a 50 MΩ resistance in series. What voltage will be across the resistance when the diode

3-5 REVERSE BIAS RESISTANCE OF DIODES

is reverse biased? Typically, silicon diode specification sheets may show that the diode has 2 μA leakage I_R at room temperature with 100 V reverse bias. Thus the diode reverse resistance may be said to be on the order of 100 V/2 μA or about 50 MΩ. The 50 V source therefore sees the reverse biased diode and the resistance as two 50 MΩ resistances in series, and 25 V is dropped across each. We may conclude that if the circuit resistance is near the value of the diode reverse resistance, the reverse biased diode cannot be assumed to be an open circuit.

In power (high-current) circuits, the trends are: an increasing popularity of silicon diodes and a decreasing use of the germanium and other types of diodes. Silicon operates well at high temperatures that destroy other semiconductor materials. Silicon diodes have low forward bias resistance and high reverse resistance and are able to conduct large forward currents.

Work the following eight examples (Exs. 3-4 through 3-11), assuming that the diodes are ideal:

Example 3-4 Referring to Fig. 3-12, if the applied voltage $E = 10$ V and terminal x is negative with respect to y (negative terminal of E is connected to x and the positive connected to y), what are the voltages across the two resistors?

Answer 3-4 $E_1 = 6$ V, $E_2 = 4$ V.—The diode in this case is reverse biased and acts like an open. Thus, as the source E sees it, R_1 and R_2 are in series. Therefore,

$$E_1 = \frac{ER_1}{R_1 + R_2} = \frac{10(300)}{300 + 200} = 6 \text{ V}$$

or $E_1 = IR_1 = 20$ mA × 300 Ω = 6 V, where $I = E/(R_1 + R_2)$. And thus $E_2 = E - E_1 = 10 - 6 = 4$ V.

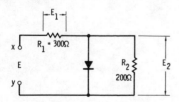

Fig. 3-12. Finding voltage drops E_1 and E_2.

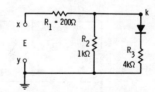

Fig. 3-13. Finding voltage at point k with respect to y.

Example 3-5 Referring to Ex. 3-4, what are the voltages E_1 and E_2 if the polarity of the 10 V source is changed? That is, $E = 10$ V, x positive with respect to y.

Answer 3-5 $E_1 = 10$ V, $E_2 = 0$ V.—With x positive with respect to y, the diode is forward biased. Since we assume that it is ideal, it acts like a short and thus shorts out resistor R_2. The full 10-V source voltage appears across R_1.

Example 3-6 Referring to Fig. 3-13, if the source $E = 12$ V and x is negative with respect to y, what is the voltage at point k with respect to ground (y)? Call it E_k.

Answer 3-6 $E_k = -10$ V.—With source E negative at x with respect to ground (y), the diode is reverse biased and it acts like an open. Thus, the branch containing the diode and R_3 is open, and as the source E sees it, R_1 and R_2 are in series. The voltage E_2 across R_2 is the voltage across the diode and is the voltage across point k and ground:

$$E_2 = \frac{ER_2}{R_1 + R_2} = \frac{12(1000)}{1200} = 10 \text{ V}$$

The other 2 V is across R_1. See Fig. 3-14. The voltage at k with respect to ground can be found by "walking" through the circuit from point k to ground, adding voltages (algebraically) as you go. For example, "walking" from k to ground through R_2, as shown with the broken line in Fig. 3-14, we "walk" from the *negative* side of R_2 to its positive side. Thus $E_k = -10$ V. Similarly, if we "walk" from k to ground in the dotted line path—that is, through R_2 and the source E—we first "walk" through R_2 from *positive* to negative 2 volts $(+2 \text{ V})$, then from *negative* to positive through the source (-12 V). The sum of these voltages is the k to ground voltage or $E_k = +2 \text{ V} - 12 \text{ V} = -10 \text{ V}$.

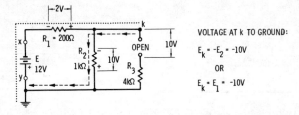

Fig. 3-14. Equivalent circuit of Fig. 3-13 for x negative 12 V with respect to y.

3-5 REVERSE BIAS RESISTANCE OF DIODES

Example 3-7 Referring to Ex. 3-6, what is the voltage at k to ground when the applied voltage E, in the circuit of Fig. 3-13, is positive 12 V at x with respect to y?

Answer 3-7 $E_k = 9.6$ V.—In this case, the diode is forward biased and acts like a short. The equivalent circuit becomes Fig. 3-15. This effectively places R_2 and R_3 in parallel, and their com-

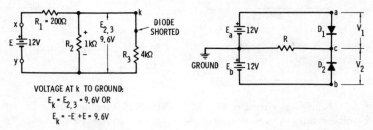

Fig. 3-15. Equivalent circuit of Fig. 3-13 for x 12 V with respect to y.

Fig. 3-16. Finding diode voltage drops and voltage at c.

bined resistance $R_{2,3} = \dfrac{R_2 R_3}{R_2 + R_3} = 800$ Ω. As seen by the source E, the total resistance is $R_t = R_1 + R_{2,3} = 1000$ Ω. The voltage across the parallel resistors is

$$E_{2,3} = \frac{E R_{2,3}}{R_t} = \frac{12(800)}{1000} = 9.6 \text{ V}$$

and therefore the voltage at k to ground is $+9.6$ V.

Example 3-8 What are the voltage drops across the diodes in Fig. 3-16, and what is the voltage at point c to ground (E_c)?

Answer 3-8 $E_c = 12$ V, $V_1 = 0$ V, $V_2 = 24$ V.—Voltage source E_a forward biases diode D_1, which acts like a short. Source E_b reverse biases diode D_2, which then acts like an open. The short through D_1 places the 12 V of E_a across R, thus making point c $+12$ V with respect to ground. And, of course, the voltage across the shorted diode is zero. Also note in Fig. 3-17 that the sum of E_a and E_b is placed across the open diode D_2 through the shorted diode D_1.

Example 3-9 If the polarities of both sources E_a and E_b are reversed in Fig. 3-16, what are the voltages across the diodes and what is the voltage from point c to ground?

Answer 3-9 $E_c = 12$ V, $V_1 = 24$ V, $V_2 = 0$ V.—In this case source E_a reverse biases diode D_1, which then acts like an open. Source E_b forward biases D_2, causing it to act like a short. The sum $E_a + E_b$ this time appears across D_1 while the shorted D_2 places the 12 V of E_b directly across R. This puts point c at 12 V with respect to ground. See Fig. 3-17.

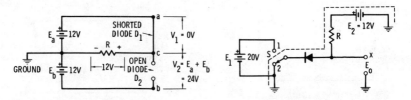

Fig. 3-17. Equivalent circuit of Fig. 3-16 with voltage distribution shown assuming the diodes are ideal.

Fig. 3-18. Finding voltage E_o with switch S in position 2, causing forward bias on diode.

Example 3-10 Referring to Fig. 3-18, what is the voltage E_o when the switch S is in position 2 as shown?

Answer 3-10 $E_o = 0$ V.—The diode is forward biased through the path indicated by the dashed line. The diode acting like a short circuit effectively shorts x to ground through the switch and terminal 2. That is, point x is at ground potential.

Example 3-11 Referring to Fig. 3-18, what is the voltage E_o when the switch S is thrown up into position 1?

Answer 3-11 $E_o = +12$ V.—In position 1, the source E_1, the diode, the resistor R, and the source E_2 are placed in series through ground. The two sources buck each other. The 20 V E_1 dominates and reverse biases the diode, which thus acts like an open. With the diode acting like an open, no current flows in R and there is no voltage drop across it. Therefore the source E_2 voltage is read across x and ground. That is, "walking" from x to ground up through R and E_2, we "walk" through only one voltage E_2 because zero voltage is across R.

REVIEW QUESTIONS

3-1. What three types of equivalents can be made of a real diode's V-I characteristics?

3-2. Briefly describe the characteristics of the ideal diode.

3-3. Briefly describe the so-called simplified diode characteristics.

3-4. Briefly describe the so-called approximated diode characteristics.

3-5. Under what circuit conditions is the assumption that the diode is ideal usually adequate?

3-6. Under what conditions is it usually better to use simplified diode characteristics instead of those of ideal diodes?

3-7. Under what conditions do approximated diode characteristics usually yield significantly more accurate results?

3-8. If the source voltage and circuit resistances are very low, how can you accurately determine the forward voltage and current values?

PROBLEMS

3-1. If using $E = 5$ V, x positive with respect to y, and simplified silicon diode characteristics, what is the voltage across R_2 in Fig. 3-12?

3-2. If the PIV rating of the diode in Fig. 3-12 is 120 V, what maximum value of the source E may be applied in reverse bias?

3-3. Suppose that $E = 50$ V in Fig. 3-13 and that it forward biases the diode. Assuming the diode is silicon with a constant 0.7-V forward drop (simplified silicon diode characteristics), what is the voltage at point k with respect to ground?

3-4. Find the voltages at k to ground and across the diode in Fig. 3-13 where $E = 200$ V, the diode is forward biased, and the diode has the characteristic in Fig. 3-8. Approximate the forward voltage drop V_F from this characteristic.

3-5. What is the voltage across R in Fig. 3-16, assuming that the diodes are silicon and they have a constant drop of 0.7 V if conducting and sources E_a and E_b are each 24 V?

3-6. Referring to Prob. 3-5, what minimum PIV rating must the nonconducting diode have?

3-7. Suppose that you have a forward biased silicon diode in a circuit like that of Fig. 3-7 in which $E = 80$ V and the diode characteristic is shown in Fig. 3-8. Approximately what value of R is needed to limit I_F to 40 mA?

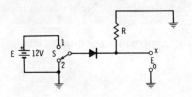

Fig. 3-19. Finding voltage E_o.   Fig. 3-20. Finding voltage E_2.

3-8. With the switch S in position 2 in Fig. 3-19, what is voltage E_o, assuming that the diode has a constant 0.7 V drop when forward biased?

3-9. What is the voltage E_o in Fig. 3-19 when switch S is in position 1, assuming that the diode has a constant 0.7 V drop when forward biased?

3-10. Suppose that the forward current of the diode in Fig. 3-19 is not to exceed 100 mA. What minimum value of R would you select?

3-11. What is voltage E_o in the circuit of Fig. 3-18 when switch S is in position 2 and the diode is a germanium type? Use the simplified germanium characteristic, that is, $V_F = 0.3$ V constantly.

3-12. What is the voltage across the diode in Fig. 3-18 when the switch S is in position 1?

3-13. Suppose that you have a germanium diode reverse biased in a circuit like that of Fig. 3-2B. What will happen to the voltage E_R if heat is applied to the diode with a soldering iron or match?

3-14. Referring to Fig. 3-20, what is the voltage E_2 if source $E = 24$ V, point a positive with respect to point b?

3-15. If the source E is reversed in Fig. 3-20, i.e., point a is made negative with respect to point b but its value is still 24 V, what is the voltage E_2?

4

Applications of Diodes

4-1 SUPERPOSITION OF DC AND AC

You have already become familiar with the term *bias voltage*, which referred to the dc voltage applied to a diode. In this and later chapters we will similarly use bias voltages with other semiconductor devices. Frequently, ac signal voltages are applied to such devices along with the dc bias voltages. Therefore, this section reviews concepts of dc and ac theory and also analyzes circuits that use both dc and ac simultaneously.

A constant dc voltage E_s of 12 V is applied to the circuit in Fig. 4-1A, and therefore the voltage E_R across the resistor is a constant 12 V as shown in Fig. 4-1B. Similarly, if a sine-wave source with a peak of 12 V is applied to a resistor as shown in

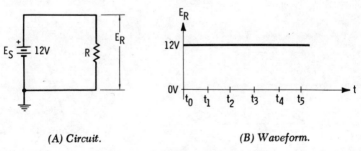

(A) *Circuit.*   (B) *Waveform.*

Fig. 4-1. Pure dc applied to resistor R.

Fig. 4-2A, the voltage e_r across the resistor is sinusoidal as shown in Fig. 4-2B. In this case, the current flows up through the resistor R half the time (on positive alternations) and down through R the other half of the time (on negative alternations).

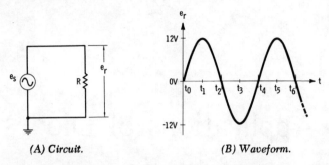

(A) Circuit.   *(B) Waveform.*

Fig. 4-2. Sine wave voltage applied to resistor R.

When both this ac source e_s and dc source E_s are in series across the resistor as shown in Fig. 4-3A, the resulting voltage waveform across R is as shown in Fig. 4-3B. This voltage waveform e_R has the ac component e_r *superimposed* on the dc component E_R. Note that the value of e_R at any instant of time t is equal to the sum of the ac component e_r and the dc component E_R at that instant (assuming that the internal resistances of the sources are negligible). For example, at time t_1 the ac component e_r is at its positive peak (12 V) and it adds to the positive dc component (12 V), giving a total of 24 V. At time of t_2, ac component $e_r = 0$ V and this added to component $E_R = 12$ V gives the sum $e_R = 12$ V. Similarly at time t_3, component $e_r = -12$ V; that is, it is its negative peak value, and adding it to $E_R = 12$ V gives the sum $-12 + 12 = 0$ V.

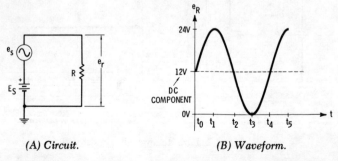

(A) Circuit.   *(B) Waveform.*

Fig. 4-3. Pure dc and sine wave ac applied simultaneously to resistor R.

4-1 SUPERPOSITION OF DC AND AC

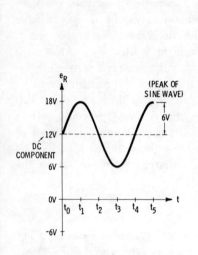

Fig. 4-4. Ac with peak value of 6 V superimposed on 12 V dc.

Fig. 4-5. Ac with peak value of 20 V superimposed on 12 V dc.

If the peak value of the ac component is smaller than the amplitude of the dc component, say for example e_r has a 6 V peak while $E_R = 12$ V in a circuit like that of Fig. 4-3A, the resulting voltage across the resistor R is shown in Fig. 4-4. If the peak value of the ac component is larger than the amplitude of the dc component, say that e_r has a 20 V peak and that $E_R = 12$ V, the voltage across R that results is shown in Fig. 4-5. Note in this case that around time t_3 the voltage across the resistor e_R is negative and therefore current flows down through R during this part of the cycle.

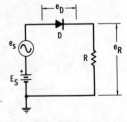

Fig. 4-6. Series diode circuit with ac and dc power supplies in series.

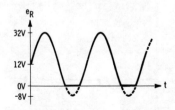

Fig. 4-7. Waveform across R of Fig. 4-6 when $e_{s(peak)} = 20$ V and $E_s = 12$ V.

Suppose that we use a series diode as shown in Fig. 4-6. What does the voltage waveform e_R look like if the peak value of $e_s =$ 20 V and $E_S = 12$ V? With these component values, e_R would have the waveform shown in Fig. 4-5 if the diode were *not* in the circuit. However, with the diode, and assuming it is ideal, waveform e_R is as shown in Fig. 4-7. There is no negative voltage across the resistor R in this case. The voltage at the top of the resistor R can be negative to ground only if current flows down through it. But this cannot happen because the reverse biased diode prevents such a current. That is, during the time of the cycle when the sum of e_s and E_S is negative, the diode is reverse biased and acts like an open. The negative sum of e_s and E_S therefore appears across the open diode as shown in Fig. 4-8.

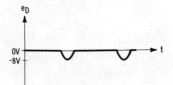

Fig. 4-8. Waveform across D of Fig. 4-6 when $e_{s(peak)} = 20$ V and $E_S = 12$ V.

Example 4-1 If the diode in Fig. 4-6 is reversed, what are the waveforms e_R and e_D? Assume $e_{s(peak)} = 20$ V and $E_S = 12$ V.

Answer 4-1 See Fig. 4-9.—In this case the diode is *reverse* biased when the sum $e_s + E_S$ is positive. The diode conducts only when this sum is negative.

Example 4-2 If dc source E_S in the circuit of Fig. 4-3 is reversed, what is the voltage waveform e_R? Use peak $e_s = 12$ V and $E_S = 12$ V.

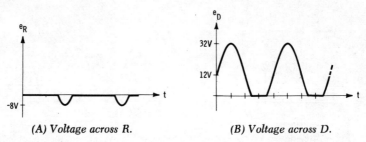

(A) *Voltage across R.*   (B) *Voltage across D.*

Fig. 4-9. Waveforms for Fig. 4-6 with reversed diode when $e_{s(peak)} = 20$ V and $E_S = 12$ V.

4-2 THE HALF-WAVE RECTIFIER

Answer 4-2 In this case the ac component e_s rides on a negative dc component as shown in Fig. 4-10.

Fig. 4-10. Waveform for Fig. 4-3A with E_S reversed and $e_{s(peak)} = 12$ V and $E_S = 12$ V.

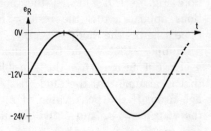

4-2 THE HALF-WAVE RECTIFIER

Rectifier circuits in general are used to convert ac to dc. This conversion can be accomplished with the simple *half-wave rectifier* circuit shown in Fig. 4-11A. During the positive alternations of the input voltage e_s, the diode is forward biased allowing current up through the resistor R. If the diode is assumed to be ideal, its forward voltage drop V_F is zero and the full source voltage e_s appears across the resistor; that is, $e_s = e_R$. (See Figs. 4-11B and 4-11C.) During the negative alternations the diode acts like an open, and

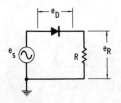

(A) *Simple half-wave rectifier circuit.*

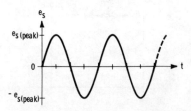

(B) *Voltage applied to rectifier.*

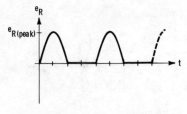

(C) *Voltage across R of Fig. 4-11A with ideal diode.*

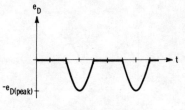

(D) *Voltage across diode shown in Fig. 4-11A.*

Fig. 4-11. Simple half-wave rectifier.

no voltage appears across the resistor R, as shown in Fig. 4-11C. However, the negative alternations do appear across the diode; note Fig. 4-11D. Thus, if the diode is ideal, all positive alternations appear across the resistor R and all negative alternations appear across the diode.

Suppose that we do not assume that the diode is ideal in Fig. 4-11A but instead use the simplified silicon diode characteristic; that is, assume that the diode has a constant of 0.7 V forward voltage drop. If the peak value of e_s is not much larger than 0.7 V, the waveforms e_R and e_D will be noticeably affected. For example, suppose that $e_{s(\text{peak})} = 5$ V. The equivalent circuit may be drawn as Fig. 4-12A, where the 0.7 V source represents the constant 0.7 V drop across the conducting diode. The input voltage waveform e_s and the resulting waveforms across the resistor and diode are shown in Figs. 4-12B, 4-12C, and 4-12D respectively. We can view this circuit as being similar to circuits covered in the last section, where we had ac superimposed on dc. In this case, the ac component e_s rides on a -0.7 V dc component as shown in Fig.

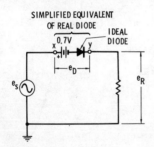

(A) Half-wave rectifier with simplified equivalent of diode.

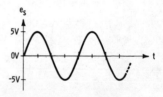

(B) Input voltage waveform e_s to rectifier circuit.

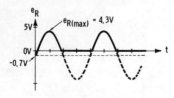

(C) Solid line is voltage across resistor R.

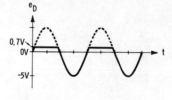

(D) Voltage across diode (terminals x and y).

Fig. 4-12. Half-wave rectifier with diode voltage drop included.

4-2 THE HALF-WAVE RECTIFIER

4-12C. The diode is reverse biased and nonconducting during portions of the input waveform shown with the broken line in Fig. 4-12C. In equation form we can show $e_R = e_s - 0.7$ V for values of e_s greater (more positive) than $+0.7$ V. At times when e_s is more negative than $+0.7$ V, e_R is zero because the diode is not forward biased and therefore not conducting. Of course, if the peak of e_s is much greater than 0.7 V, we can, for most practical purposes, ignore this 0.7-V diode forward voltage drop.

With few exceptions, we may assume that the reverse leakage current I_R is negligible. An exception might be when a germanium diode is used in a circuit like that in Fig. 4-11A and the diode becomes too hot. In such a case, the reverse leakage current I_R, down through R, on negative alternations, becomes large enough to produce a waveform e_R as shown in Fig. 4-13.

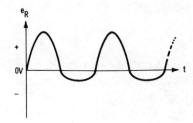

Fig. 4-13. Output waveform of half-wave rectifier when the diode has high leakage current.

Note that the output voltage e_R of the half-wave rectifier (Fig. 4-11A) is a pulsating dc voltage as shown in Fig. 4-11C. This voltage is dc because it never changes polarity; that is, it is either a positive value or zero but never negative. Thus, the output of a half-wave rectifier (Fig. 4-11C) has an average dc value which can be determined with either of the following equations, if the diode is assumed ideal:

$$E_{dc} = \frac{e_{s(peak)}}{\pi} \qquad (4\text{-}1A)^*$$

or

$$E_{dc} \cong \frac{0.637 e_{s\ (peak)}}{2} \qquad (4\text{-}1B)$$

where

---

*See Appendix A-1.

$e_{s(peak)}$ is the peak of the sine wave source voltage,
E_{dc} is the average dc voltage across resistance R as would be measured with a dc-reading voltmeter.

Example 4-3 If in Fig. 4-11A, the sine wave source voltage $e_s = 35.35$ V rms, about what average dc voltage E_{dc} would you expect to measure across R?

Answer 4-3 $E_{dc} \cong 15.92$ V or about 16 V.—Knowing the effective rms value of a sine wave, we can determine its peak value with the equation

$$e_{s(peak)} = e_s(\sqrt{2}) \cong e_s(1.414)$$

Thus in this case,

$$e_{s(peak)} \cong 35.35(1.414) \cong 50 \text{ V}$$

And by Eq. 4-1A,

$$E_{dc} \cong \frac{50}{3.14} \cong 15.92 \text{ V}$$

The pulsating (varying amplitude) waveform in Fig. 4-11C has limited usefulness. Most practical uses of dc voltage require that it be as unvarying as possible. The variation present in the output of a rectifier or any dc power supply is called *ripple*. Generally, the smaller the ripple, the better is the power supply. Capacitors are used with rectifiers to reduce the ripple. We may get an idea of how this is done by first analyzing a simple circuit like that of Fig. 4-14A. The resistor has been replaced with a capacitor C. The input voltage waveform e_s is shown with the broken line in Fig. 4-14B. Assuming that the capacitor C is initially uncharged, it charges to the peak value of the source voltage $e_{s(peak)}$ in the first positive quarter cycle (in time t_0 to t_1). After this, the voltage across the capacitor remains constant even though e_s continues to vary, assuming that the capacitor C has infinite or very large leakage resistance. In other words, once the capacitor C is charged by a momentary counterclockwise current through the forward biased diode, it has no discharge path. That is, the discharge current path must be clockwise but the diode is reverse biased to such a discharge current. Therefore, the capacitor remains charged as shown by the solid line in Fig. 4-14B.

It is interesting to note the resulting voltage across the diode (Fig. 4-14C). The diode is forward biased for the first quarter

4-2 THE HALF-WAVE RECTIFIER 55

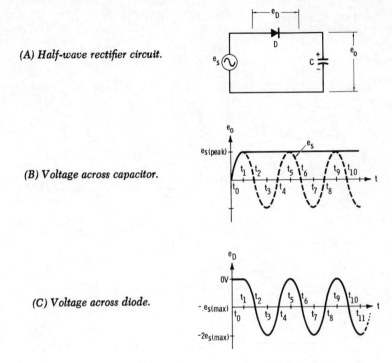

(A) Half-wave rectifier circuit.

(B) Voltage across capacitor.

(C) Voltage across diode.

Fig. 4-14. Use of capacitor to reduce ripple of half-wave rectifier.

cycle and acts like a short, assuming that it is ideal. Thus, between t_0 and t_1 voltage $e_D = 0$ V. After time t_1 the diode is held in reverse bias by the voltage across the capacitor C except at the instants t_5, t_9, etc. At these instants e_s is at its positive peaks and, as the diode sees it, $+e_{s(\text{peak})}$ and e_o are in series bucking; thus being equal in magnitude, they cancel each other out. For example, suppose that the voltage source has a peak $e_{s(\text{peak})} = 20$ V. Therefore the capacitor voltage $e_o = 20$ V too because it was charged to this value on the first quarter cycle. However, the positive $e_{s(\text{peak})}$ "wants" to forward bias the diode (cause counterclockwise current) but the capacitor voltage e_o "wants" to reverse bias the diode (cause clockwise current). These equal but opposing effects cancel each other at instants t_5, t_9, etc. Note in Fig. 4-14C that $e_D = 0$ V at these instants.

Also an important point to note is that the diode reverse voltage is twice the peak of the input voltage $-2e_{s(\text{max})}$ at instants t_3, t_7, t_{11}, etc. because e_s and e_o are series aiding then. This means that the

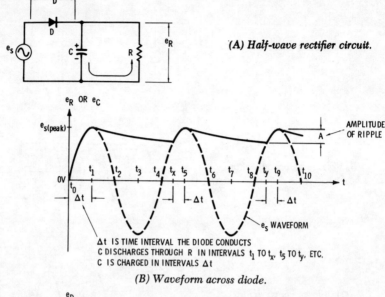

(A) Half-wave rectifier circuit.

(B) Waveform across diode.

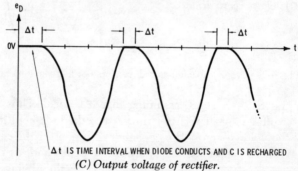

Δt IS TIME INTERVAL WHEN DIODE CONDUCTS AND C IS RECHARGED
(C) Output voltage of rectifier.

Fig. 4-15. Use of load resistor in half-wave rectifier.

diode PIV (peak inverse voltage) rating must be at least twice the input peak. So if $e_{s(\max)} = 20$ V, the diode PIV rating must be 40 V or more.

If we compare the output waveform of Fig. 4-11C with that of Fig. 4-14B, it is obvious that the capacitor has been quite effective in removing the ripple (variations). Capacitors are called *filter components*, when used in this way with rectifiers or with any dc power supplies in general, because they help to filter out the ripple from the dc output voltage.

4-2 THE HALF-WAVE RECTIFIER

Actually, the capacitor C alone as a load, as shown in Fig. 4-14A, is not very practical. Rectifier circuits are usually required to supply dc current to a resistive load. The circuit in Fig. 4-15A is a more practical half-wave rectifier. It uses both a filter capacitor C and a resistive load R. With this circuit, as you will see, we get an output that is neither a pulsating waveform like that of Fig. 4-11C nor a constant dc like that of Fig. 4-14B, but instead a compromise between the two (Fig. 4-15B).

As before, the capacitor C is charged to $e_{s(peak)}$ in the first quarter cycle (between times t_0 and t_1) and cannot discharge after that through the reverse biased diode. However, the capacitor can discharge through R in the path shown with the arrow in Fig. 4-15A. This discharge causes e_R (voltage across C and R) to decrease between positive peaks of e_s, as shown in Fig. 4-15B. Since the

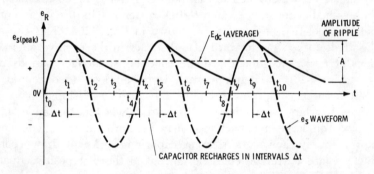

(A) Waveform across resistor of Fig. 4-15A where R value has been reduced.

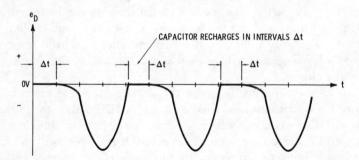

(B) Waveform across diode of Fig. 4-15A where R value has been reduced.

Fig. 4-16. Reduced value of R causes greater discharge of capacitor in intervals t_1 to t_x, t_5 to t_y, etc.

voltage across the capacitor e_R decreases to values less than $e_{s(\text{peak})}$, the diode becomes forward biased, allowing a recharge of C to voltage $e_{s(\text{peak})}$ on each positive peak of the input voltage. In this case, C starts to recharge at instants t_x, t_y, etc., and the diode is forward biased at the time intervals Δt. The voltage waveform across the diode is similar to that of Fig. 4-14C except that its peaks are clipped off as shown in Fig. 4-15C.

This output voltage waveform in Fig. 4-15B results with a relatively large resistance R in the circuit of Fig. 4-15A. Obviously, a ripple component has been introduced in spite of C. The solid line waveform in Fig. 4-15B shows some variations in the dc output. Thus, with resistance in the output we no longer have a pure (unvarying) dc output like that in Fig. 4-14B.

The amplitude A of the ripple component becomes larger if the resistance R in a circuit like that of Fig. 4-15A is reduced. That is, with a smaller resistance value R, the capacitor will discharge more between positive peaks of e_s, as shown in Fig. 4-16A. This causes the diode to become forward biased sooner in the positive alternations at t_x, t_u, etc., and the diode conducts for longer periods Δt (see Fig. 4-16B). Therefore, with increased load (reduced load resistance) on a dc power supply, the amount of ripple may increase. If a larger capacitance C is used, it will tend to reduce the ripple. Too large a capacitance may be hard on the diode and power supply, however. A larger capacitance C will draw (conduct) a larger average charging current in the first positive quarter cycle than will a smaller capacitance. Such high initial charging currents flow through the forward biased diode and the source e_s. This may destroy the diode by exceeding its maximum current capability and/or damage the source e_s. A surge-current limiting resistor is usually used in series with the diode and the source e_s when a large capacitor C is used across the load resistance R of a rectifier circuit.

Since the capacitor C across the load resistance R tends to remove the variation in e_R, Eqs. 4-1A and 4-1B for the waveform in Fig. 4-11C do not apply here. That is, in the case of the circuit in Fig. 4-15A, the average dc voltage across R may, for most practical purposes, be assumed equal to the peak of the source voltage $e_{s(\text{peak})}$:

$$E_{\text{dc}} \cong e_{s(\text{peak})} \qquad (4\text{-}2)$$

4-3 THE FULL-WAVE RECTIFIER

If resistance R and/or capacitance C in Fig. 4-15A is made too small, Eq. 4-2 above becomes inaccurate. As a "rule of thumb," if the frequency of e_s is 60 Hz, Eq. 4-2 gives good accuracy if the time constant $(R \times C)$ of the load R and capacitance C is greater than 25 ms.

Example 4-4 Suppose that you have a circuit like that in Fig. 4-15A, in which the source voltage is a sine wave with $e_{s(\text{peak})} = 80$ V, $C = 20$ μF, and $R = 10$ kΩ. By placing a dc voltmeter across R, what voltage would you expect to read? If the capacitor C is disconnected, what will you read?

Answer 4-4 About 80 V with C in the circuit and about 25.4 V with C disconnected.—Since the time constant $R \times C = 200$ ms, which is much larger than 25 ms, Eq. 4-2 applies. Thus, across R, we will read $E_{\text{dc}} = e_{s(\text{peak})} = 80$ V. When the capacitor is disconnected, this circuit becomes a simple half-wave rectifier as shown in Fig. 4-11A. Therefore, Eqs. 4-1A and 4-1B apply, and the meter across R reads

$$E_{\text{dc}} = \frac{e_{s(\text{peak})}}{\pi} \cong \frac{80}{3.14} \cong 25.4 \text{ V}$$

Example 4-5 If in the circuit described in the last example, the 20-μF capacitor C is reconnected and the 10 kΩ load R is replaced with a 250 Ω resistance, what would you expect to read on the dc voltmeter?

Answer 4-5 Less than 80 V.—In this case $R \times C$ is less than 25 ms, which means that the capacitor discharges significantly through the 250 Ω in the time between positive peaks. This increases the ripple amplitude and lowers the average dc across R.

4-3 THE FULL-WAVE RECTIFIER

In the previous section we studied the half-wave rectifier. The reason why it is so called is obvious when we look at the input and output waveforms of the half-wave rectifier: Figs. 4-11B and 4-11C. The output waveform has half the alternations of the input waveform, and current flows through the load only during positive alternations. The full-wave rectifier, as its name implies, has load current during both negative and positive alternations of the input voltage.

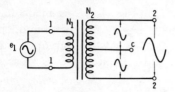 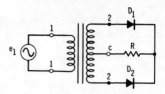

Fig. 4-17. Transformer with secondary center tapped.

Fig. 4-18. Transformer used in full-wave rectifier.

The full-wave rectifier is used with a transformer whose secondary is provided with a center tap like that shown in Fig. 4-17. When a sine wave voltage is applied to the primary windings terminals 1-1, a sine wave voltage appears across the entire secondary (terminals 2-2) by induction. The voltage from either end of the secondary to the center tap c has half the amplitude of the voltage across the entire secondary. The amplitude of the secondary voltage is determined by the amplitude of the input voltage e_1 and the turns ratio of the transformer. The *turns ratio* may be defined as the ratio of turns in the higher voltage windings to the turns in the lower voltage windings. If we assume close coupling between primary and secondary windings and that losses are negligible, which is quite reasonable when working with power transformers, the following relationship exists:

$$\frac{e_2}{e_1} = \frac{N_2}{N_1}$$

where
  e_1 is the input or applied voltage,
  e_2 is the induced voltage across terminals 2-2,
  N_1 is the number of turns in the primary,
  N_2 is the number of turns in the secondary.

Algebraically manipulating this equation, we can show that

$$e_2 = \frac{N_2}{N_1} e_1 \tag{4-3}$$

If two diodes and a load resistance R are added to the secondary as shown in Fig. 4-18, we have a full-wave rectifier. The voltage induced into the secondary forward biases diode D_1 and reverse biases diode D_2 on positive alternations. On the negative alternations of the induced voltage, D_1 is reverse biased while D_2 is forward biased. For example, on positive alternations, the top terminal 2 of

4-3 THE FULL-WAVE RECTIFIER

the secondary is positive with respect to the bottom terminal 2 as shown in Fig. 4-19A. This makes the center tap c negative with respect to the top of the secondary and positive with respect to the bottom. The top half of the secondary N_a acts like a voltage source forward biasing D_1 and causing current I to flow as shown. The bottom half, N_b, reverse biases D_2 and no current flows out of the turns N_b. Since the two halves of the secondaries act like voltage sources, the equivalent of Fig. 4-19A is Fig. 4-19C. In the latter, $e_a + e_b$ at any instant is equal to the induced voltage across 2-2 at the same instant. Note that voltage $e_a + e_b$ is in series across reverse biased diode D_2 via conducting diode D_1. This means that the diodes in this circuit must have a PIV rating exceeding the peak of the entire secondary voltage, $e_{a(peak)} + e_{b(peak)}$.

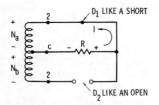

(A) *Current for positive alternation.*   (B) *Current for negative alternation.*

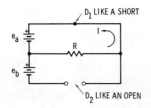 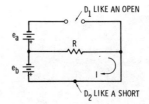

(C) *Voltage for positive alternation.*   (D) *Voltage for negative alternation.*

Fig. 4-19. Operation of transformer in full-wave rectifier circuit.

On the negative alternations, the top terminal 2 is negative with respect to the bottom terminal 2, thus making the center tap c positive with respect to the top of the secondary and negative with respect to the bottom as shown in Fig. 4-19B. Both halves of the secondary N_a and N_b act like individual voltage sources e_a and e_b as shown in Fig. 4-19D. In the latter, e_b forward biases D_2 causing current I to flow while D_1 is reverse biased and has $e_a + e_b$ volts across it.

Note that regardless of the polarity of the secondary induced voltage, Figs. 4-19A or 4-19B, the current is from left to right through the resistance R, which means that this current is dc and therefore rectification (conversion of ac to dc) has taken place.

The full-wave rectifier is often drawn as in Fig. 4-20A. The output voltage e_o has the heavy lined waveform shown in Fig. 4-20B. The full induced secondary voltage is shown with the broken line. The peak output $e_{o(peak)}$ at instants t_1, t_3, t_5, etc. is half the peak of the full secondary voltage. This is a disadvantage of the full-wave rectifier; that is, we can't rectify the full secondary voltage because only half of the secondary is working (supplying current I to the load R) at any instant as was shown in Fig. 4-19.

Since the output voltage of the full-wave rectifier (Fig. 4-20B), never changes polarity, it has an average dc value that can be determined with either of the following equations:

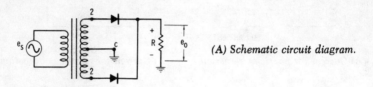

(A) Schematic circuit diagram.

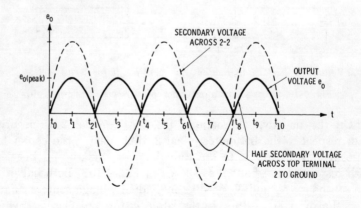

(B) Voltage waveforms.

Fig. 4-20. Basic full-wave rectifier.

4-3 THE FULL-WAVE RECTIFIER

$$E_{dc} = \frac{2e_{o(peak)}}{\pi} \quad (4\text{-}4\text{A})^*$$

or

$$E_{dc} \cong 0.637 e_{o(peak)} \quad (4\text{-}4\text{B})$$

where

E_{dc} is the average dc voltage across resistance R as would be measured with a dc voltmeter,

$e_{o(peak)}$ is the peak voltage across either end of the secondary with respect to the center tap c,

the diode is assumed ideal.

Example 4-6 If the transformer in Fig. 4-20A is 4:1 step-up and the sine wave source voltage e_s has a peak of 40 V, that is, $e_{s(peak)} = 40$ V, what is the output voltage $e_{o(peak)}$ and what average dc voltage would you expect to read across R?

Answer 4-6 $e_{o(peak)} = 80$ V, $E_{dc} \cong 51$ V.—A transformer 4:1 step-up ratio means that there are four times more turns in the secondary than in the primary; thus $N_2/N_1 = 4/1$. Since the voltage applied to the primary has a peak of 40 V, the peak voltage induced into the secondary, by Eq. 4-3 is

$$e_{2(peak)} = \frac{4}{1} \times 40 = 160 \text{ V}$$

Thus, from either end to the center tap we have 80 V peak, which is $e_{o(peak)}$, ignoring the drop across the diode. By using either Eq. 4-3A or Eq. 4-3B we find that

$$E_{dc} \cong \frac{2(80)}{3.14} \cong 51 \text{ V}$$

Example 4-7 Suppose that the transformer in circuit Fig. 4-20A is step-down 3:1, the source $e_s = 120$ V and is sinusoidal, and $R = 2000 \, \Omega$. What PIV rating should the diodes have? What is the peak value of the output voltage e_o, and what is the peak current $i_{R(max)}$ through the load R?

Answer 4-7 PIV = 56.7 V or more, $e_{o(max)} \cong 28.3$ V, $i_{R(max)} \cong 14.1$ mA.—Unless otherwise specified, the 120 V ac input to the primary is assumed to be an rms (effective) value. Therefore, the input

---

*See Appendix A-2.

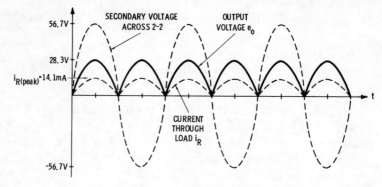

(A) *Waveforms of full-wave rectifier circuit.*

(B) *Full-wave rectifier output for filter C across load R and R×C relatively large.*

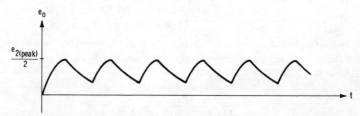

(C) *Full-wave rectifier output for filter C and across load R and R×C relatively small.*

Fig. 4-21. Effect of $R \times C$ on full-wave rectifier output.

voltage $e_{s(\text{peak})} = 120\sqrt{2} \cong 120 \times 1.414 \cong 170$ V. With a step-down turns ratio of 3:1, the full peak secondary voltage is $170/3 \cong 56.7$ V. In other words, by Eq. 4-3. $e_2 = (N_2/N_1)e_1 = 1/3(170) \cong 56.7$ V, where turns ratio 3:1 implies $3/1 = N_1/N_2$ or $N_2/N_1 = 1/3$. Thus, from either end of the secondary to the center tap c, we have $56.7/2 \cong 28.3$ V peak, and the peak of the output voltage $e_{o(\text{peak})} \cong 28.4$ V if the diode is assumed ideal. (See Fig. 4-21A.) The PIV rating of

4-3 THE FULL-WAVE RECTIFIER

the diode must be at least equal to or exceed the end-to-end peak secondary voltage, which in this case is about 56.7 V. With a maximum voltage across R equal to about 28.3 V, by Ohm's law the peak current through R is $I_{R(peak)} = e_{o(peak)}/R = 28.3 \text{ V}/2 \text{ k}\Omega = 14.1 \text{ mA}$.

As with the half-wave circuit, a capacitor across the load R of the full-wave rectifier tends to remove the variations from the output voltage e_o. That is, if a capacitor C is placed across R in the circuit of Fig. 4-20 A, the output e_o will look much like the waveforms in Figs. 4-21B and 4-21C, depending on how large the values of R and C are. If the values of R and/or C are large such as to give a time constant larger than 25 ms when input $f = 60$ Hz, the waveform e_o will be more like that of Fig. 4-21B, which is nearly an unvarying dc with an average value about equal to half of the peak of the secondary voltage. That is,

$$E_{dc} \cong \frac{e_{2(peak)}}{2} \qquad (4\text{-}5)$$

Example 4-8 If a capacitor C is placed across R in the circuit described in the previous example, what dc voltage would

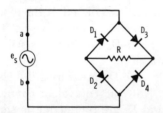

(A) Simple bridge rectifier circuit.

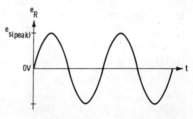

(B) Waveform of input voltage e_s.

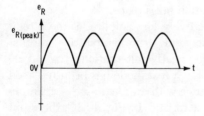

(C) Voltage across R where $e_{s(peak)}$ is much larger than forward voltages of the conducting diodes.

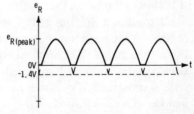

(D) Voltage across R where $e_{s(peak)}$ is not much larger than forward voltages of the conducting diodes.

Fig. 4-22. Bridge rectifier and waveforms.

you expect to read across R, assuming that the time constant is greater than 25 ms ($R \times C > 25$ ms).

Answer 4-8 $E_{dc} \cong 28.3$ V.—Since the peak of the full secondary voltage $e_{2(peak)}$ was found to be 56.7 V, E_{dc} is half this by Eq. 4-5.

4-4 THE BRIDGE RECTIFIER

Fig. 4-22A shows a simple *bridge rectifier*. It provides full-wave rectification of the input ac voltage e_s. It differs from the full-wave rectifier covered in the last section in obvious ways. The bridge rectifier does not require a center-tapped transformer but uses four instead of two diodes. If the voltage waveform of Fig. 4-22B is applied to the bridge rectifier, the output voltage waveform e_R is as shown in Fig. 4-22C with voltage $e_{R(peak)} = e_{s(peak)}$ if the diodes are assumed ideal. Where $e_{s(max)}$ is not much larger than the total forward drops across the conducting diodes, the output voltage e_R will look more like that of Fig. 4-22D. The origin of the -1.4 V will be explained in the following paragraphs.

During the positive alternations of the input e_s in the circuit of Fig. 4-22A, point a is positive with respect to b. Diodes D_2 and D_3 are forward biased while D_1 and D_4 are reverse biased. If the diodes are assumed ideal, Fig. 4-23A is an equivalent for these positive alternations. Note that the diodes D_2 and D_3 are in series with R and that i_R flows from left to right through R.

On the negative alternations, D_1 and D_4 are forward biased while D_2 and D_3 are reverse biased. Now the equivalent circuit is given in Fig. 4-23B. In this case D_1 and D_4 are conducting in series with R and this conduction of i_R is again from left to right through R. Regardless of the polarity of the input voltage, the polarity across R is always the same, thus rectification takes place.

If the input voltage e_s has relatively low amplitude, the circuit of Fig. 4-23C is a more accurate equivalent, on positive alternations, than Fig. 4-23A. Similarly on negative alternations, Fig. 4-23D is more accurate than Fig. 4-23B. Each real diode has been replaced with a constant 0.7 V source in series with an ideal diode—a simplification of a silicon diode V-I characteristic. In Fig. 4-23C, the input voltage must be more positive than 1.4 V before diodes D_2 and D_3 conduct current i_R. Similarly, in Fig. 4-23D, the source must be more negative than -1.4 V before diodes D_1 and D_4 conduct i_R. This causes the flat (zero volt) portions between pulses of the out-

4-4 THE BRIDGE RECTIFIER

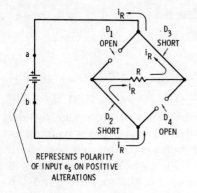

(A) *Circuit for positive alternation of input voltage much larger than 0.7 V.*

(B) *Circuit for negative alternation of input voltage much larger than 0.7 V.*

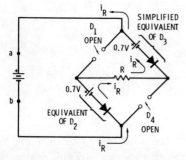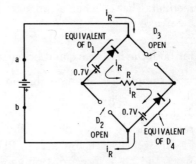

(C) *Diode simplified to 0.7-V source in series with ideal diode: for positive alternation of low input voltage.*

(D) *Diode simplified to 0.7-V source in series with ideal diode: for negative alternation of low input voltage.*

Fig. 4-23. Simplified bridge circuit operation.

put e_R as shown in Fig. 4-22D. That is, since two diodes always conduct in series and each has 0.7 V forward drop, the output voltage is $e_R = e_s - 1.4$ V when e_s has an amplitude large enough to forward bias the diodes. This 1.4-V total drop across the two conducting diodes may be ignored (real diodes may be assumed ideal) when the peak value of e_s is much larger than 1.4 V. On the other hand, if e_s is relatively small, the 1.4-V diode drop significantly affects the output e_R waveform.

The average dc voltage across R, in the circuit of Fig. 4-22A, may be found with the equations

$$E_{dc} = \frac{2e_{R(peak)}}{\pi} \qquad (4\text{-}6A)$$

or

$$E_{dc} \cong 0.637 e_{R(peak)} \qquad (4\text{-}6B)$$

where $e_{R(peak)} = e_{s(peak)}$, assuming that the diodes are ideal.

Obviously, the output voltage waveforms of the bridge rectifier and the full-wave rectifier are similar. Compare the waveforms in Figs. 4-20B and 4-22C. Thus, if we put a capacitor across the load R in the bridge rectifier circuit of Fig. 4-22A, we can expect a voltage across R with waveforms much like the ones in Figs. 4-21B and 4-21C. If the capacitor C and the load R have a time constant ($R \times C$) greater than about 25 ms, we can find the average dc voltage across R with Eq. 4-2, which we used with the half-wave rectifier. That is, when a capacitor is across the load R.

$$E_{dc} \cong e_{s(peak)} \qquad (4\text{-}2)$$

for both the half-wave and bridge rectifiers.

Example 4-9 Suppose that in a circuit like that of Fig. 4-22A, the source voltage e_s has a peak value of 150 V. What does the output waveform look like, what is its peak value, and what PIV ratings should the diodes have?

Answer 4-9 See Fig. 4-24A, PIV = 150 V or more.—With a 150 V peak, the 1.4 V forward drop across both diodes is insignificant and the diodes may be assumed to be ideal. The PIV rating of each diode must exceed the peak of the input voltage $e_{s(peak)}$ because nearly the full $e_{s(peak)}$ value appears across the nonconducting diodes twice each cycle. Note in Fig. 4-23A that the input voltage is across the nonconducting D_1 through the conducting D_2. Similarly, the full input is across D_4 via the conducting D_3. On the other alternation in Fig. 4-23B, we see similarly that the full input is across nonconducing D_3 through conducting D_4, and the input is across D_2 through D_1.

Example 4-10 Referring to the circuit described in the previous example, if a dc voltmeter is placed across R, (a) what reading would you expect? (b) If R is shunted with a capacitor C, such that $R \times C > 25$ ms, what will be the voltage value that is read on the voltmeter?

4-4 THE BRIDGE RECTIFIER

Answer 4-10 (a) About 95.5 V, (b) About 150 V.—Without the filter capacitor, E_{dc} is found with Eq. 4-6A or Eq. 4-6B. In this case, then,

$$E_{dc} \cong 0.637 \times 150 \cong 95.5 \text{ V}$$

With the filter C, as indicated by Eq. 4-2, E_{dc} is nearly the peak of the input, or 150 V in this case.

Example 4-11 Repeat Ex. 4-9 but use a source voltage $e_{s(peak)} = 5$ V.

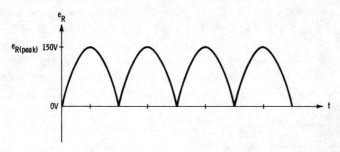

(A) For source $e_s \gg V_F$ each diode.

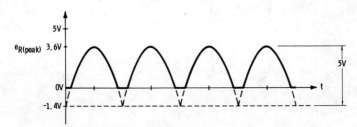

(B) For silicon diodes and $e_{s(peak)} = 5$ V.

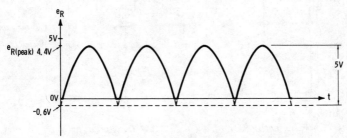

(C) For germanium diodes and $e_{s(peak)} = 5$ V.

Fig. 4-24. Output waveforms of bridge rectifier of Fig. 4-22A.

Answer 4-11 See Fig. 4-24B, PIV about 5 V.—On the positive alternations, the diodes D_2 and D_3 do not conduct until the amplitude of the input voltage e_s exceeds 1.4 V, the total forward drop of both these diodes. Similarly, on negative alternations of e_s diodes D_1 and D_2 do not conduct until the amplitude of the negative alternations exceeds the 1.4-V total drop of these diodes. This causes i_R and e_R to be zero between positive alternations as shown. Note also that since the conducting diodes have a total 1.4-V drop, the output peak $e_{R(peak)}$ is less than the input peak $e_{s(peak)}$ by 1.4 V. Note in Fig. 4-23C that nonconducting diode D_1 has the input voltage across it minus the 0.7-V drop across conducting D_2. Thus, the PIV of D_1 is $5 - 0.7 = 4.3$ V. To be on the safe side, use a 5-V or greater PIV rating. Actually, the PIV ratings available are typically over 50 V even with the least expensive diodes. So the PIV that the diode is expected to withstand becomes an important point to consider only if we expect reverse voltages up to 50 V or more.

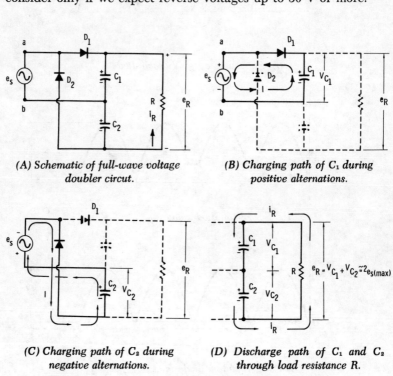

(A) Schematic of full-wave voltage doubler circut.

(B) Charging path of C_1 during positive alternations.

(C) Charging path of C_2 during negative alternations.

(D) Discharge path of C_1 and C_2 through load resistance R.

Fig. 4-25. Operation of full-wave voltage doubler circuit.

4-5 VOLTAGE DOUBLERS

Example 4-12 Repeat Ex. 4.9 but assume that the diodes are germanium and $e_{s(peak)} = 5$ V.

Answer 4-12. See Fig. 4-24C, PIV about 5 V.—Just as in Ex. 4-5, the diodes are not assumed ideal because $e_{s(peak)}$ is not much greater than the total forward drop of the conducting diodes. In this case, however, we assume that each conducting diode has a constant 0.3-V drop.

4-5 VOLTAGE DOUBLERS

The voltage doubler circuit (Fig. 4-25A) is used to obtain a dc output voltage e_R that is approximately twice the peak value of the input voltage e_s. The positive alternations of e_s charge the capacitor C_1 through the path shown in Fig. 4-25B. As with the half-wave rectifier (Fig. 4-14A), the capacitor C_1 charges to the peak value of e_s. In a similar way capacitor C_2 is charged to $e_{s(peak)}$ on negative alternations in the path shown in Fig. 4-25C. The sum of the voltages across the two capacitors is across the load resistance R; that is, $e_R = V_{C1} + V_{C2}$.

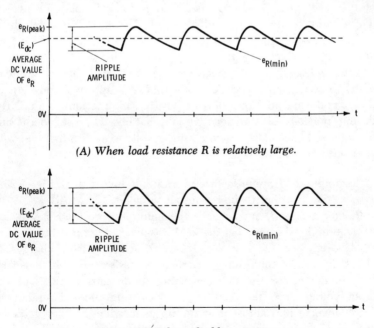

Fig. 4-26. Voltage doubler outputs.

If the resistance R in Fig. 4-25A is relatively large so that i_R is small, the output e_R will have waveform of Fig. 4-26A. The peak output voltage is $e_{R(peak)} \cong 2e_{s(peak)}$ and occurs in the output at the same instants the input e_s reaches each its peaks. Between peaks of e_s, the output e_R drops off because the capacitors are discharging through R. The average dc value of e_R, which is the value that would be read with a dc voltmeter place across R, is between $e_{R(peak)}$ and $e_{R(min)}$ as shown with the broken line in Fig. 4-26A.

If the resistance R is decreased, discharge current i_R becomes larger. Thus, the capacitors will discharge more between peaks as shown in Fig. 4-2B. This not only increases the ripple in the output but also lowers the average dc value of e_R.

If resistance R is very large or an open, the output voltage e_R will have negligible ripple. The capacitors will each charge to $e_{s(peak)}$ and stay charged and the output waveform would look much like that of Fig. 4-14B.

The PIV rating of each diode should equal or exceed twice the input peak, that is, $PIV \geq 2e_{s(peak)}$. The reason why this is so can be seen by referring to the circuit in Fig. 4-25A. During a portion of the positive alternations, diode D_1 is forward biased and acts like a short, assuming it is ideal. Diode D_2 at the same instant is reverse biased. Imagine D_1 replaced with a short circuit. The output voltage e_R is effectively placed in parallel with D_2. Thus, this diode reverse voltage $V_{D2} = e_R$. Since we already found that e_R approximately equals $2e_{s(peak)}$ we can say that V_{D2} approximately equals $2e_{s(peak)}$ during portions of the positive alternations. Similarly, during portions of the negative alternations D_2, the conducting diode, acts like a short, and the output voltage e_R is in parallel with nonconducting diode D_1.

Example 4-13 Suppose that in a circuit like that of Fig. 4-25A, $e_{s(peak)} = 90$ V. What is the dc output voltage E_{dc} as read with a dc voltmeter across R, and what is the minimum required PIV rating of each diode? Assume that the load R is a very high resistance.

Answer 4-13 $E_{dc} \cong PIV \cong 180$ V.—Each capacitor charges to about 90 V and their sum is across the resistor R, which is also the voltage across the nonconducting diode during peaks of the input voltage e_s. That is, the nonconducting diode has about a 180 V reverse bias across it at the instance when e_s is at its peaks and when the other diode is conducting (forward biased).

4-5 Voltage Doublers

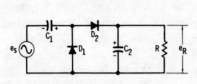

(A) Schematic for half-wave voltage doubler.

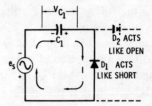

(B) Charge path for C_1 on negative alternations.

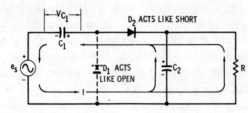

(C) Charge path for C_2 on positive alternations.

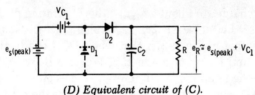

(D) Equivalent circuit of (C).

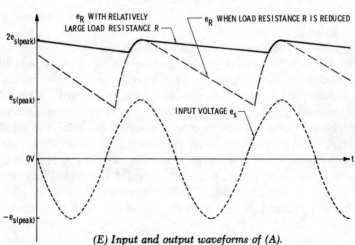

(E) Input and output waveforms of (A).

Fig. 4-27. Half-wave voltage doubler.

Fig. 4-27A shows a voltage doubler of a different type. It too has a dc output voltage e_R equal to about twice the input peak voltage. In this case it is convenient to start our analysis on the negative alternation of the input e_s, as shown in Fig. 4-27B. On this negative alternation D_1 is forward biased and D_2 is reverse biased. This, current I flows, as shown, through D_1, charging the capacitor C_1 to the peak of e_s. That is, after the first negative alternation, the capacitor C_1 is charged to voltage V_{C1}, which is equal to $e_{s(\text{peak})}$, neglecting the drop across D_1. During portions of the positive alternations, D_1 is reverse biased and D_2 is forward biased. Current I flows in the paths in Fig. 4-27C. The capacitor C_2 is charged to a voltage equal to the peak of the input $e_{s(\text{peak})}$ plus the voltage on C_1. This may be more obvious in the equivalent circuit in Fig. 4-27D. The voltage across C_1 is shown as a voltage source V_{C1}. It is in series aiding with $e_{s(\text{peak})}$. Their sum $e_{s(\text{peak})} + V_{C1}$ is the voltage to which capacitor C_2 charges. Since V_{C1} is about equal to $e_{s(\text{peak})}$, capacitor C_2 charges to about $2e_{s(\text{peak})}$, which is the output voltage e_R.

Since C_2 charges to about $2e_{s(\text{peak})}$, the PIV rating of both diodes must likewise be at least $2e_{s(\text{peak})}$ volts. The reason why can be seen by noting in Fig. 4-27A that when D_1 is conducting (acting like a short) and D_2 is nonconducting (reverse biased), the capacitor C_2 full voltage e_R is placed across D_2 via conducting D_1. Similarly, when D_2 is conducting and D_1 is not, as is the case on positive peaks of e_s, the full voltage e_R is placed across D_1 via conducting D_2. Thus both diodes have PIV $\cong 2e_{s(\text{peak})} \cong e_R$.

Input and output waveforms for this doubler circuit are shown in Fig. 4-27E. With a relatively large load resistance R there is little discharge of the voltage on C_2 between peaks of the positive alternations, and the output e_R is nearly pure dc with negligible ripple, as shown. When the load resistance R is decreased, C_2 loses more charge through R between the positive peaks of the input, and the output e_R picks up more ripple and has a lower dc average voltage value E_{dc}. See the dashed line waveform in Fig. 4-27E.

Both types of doublers, circuits in Figs. 4-25A and 4-27A, give an average dc output voltage E_{dc} that is about twice the peak of the source e_s. That is, a dc voltmeter across R in either circuit will read

$$E_{dc} \cong 2e_{s(\text{peak})} \quad (4\text{-}7)$$

4-6 RECTIFIER ASSEMBLIES

if the load R or the total capacitance across R is not too small. More specifically, the time constant of the capacitor(s) discharge path should not, as a "rule of thumb," be shorter than 25 ms if e_s is a 60 Hz source.

Example 4-14 If C_1 and C_2, shown in Fig. 4-25A, are each equal to 40 μF, and the source frequency is 60 Hz, what is the smallest load resistance R that you would use to obtain maximum dc output voltage and minimum ripple amplitude? (Use the rule of thumb.)

Answer 4-14 About 1250 Ω.—The two capacitors discharge in series through the load R. Their total capacitance is one-half the value of either capacitor or 40 μF/2 = 20 μF in this case (remember that capacitors in series add like resistors in parallel). Since our minimum time constant is 25 ms, that is, 25 ms = RC_t, then

$$R = \frac{25 \text{ ms}}{C_t} = \frac{25 \text{ ms}}{20 \ \mu\text{F}} = 1250 \ \Omega$$

4-6 RECTIFIER ASSEMBLIES

The trends in electronics are toward compactness and miniaturization. *Rectifier assemblies*, which are actually sealed packages

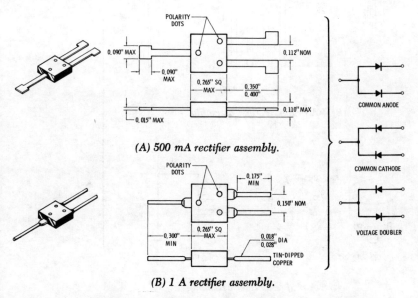

(A) *500 mA rectifier assembly.*

(B) *1 A rectifier assembly.*

Fig. 4-28. Three-lead diode assemblies and their typical contents.

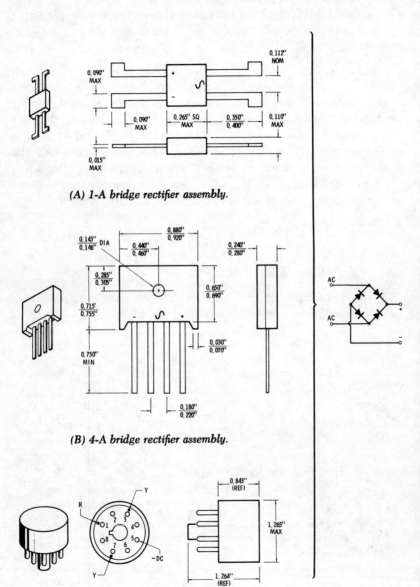

(A) 1-A bridge rectifier assembly.

(B) 4-A bridge rectifier assembly.

(C) 5-A bridge rectifier assembly.

Fig. 4-29. Four-lead diode assemblies and their contents.

4-6 RECTIFIER ASSEMBLIES

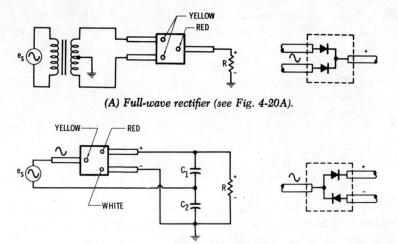

(A) *Full-wave rectifier (see Fig. 4-20A).*

(B) *Doubler (see Fig. 4-25A).*

Fig. 4-30. Examples of diode applied assemblies.

containing diodes in a variety of possible connections, are manufactured to meet the demands of such trends and to offer other conveniences such as economy, simplified assembly of equipment, etc. Figs. 4-28 and 4-29 show some examples of such packages and what they contain. These assemblies are used in full-wave, bridge, and doubler rectifier circuits. Examples of full-wave and doubler circuits are shown in Fig. 4-30. The leads are usually identified by color dots or embossed printing on the package. For example, as shown in Fig. 4-30, *yellow* dots identify ac input leads, *red* the positive dc lead, and *white* the negative dc lead. Sometimes the symbol $\sim$ identifies the ac leads and simply plus ($+$) and minus ($-$) signs are used for the dc leads.

The physical size of the assemblies depends a lot on their current ratings. A high-current package is usually larger than one rated for a smaller current. The manufacturers also rate them in terms of rms maximum input voltage. The user thus has a good idea of the maximum rms ac input voltage that he can use and the maximum current average dc output current he can safely draw from the rectifier. The maximum current rating dictates the minimum load resistance that can be used for a given ac input voltage. For example, suppose that a full-wave rectifier assembly is rated: maximum rms ac input voltage, 25 V; maximum dc output current, 2 A. You are to use this assembly in a circuit like that of Fig. 4-30A in which the transformer secondary voltage is 20 V rms end-to-end. The load resistance may at times be as low as 1000 Ω. Are the specifications of the package exceeded? No. Obviously the actual ac voltage applied to the ac input terminals is 5 V less than the allowable maximum specified. The dc current through a 1 kΩ load can be determined by first finding the average dc voltage across the load R. This is done with Eq. 4-4A or Eq. 4-4B. Thus, in this case

$$E_{dc} = \frac{2e_{o(peak)}}{\pi} \cong \frac{2(14.14)}{3.14} \cong 9 \text{ V}$$

where $e_{o(peak)}$ is half of the peak secondary voltage, neglecting the voltage drops across the diodes within the assembly. With $E_{dc} \cong$ 9 V across R, the dc current through it is

$$I_{dc} = \frac{E_{dc}}{R} \cong \frac{9 \text{ V}}{1 \text{ k}\Omega} = 9 \text{ mA}$$

which is far less than the 2 A maximum allowed.

Example 4-15 The full-wave rectifier shown in Fig. 4-30A has a positive output. If a negative output is required instead, what change(s) should be made?

Answer 4-15 The simplest change is to replace the rectifier assembly, which has a *red* dot or a plus sign identifying the dc lead, with an assembly that has a *white* dot or negative sign. That is, replace the common cathode assembly with a common anode type (see Fig. 4-28). It should be pointed out that the color code described here is not universal but may vary with different manufacturers.

Example 4-16 If e_s, in the circuit of Fig. 4-30B, is 25 V rms and sinusoidal, what would you expect to read with a dc voltmeter across R?

Answer 4-16 About 70.7 V.—Since $e_s = 25$ V rms, then $e_{s(peak)} \cong$ 25(1.414) = 35.35 V. Eq. 4-7 applies to this doubler circuit. Therefore,

$$E_{dc} \cong 2e_{s(peak)} = 2(35.35) = 70.7 \text{ V}$$

The bridge assemblies, shown in Fig. 4-29, are easy to use. Just apply ac to the proper two terminals and take dc off the other two.

Note that the leads are clearly identified with symbols or colored dots.

Example 4-17 A 120-V rms sine wave is applied to the ac terminals of a bridge assembly. (a) What voltage waveform would you expect on a resistor placed across the other two (output) terminals and what dc average value would this voltage have? (b) If a resistor and capacitor are connected in parallel with each other and then across the output terminals, what voltage waveform would you expect across them and what would the dc average value of this voltage be? (Assume that the time constant of this resistance and capacitance is much greater than 25 ms.)

Answer 4-17 (a) See Fig. 4-22C, $E_{dc} \cong 108$ V, (b) see Fig. 4-21B, $E_{dc} \cong 170$ V.—When the load on a bridge rectifier is resistance only, the average dc voltage across this load can be found with Eq. 4-6A or Eq. 4-6B. Thus,

$$E_{dc} \cong 0.637(170) \cong 108 \text{ V}$$

where $170 \cong 1.414(120)$.

When the load resistance has a shunt capacitor, the capacitor charges to the peak of the input voltage, which is 170 V in this case. The voltage across this RC combination looks much like the waveform in Fig. 4-21B except that its peak is $e_{s(peak)}$ or 170 V instead of $e_{2(peak)}/2$.

4-7 LOGIC CIRCUITS

Until now we have used diodes in a variety of circuits that convert ac to dc. Another common application of diodes is in logic circuits which are, sort of, decision-making circuits used in digital computers and in all kinds of electronic automated equipments. Among the most basic of these decision-making circuits are the AND gate and the OR gate, which can be constructed in a number of different ways.

To get an idea what these circuits are we can look at a couple of simple versions that don't use diodes, such as is shown in Fig. 4-31. The circuit in Fig. 4-31A has two switches, A and B, a lamp, and a voltage source in series. Obviously, the lamp will not light (turn on) unless both switches A *and* B are closed; thus it is called an AND gate. The circuit in Fig. 4-31B is a simple OR gate because the lamp will light if either switch *A or B* is closed. Such simple AND and OR gates, though useful in some industrial applications, are inadequate for use in more sophisticated equipments. The more flexible and interesting diode AND gates and OR gates are examples of more sophisticated logic circuits.

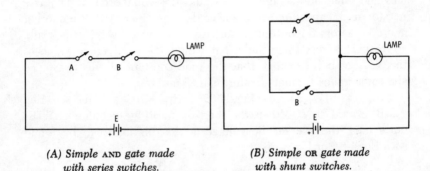

(A) *Simple* AND *gate made with series switches.*

(B) *Simple* OR *gate made with shunt switches.*

Fig. 4-31. Switch logic circuits.

A typical diode AND gate is shown in Fig. 4-32A. It has two inputs, *A* and *B,* and one output *X*. The voltages applied to the inputs are either of two values, say in this case 0 V and 12 V with respect to ground. The output then will be either of the same two levels, assuming that the diodes are ideal. If both inputs *A* and *B* are 12 V to ground, the output at *X* is 12 V to ground. If either or both inputs is 0 V to ground, the voltage at *X* is 0 V. Thus, the output is 12 V only if input *A* and input *B* are 12 V to ground.

Fig. 4-32B shows a more detailed version of the diode AND gate. Note that the switches *A* and *B* can be put in position 1 (up) for an input of 12 V to ground. Switches in position 2 (down) give 0 V to ground. Of course, in this case both switches are down and the 0 V is at both inputs *A* and *B*. We can see that the positive terminal of the 12 V source is open; thus, this source is not even in the circuit. The 20 V source, however, does forward bias both diodes via ground, and a current conducting path exists as shown with the dotted line. Assuming the diodes are ideal, they act like shorts, effectively connecting point *X* to ground, which is to say that point *X* is 0 V to ground.

Now suppose that we put the switch *A* in position 1 (up) as shown in Fig. 4-32C. Note now that the 12 V source is in the

4-7 Logic Circuits

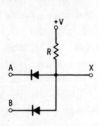

(A) Simple two-input diode AND gate.

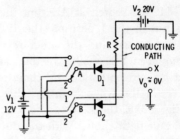

(B) With 0 V at each input, resulting in 0 V output.

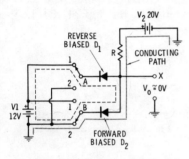

(C) With 12 V at A and 0 V at B, resulting in 0 V output.

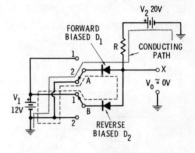

(D) With 0 V at A and 12 V at B, resulting in 0 V output.

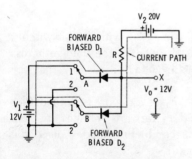

(E) Both inputs at 12 V, resulting in 12 V output.

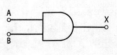

(F) Commonly used logic symbol for AND gate.

Fig. 4-32. Diode AND gate.

circuit and it reverse biases D_1 but forward biases D_2 in the dashed line path shown. Since D_2 is still conducting, point X is still at ground potential (0 V). Therefore, we see that though one of the inputs is 12 V to ground the output is still 0 V.

If the switch at A is put in the 2 position and the switch at B is put in the 1 position, the 12 V source again sees one diode forward biased and the other reverse biased in the dotted line path shown in Fig. 4-32D. In this case, the conducting D_1 puts point X at ground potential, and therefore the output voltage is still 0 V.

Finally, when both switches are up in position 1, as shown in Fig. 4-32E, the 12 V source tends to reverse bias both diodes in the circuit path shown with the dotted line. However, the 20 V source tends to forward bias the diodes in the same path. That is, V_1 and V_2 are bucking each other. The 20-V source, being larger, dominates and forward biases both diodes, which places point X at 12 V to ground. That is, point X is connected to the top of the 12 V source through the conducting diodes.

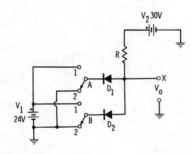

Fig. 4-33. AND gate problem.

Example 4-18 In the circuit of Fig. 4-33 with both switches down as shown, what is the voltage V_o, what is the voltage across the resistor R, and what are the diode voltages? Assume that the diodes are ideal.

Answer 4-18 $V_o = 0$ V, $E_R = 30$ V, $E_{D1} = E_{D2} = 0$ V.—Both diodes are forward biased in this case and their forward voltages are zero and point X is 0 V to ground. The conducting diodes place the full 30 V of source V_2 across R to ground.

Example 4-19 If in Fig. 4-33 the switch at A is up and the switch at B is down, what are the voltages V_o, E_R, E_{D1} and E_{D2}? Assume that the diodes are ideal.

Answer 4-19 $V_o = 0$ V, $E_R = 30$ V, $E_{D1} = 24$ V, $E_{D2} = 0$ V.—Now only D_2 is forward biased. Diode D_1 is reverse biased by the 24 V of V_1; that is, V_1 is placed across D_1 through conducting diode D_2 and the switch at input A.

4-7 Logic Circuits

Example 4-20 Suppose that in Fig. 4-33, both switches are up. What are the voltages V_o, E_R, E_{D1}, and E_{D2}, assuming the diodes are ideal?

Answer 4-20 $V_o = 24$ V, $E_R = 6$ V, $E_{D1} = E_{D2} = 0$ V.—When both switches are up in position 1, V_1 and V_2 are bucking each other from the diode point of view. Since V_1 is smaller than V_2 source V_1 dominates and forward biases both diodes, resulting in 0 V across

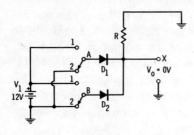

(A) Diode OR gate with 0 V at inputs A and B.

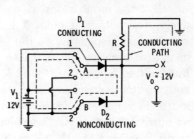

(B) With 12 V at input A and 0 V at input B.

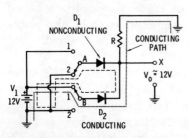

(C) With 0 V at input A and 12 V at input B.

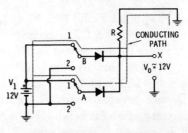

(D) With 12 V at input A and 12 V at input B.

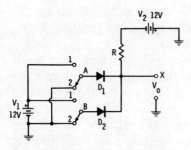

(E) With voltage source V_2.

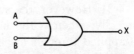

(F) Logic symbol for OR gate.

Fig. 4-34. Diode OR gate.

each of them. As the resistor R sees it (looking from the terminals of R), V_1 and V_2 are bucking each other. Their difference, 6 V, appears across R. Since both diodes act like short circuits, point X is effectively connected to the positive terminal of V_1, thus putting it at 24 V to ground.

We have a diode OR gate in Fig. 4-34A. It has two inputs, which may be either 0 V or 12 V to ground, depending on whether the switch at the input is down (position 2) or up (position 1) respectively. The output at point X is 0 V *or* 12 V if the diodes are assumed ideal, depending on the input switch positions. In this diode OR gate, if both inputs are 0 V to ground, the output at X is 0 V too. But if either input A or B is 12 V, the output is 12 V. In Fig. 4-34A, both switches are down and therefore both inputs are 0 V. The source V_1 is not even in the circuit; thus neither diode conducts and no current flows in R. The voltage at X to ground is the voltage across R and is certainly 0 V in this case.

If the switch at A is thrown in the up position as shown in Fig. 4-34B, source V_1 is placed in the circuit and forward biases D_1. The conducting D_1, acting like a short, places the positive terminal of V_1 at point X and thus $V_o \cong 12$ V. The conducting D_1 also places V_1 across diode D_2 in reverse bias, via the dashed line path, and therefore its reverse voltage $E_{D2} = 12$ V.

We can similarly analyze the circuit shown in Fig. 4-34C. Here input A is 0 V and input B is 12 V. The 12 V of V_1 forward biases D_2 but reverse biases D_1 in the dashed line path. Once D_2 is forward biased it conducts current in the dotted line path and also it effectively connects point X to the positive terminal of V_1. Of course, then point X is 12 V to ground.

Finally, in the case where both switches are up as shown in Fig. 4-34D, both diodes are forward biased and conduct. In this case we can trace from output point X through either diode to V_1. That is, the voltage at X is still 12 V to ground.

It may appear to you that these diode logic circuits are needlessly complicated when compared to the simple AND and OR gates in Fig. 4-31. However, the switches in Fig. 4-31 must be operated mechanically, whereas the inputs to the diode logic circuits need not be applied through mechanical switches. That is, the voltages to inputs A and B of the diode logic circuits may come directly from the outputs of other logic circuits and not necessarily from mechanical switches, as shown in Figs. 4-32 and 4-34.

4-7 LOGIC CIRCUITS

Frequently, the diode OR gate is used including a voltage source V_2 as shown in Fig. 4-34E. This source V_2 is sometimes necessary as part of the circuitry that is connected to the right of output terminal X.

Example 4-21 Referring to Fig. 4-34E, assuming the diodes are ideal, find the voltages V_o, E_R, E_{D1}, and E_{D2} with each of the following conditions:

(a) Both switches are down (position 2).
(b) Switch at input A is up and switch at input B is down.
(c) Switch at input A is down and switch at input B is up.
(d) Both switches are up (position 1).

Answer 4-21 (a) $V_o = 0$ V, $E_R = 12$ V, $E_{D1} = 0$ V, $E_{D2} = 0$ V.— The equivalent in this case is shown in Fig. 4-35A. Both diodes are forward biased by V_2 and conduct in the paths shown. Source V_1 is not actively part of the circuit and the full 12 V of V_2 appears across resistor R. The output point X is effectively grounded through the conducting diodes. (b) $V_o = 12$ V, $E_R = 24$ V, $E_{D1} = 0$ V, $E_{D2} =$

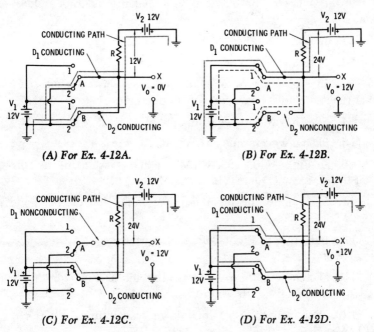

(A) For Ex. 4-12A.   (B) For Ex. 4-12B.

(C) For Ex. 4-12C.   (D) For Ex. 4-12D.

Fig. 4-35. Circuits for example problems.

12 V.—In this case the equivalent circuit is shown in Fig. 4-35B. Diode D_1 is forward biased by both V_1 and V_2, which are in series aiding in the conducting path shown. The sum of V_1 and V_2 is thus across R. Note that D_2 is held in reverse bias by V_1 through conducting D_1 in the dashed line path shown. The switch at input A connects output X to the top of V_1 via diode D_1. Therefore the output is 12 V. (c) $V_o = 12$ V, $E_R = 24$ V, $E_{D1} = 12$ V, $E_{D2} = 0$ V.— We now have the circuit of Fig. 4-35C. Diode D_2 is forward biased by the series sources V_1 and V_2. The sum of these voltages, 24 V, appears across R. Diode D_1 is held in reverse bias by V_1 through conducting D_{D2}. Also the 12 V of V_1 is at the output X through the switch at input B and conducting D_2. (d) $V_o = 12$ V, $E_R = 24$ V, $E_{D1} = 0$ V, $E_{D2} = 0$ V.—Now note the equivalent circuit in Fig. 4-35D. In this case both diodes are forward biased by V_1 and V_2 via the conducting paths shown. The diodes and switches connect the top of V_1 to the output point X. The sum of V_1 and V_2 appears across R.

PROBLEMS

In the following problems assume that the diode or diodes are ideal:

4-1. Referring to Fig. 4-36A, sketch voltage waveform e_o.

4-2. Sketch waveform e_{R1} in Fig. 4-36A.

4-3. In Fig. 4-36A, what is the maximum instantaneous current that the diode must be capable of conducting?

4-4. Sketch waveform e_o in Fig. 4-36B.

4-5. Sketch waveform e_D in Fig. 4-36B.

4-6. In Fig. 4-36B, what is the maximum instantaneous current that the diode must be capable of conducting?

4-7. What must the minimum PIV rating be for the diode in Fig. 4-36B?

4-8. Referring to Fig. 4-36C, sketch waveform e_R, where $e_{s(peak)} = 20$ V and $E_s = 5$ V.

4-9. Sketch waveform e_D of the circuit of Fig. 4-36C, where $e_{s(peak)} = 20$ V and $E_s = 5$ V.

4-10. Referring to Fig. 4-36C, sketch waveform e_R, where $e_{s(peak)} = 12$ V and $E_s = 20$ V.

Problems

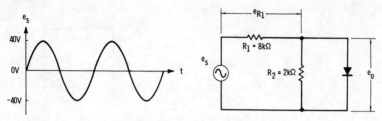

(A) For Probs. 4-1, 4-2, and 4-3.

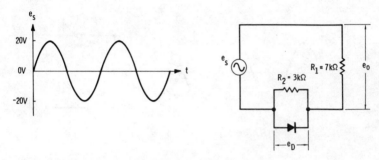

(B) For Probs. 4-4, 4-5, 4-6, and 4-7.

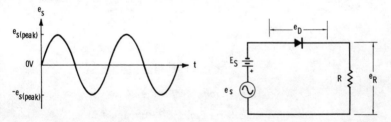

(C) For Probs. 4-8, 4-9, 4-10, 4-11, and 4-12.

Fig. 4-36. Circuits for Probs. 4-1 through 4-12.

4-11. Using the voltages given in Prob. 4-10, sketch waveform e_D.

4-12. What must the minimum PIV rating be for the diode in Fig. 4-36C if $e_{s(\text{peak})} = 12$ V and $E_S = 20$ V?

4-13. Referring to Fig. 4-37A, sketch voltage waveform e_{xy}, which is the voltage across points x and y.

4-14. Sketch waveform e_R in Fig. 4-37A.

APPLICATIONS OF DIODES

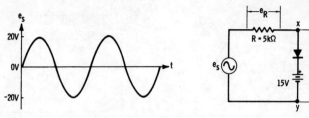

(A) For Probs. 4-13, 4-14, 4-15, and 4-16.

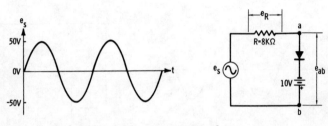

(B) For Probs. 4-17, 4-18, and 4-19.

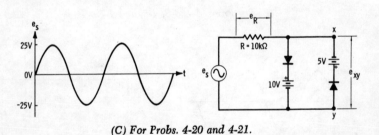

(C) For Probs. 4-20 and 4-21.

Fig. 4-37. Circuits for Probs. 4-13 through 4-21.

4-15. What is the maximum instantaneous current that the diode in Fig. 4-37A must be capable of conducting?

4-16. What must be the minimum PIV rating of the diode in Fig. 4-37A?

4-17. Referring to Fig. 4-37B, sketch waveform e_{ab}.

4-18. Sketch waveform e_R in Fig. 4-37B.

4-19. What is the minimum required PIV rating of the diode in Fig. 4-37B.

4-20. Sketch waveform e_{xy} in Fig. 4-37C.

PROBLEMS

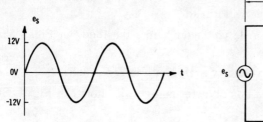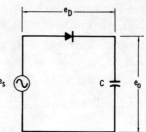

(A) For Probs. 4-22, 4-23, 4-24, 4-25, 4-26, 4-33, and 4-34.

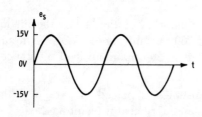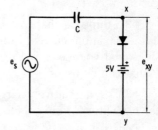

(B) For Probs. 4-27 and 4-28.

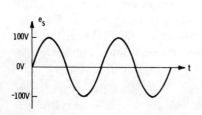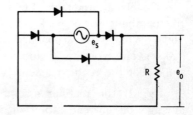

(C) For Probs. 4-29 and 4-30.

Fig. 4-38. Circuits for Probs. 4-22 through 4-30 and 4-33 and 4-34.

4-21. Sketch waveform e_R in Fig. 4-37C.

4-22. Sketch waveform e_o in Fig. 4-38A.

4-23. Sketch waveform e_D in Fig. 4-38A.

4-24. What must be added to a circuit like that shown in Fig. 4-38A to prevent excessive initial current; that is, when e_s is first applied and capacitor C is charging, how is the charging current kept below the maximum current rating of the diode?

90   APPLICATIONS OF DIODES

4-25. If a dc voltmeter is placed across the capacitor C in Fig. 4-38A, what dc voltage will it read?

4-26. What will a dc voltmeter across the diode in Fig. 4-38A read?

4-27. Sketch the voltage waveform e_{xy} in Fig. 4-38B.

4-28. What will a dc voltmeter across capacitor C in Fig. 4-38B read?

4-29. Referring to Fig. 4-38C, sketch waveform e_o.

4-30. What is the minimum required PIV rating of each diode in Fig. 4-38C?

4-31. Suppose that in a circuit like that of Fig. 4-11A the diode has a maximum current rating of 500 mA and the peak of the input sinusoidal voltage e_a is 18 V. What is the minimum load resistance R that can be used?

4-32. If a relatively large resistance is placed in parallel with the capacitor C in Fig. 4-38A, sketch the typical waveform e_o.

4-33. What is the minimum required PIV rating of the diode in Fig. 4-38A?

4-34. In a circuit like Fig. 4-30A the source voltage e_s is 120 V rms or about 170 V peak. The transformer turns ratio is 5:1 step-down. What is the secondary end-to-end voltage? Sketch the voltage waveform across R and indicate its maximum amplitude. Assume that the rectifier assembly is a common cathode type.

4-35. If the rectifier assembly in Prob. 4-34 has a maximum average dc current rating of 500 mA, what minimum load resistance will you use?

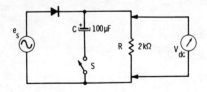

Fig. 4-39. Circuit for Probs. 4-38 through 4-41.

4-36. If a sine wave voltage with a 50-V peak is applied to a circuit like that of Fig. 4-30B, what is the average dc voltage across R if $RC_t > 100$ ms?

4-37. Referring to Prob. 4-36, if the doubler assembly has a maximum average dc current rating of 1 A, what minimum load resistance can you use?

4-38. If in a circuit as shown in Fig. 4-39 the source $e_s = 20$ V rms and is sinusoidal, about what voltage is read by the dc voltmeter V_{dc} when the switch S is open?

4-39. If the circuit in Fig. 4-39 has a sine wave source e_s with a peak of 40 V, about what voltage is indicated by the dc voltmeter V_{dc} when the switch S is open?

4-40. What will the dc voltmeter read in the circuit described in Prob. 4-38 if the switch S is closed?

4-41. What will the dc voltmeter read in the circuit of Fig. 4-39 if the switch S is closed with $e_s = 80$ V (peak) and is sinusoidal?

4-42. In the circuit shown in Fig. 4-40, the source is sinusoidal with a peak voltage $e_{s(peak)} = 30$ V. The transformer has a turns ratio of 5:1 step-up. Assuming that the transformer is ideal, about what dc voltage is read by meter V_{dc}? (Assume switch S is open.)

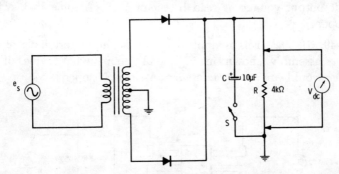

Fig. 4-40. Circuit for Probs. 4-42 through 4-45.

4-43. The voltage applied to the primary in the circuit shown in Fig. 4-40 is 120 V rms. The transformer is ideal and has a step-down turns ratio of 4:1. About what dc voltage is read on the meter V_{dc} when the switch S is open?

4-44. If the switch S is closed in the circuit of Fig. 4-40, what voltage is read on V_{dc} if $e_{s(peak)} = 30$ V, and the transformer turns ratio is 5:1 step-up?

4-45. With the switch S closed in the circuit of Fig. 4-40, what is the voltage read on V_{dc} if $e_s = 120$ V rms and the turns ratio of the transformer is 4:1 step-down?

4-46. The circuit in Fig. 4-41 has a voltage-doubler diode assembly wired in. If $e_s = 120$ V rms and is sinusoidal, about what voltage is read by meter V_{dc}?

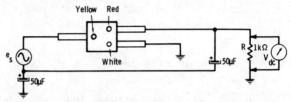

Fig. 4-41. Circuit for Probs. 4-46 and 4-47.

4-47. If the sine wave source has a peak value $e_{s(peak)} = 12$ V in the circuit shown in Fig. 4-41, what dc voltage would you expect to read on meter V_{dc}?

4-48. A bridge rectifier assembly is used in the circuit shown in Fig. 4-42. If the sine wave source has a peak $e_{s(peak)} = 60$ V, what is the dc output voltage as read by meter V_{dc}? (Assume that switch S is open.)

4-49. If 80 V rms is applied to the ac input terminals of the bridge assembly shown in Fig. 4-42, about what value of dc voltage would you expect across R? (Assume S is open).

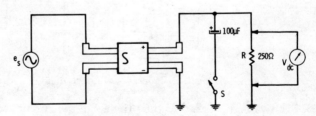

Fig. 4-42. Circuit for Probs. 4-48 through 4-51.

4-50. What is the dc voltage across R when switch S is closed in the circuit described in Prob. 4-48?

4-51. What is the dc voltage across R when S is closed in the circuit described in Prob. 4-49?

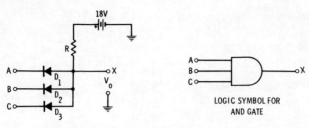

(A) *Three-input* AND *gate circuit and symbol.*

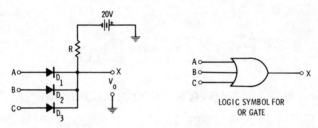

(B) *Three-input* OR *gate circuit and symbol.*

Fig. 4-43. AND and OR gates for Probs. 4-52 through 4-58.

4-52. Fig. 4-43A shows a three-input AND gate. If all three inputs are 0 V with respect to ground, what are voltages V_o, E_{D1}, E_{D2}, E_{D3}, and E_R?

4-53. If 12 V to ground is applied to input A in Fig. 4-43A, while inputs B and C are 0 V to ground, what are voltages V_o, E_{D1}, E_{D2}, E_{D3}, and E_R?

4-54. If all three inputs in Fig. 4-43A are 12 V to ground, what are voltages V_o, E_{D1}, E_{D2}, E_{D3}, and E_R?

4-55. Fig. 4-43B shows a three-input OR gate. If all three inputs are 0 V to ground, what are voltages V_o, E_{D1}, E_{D2}, E_{D3}, and E_R?

4-56. If inputs A, B, and C are 0 V, 12 V, and 0 V respectively in Fig. 4-43B, what are voltages V_o, E_{D1}, E_{D2}, E_{D3} and E_R?

4-57. If all inputs are 12 V to ground in Fig. 4-43B, what are voltages V_o, E_{D1}, E_{D2}, E_{D3}, and E_R?

4-58. In a circuit like that in Fig. 4-11A, sketch e_R where $e_{s(max)} = 5$ V and $R = 200$ kΩ. Assume that the diode is a germanium type which has a reverse leakage of 0.1 mA with a reverse voltage of 50 V.

5

The Zener Diode

5-1 V-I CHARACTERISTICS OF THE ZENER DIODE

In the last chapter we used reverse bias voltages that did not exceed the PIV rating of the diodes. Reverse voltages greater than the PIV rating of a diode may cause the diode to *break down*. Breakdown occurs when a diode under excessive reverse bias loses its high reverse resistance and starts to conduct a large reverse current I_R. This large reverse current causes high power dissipation in the diode which may destroy it. The reverse voltage at which this loss of reverse resistance occurs is called the *breakdown voltage* in ordinary diodes, or it is called the *zener voltage* V_Z in the case of zener diodes. The cause of the increased conductivity of the diode at the breakdown or zener voltage is an increased number of minority carriers. These carriers are produced by the reverse voltage (electromotive force) which tears valence electrons away from their parent atoms. The electrons and holes thus produced become available to support current.

The zener diode differs from the ordinary diode in that it is specifically designed to operate in reverse bias such that V_Z voltage is reached. That is, the zener diode is used so that reverse bias breakdown does occur, but the zener is not ruined by this, provided that its maximum reverse current value is not exceeded. We may refer to this part of the zener characteristic as the *zener region* (see Fig. 5-1). When operating in this region, large changes in reverse current may occur for relatively small changes in reverse

5-1 V-I CHARACTERISTICS OF THE ZENER DIODE

voltage. Note on the characteristic (Fig. 5-1A) that at zener voltage V_Z a reverse current I_{R1} flows. If the reverse current increases to I_{R2} (a large change) the reverse voltage increases to V_Z' (a small change). For most practical purposes, the change in zener voltage is considered negligible. That is, the zener voltage V_Z of a zener diode may usually be assumed a constant value even though the reverse current changes significantly. Thus the actual zener characteristics may be simplified to Fig. 5-2. These characteristics imply

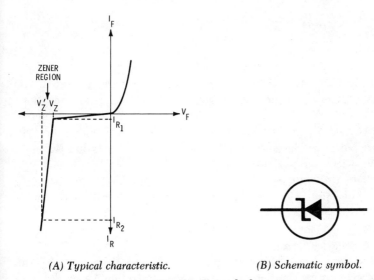

(A) Typical characteristic.　　　(B) Schematic symbol.

Fig. 5-1. Zener diode.

that the zener acts as a short when forward biased, then acts as an open when reversed biased until voltage V_Z is reached, whereupon V_Z is constant with large changes in I_R.

Manufacturers usually specify the V_Z voltage and the maximum power that the zener can safely dissipate at 25°C (about room temperature). This tells you the maximum reverse current $I_{R(\max)}$ that the zener can safely conduct. For example, if the manufacturer specifies $V_Z = 50$ V, $P_{\max} = 500$ mW at 25°C, the power equation $P = EI$ can be solved for I and thus

$$I_{\max} = P/E = 500 \text{ mW}/50 \text{ V} = 10 \text{ mA}$$

We can acquire an understanding of how the zener behaves by starting with a simple circuit like that of Fig. 5-3 and using ideal

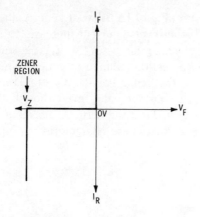

Fig. 5-2. Ideal zener diode characteristic.

zener characteristics. Note that the source E is variable and that it reverse biases the zener D. Suppose that this zener specifications are: $V_z = 25$ V, $P_{max} = 1$ W at 25°C. When $E = 0$ V, of course both V_o and E_R are 0 V too. When $E = 10$ V, the zener is reverse biased but is not in zener conduction. That is, it acts like an open and $V_o = 10$ V while $E_R = 0$ V.

When the source is increased so that $E = 25$ V, the zener is ready to go into zener conduction. However, I_R is still 0 A because the full source voltage is across the zener; that is, $V_o = 25$ V, and no voltage is yet across R, i.e., $E_R = 0$ V. With no voltage across R it cannot conduct current I_R.

Now if source E is increased to 30 V, the voltage across the zener remains constant at 25 V. Thus, the remaining 5 V is dropped across R. Therefore $V_o = 25$ V and $E_R = 5$ V. The reverse current I_R is the current through R and by Ohm's law $I_R \cong 5\text{ V}/250\text{ }\Omega = 20$ mA.

If the source E is increased further to 40 V, the zener voltage remains constant or $V_o = 25$ V. The remaining 15 V is across R and therefore $I_R = 15\text{ V}/250\text{ }\Omega = 60$ mA.

Actually, this last current exceeds the maximum the diode can

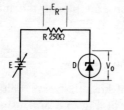

Fig. 5-3. Simple zener diode circuit.

5-2 POWER AND DERATING

handle. With $I_R = 60$ mA, the power consumed by the zener is $P = V_z I_R = 25$ V $\times 60$ mA $= 1.5$ W. This exceeds the maximum specified power of 1 W. To avoid ruining the zener, determine its maximum safe current if it is not specified. In this case we find it is $I_{R(max)} = P/V_z = 1$ W/25 V $= 40$ mA. Thus, 40 mA is the maximum current allowable in R and since R is a fixed 250 Ω, the maximum voltage across it is $E_{R(max)} = I_{R(max)} \times 250$ Ω $= 10$ V. Therefore the source E cannot be adjusted over 35 V if destruction of the zener is to be avoided.

Example 5-1 Suppose that in a circuit like that of Fig. 5-3, the zener has characteristics $V_z = 50$ V, $P_{max} = 6$ W. What maximum reverse current can the zener safely conduct?

Answer 5-1 $I_{max} = 120$ mA.—Solving the power equation $P = EI$ for I, we have $I_{max} = P_{max}/E = 6$ W/50 V $= 120$ mA.

Example 5-2 If the zener in Ex. 5-1 is used in a circuit like that of Fig. 5-3 where the source $E = 35$ V, what is the zener current?

Answer 5-2 $I = 0$.—This zener goes into zener conduction with a reverse voltage of 50 V. With 35 V reverse bias, the zener acts like an open, assuming the ideal zener characteristic of Fig. 5-2, and no current flows.

Example 5-3 Referring to the zener in Ex. 5-1 above, if the source E is increased to 75 V in Fig. 5-3, what is the circuit current?

Answer 5-3 $I = 100$ mA.—With source $E = 75$ V, the zener goes into zener conduction and the voltage across it $V_o = 50$ V. The remaining 25 V appears across the resistance R. Thus, by Ohm's law, $I = E_R/R = 25$ V/250 Ω $= 100$ mA.

Example 5-4 Referring to Ex. 5-3, if voltage E is to remain 75 V but resistance R is to be increased so that the current is reduced to 40 mA, what value of R would you use?

Answer 5-4 $R = 625$ Ω.—With $E = 75$ V and $V_o = 50$ V, 25 V is across R. If the current through R is to be limited to 40 mA, its value, by Ohm's law, must be

$$R = \frac{25 \text{ V}}{40 \text{ mA}} = 625 \text{ Ω}$$

5-2 POWER AND DERATING

The power that a zener can safely dissipate decreases as the surrounding temperature (ambient temperature) increases. Thus, for example, a zener that has a maximum power dissipation capability of 1 W at 25°C may not be capable of 1-W dissipation at temperatures higher than 25°C. To help the designer using zeners avoid ruining them, zener manufacturers supply *derating* curves as shown in Fig. 5-4. The curve in Fig. 5-4 shows that a zener

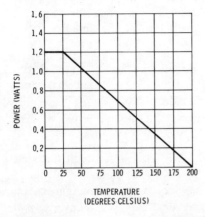

Fig. 5-4. Power rating vs ambient temperature derating curve.

with a maximum power rating of 1.2 W can safely dissipate this maximum power at temperatures 0°C to 25°C. At temperatures above 25°C, the power capability decreases. At 50°C the power capability drops to about 1 W. At 100°C it drops to 0.7 W, etc.

Example 5-5 Suppose that a zener with $V_Z = 100$ V and whose derating curve is Fig. 5-4, is used in a circuit like that of Fig. 5-3 in which $R = 20$ kΩ. The circuit is to be able to work in an environment with an ambient temperature as high as 100°C. What is the maximum source E that can be used?

Answer 5-5 $E = 240$ V maximum.—From the curve of Fig. 5-4 we see that at 100°C, this zener can safely dissipate about 0.7 W. Since $V_Z = 100$ V, the zener can conduct $I_{max} = 0.7$ W/100 V = 7 mA. With this maximum current, the maximum voltage across R is $E_{R(max)} = I_{max}R = 7$ mA × 20 kΩ = 140 V. The source E is the sum of E_R and V_Z. Therefore in this case the source E may safely be as high as $140 + 100 = 240$ V.

Instead of supplying derating curves, the manufacturer may specify a *derating factor*. The derating factor may be given in milliwatts per degree Celsius. For example, if the derating factor given is 5 mW/°C, it means that the zener power dissipating capability

P_D decreases 5 mW for every degree Celsius increase in temperature.

Example 5-6 If a certain zener has $P_D = 1$ W at 25°C, what is its P_D at 75°C if its derating factor is 5 mW/°C?

Answer 5-6 $P_D = 0.75$ W.—The temperature increase here is $75 - 25$ or 50°C. Since P_D decreases 5 mW for every degree increase, the total decrease of P_D is $5 \times 50 = 250$ mW or 0.25 W. Thus the new $P_D = 1 - 0.25 = 0.75$ W.

5-3 THE ZENER AS A VOLTAGE REGULATOR

A popular application of zener diodes is in *voltage regulators*. Voltage regulators used with dc power supplies help keep the voltage across the load constant or nearly constant. An example of this kind of application is shown in Fig. 5-5. The zener is reverse biased by voltage E and is in zener conduction. The zener voltage V_Z is the load voltage E_o. Since V_Z is constant, assuming ideal characteristics, with varying current through the zener, it is possible to have E_o constant even though the load R_L and the output of E of the dc power supply may vary.

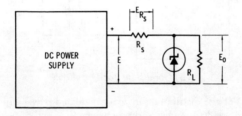

Fig. 5-5. Zener diode regulated power supply.

For example, suppose that $V_Z = 30$ V and E fluctuates from 40 to 55 V. When $E = 40$ V, E_o is held constant at 30 V by the zener and the remaining 10 V appears across the resistor R_s. When E increases to 55 V, E_o remains 30 V, and 25 V in this case is across R_s. Thus, the fluctuations of E cause fluctuation in E_{Rs} but E_o remains constant.

Example 5-7 Suppose that in a circuit like that of Fig. 5-5, voltage E fluctuates between 40 and 55 V, $R_s = 1$ kΩ, $R_L = 4$ kΩ, and $V_Z = 30$ V. Find the maximum and minimum currents through

R_s and the zener, the load current I_L, and the maximum power the zener will dissipate.

Answer 5-7 $I_{Rs(max)} = 25$ mA, $I_{Rs(min)} = 10$ mA, $I_{Z(max)} = 17.5$ mA, $I_{Z(min)} = 2.5$ mA, $I_L = 7.5$ mA, $P_{Z(max)} = 525$ mW.—Assuming ideal zener characteristics, the load voltage is a constant 30 V. Since $R_L = 4$ kΩ constant, a constant load current I_L flows whose value is $I_L = 30$ V/4 kΩ = 7.5 mA.

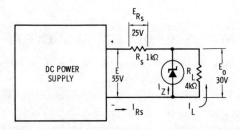

(A) Conditions for E = 55 V.

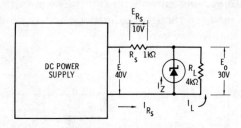

(B) Conditions for E = 40 V.

Fig. 5-6. Voltage distribution in zener regulated power supply where $V_Z = 30$ V and E varies between 40 V and 55 V.

When Supply Output Voltage E = 55 V, as Shown in Fig. 5-6A:

$E_{Rs} = 55 - 30 = 25$ V

$I_{RS(max)} = 25$ V/1 kΩ = 25 mA

$I_{Z(max)} = I_{Rs} - I_L = 25 - 7.5$
$\phantom{I_{Z(max)}} = 17.5$ mA

$P_{Z(max)} = V_Z I_{Z(max)} = 30(17.5$ mA$)$
$\phantom{P_{Z(max)}} = 525$ mW

When Supply Output Voltage E = 40 V, as Shown in Fig. 5-6B:

$E_{Rs(min)} = 40 - 30 = 10$ V

$I_{RS(min)} = 10$ V/1 kΩ = 10 mA

$I_{Z(min)} = 10 - 7.5 = 2.5$ mA

$P_{Z(min)} = 30(2.5$ mA$) = 75$ mW

5-3 THE ZENER AS A VOLTAGE REGULATOR

Since the zener must be capable of at least 525 mW, a 3/4 W or 1 W zener could be used, depending on the operating temperature and available sizes.

Often the output E of the dc power supply in a circuit like that of Fig. 5-5 is constant but the load resistance R_L varies. The zener again will keep the load voltage E_o constant. In such circuits the voltage drop across the series resistor R_s and the current through it are constant. As the load R_L varies, the currents in the zener and load vary. If the load current I_L increases, the zener current I_Z decreases by the same amount, provided the V_Z voltage across the zener is maintained. That is, if the zener is to act like a voltage regulator, it must be operated in its zener region; see Figs. 5-1 and 5-2.

Example 5-8 Suppose that in a circuit like that of Fig. 5-5, the supply voltage $E = 90$ V constantly, $R_s = 5$ kΩ, zener voltage $V_Z = 60$ V, and the load resistance R_L varies between 12 kΩ and infinite ohms (an open). What are the maximum and minimum load currents, the maximum and minimum zener currents, and the maximum power dissipation of the zener?

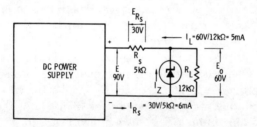

(A) Voltage and current distribution for $R_L = 12$ kΩ.

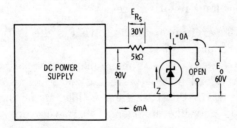

(B) Voltage and current distribution for R_L open.

Fig. 5-7. Zener regulated power supply with R_L varying between 12 kΩ and open.

Answer 5-8 $I_{L(max)} = 5$ mA, $I_{L(min)} = 0$, $I_{Z(max)} = 6$ mA, $I_{Z(min)} = 1$ mA, $P_{Z(max)} = 360$ mW. The calculations are as follows:

When the Load Resistance $R_L = 12$ kΩ, as Shown in Fig. 5-7A:

$E_{Rs} = E - E_o = 90 - 60 = 30$ V

$I_{Rs} = 30$ V/5 k$\Omega = 6$ mA

$I_{L(max)} = E_o/R_L = 60$ V/12 k$\Omega = 5$ mA

$I_{Z(min)} = I_{Rs} - I_L = 6 - 5 = 1$ mA

$P_{Z(min)} = V_Z I_{Z(min)} = 60(1 \text{ mA}) = 60$ mW

When the Load Resistance $R_L = \infty$ Ω, as Shown in Fig. 5-7B:

$E_{Rs} = 90 - 60 = 30$ V

$I_{Rs} = 30$ V/5 k$\Omega = 6$ mA

$I_{L(min)} = 0$

$I_{Z(max)} = I_{Rs} - I_L = 6 - 0 = 6$ mA

$P_{Z(max)} = V_Z I_{Z(max)} = 60(6 \text{ mA}) = 360$ mW

Note that when the load current is maximum, the zener current is minimum, and vice versa.

Example 5-9 Referring to Ex. 5-8, what minimum load resistance R_L can we use in the circuit but still maintain voltage regulation?

Answer 5-9 $R_{L(min)} = 10$ kΩ.—With source $E = 90$ V constant, the voltage E_{Rs} must be 30 V constant if the output is to be kept 60 V constant. Thus, the current through R_s must be 6 mA constantly if we need a constant 30 V across it. This sets the limit on load current I_L. In this case, therefore, the maximum load current is 6 mA at 60 V. By Ohm's law, then, the minimum value of load resistance is $R_L = 60$ V/6 mA $= 10$ kΩ. If a load R_L smaller than 10 kΩ is used, the current through it would be greater than 6 mA, thus drawing more than 6 mA through R_s. This will cause E_{Rs} to be larger than 30 V and E_o to be smaller than 60 V. The zener comes out of zener conduction and voltage regulation ceases.

Example 5-10 Referring to Ex. 5-8, what is the load voltage E_o if the load R_L is reduced to 7 kΩ?

Answer 5-10 $E_o = 52.5$ V.—As mentioned in the last example, if the load R_L is reduced below 10 kΩ, the zener comes out of conduction. The zener in this case then is like an ordinary reverse biased diode and acts like an open. The power supply therefore sees R_s and R_L in series. Solving as a simple series circuit,

$$E_o = \frac{ER_L}{R_L + R_s} = \frac{90 \times 7}{12} = 52.5 \text{ V}$$

5-3 THE ZENER AS A VOLTAGE REGULATOR

Example 5-11 Suppose that in a circuit like that of Fig. 5-5, the supply voltage E varies between 100 V and 140 V, $V_Z = 90$ V and the load $R_L = 4.5$ kΩ constantly. What power must the zener be capable of dissipating and what value of R_s would you use that will provide proper voltage regulation but will provide the minimum possible power dissipation in the zener?

Answer 5-11 $P_Z = 7.2$ W, $R_s = 500$ Ω.—If a smaller value than 500 Ω of R_s is used, the zener must be capable of dissipating more than 7.2 W. If R_s is made larger than 500 Ω, the zener voltage drops below 90 V when E drops as low as 100 V; that is, regulation ceases. We should reason as follows: If voltage regulation is to be maintained, the load current I_L will be constant because R_L is constant. Thus, when the supply voltage E varies, causing E_{Rs} and I_{Rs} to vary, the zener current must vary too. Minimum zener current will flow when supply E is minimum. On the other hand, the zener current is maximum when E is maximum. Assuming an ideal zener characteristic (Fig. 5-2), we can use a zener current of about zero and still maintain V_Z across the zener. Therefore we may allow the zener current to drop to as low as 0 A when the supply voltage E drops to its minimum 100 V. As E fluctuates upward, I_Z can only increase from zero.

Specifically in this case, load current $I_L = 90$ V/4.5 kΩ = 20 mA constantly. (a) When supply $E = 100$ V, $E_{Rs(min)} = 10$ V. Since at this minimum E we can make $I_Z = 0$, we can make I_{Rs} as low as the load current 20 mA. And R_s can be made as large as 10V/20 mA = 500 Ω. (b) When supply $E = 140$ V, $E_{Rs(max)} = 50$ V. With $R_s = 500Ω$, $I_{Rs(max)} = 50$ V/500Ω = 100 mA. Since the load draws a constant 20 mA, the zener must conduct the other 80 mA. That is, $I_{Z(max)} = 80$ mA and the maximum power dissipation is $P_{(max)} = V_Z I_{Z(max)} = 90(80 \text{ mA}) = 7.2$ W.

In practice, it would be wise to use an R_s value slightly smaller than 500 Ω in this case, say 490 Ω or so. This will cause $I_{Z(min)}$ to be a little larger than zero, which tends to prevent the zener from coming out of zener conduction. That is, in practice it takes a small amount of current to hold the zener voltage at V_Z. Note in Fig. 5-1A that with currents less than I_{R1}, the reverse voltage drops below V_Z. This minimum current is called the *holding current*. In Fig. 5-1A, the holding current is about current value I_{R1}.

Example 5-12 Suppose that in a circuit like that of Fig. 5-5, the supply voltage $E = 40$ V constant. The load voltage is to be maintained at 25 V while the load R_L varies between 1 kΩ and 5 kΩ. Choose an R_s value that will maintain regulation and will keep power dissipation in the zener at the minimum possible value.

Answer 5-12 $R_s = 600$ Ω.—If R_s is made larger than 600 Ω, the circuit will not regulate and the output voltage E_o will drop below 25 V when R_L fluctuates to a low of 1 kΩ. With $R_s = 600$ Ω, the zener will dissipate a maximum of 0.5 W. If R_s smaller than 600 Ω is used, the zener at times will dissipate more than 0.5 W. When solving this problem, you could reason as follows: With $E = 40$ V constantly and $E_o = 25$ V constantly, voltage $E_{Rs} = 15$ V constantly. Thus, current I_{Rs} is constant too. And when the load current I_L varies, zener current I_Z varies. Zener current is maximum when load current is minimum and vice versa.

When $R_L = 1$ kΩ, maximum load current will flow, which is $I_{L(max)} = 25$ V/1 kΩ = 25 mA. If we make the minimum zener current zero by using ideal zener characteristics, the current I_{Rs} must be 25 mA. Since $E_{Rs} = E - E_o = 40 - 25 = 15$ V, resistance $R_s = 15$ V/25 mA = 600 Ω.

When $R_L = 5$ kΩ, the load current is minimum $I_{L(min)} = 25$ V/5 kΩ = 5 mA. Since I_{Rs} is a constant 25 mA, the zener conducts the other 20 mA. The maximum zener power dissipation is $P_{max} = V_Z I_{Z(max)} = 25 \times 20$ mA = 0.5 W.

As in Ex. 5-11, a slightly smaller R_s than the calculated value is used to avoid minimum zener currents below the holding current value.

So far we have considered regulated power supplies where either the power supply voltage E varied or the load resistance R_L varied. Frequently, in practice, however, both E and R_L will vary. Thus, both varying quantities must be considered when choosing the proper R_s and zener power capability.

Example 5-13 In a circuit like that of Fig. 5-5, $R_s = 470$ Ω and the zener $V_Z = 12$ V. The load fluctuates between 1 kΩ and 3 kΩ. The supply voltage E fluctuates between 18 V and 22 V. Find the maximum and minimum currents in R_s, in the zener, and in the load R_L, and the maximum power dissipation in the zener.

Answer 5-13 $I_{Rs(max)} = 20.8$ mA, $I_{Rs(min)} = 12.5$ mA, $I_{Z(max)} = 16.8$ mA, $I_{Z(min)} = 0.5$ mA, $I_{L(max)} = 12$ mA, $I_{L(min)} = 4$ mA, $P_{max} = 201.6$ mW.—The calculations are as follows:

5-3 THE ZENER AS A VOLTAGE REGULATOR

When $E = 18$ V and $R_L = 1$ kΩ:

$E_{Rs} = 6$ V

$I_{Rs(min)} = 6$ V/480 Ω $= 12.5$ mA

$I_{L(max)} = 12$ V/1 kΩ $= 12$ mA

$I_{Z(min)} = I_{Rs} - I_L = 12.5 - 12 = 0.5$ mA

$P_Z = (12\ \text{V})(0.5\ \text{mA}) = 6$ mW

When $E = 18$ V and $R_L = 3$ kΩ:

$E_{Rs} = 6$ V

$I_{Rs(min)} = 6$ V/480 Ω $= 12.5$ mA

$I_{L(min)} = 12$ V/3 kΩ $= 4$ mA

$I_Z = 12.5 - 4 = 8.5$ mA

$P_Z = (12\ \text{V})(8.5\ \text{mA}) = 104$ mW

When $E = 22$ V and $R_L = 1$ Ω:

$E_{Rs} = 10$ V

$I_{Rs(max)} = 10$ V/480 Ω $= 20.8$ mA

$I_{L(max)} = 12$ V/1 kΩ $= 12$ mA

$I_Z = 20.8 - 12 = 8.8$ mA

$P_Z = 12\ \text{V}(8.8\ \text{mA}) = 105.6$ mW

When $E = 22$ V and $R_L = 3$ kΩ:

$E_{Rs} = 10$ V

$I_{Rs(max)} = 10$ V/480 Ω $= 20.8$ mA

$I_{L(min)} = 12$ V/3 kΩ $= 4$ mA

$I_{Z(max)} = 20.8 - 4 = 16.8$ mA

$P_{Z(max)} = 12\ \text{V}(16.8\ \text{mA}) = 201.6$ mW

Example 5-14 Referring to Ex. 5-13, what maximum value of R_s can be used with the E and R_L variations given? Assume ideal zener characteristics.

Answer 5-14 $R_s = 500$ Ω.—With values of R_s larger than 500 Ω, this circuit will not regulate when E and R_L are minimum. As shown in the previous example, the zener current I_Z dropped to its minimum value when E and R_L were minimum. We can design to make $I_{Z(min)}$ smaller than 0.5 mA. In fact, with ideal zener characteristics we can theoretically use $I_{Z(min)} = 0$. Thus, when $E = 18$ V and $R_L = 1$ kΩ, the voltage $E_{Rs} = 6$ V, and the load current $I_L = 12$ V/1 kΩ $= 12$ mA. Since we assume $I_{Z(min)} = 0$, then $I_{Rs} = I_Z + I_L = 0 + 12$ mA $= 12$ mA. Thus, the maximum value of $R_s = 6$ V/12 mA $= 500$ Ω.

When assuming ideal zener characteristics, the maximum value of R_s that can be used in a circuit like that of Fig. 5-5 so that the zener regulates all of the time can be found by the formula

$$R_{s(max)} = \frac{E_{min} - V_Z}{I_{L(max)}} \quad (5\text{-}1)$$

where

E_{min} is the minimum output voltage from the dc power supply,

V_Z is the zener voltage, which is also the output voltage,
$I_{L(\max)}$ is the maximum load current.

If we would have used this formula to solve Ex. 5-14, we would have worked as follows:

$$R_{s(\max)} = \frac{E_{\min} - V_Z}{I_{L(\max)}} = \frac{18 \text{ V} - 12 \text{ V}}{12 \text{ mA}} = 500 \text{ }\Omega$$

In practice, when using this formula, the next lower value of resistor available is used instead of the calculated one. That is, in the circuit in Ex. 5-14, about 490 Ω or 480 Ω or even 470 Ω would be used. This is done to avoid a minimum zener current lower than the zener holding current, and it does not appreciably add to the zener power dissipation.

The maximum power that the zener must be capable of dissipating can be determined with the formula

$$P_{Z(\max)} = V_Z \left(\frac{E_{\max} - V_Z}{R_s} - I_{L(\min)} \right) \qquad (5\text{-}2)$$

where
$E_{\max}$ is the maximum output voltage from the dc power supply,
V_Z is the zener or output voltage,
$I_{L(\min)}$ is the minimum load current,
R_s is the resistance in series with the zener and load R_L.

Example 5-15 Suppose that you are to design the zener regulator portion of a circuit like that of Fig. 5-5. The voltage E varies between 80 V and 110 V, R_L varies between 2 kΩ and 10 kΩ and V_Z = 50 V. (a) What are the theoretical maximum R_s and minimum P_Z if the zener is assumed ideal? (b) What are practical maximum R_s and minimum P_Z values?

Answer 5-15 (a) $R_{s(\max)} = 1.2$ kΩ, $P_{Z(\max)} \cong 2.25$ W, (b) $R_s = 1.1$ kΩ, $P_{Z(\max)} = 2.475$ W.—In practice, a 2.5-W to 3-W zener would be used if the ambient temperature is about 25°C. Larger than a 3-W zener would be used at higher temperatures. Answers (a) were found as follows: By Eq. 5-1

$$R_s = \frac{80 \text{ V} - 50 \text{ V}}{25 \text{ mA}} = \frac{30 \text{ V}}{25 \text{ mA}} = 1.2 \text{ k}\Omega$$

By Eq. 5-2

$$P_{Z(\max)} = 50 \left(\frac{110 \text{ V} - 50 \text{ V}}{1.2 \text{ k}\Omega} - \frac{50 \text{ V}}{10 \text{ k}\Omega} \right) = 50 \left(\frac{60 \text{ V}}{1.2 \text{ k}\Omega} - 5 \text{ mA} \right)$$
$$= 50(50 \text{ mA} - 5 \text{ mA}) = 2.25 \text{ W}$$

5-4 PRACTICAL CONSIDERATIONS

In previous examples, we assumed that the zener voltage V_Z is constant though the current in the zener changes. While this assumption is adequate for most practical purposes, we may sometimes have to consider the more accurate facts. One fact is that the zener voltage *does* change with changes in zener current as we noted in Fig. 5-1A. When conducting in the zener region, the zener displays resistive properties. That is, as the zener current increases, the voltage drop across it increases, too, like in a resistive circuit where voltage is proportional to current. For conduction in the zener region, the real zener and its equivalent circuit are shown in Fig. 5-8A. Compare this to the ideal zener and its equiv-

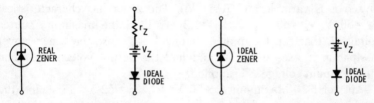

(A) *Real zener and its equivalent circuit.* (B) *Ideal zener and its equivalent circuit.*

Fig. 5-8. Zener diodes and equivalent circuits when in zener conduction.

alent in Fig. 5-8B. The voltage source V_Z represents the voltage at which the zener goes into zener conduction. The resistance r_Z represents the zener resistance which accounts for the increased voltage drop when current is increased. The value of r_Z is usually supplied by the manufacturer, or it may be determined from the zener characteristics. For example, if in Fig. 5-1A, $V_Z = 50$ V, $V_Z' = 50.8$ V, $I_{R1} = 0.5$ mA, and $I_{R2} = 10.5$ mA, the zener resistance is

$$r_Z = \frac{\text{change in zener voltage}}{\text{change in zener current}} \cong \frac{(50.8 - 50)\text{V}}{(10.5 - 0.5)\text{mA}} = \frac{0.8 \text{ V}}{10 \text{ mA}} = 80 \text{ }\Omega$$

Fig. 5-9. Equivalent of zener circuit of Fig. 5-3.

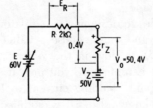

Further suppose that in a circuit like that of Fig. 5-3, the zener characteristics are: $V_Z = 50$ V, $r_Z = 80$ Ω. If $E = 60$ V and $R = 2$ kΩ, what is voltage V_o? Replacing the zener with its equivalent, we get the circuit of Fig. 5-9. Since the zener drop is approximately 50 V when in zener conduction, there is about 10 V across resistor R. Thus, by Ohm's law

$$I \cong \frac{E_R}{R} = \frac{10 \text{ V}}{2 \text{ k}\Omega} = 5 \text{ mA}$$

With this current, the voltage drop across the zener resistance is $Ir_Z = (5 \text{ mA})(80 \text{ }\Omega) = 0.4$ V. Thus, V_o, the actual voltage across the zener, is $V_Z + Ir_Z = 50 + 0.4 = 50.4$ V.

Example 5-16 Suppose that in a circuit like that of Fig. 5-3, the source E is variable and $R = 1$ kΩ. The zener has characteristics: $V_Z = 30$ V, $r_Z = 50$ Ω, $P_{Z(max)} = 1.2$ W. Find the maximum source voltage E that can be applied that will not cause the zener power dissipation to exceed 1.2 W, and find the output voltage V_o when the maximum voltage E is applied.

Answer 5-16 Maximum $E \cong 70$ V, $V_{o(max)} = 32$ V.—First find the maximum zener current. Since the zener voltage is about 30 V, $I_{max} = P_{max}/V_Z = 1.2$ W/30 V $\cong 40$ mA. With this maximum current, we can determine the maximum voltage across R. So $E_{R(max)} = I_{R(max)}R \cong 40$ mA$(1$ kΩ$) = 40$ V. The source $E = E_R + V_Z \cong 40 + 30 = 70$ V. The voltage across the zener resistance is $Ir_Z = 40$ mA$(50 $ Ω$) = 2$ V. Thus $V_o = V_Z + Ir_Z = 30 + 2 = 32$ V.

The manufacturers may also list zener holding currents I_H. Typically, it may be about 0.25 mA for zeners rated for 1 W or less, 0.5 mA for zeners rated 1 W to 5 W, 1 mA for zeners rated 10 W, and may be as high as 5 mA for 50 W zeners. These holding currents I_H also depend on the zener voltage V_Z. For example, a zener with a lower V_Z may require a larger holding current than one with a larger V_Z, even though the power ratings of the two zeners are the same.

If you are designing a regulator for a power supply like that in Fig. 5-5, and must choose a value of R_s that will allow a holding current I_H to flow in the zener when I_L is maximum and power supply E is minimum, you may use Eq. 5-1. Simply add the specified minimum holding current to $I_{L(max)}$. Thus, if the holding current I_H is known, Eq. 5-1 is modified to

5-4 Practical Considerations

$$R_s = \frac{E_{min} - V_Z}{I_{Lmax} + I_H} \quad (5\text{-}1A)$$

Example 5-17 Suppose that you are to design the regulator for a circuit like that of Fig. 5-5 where the dc supply voltage E varies between 40 V and 50 V, the load R_L varies between 10 kΩ and an open, and the load voltage is to be regulated at about 30 V. Find the necessary value of R_s that will provide a minimum zener current of 0.5 mA. Also find the maximum power dissipation in the zener.

Answer 5-17 $R_{s(max)} = 2.86$ kΩ, $P_{Z(max)} = 1.05$ W.—Using Eq. 5-1A we get

$$R_{s(max)} = \frac{E_{(min)} - V_Z}{I_{L(max)} + 0.5 \text{ mA}} = \frac{(40-30) \text{ V}}{(3+0.5) \text{ mA}} = \frac{10 \text{ V}}{3.5 \text{ mA}} = 2.86 \text{ kΩ}$$

where

$$I_{L(max)} = \frac{V_Z}{R_{L(min)}} = \frac{30 \text{ V}}{10 \text{ kΩ}} = 3 \text{ mA}$$

The maximum power dissipated by the zener is found by Eq. 5-2:

$$P_{Z(max)} = V_Z \left(\frac{E_{(max)} - V_Z}{R_s} - I_{L(min)} \right)$$

$$= 30 \left(\frac{(40-30)\text{V}}{2.86 \text{ kΩ}} - 0 \right) = 1.05 \text{ W}$$

where $I_{L(min)} = 0$ because the load may at times be open.

In practice a zener with a power rating a little larger than 1.05 W would be used if the ambient temperature is about 25°C. Availability would probably dictate a 1.2 W or 1.5 W zener.

Example 5-18 Referring to Ex. 5-17, if the specification sheet for the available zener lists the characteristics, $V_Z = 30$ V, $r_Z = 100$ Ω, what will the load voltage variations be as the load R_L varies from an open to 10 kΩ?

Answer 5-18 E_o varies from about 30 V to 30.7 V as the load varies from an open to 10 kΩ.—The circuit was designed for an $I_{Z(min)} = 0.5$ mA, thus the minimum drop across r_Z is $(0.5 \text{ mA})(100 \text{ Ω}) = 0.05$ V. Since the output voltage $E_o = V_Z + Ir_Z$, we have a minimum output voltage of 30 V + 0.05 V = 30.05 V. Call it 30 V. for all practical purposes. The maximum zener current flows when the voltage E is maximum and the load current I_L is minimum.

With maximum input, $E = 50$ V, the current $I_{Rs(max)} = E_{Rs}/R_s = 20$ V$/2.86$ k$\Omega = 7$ mA. Since $I_{Rs(max)} = I_{Z(max)} + I_{L(min)}$ and since $I_{L(min)} = 0$, we find in this case that $I_{L(max)} = I_{Rs(max)} = 7$ mA. Thus the maximum drop across r_Z is $(7$ mA$)(100 \ \Omega) = 0.7$ V. Therefore

$$E_{o(max)} = V_Z + 0.7 \text{ V} = 30.7 \text{ V}$$

5-5 THE ZENER AS A VOLTAGE LIMITER

Consider now a zener used in a circuit like that of Fig. 5-10A. Suppose that the applied voltage is sinusoidal as shown in Fig. 5-10B and has a peak value $e_{max} = 20$ V. What is the output voltage e_o waveform if the zener voltage $V_Z = 30$ V? In this case, the peak of the input voltage is not as large as the zener voltage V_Z and the zener never goes into zener conduction. In this case, the zener acts like an ordinary diode and it half-wave rectifies the ac input voltage. On the positive alternations of the input voltage e_{in}, the zener is reverse biased and essentially acts as an open, assuming

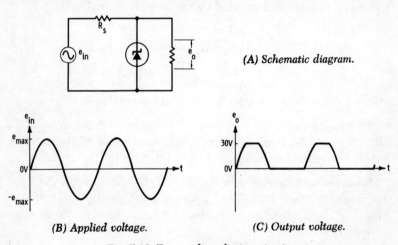

Fig. 5-10. Zener voltage limiter circuit.

that R_s is much smaller than the reverse resistance. Thus the full positive alternations appear across the diode and the output. On negative alternations, the zener is forward biased and acts like a short if we assume that this is an ideal zener. Therefore the voltage across the zener which is the ouput voltage e_o is 0 V on these

5-5 THE ZENER AS A VOLTAGE LIMITER

alternations or at most a few tenths of a volt if simplified characteristics are used.

Consider now applying e_{in} that has a peak value of 40 V. What is the output voltage waveform? In this case, the input exceeds the zener voltage V_Z and we have zener conduction during portions of the positive alternations. That is, as e_{in} rises above 30 V on positive alternations, zener conduction occurs and e_o remains nearly a constant 30 V. Voltage waveform e_o is shown in Fig. 5-10C, which is clipped at 30 V.

What does the output voltage e_o waveform look like for the circuit of Fig. 5-11 if the input voltage e_{in} is as shown in Fig. 5-10B, where $e_{max} = 20$ V and $V_Z = 30$ V for both zeners? The output voltage in this case would be a sine wave. Neither zener ever goes into conduction because e_{max} of the input is less than the zener voltage. On positive alternations D_2 tends to be forward biased, but D_1 is reverse biased and acts like an open. On the negative alternations D_1 is forward biased, but D_2 is reverse biased, causing the branch containing both diodes to act like an open on these alternations, too. Thus, as the input e_{in} see it, R_s and R_L are in series. The voltage e_o across R_L is the input e_{in} minus the drop across R_s. That is, e_o has the same waveform as e_{in}, but is reduced in amplitude.

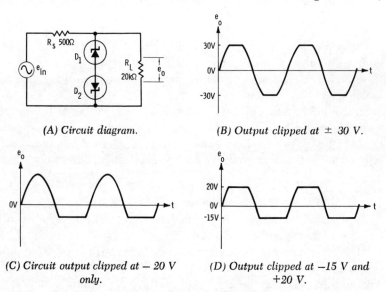

(A) Circuit diagram.

(B) Output clipped at ± 30 V.

(C) Circuit output clipped at − 20 V only.

(D) Output clipped at −15 V and +20 V.

Fig. 5-11. Two-zener voltage limiter.

If input voltage e_{in} to the circuit in Fig. 5-11A is increased so that $e_{max} = 40$ V, zener D_1 goes into zener conduction for portions of the positive alternations. D_2 is forward biased during those times and acts like a short. Therefore the output voltage e_o, which is the voltage across zener D_1, cannot have an amplitude greater than the zener voltage. Similarly during portions of the negative alternations, D_2 goes into zener conduction while D_1 is forward biased. Thus e_o has a waveform shown in Fig. 5-11B, which looks like the input waveform but is clipped at ±30 V.

Example 5-19 Suppose that in a circuit like Fig. 5-11A voltage e_{in} has a peak of 40 V and the V_Z of both zeners is 30 V, but in this case both R_s and R_L are 20 kΩ. What does output e_o waveform look like?

Answer 5-19 Voltage e_o is an unclipped sine wave with a peak value of 20 V. In this case, neither zener goes into zener conduction because half of the input voltage is dropped across R_s. Thus, even when e_{in} is at its peak 40 V, half of it, 20 V, is dropped across R_s, causing $e_{o(max)}$ never to be greater than 20 V. It takes 30 V to bring the reverse biased diode into zener conduction. Therefore the branch containing the zeners always appears open, and, as e_{in} sees it, R_s and R_L are in series during all portions of the input cycle.

Example 5-20 Suppose that in a circuit like that of Fig. 5-11A with resistance values as shown, sine wave input $e_{in(max)} = 30$ V, $V_Z = 40$ V for D_1, and $V_Z = 20$ V for D_2. What does waveform e_o look like?

Answer 5-20 See Fig. 5-11C.—Since V_Z of D_1 exceeds the input peak, D_1 never goes into zener conduction and acts like an open during the positive alternations. The negative alternations are clipped at −20 V, however, due to the zener conduction of D_2.

If in Fig. 5-11A, e_{in} has a peak of 30 V, $V_Z = 20$ V for D_1 and $V_Z = 15$ V for D_2, the output waveform e_o looks like Fig. 5-11D. The positive alternations are clipped at 20 V and the negative alternations are clipped at −15 V.

The zeners in Fig. 5-11A are said to be *back-to-back*. They are often used this way to protect a load R_L from excessive voltage. For example, R_L may be a speaker and e_{in} the output signal from a hi-fi or stereo. The back-to-back zeners clip high-amplitude bursts

5-6 ZENERS AND RECTIFIERS

from the hi-fi and thus protect the speaker from damage. In this application, the zeners are chosen with V_Z values large enough so that signals with normal amplitude are not clipped.

5-6 ZENERS AND RECTIFIERS

If a sine wave voltage like that of Fig. 5-10B is applied to a circuit like that in Fig. 5-12A, the output voltage e_o will have a waveform like that in Fig. 5-12B, provided that the peak of the input voltage is larger than the zener V_Z. For example, if e_{in} has a peak of 50 V and the zener $V_Z = 30$ V, the output e_o is clipped at

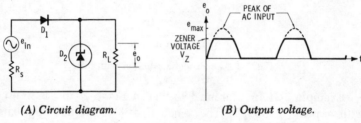

(A) Circuit diagram.   (B) Output voltage.

Fig. 5-12. Zener rectifier-limiter circuit.

30 V. The ordinary diode D_1 is forward biased on the positive alternations while the zener D_2 is reverse biased. In the case, D_2 acts like an open for e_{in} voltages up to 30 V, the D_2 goes into zener conduction and thus holds e_o at nearly a constant 30 V. The difference between e_{in} and e_o, when the zener is conducting, appears across R_s. Component R_s may be a resistor placed in the circuit by the designer or it may simply represent the internal resistance of the source of voltage e_{in}. In our analysis, we may assume that its resistance is much smaller than resistance R_L.

Thecircuit in Fig. 5-12A is often used in practice but in modified form (Fig. 5-13A). In the latter, a filter capacitor C is used to remove the ripple, resulting in the output waveform of Fig. 5-13B. The capacitor C is charged on the first positive alternation to the zener voltage V_Z. In between the positive alternations the capacitor tends to hold the voltage e_o relatively constant. The output e_o is more or less constant, depending on the values of R_L and C. A larger resistance R_L and/or larger capacitance C tends to make e_o more constant; that is, it reduces ripple.

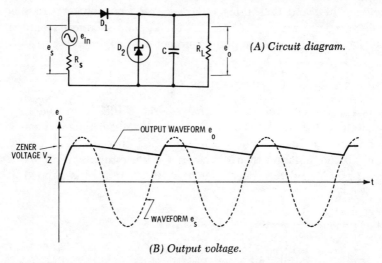

(B) Output voltage.

Fig. 5-13. Zener rectifier-limiter with filter.

Example 5-21 In a circuit like that of Fig. 5-13, e_{in} is sinusoidal with a peak of 155 V. The zener $V_Z = 90$ V. What approximate dc voltage would you read across the load R_L?

Answer 5-21 Output e_o is about 90 V dc.—The capacitor charges to the zener voltage V_Z, which decays only a little between positive alternations if capacitance C and resistance R_L are relatively large.

PROBLEMS

5-1. Referring to typical zener characteristics Fig. 5-1A, if V_Z, V_Z', I_{R1} and I_{R2} are 11V, 11.4 V, 0.5 mA, and 9 mA respectively, what is the zener resistance r_Z?

5-2. Referring to typical zener characteristics (Fig. 5-1A) if V_Z, V_Z', I_{R1}, and I_{R2} are 11 V, 11.4 V, 0.5 mA, and 9 mA respectively, what is the zener resistance r_Z?

5-3. What does an equivalent circuit of the zener described in Prob. 5-1 look like if the zener is in zener conduction?

5-4. What does an equivalent circuit of the zener described in Prob. 5-2 look like if the zener is in zener conduction?

5-5. If in a circuit like that of Fig. 5-3, the source E varies between 30 V and 45 V, and $R = 5$ kΩ, find the minimum and maxi-

PROBLEMS

mum circuit currents and output voltage V_o. Assume that the zener is ideal with $V_z = 20$ V.

5-6. If in a circuit like that of Fig. 5-3, the source E varies between 16 V and 21 V, and $R = 5$ kΩ, find the minimum and maximum circuit currents and output voltage V_o. Assume that the zener is ideal with $V_z = 11$ V.

5-7. If in a circuit like that of Fig. 5-3, the source E varies between 30 V and 45 V, $R = 5$ kΩ, $V_z = 20$ V, and $r_z = 182$ Ω, find the minimum and maximum values of the output voltage V_o. Account for r_z's effect.

5-8. If in a circuit like that of Fig. 5-3, the source E varies between 16V and 21 V, $R = 5$ kΩ, $V_z = 11$ V, and $r_z = 47$ Ω, find the minimum and maximum values of the output voltage V_o. Account for r_z's effect.

5-9. A certain zener has a derating curve as shown in Fig. 5-4. What power can it dissipate at 125°C?

5-10. If a zener has the derating curve of Fig. 5-4, what power can it dissipate at 138°C?

5-11. A certain zener is rated 1 W at 25°C and it has a derating factor 7 mW/°C. What power is it capable of dissipating at 45°C?

5-12 A certain zener is rated 10 W at 25°C and has a derating factor 80 mW/°C. What power is it capable of dissipating at 100°C?

5-13. Suppose that in a circuit like that of Fig. 5-5, the load is a constant 10 kΩ, E_o is to be regulated at 80 V, and the power supply E varies between 90 V and 120 V. Using ideal zener characteristics, what maximum value of R_s can be used to provide the least possible power dissipation in the zener and maintain regulation?

5-14. Suppose that in a circuit like that of Fig. 5-5, the load is a constant 3 kΩ, E_o is to be a regulated 15 V, and the power supply E varies between 20 V and 25 V. Using ideal zener characteristics, what maximum value of R_s can be used to provide the least possible power dissipation in the zener and maintain regulation of E_o?

5-15. Rework Prob. 5-13 where the minimum zener holding current, as specified by the manufacturer, is 1 mA.

5-16. Rework Prob. 5-14 where the minimum zener holding current is specified as 0.5 mA.

5-17. In a circuit like that of Fig. 5-5, the load R_L is constantly an open, E_o is to be a regulated 12 V, and the power supply E varies between 16 V and 20 V. If the minimum zener holding current is specified as 1 mA, what maximum value of R_s can you use that will cause the least possible power dissipation in the zener, yet will maintain regulation of E_o?

5-18. In a circuit like that of Fig. 5-5, the load R_L is constantly open, E_o is to be a regulated 150 V, and the power supply E varies between 170 V and 200 V. If the specified minimum zener holding current is 1 mA, what maximum value of R_s can you use that will cause the least possible power dissipation in the zener, yet will maintain regulation of E_o?

5-19. In a circuit like that of Fig. 5-5, the load R_L is constantly an open, E_o is to be regulated at 12 V, the voltage source E varies between 16 V and 20 V, and $r_Z = 50$ Ω. Accounting for r_Z effect and referring to Prob. 5-17 and its answer, what are the maximum and minimum values of E_o?

5-20. In a circuit like that of Fig. 5-5, the load R_L is constantly an open, E_o is to be regulated at 150 V, the source voltage E varies between 170 V and 200 V, and $r_Z = 300$ Ω. Referring to Prob. 5-18 and its answer, what are the maximum and minimum values of E_o, accounting for r_Z effect.

5-21. In a circuit like that of Fig. 5-5, the power supply voltage $E = 20$ V constantly, the load R_L varies between 300 Ω and 1.2 kΩ, and E_o is to be regulated at 12 V. Assuming that the zener is ideal, what maximum value of R_s can you use that will cause the least possible dissipation in the zener and will maintain regulation of E_o? What is the maximum power dissipation in the zener?

5-22. In a circuit like that of Fig. 5-5, the power supply voltage $E = 120$ V constantly, the load R_L varies between 3 kΩ and an open, and E_o is to be regulated at 90 V. Assuming that the zener is ideal, what maximum value of R_s can you use that will cause the least possible power dissipation in the zener and will maintain regulation of E_o? What is the maximum power dissipation in the zener?

5-23. Rework Prob. 5-21 using a minimum zener holding current of 2 mA; that is, the zener is not ideal.

5-24. Rework Prob. 5-22 using a minimum zener holding current of 3 mA.

5-25. In a circuit like that of Fig. 5-5, the power supply voltage E varies between 120 V and 140 V, the load R_L varies from 25 kΩ to an open, and the load voltage E_o is to be regulated at 100 V. What maximum value of R_s can you use that will cause the least possible power dissipation and maintain a minimum holding current of 1 mA in the zener? What is the maximum power dissipation?

5-26. In a circuit like that of Fig. 5-5, the power supply voltage E varies between 30 V and 32 V, the load R_L draws currents between 0 A to 20 mA at 24 V. What maximum value of R_s can you use that will cause the least possible power dissipation and will maintain a minimum holding current of 2 mA in the zener? What is the maximum power dissipation?

5-27. Referring to Prob. 5-25, if the zener $r_Z = 120$ Ω, what are the maximum and minimum values of E_o when the input E varies from 120 V to 140 V?

5-28. Referring to Prob. 5-26, if the zener $r_Z = 50$ Ω, what are the maximum and minimum values of E_o when the input E varies from 30 V to 32 V?

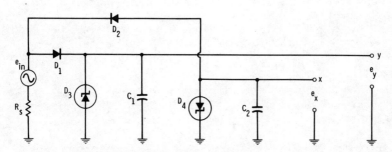

Fig. 5-14. Power supply provides negative and positive dc output voltages with respect to ground.

5-29. In the circuit of Fig. 5-14, the voltage e_{in} is sinusoidal with a peak of 160 V. The zeners D_3 and D_4 have V_Z values 100 V and 120 V respectively. What are the output voltages at points x

and y with respect to ground, and what are the PIV values across diodes D_1 and D_2?

5-30. In the circuit of Fig. 5-14, the voltage e_{in} is sinusoidal with a peak of 18 V. Zeners D_3 and D_4 have V_Z values 12 V and 8 V respectively. What are the output voltages at points x and y with respect to ground, and what are the PIV values across the diodes D_1 and D_2?

6

The Junction Transistor

In modern equipment, transistors are often used where vacuum tubes were used years ago. While tubes are holding their own in some areas of electronics, they have some significant disadvantages compared to transistors. Transistors are generally more rugged, more efficient, smaller, and usually less expensive per unit. They also have simpler power supply and circuit wiring requirements. These are important facts in the design of portable equipment such as radio receivers, hearing aids, and transmitter-receiver sets for aircraft, boats, and land vehicles. Tubes may still have the edge where the power and frequency requirements are very high, but they're "losers" in almost all other areas.

In amplifiers, transistors enable us to increase the amplitude of small electrical signals. That is, small ac voltages or currents may be amplified to larger ac voltages or currents. Also, a small *change* in a dc value may be amplified to a larger *change* of dc. As a switch, the transistor may be used to turn a circuit current on or off with no arcing, which is often a problem with mechanical switches.

6-1 THE PNP AND NPN JUNCTIONS

The junction transistor has *two* junctions of p and n semiconductor materials, and for this reason is also called a *bipolar* semiconductor and is to be distinguished from unipolar devices, which are covered later.

If an n material is "sandwiched" between to p materials, it is a *pnp* transistor (Fig. 6-1A). On the other hand, if a p material is "sandwiched" between two n materials, it is an *npn* transistor (Fig. 6-1B). Leads are connected to each of the semiconductors and they are called *emitter, base,* and *collector.* The base material is always "sandwiched" between the emitter and collector and is considerably thinner than either. The base is also *lightly doped.* This means that relatively few impurity atoms have been added to the base material, thus providing it with few majority carriers. Schematic symbols of pnp and npn transistors are shown in Figs. 6-1C and 6-1D.

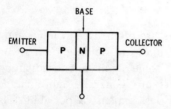

(A) Construction of pnp transistor.   (B) Construction of npn transistor.

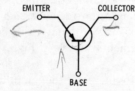

(C) Schematic symbol of pnp transistor.   (D) Schematic symbol of npn transistor.

Fig. 6-1. Basic junction transistors.

In almost all practical applications of transistors, the collector-base junction is reverse biased as shown in Figs. 6-2A and 6-2B. Note that the collector and base form a diode junction. Therefore the term "reverse bias" has the same meaning here as it did in diode theory in Chap. 2.

With the emitter lead hanging open as shown, only a small leakage current flows, which is called I_{CBO} or is abbreviated I_{CO}. The subscript *CBO* means Collector-to-Base leakage with emitter Open. We will, for most practical purposes, assume that this leakage I_{CO} is negligible, just as we usually assume that the reverse

6-1 THE PNP AND NPN JUNCTIONS

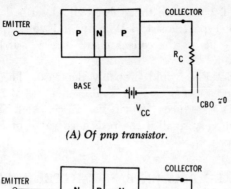

(A) Of pnp transistor.

(B) Of npn transistor.

Fig. 6-2. Reverse bias of collector-base junction.

bias leakage of an ordinary diode is about zero. An exception might be where the operating temperature is quite high and the transistors are the germanium type. As with diodes in reverse bias, high temperatures cause thermally generated electron-hole pairs and thus minority carriers, especially in germanium semiconductors.

Example 6-1 Suppose that in a circuit like that of Fig. 6-2, the collector supply voltage $V_{CC} = 20$ V and collector resistance $R_C = 5$ kΩ. What voltage would you expect to read across the collector and base leads if the emitter is open as shown?

Answer 6-1 The collector-to-base voltage $V_{CB} \cong 20$ V.—In a properly working transistor we expect negligible leakage I_{CBO}. Thus, the voltage drop across R_C should be negligible. Therefore, with emitter open, we should read the full source V_{CC} across the collector and base leads.

The transistor is not definitely okay if $V_{CB} \cong V_{CC}$, however. It may have been previously ruined by excessive current so that the collector-base junction acts like a permanent open which also causes $V_{CB} \cong V_{CC}$. Needless to say, a transistor with an open collector-base junction will not work properly.

Example 6-2 Suppose that the circuit described in Ex. 6-1 has a germanium type transistor. What would you expect to happen to the voltage across the collector and base V_{CB} if heat is applied to the transistor?

Answer 6-2 V_{CB} would probably decrease.—The heat will significantly increase I_{CO} and thus also increase the voltage drop across R_C. This reduces V_{CB} if V_{CC} is constant. The change in V_{CB} would be very small if the resistance value of R_C is small, say about 100 Ω or less.

6-2 FORWARD BIAS ON BASE-EMITTER JUNCTIONS AND ITS EFFECTS

In practice, the emitter-base junction of the transistor is usually forward biased as shown in Fig. 6-3. This junction acts much like an ordinary forward biased diode. A forward bias current I_E flows

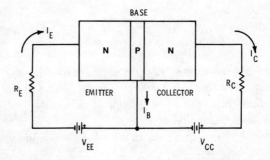

(A) Typical bias polarities of npn transistor.

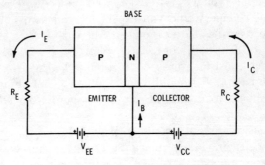

(B) Typical bias polarities of pnp transistor.

Fig. 6-3. Bias of base-emitter junction.

6-2 FORWARD BIAS ON BASE-EMITTER JUNCTION AND ITS EFFECTS

in the emitter lead. Its magnitude depends largely on the values of V_{EE} and R_E. That is, V_{EE}, R_E, and the forward biased emitter-base junction are in series, and since the emitter-to-base voltage V_{EB} is in the order of a few tenths of a volt, which may usually be considered negligible, most of the source V_{EE} is dropped across R_E. Thus, by Ohm's law, the current is

$$I_E \cong \frac{V_{EE}}{R_E} \tag{6-1}$$

In other words, since the emitter-base junction is a diode, it has typical forward bias characteristics much like Figs. 2-4 and 2-5, depending on whether the transistor is a silicon or a germanium type respectively. Using simplified characteristics, we assume that the emitter-to-base forward voltage V_{EB} is constant at about 0.7 V for silicon transistors or about 0.3 V for germanium transistors. In practice, V_{EE} is usually much larger than V_{EB}, and we can disregard this forward drop, which makes Eq. 6-1 valid and very useful.

Interesting things happen when the emitter-base junction is forward biased. Not only does it cause an emitter current I_E flow but also causes a significant collector current I_C. In fact, the emitter and collector currents are very nearly equal over wide range of I_E variations. That is, when $I_E = 0$, $I_C \cong 0$, and if $I_E = 1$ mA, $I_C \cong 1$ mA, etc. Therefore we can say that

$$I_E \cong I_C \tag{6-2}$$

For example, suppose that in a circuit like that of Fig. 6-3B, $V_{EE} = 20$ V and $R_E = 10$ kΩ. What are the emitter and collector currents? By Eq. 6-1, we find that

$$I_E \cong \frac{20 \text{ V}}{10 \text{ k}\Omega} = 2 \text{ mA}$$

Since I_C is approximately equal to I_E, according to Eq. 6-2, $I_C \cong 2$ mA.

Resistance R_C sees the collector as a constant-current source. That is, even though the resistance of R_C is varied within limits, the current I_C through it remains nearly constant.

The cause of collector current dependence on emitter current can be explained as follows: If the emitter-base junction of an npn transistor is forward biased, the majority carriers in the emitter and base materials are repelled by V_{EE} toward the junction as

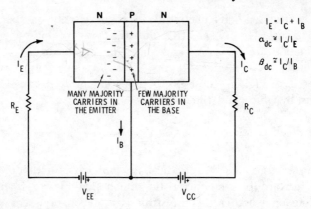

Fig. 6-4. Current relationships in npn transistor.

shown in Fig. 6-4. At the junction the electrons from the n-type material tend to recombine with holes from the p-type material. However, since the base material is lightly doped, there are far more electrons than holes converging at the junction. Thus, as electrons move from left to right toward the junction, only a few recombine with holes. The rest of the electrons flow across the thin base material and across the collector-base junction. These electrons eventually flow out of the collector and support collector current I_C. Since most of the electrons from the emitter flow across the base to the collector, the collector current I_C is very nearly equal to emitter current I_E. The small difference between emitter and collector currents is the base current I_B. That is,

$$I_E - I_C = I_B \quad \text{or} \quad I_E = I_C + I_B \qquad (6\text{-}3)$$

For example, if $I_E = 2$ mA and $I_C = 1.96$ mA, the base current $I_B = 2 - 1.96 = 0.04$ mA or 40 μA. Or if $I_C = 2.94$ mA and $I_B = 60$ μA, the emitter current $I_E = 2.94$ mA $+ 0.06$ mA $= 3$ mA. In all good-quality junction transistors, the base current I_B is very small compared to I_E and I_C.

6-3 MORE ON HOW I_E, I_C, AND I_B ARE RELATED

If current I_E is varied, the currents I_C and I_B tend to vary by the same percentage. For example, suppose that initially $I_E = 2$ mA, $I_C = 1.96$ mA, and $I_B = 40$ μA and that emitter current is increased by 50 percent so that $I_E = 3$ mA. The collector and base currents will

6-3 MORE ON HOW I_E, I_C, AND I_B ARE RELATED

likewise increase by about 50 percent so that $I_C = 2.94$ mA and $I_B = 60$ μA, provided that R_C is not too large and V_{CC} is not too small in a circuit like that of Fig. 6-3. The effects of R_C and V_{CC} are covered in a later chapter. In general, we can say that the ratios I_C/I_E and I_C/I_B tend to be constant even though the individual currents may vary. In the preceding example we initially had

$$\frac{I_C}{I_E} = \frac{1.96 \text{ mA}}{2 \text{ mA}} = 0.98 \quad \text{and} \quad \frac{I_C}{I_B} = \frac{1.96 \text{ mA}}{0.04 \text{ mA}} = 49$$

After the increases the ratios became

$$\frac{I_C}{I_E} = \frac{2.94 \text{ mA}}{3 \text{ mA}} = 0.98 \quad \text{and} \quad \frac{I_C}{I_B} = \frac{2.94 \text{ mA}}{0.06 \text{ mA}} = 49$$

In this case the ratios remained constant, and for most practical purposes we may assume that they always do. Since these ratios are practically constants, they become useful factors when designing and analyzing transistor circuits. If the leakage current I_{CO} is negligible, the ratio I_C/I_E may be called the *dc alpha* a_{dc} or h_{FB} of the transistor. The ratio I_C/I_B may be called the *dc beta* β_{dc} or h_{FE} of the transistor. In equation form these relationships are

$$a_{dc} = h_{FB} \cong \frac{I_C}{I_E} \tag{6-4}$$

$$\beta_{dc} = h_{FE} \cong \frac{I_C}{I_B} \tag{6-5}$$

These factors are called *parameters* of the transistor and are supplied by the manufacturer.

In good-quality transistors a_{dc} is almost equal to 1 and β_{dc} is 50 or more. If the β_{dc} of a transistor is known, its a_{dc} can be found with the equation

$$a_{dc} = \frac{\beta_{dc}}{\beta_{dc} + 1} \tag{6-6}$$

If the a_{dc} of the transistor is known, its β_{dc} can be determined by the equation

$$\beta_{dc} = \frac{a_{dc}}{1 - a_{dc}} \tag{6-7}$$

For example, suppose that in a circuit like that of Fig. 6-3A, $I_E = 4$ mA and the transistor $a_{dc} = 0.99$. From this information we can find I_C, I_B, and β_{dc} as follows: Solving Eq. 6-4 for I_C we get $I_C = \alpha_{dc} I_E$. Therefore,

$$I_C = 0.99(4 \text{ mA}) = 3.96 \text{ mA}$$

Now solving Eq. 6-3 for I_B, we get $I_B = I_E - I_C$. Therefore,

$$I_B = 4 - 3.96 = 0.04 \text{ mA} = 40 \text{ } \mu\text{A}$$

By Eq. 6-7 we find that

$$\beta_{dc} = \frac{0.99}{1 - 0.99} = 99$$

We can also show that by Eq. 6-5

$$\beta_{dc} = \frac{3.96}{40 \text{ } \mu\text{A}} = 99$$

We can in most cases assume that the α_{dc} of the transistor is 1 and that therefore collector and emitter currents are equal.

Example 6-3 In a circuit like that of Fig. 6-5, $V_{EE} = 12$ V, $V_{CC} = 20$ V, $R_E = 3$ kΩ, and $R_C = 2$ kΩ. Find the approximate values of I_E and I_C.

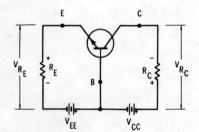

Fig. 6-5. Simple transistor circuit.

Answer 6-3 $I_E \cong I_C \cong 4$ mA.—Assuming that the voltage across the forward biased emitter-base junction is negligible we can use Eq. 6-1. Thus,

$$I_E \cong \frac{12 \text{ V}}{3 \text{ k}\Omega} = 4 \text{ mA}$$

Assuming $\alpha_{dc} = 1$, we can say that I_C is about 4 mA, too.

Example 6-4 Referring to Ex. 6-3, what is the voltage drop across R_C and what is the voltage across the collector and base terminals (voltage across points C and B)?

6-3 MORE ON HOW I_E, I_C, AND I_B ARE RELATED

Answer 6-4 $V_{Rc} \cong 8$ V, $V_{CB} \cong 12$ V.—Since in our answer to Ex. 6-2 we found that $I_C \cong 4$ mA, we find the voltage across R_C by Ohm's law: $V_{Rc} \cong I_C R_C = (4 \text{ mA} \times 2 \text{ k}\Omega) = 8$ V. By Kirchhoff's law the voltage across the collector and base (V_{CB}) plus the voltage V_{Rc} must be equal to V_{CC}:

$$V_{CC} = V_{CB} + V_{Rc} \tag{6-8}$$

Therefore, by rearranging, we find

$$V_{CB} = V_{CC} - V_{Rc} \cong 20 - 8 = 12 \text{ V}$$

At this time, we can determine the approximate power dissipated in the transistor, which is sometimes called the *collector power* P_C. It is simply the product of the collector-to-base voltage V_{CB} and the collector current I_C. Thus, in Ex. 6-3,

$$\begin{aligned} P_C &\cong V_{CB} I_C \\ P_C &\cong (12 \text{ V})(4 \text{ mA}) = 48 \text{ mW} \end{aligned} \tag{6-9}$$

Since the emitter-to-base voltage V_{EB} is considered negligible, we may say that the emitter and base leads are at practically the same potential. Therefore, the voltage from collector to base, V_{CB}, must be about equal to the collector-to-emitter voltage V_{CE}; that is,

$$V_{CB} \cong V_{CE}$$

Substituting this into Eq. 6-9, we can show that the collector power can also be found with the equation

$$P_C \cong V_{CE} I_C \tag{6-10}$$

Example 6-5 Suppose that in a circuit like that of Fig. 6-5, $V_{EE} = V_{CC} = 20$ V, $R_E = 10$ kΩ and $R_C = 7$ kΩ. What are the values of I_E, I_C, and P_C?

Answer 6-5 $I_E \cong I_C \cong 2$ mA, $P_C \cong 12$ mW.—Since the forward voltage across the emitter-base junction is negligible, $I_E \cong V_{EE}/R_E = 20$ V/10 kΩ = 2 mA. Thus, $I_C \cong 2$ mA also. By Ohm's law, then, $V_{Rc} \cong I_C R_C \cong (2 \text{ mA})(7 \text{ k}\Omega) = 14$ V, and therefore $V_{CB} = V_{CC} - V_{Rc} = 20 - 14 = 6$ V. Finally $P_C \cong V_{CB} I_C \cong (6 \text{ V})(2 \text{ mA}) = 12$ mW.

REVIEW QUESTIONS

6-1. What are some uses of transistors?

6-2. What other term is used to describe the junction transistor?

6-3. How is the collector-base junction biased?

6-4. How is the emitter-base junction biased?

6-5. What is the approximate dc voltage drop across the emitter-base junction while emitter current is flowing, if the transistor is a silicon type?

6-6. What do the symbols I_{CBO} and I_{CO} mean?

6-7. How are the values of I_E, I_C, and I_B related?

6-8. What symbol is used to equal approximately the ratio I_C/I_E?

6-9. What symbol is used to equal approximately the ratio I_C/I_B?

6-10. What might cause a change in leakage across a reverse biased collector-base junction?

6-11. What do the symbols V_{EB}, V_{CB}, and V_{CE} mean?

6-12. How are the collector-to-base voltage V_{CB} and the collector-to-emitter voltage V_{CE} related?

PROBLEMS

6-1. In a circuit like that of Fig. 6-5, $V_{EE} = V_{CC} = 24$ V, $R_E = 10$ kΩ and $R_C = 7.5$ kΩ. Find the emitter and collector currents, the collector-to-base voltage V_{CB}, and the power dissipated by the transistor. Assume that $h_{FB} \cong 1$.

6-2. If $V_{EE} = 15$ V, $V_{CC} = 24$ V, $R_E = 10$ kΩ, and $R_C = 6$ kΩ in a circuit like that of Fig. 6-5, what are the emitter and collector currents, the collector-to-base voltage V_{CB}, and the power dissipated by the transistor? Assume that $\alpha_{dc} \cong 1$.

6-3. Suppose that the manufacturer lists $h_{FE} = 50$ for the transistor used in Prob. 6-1. What is the more exact value of the transistor h_{FB}?

6-4. Suppose that the manufacturer lists $h_{FE} = 200$ for the transistor used in Prob. 6-2. What is the more exact value of the transistor h_{FB}?

6-5. Referring to Prob. 6-1 and the circuit in Fig. 6-5, if the transistor $\alpha = 0.98$, what is the base current I_B?

Problems

6-6. Referring to Prob. 6-2 and the circuit in Fig. 6-5, if the transistor $\beta = 200$, what is the base current I_B?

6-7. In a circuit like that of Fig. 6-5, $V_{EE} = 12$ V, $V_{CC} = 18$ V, and $R_C = 5$ kΩ. If the required collector-to-base voltage V_{CB} is 8 V, what value of R_E would you use?

6-8. If in a circuit like that of Fig. 6-5, $V_{EE} = 30$ V, $V_{CC} = 40$ V, and $R_C = 50$ kΩ, what value of R_E would you use if the required value of voltage across the collector and base V_{CB} is 15 V?

6-9. Referring to Fig. 6-6, if $R_C = 5$ kΩ and V_{CB} is to be 12 V, what value of R_E would you use? The given circuit voltages are with respect to ground.

6-10. In the circuit in Fig. 6-6, resistance $R_C = 5$ kΩ and the required value of V_{CB} is 5 V, what value of R_E will you use?

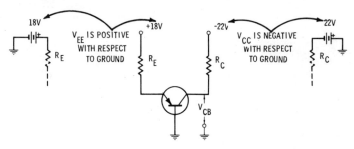

Fig. 6-6. Grounded-base npn transistor circuit.

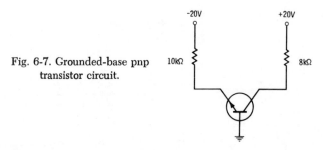

Fig. 6-7. Grounded-base pnp transistor circuit.

6-11. Referring to Prob. 6-9, what is the collector power dissipation, P_C?

6-12. Referring to Prob. 6-10, what is the collector power dissipation, P_C?

6-13. Is the transistor in Fig. 6-5 a pnp or an npn?

6-14. Is the transistor in Fig. 6-6 a pnp or an npn?

6-15. Find the values of V_{CB} and P_C in the circuit of Fig. 6-7.

6-16. If the collector voltage source V_{CC} in Fig. 6-7 is increased from 20 V to 25 V, what are the new values of V_{CB} and P_C?

7

The Common-Base Connection

Actually we have already considered junction transistors in common-base connection, sometimes called *grounded-base connection*, in the last chapter (Figs. 6-3, 6-4, 6-5, and 6-7). We learned that the transistor in this kind of connection acts like a constant-current source, within limits, as the load resistance R_C sees it. That is, in a circuit like Fig. 6-7, I_C tends to remain constant, dependent only on I_E even though R_C and V_{CB} vary. We can now pursue methods for finding the limits in which this is true.

7-1 CUTOFF AND SATURATION DEFINED

Suppose that in a circuit like that shown in Fig. 7-1 we intend to vary I_E from zero to larger values. Will I_C always follow the changes in I_E according to Eq. 6-2? The answer is *no*. There is an upper limit to which I_C can increase, and it is dictated largely by the circuit component values. When I_C reaches its limit, it will not increase further even though I_E does. If I_E continues to increase but I_C does not follow, the collector is said to be *saturated*. It is caused by the fact that at some larger I_C value, the voltage V_{CB} becomes zero and no longer reverse biases the collector-base junction. Note in Fig. 7-1 that by Kirchhoff's law, the sum of voltages V_{CB} and $I_C R_C$ is equal to V_{CC}. Thus, as I_C increases, the voltage across R_C increases, causing V_{CB} to decrease. If I_C is increased to the point where $I_C R_C$ becomes equal to the source V_{CC}, voltage V_{CB} becomes zero and normal transistor action ceases and the collector is said to be saturated.

For example, suppose that in Fig. 7-1 we want to determine the range in which I_E can be varied so that always $I_E \cong I_C$. Of course, if R_E has infinite resistance, or if the emitter lead is open, then $I_E = 0$. And if leakage I_{CO} is negligible, I_C will be about zero by Eq. 6-2; that is, $I_E \cong I_C \cong 0$. Thus the minimum possible I_C is about zero. When I_C is zero, the voltage drop across R_C is about zero, too, causing the full source voltage V_{CC} to appear across the collector and base. That is, in this case, when $I_C \cong 0$, then $V_{CB} \cong V_{CC} = 30$ V and the transistor is said to be *cut off*. In other words, a transistor is in *cutoff* when its emitter current is reduced to zero, which causes the collector current to drop to about zero, too, which in turn causes the entire collector source voltage V_{CC} to appear across the collector and base leads.

We can increase I_E, and consequently I_C, by decreasing the resistance of R_E. As we do, the voltage across R_C increases and V_{CB} decreases. Current I_C can be increased with increases of I_E until the

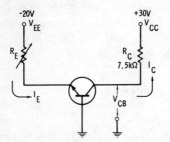
Fig. 7-1. Basic common-base transistor circuit.

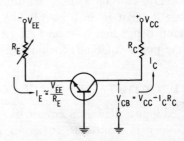
Fig. 7-2. Example of common-base circuit.

voltage across R_C is about equal to the source V_{CC} or 30 V in this case. The current that causes this maximum voltage across R_C is called the *collector saturation current* $I_{C(\text{sat})}$. Its value can be determined by the equation

$$I_{C(\text{sat})} \cong \frac{V_{CC}}{R_C} \qquad (7\text{-}1)$$

For the circuit in Fig. 7-1

7-1 CUTOFF AND SATURATION DEFINED

$$I_{C(\text{sat})} \cong \frac{30 \text{ V}}{7.5 \text{ k}\Omega} = 4 \text{ mA}$$

Therefore, in this circuit, $I_E \cong I_C$ for the range from $I_E = 0$ to $I_E \cong 4$ mA. If we want to increase I_E, and consequently I_C to about 4 mA, the resistance R_E must be reduced to 20 V/4 mA = 5 kΩ, by Eq. 6-1.

Example 7-1 Suppose that in a circuit such as that in Fig. 7-2, $V_{EE} = -21$ V, $V_{CC} = 24$ V, and $R_C = 8$ kΩ. What is the collector saturation current $I_{C(\text{sat})}$, and what value of R_E will cause the collector to saturate?

Answer 7-1 $I_{C(\text{sat})} \cong 3$ mA, $R_E \cong 7$ kΩ.—The collector is saturated when the voltage across R_C is about equal to the source V_{CC}. Therefor by Eq. 7-1

$$I_{C(\text{sat})} \cong \frac{24 \text{ V}}{8 \text{ k}\Omega} = 3 \text{ mA}$$

The emitter current I_E that causes saturation is about 3 mA too. This $I_E \cong 3$ mA flows when $R_E \cong 21$ V/3 mA $\cong 7$ kΩ, by Eq. 6-1 when solved for R_E.

Example 7-2 Suppose that in a circuit like that of Fig. 7-2, $V_{EE} = -20$ V, $V_{CC} = 20$ V, and that the manufacturer of the transistor specifies that the collector current should never exceed 10 mA. What value of collector resistance could you use that would certainly prevent over 10 mA of collector current?

Answer 7-2 $R_C = 2$ kΩ.—A resistance R_C can be chosen so that the full source V_{CC} is dropped across it when the maximum allowed collector current flows through it. Thus, by Ohm's law

$$R_C = \frac{20 \text{ V}}{10 \text{ mA}} = 2 \text{ k}\Omega$$

Example 7-3 Suppose $V_{EE} = 12$ V and $V_{CC} = 12$ V in a circuit like that of Fig. 7-2, and you are required to make $V_{CB} = 6$ V and $I_E = 1$ mA. What values of R_E and R_C will you use?

Answer 7-3 $R_E = 12$ kΩ, $R_C = 6$ kΩ.—Solving Eq. 6-1 for R_E we have $R_E \cong V_{EE}/I_E$, and since I_E is specified at 1 mA, $R_E \cong 12$ V/1 mA $= 12$ kΩ. With the collector-to-base voltage specified at 6 V, which means that the collector is not saturated in this case, the other 6 V of the 12 V source V_{CC} is dropped across R_C. Since I_E is 1 mA, current I_C must be about 1 mA too. Thus, by Ohm's law $R_C \cong 6$ V/ 1 mA = 6 kΩ.

Example 7-4 Referring to the circuit described in Ex. 7-3, suppose that R_E and R_C remain constant at 12 kΩ and 6 kΩ respectively but the source V_{EE} is varied from -18 V to -6 V. What maximum and minimum values of I_E, I_C and V_{CB} will be caused by these V_{EE} variations?

Answer 7-4 When $V_{EE} = -18$ V, $I_E \cong I_C \cong 1.5$ mA, and $V_{CB} \cong 3$ V. When $V_{EE} = -6$ V, $I_E \cong I_C \cong 0.5$ mA, and $V_{CB} \cong 9$ V.—When $V_{EE} = -18$ V, nearly this full voltage is across R_E, and therefore $I_E \cong V_{EE}/R_E = 18$ V/12 kΩ = 1.5 mA, and $V_{CB} = V_{CC} - I_C R_C \cong 12 - (1.5 \text{ mA})(6 \text{ k}\Omega) = 12 - 9 = 3$ V. When $V_{EE} = -6$ V, resistor R_E has about 6 V dropped across it and $I_E \cong 6$ V/12 kΩ = 0.5 mA, and $V_{CB} = 12 - (0.5 \text{ mA})(6 \text{ k}\Omega) = 12 - 3 = 9$ V.

7-2 COLLECTOR CHARACTERISTICS AND LOAD LINES

The collector characteristics of the common-base transistor circuit are shown in Fig. 7-3. They indicate the dependence of I_C on I_E and the independence of I_C on V_{CB}. For example, if $V_{CB} = 6$ V

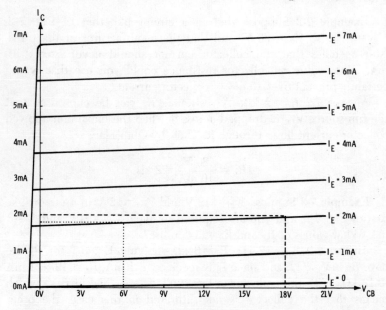

Fig. 7-3. Collector characteristics of common-base transistor circuit.

7-2 COLLECTOR CHARACTERISTICS AND LOAD LINES

and $I_E = 2$ mA, we can note that $I_C \cong 2$ mA as projected with the dotted line in Fig. 7-3. If V_{CB} is increased, say to 18 V, while I_E is still equal to 2 mA, we project with the dashed line and note that I_C is still approximately equal to 2 mA. Since a large change in voltage V_{CB} has very little effect on the current I_C, we may assume that it has none for most practical purposes. With this assumption we may use simplified collector characteristics like those in Fig. 7-4, which are sometimes called *ideal common-base characteristics*. In these ideal characteristics the I_E curves are perfectly horizontal, thus ignoring the fact that large changes in V_{CB} has a slight effect on I_C.

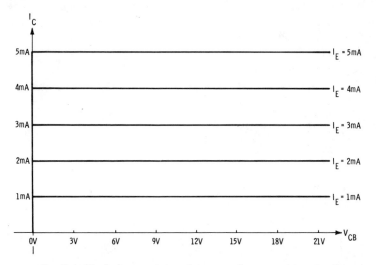

Fig. 7-4. Ideal characteristics of common-base transistor circuit.

Collector characteristics can be obtained with the aid of the circuit of Fig. 7-5. We start by adjusting R_1 so that $I_E = 1$ mA as read on milliammeter M_1 while the slide on R_2 is in the left of center position (low reading on voltmeter M_3). We then read meters M_2 and M_3 and plot a I_C vs V_{CB} point on graph paper. Now the slide of R_2 is moved a little to the right (V_{CB} is increased). If I_E changes, R_1 is adjusted to make it 1 mA again. Though V_{CB} increases noticeably, the increase in I_C is usually very small. Nevertheless, another I_C vs V_{CB} point is plotted. This process is repeated until several points are obtained; that is, V_{CB} is increased by increments and I_E is readjusted to 1 mA, if necessary, each time. The

I_C and V_{CB} values obtained after each increase are the coordinates of one point. A smooth curve drawn through several such points is nearly a horizontal line like the one marked $I_E = 1$ mA in Fig. 7-3.

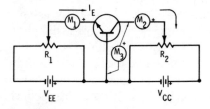

Fig. 7-5. Circuit used to plot collector characteristics of common-base circuit.

The curves marked $I_E = 2$ mA or $I_E = 3$ mA are similarly plotted except that we use R_1 to hold the reading on M_1 constant at 2 mA or 3 mA respectively while R_2 is increased by increments to obtain I_C vs V_{CB} points.

When the collector resistors R_C and the sources V_{CC} are known in circuits like those of Figs. 7-1 and 7-2, we can sketch *load lines*

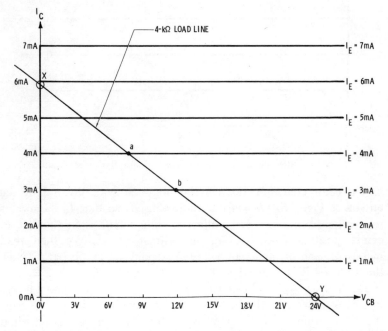

Fig. 7-6. Load line drawn on the ideal collector characteristics of a common-base circuit.

7-2 COLLECTOR CHARACTERISTICS AND LOAD LINES

on the transistor collector characteristics. Load lines, as you will see, will show us *graphically* the variations in I_C and V_{CB} when I_E varies. For example, suppose that in Fig. 7-2, source $V_{CC} = 24$ V and $R_C = 4$ kΩ. The load line, in this case, is drawn on the collector characteristics as shown in Fig. 7-6. Note that it crosses the vertical I_C axis at 6 mA (point X) and that it crosses the horizontal axis V_{CB} at 24 V (point Y). Whenever a single resistor R_C is used as the load on the collector, as in the case of Fig. 7-2, the load line always crosses the I_C axis at the saturation current value. Thus, in this example we can show by Eq. 7-1 that point X is at

$$I_{C(\text{sat})} \cong \frac{V_{CC}}{R_C} = \frac{24 \text{ V}}{4 \text{ k}\Omega} = 6 \text{ mA}$$

Also the load line crosses the V_{CB} axis at the collector source V_{CC} value, which in this example is 24 V.

This load line shows us that if $I_E = 6$ mA the curve marked $I_E = 6$ mA crosses the load line at coordinates $I_C = 6$ mA, $V_{CB} = 0$, which is the saturation point X.

Similarly, if $I_E = 4$ mA, the 4-mA curve crosses the load line at point a, which has coordinates $I_C = 4$ mA, $V_{CB} = 8$ V.

When $I_E = 3$ mA, the 3-mA curve crosses the load line at point b, which has coordinates $I_C = 3$ mA, $V_{CB} = 12$ V.

Example 7-5 Suppose that in a circuit like that of Fig. 7-2, $V_{EE} = -12$ V, $V_{CC} = 12$ V, $R_E = 12$ kΩ, and $R_C = 6$ kΩ. Solve for I_E, sketch a load line on the characteristics in Fig. 7-6, and graphically determine V_{CB}.

Answer 7-5 $I_E \cong 1$ mA, see Fig. 7-7. $V_{CB} \cong 6$ V.—By Eq. 6-1 we solve for emitter current $I_E \cong 12$ V/12 kΩ = 1 mA. The current at point X in Fig 7-7 is the saturation current. Thus, by Eq. 7-1 its value is $I_{C(\text{sat})} \cong 12$ V/6 kΩ = 2 mA. The load line also crosses point Y, which is at 12 V and is equal to the V_{CC} voltage. Note that the $I_E = 1$ mA curve crosses the load line at point Q, which is directly above $V_{CB} = 6$ V. Also note that these answers are consistent with the ones solved mathematically in Ex. 7-3, which used the same circuit.

Example 7-6 If in the circuit of Fig. 7-2, $V_{EE} = -12$ V, $V_{CC} = 12$ V, $R_E = 24$ kΩ, and $R_C = 6$ kΩ, sketch a load line and graphically determine V_{CB} on the typical collector characteristics in Fig. 7-3 or the ideal characteristics of Fig. 7-4.

Answer 7-6 $I_E \cong 0.5$ mA, $V_{CB} \cong 9$ V.—Since the load in the collector R_C and the source V_{CC} are the same as in the previous example, the load line is the same as shown in Fig. 7-7. In this case

$$I_E \cong \frac{V_{EE}}{R_E} \cong \frac{12 \text{ V}}{24 \text{ k}\Omega} = 0.5 \text{ mA}$$

If an $I_E = 0.5$ mA curve is not given, we can estimate one. In Fig. 7-7 we would estimate an $I_E = 0.5$ mA curve being halfway between the 1 mA curve and the V_{CB} axis. This estimated curve crosses the load line at point S, which is above $V_{CB} \cong 9$ V.

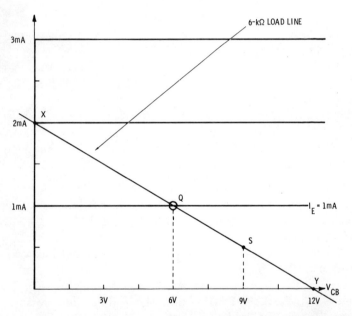

Fig. 7-7. Load line when $V_{CC} = 12$ V and $R_C = 6$ kΩ in circuit of Fig. 7-2.

Point Q for the circuit described in Ex. 7-5 is called the *operating point* of that circuit. Similarly, point S is the *operating point* for the circuit in Ex. 7-6. Knowing how to find the operating point will come in handy when we analyze and design amplifiers in later sections. Since the operating point has dc (steady values) of collector current I_C and collector voltage V_{CB} as its coordinates, it is often called the *quiescent point*, which means *steady-state point*.

7-3 SUPERPOSITION OF DC AND AC SIGNALS IN THE COMMON-BASE AMPLIFIER

We have to this point considered the dc values I_E, I_C, I_B, V_{CB}, etc., and how they are established in the common-base circuit. When the common-base circuit is used as an amplifier, ac signals are superimposed on the dc values. In this section we will consider a common-base circuit used as an amplifier, its input as dc "sees" it, its input as ac signal "sees" it, its output as dc "sees" it, and its load as ac signal "sees" it. In other words, the amplifier can be analyzed by drawing its *dc equivalent circuit* to determine the dc voltage and current distribution and by drawing its *ac signal equivalent circuit* to determine the ac signal voltage and current distribution. We will also note the graphs of the voltages and currents and that the symbols used are carefully chosen to identify where in the circuit a specific voltage or current is and whether it is pure dc, pure ac, or ac superimposed on dc.

A simple common-base amplifier is shown in Fig. 7-8. An ac signal voltage v_s is applied to the input (emitter circuit) through a coupling capacitor C_1. For the present we will assume that the capacitor reactance is negligible at the frequency of v_s. The purpose of the capacitor is to block dc current through the source v_s. That is, as far as the dc source V_{EE} "sees" it, the input looks like the circuit of Fig. 7-9. The forward biased emitter-base diode junction is shown as a diode and we may assume that it has only a few tenths of a volt dc drop. As in the previous chapter $I_E \cong V_{EE}/R_E$ by Eq. 6-1.

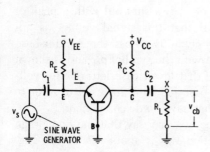
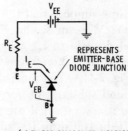

Fig. 7-8. Common-base amplifier where v_s is the input signal and v_{cb} the output signal.

Fig. 7-9. Dc equivalent of the input of the common-base amplifier shown in Fig. 7-8.

THE COMMON-BASE CONNECTION

From the point of view of the signal source v_s, the input to the circuit in Fig. 7-8 looks like the circuit of Fig. 7-10A or Fig. 7-10B. In Fig. 7-10A the source v_s "sees" the dc source V_{EE} as a short to ground if the internal resistance of V_{EE} is negligible. This places the emitter-base diode junction in parallel (shunt) with R_E. The parallel combination of the diode junction and R_E is in series with C_1 and v_s. The diode junction offers resistance to ac signal current through it. This ac resistance is called the *dynamic emitter resistance* r_e'. Its value is relatively small—typically 100 Ω or less. Since signal "sees" resistance r_e' looking into the emitter, and if the reactance of C_1 is assumed negligible, the ac equivalent circuit is more simply shown as Fig. 7-10B. Note that R_E and r_e' are in parallel. Since R_E is usually in the order of thousands of ohms, and r_e' is typically 100 Ω or less, most of the signal current that flows out of the source v_s will flow through r_e'. That is, the signal current drawn by the dc bias resistor R_E is considered negligible.

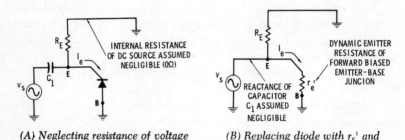

(A) *Neglecting resistance of voltage source V_{EE}.*

(B) *Replacing diode with r_e' and neglecting C_1 reactance.*

Fig. 7-10. Ac equivalent of input of common-base amplifier.

Now suppose that the output of v_s is a sinusoid with a peak of 2 mV. We can see in Fig. 7-10B that this signal voltage is applied directly to r_e', which is to say it is across the emitter-base junction. It is therefore referred to with the symbol v_{eb} and is shown in Fig. 7-11A. You must note here that lower case letters v for voltage and subscript eb for emitter-base junction are used to refer to pure ac values, which are values that have no dc component.

If the transistor is a silicon type, we may say that the dc voltage drop across the emitter-base junction V_{EB} is about 0.7 V (see Fig. 7-11B). Note that capital (upper case) letter V for voltage and subscript letters EB for emitter-base junction, are used in reference to pure dc values, which are values that have no ac component.

7-3 SUPERPOSITION OF DC & AC SIGNALS IN THE AMPLIFIER

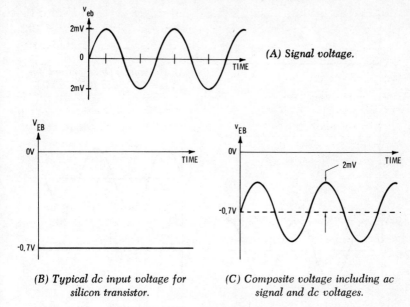

(A) Signal voltage.

(B) Typical dc input voltage for silicon transistor.

(C) Composite voltage including ac signal and dc voltages.

Fig. 7-11. Voltages at emitter-base junction.

The actual voltage across the emitter-base junction has both the ac and dc values superimposed. This composite voltage is shown in Fig. 7-11C and is referred to with the symbol v_{EB}. The lower case v for voltage and upper case subscript letters EB for emitter-base junction are used when referring to ac signals with a dc component, that is, when ac is superimposed on dc.

The dc and ac signal currents may similarly be sketched. For example, suppose that $V_{EE} = -20$ V, $V_{CC} = 24$ V, $R_C = 10$ kΩ and $R_E = 20$ kΩ in the circuit of Fig. 7-8. Then the dc emitter current $I_E \cong 20$ V/20 kΩ $= 1$ mA, which may be plotted as in Fig. 7-12A.

The amount of emitter signal current i_e depends on the value of r_e'. That is, by Ohm's law $i_e \cong v_s/r_e'$ in the equivalent circuit of Fig. 7-10B. Later you will learn to estimate the values of r_e' but for this example let's suppose that it is about 100 Ω. Since the peak value of v_s is given as 2 mV, we find that

$$i_{e(\text{peak})} \cong \frac{v_{s(\text{peak})}}{r_e'} = \frac{2 \text{ mV}}{100 \text{ Ω}} = 0.02 \text{ mA} = 20 \text{ μA}$$

This ac signal component i_e is shown in Fig. 7-12B. In the actual circuit of Fig. 7-8, i_e is superimposed on the dc component current

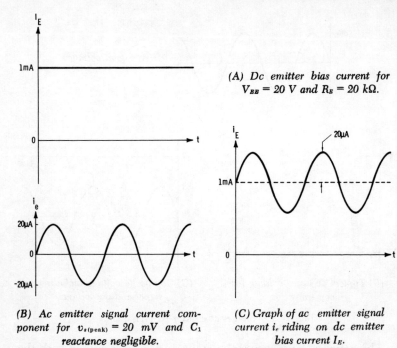

(A) Dc emitter bias current for $V_{BB} = 20$ V and $R_E = 20$ kΩ.

(B) Ac emitter signal current component for $v_{s(peak)} = 20$ mV and C_1 reactance negligible.

(C) Graph of ac emitter signal current i_e riding on dc emitter bias current I_E.

Fig. 7-12. Emitter currents of common-base amplifier of Fig. 7-8.

I_E, as shown in Fig. 7-12C. Their composite is called i_E. Note again how upper and lower case letters are used. The dc component is I_E, the ac component is i_e, and the composite signal is i_E.

The collector currents of the circuit of Fig. 7-8 are shown in Fig. 7-13. Using the value given in the last example, $I_E \cong 1$ mA in the circuit of Fig. 7-8, and therefore $I_C \cong 1$ mA. As shown in Fig. 7-14. I_C is dc and flows up through R_C and the V_{CC} supply. It cannot flow through R_L because of the dc blocking capacitor C_2. Since $I_C \cong 1$ mA, it causes $I_C R_C = (1 \text{ mA})(10 \text{ k}\Omega) = 10$ V drop across R_C. Thus, the dc voltage across the collector and base is $V_{CB} \cong 24 - 10 = 14$ V. See Figs. 7-14B and 7-14C.

Ac equivalent circuits of the output of the common-base amplifier of Fig. 7-8 are shown in Fig. 7-15. Note in Fig. 7-15A that the transistor symbol is replaced with a symbol of a constant-current source whose output is the collector signal current i_c, the dc source V_{CC} is replaced with a short, assuming that its internal resistance is about zero ohms, and capacitor C_2 is replaced with a

7-3 SUPERPOSITION OF DC & AC SIGNALS IN THE AMPLIFIER

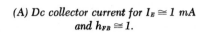

(A) Dc collector current for $I_E \cong 1$ mA and $h_{FB} \cong 1$.

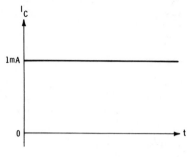

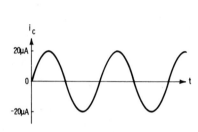

(B) Ac collector current component for $i_{e(peak)} \cong 20$ μA and $h_{fb} \cong 1$.

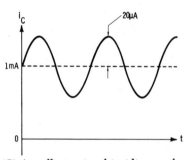

(C) Ac collector signal i_c riding on the dc collector current I_C.

Fig. 7-13. Collector currents of common-base amplifier of Fig. 7-8.

short, assuming that its reactance is negligible at the frequency of the signal. The term "constant-current source" here means that the impedance of transistor (between points B and C) is large compared with the load impedance; thus, the rms value of the load current is constant, regardless of changes in the load. Fig. 7-15B shows that R_C and R_L are parallel resistances as the constant-current source (the transistor) sees it. Therefore, the constant-current source works into a load whose total resistance r_L may be determined as follows:

$$r_L = \frac{R_C R_L}{R_C + R_L} \qquad (7\text{-}2)$$

The collector signal current i_c flows through r_L as shown in Fig. 7-15C. The value of i_c is approximately equal to i_e just as I_C is approximately equal to I_E. These relationships may be shown

$$\frac{i_c}{i_e} = a = h_{fb} \cong 1 \qquad \text{and therefore} \qquad i_c \cong i_e$$

just as

$$\frac{I_C}{I_E} = \alpha_{dc} = h_{FB} \cong 1 \quad \text{and} \quad I_C \cong I_E$$

The symbols a without a subscript and h_{fb} with lower case letters in subscript imply ac alpha, that is, the ratio of ac collector current i_c to the ac emitter current i_e. The symbols h_{FB} and h_{fb} are often called dc and ac *current transfer ratios,* respectively, of the common-base amplifier.

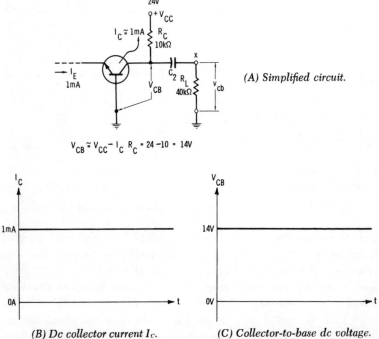

Fig. 7-14. Dc output current path of common-base amplifier of Fig. 7-8.

In this example we found that the peak value of the emitter signal is about 20 μA (Fig. 7-12B) and since $i_e \cong i_c$, we may say that in Fig. 7-15, $i_{c(\text{peak})} \cong 20$ μA. Also since the values of R_C and R_L are known, the total resistance into which i_c flows is

$$r_L = \frac{10 \times 40}{50} = 8 \text{ k}\Omega$$

by Eq. 7-2.

7-3 SUPERPOSITION OF DC & AC SIGNALS IN THE AMPLIFIER

We may now see a very important point. Even though the ratio of output to input current i_c/i_e, which is the current gain, is slightly less than 1, we can have considerable voltage gain A_e, which means

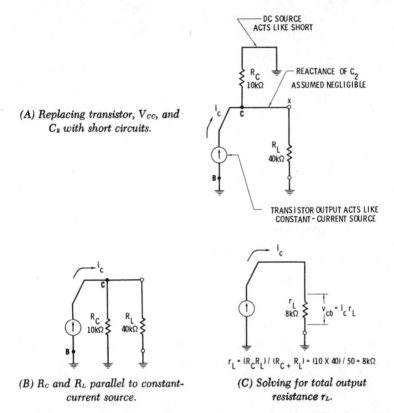

(A) *Replacing transistor, V_{CC}, and C_2 with short circuits.*

(B) *R_C and R_L parallel to constant-current source.*

(C) *Solving for total output resistance r_L.*

Fig. 7-15. Ac equivalent circuits of output of common-base amplifier of Fig. 7-8.

that the output signal voltage is larger than the input voltage. This is possible because the input signal current i_e works into a low resistance r_e', which is about 100 Ω in this case, but the output current i_c works into a much larger resistance r_L, which is 8 kΩ. That is, since r_L is larger than r_e', the output signal voltage v_{cb} is larger than the input voltage v_{eb} even though input and output currents are about equal.

By Ohm's law we can solve for the peak of the output voltage:

$$v_{cb(\text{peak})} = i_{c(\text{peak})} r_L \cong (20\ \mu\text{A})(8\ \text{k}\Omega) = 160\ \text{mV}$$

Previously we found that the input voltage $v_{eb(peak)} = 2$ mV; see Fig. 7-11A. The voltage gain is defined as the ratio of output voltage to the input voltage. Therefore, in this case,

$$\text{voltage gain } A_e = \frac{v_{cb(peak)}}{v_{eb(peak)}} = \frac{160 \text{ mV}}{2 \text{ mV}} = 80$$

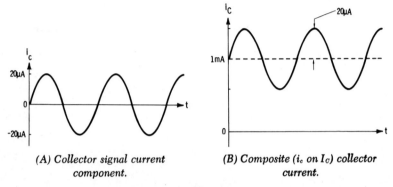

(A) Collector signal current component.

(B) Composite (i_c on I_C) collector current.

Fig. 7-16. Collector currents.

In other words, this common-base amplifier has an output signal voltage that is 80 times larger than the input voltage.

The signal collector current component i_c is shown in Fig. 7-16A. The composite collector current i_C is shown in Fig. 7-16B, which shows the ac signal component i_c superimposed on the dc collector current component I_C. Similarly, the signal output voltage v_{cb} is

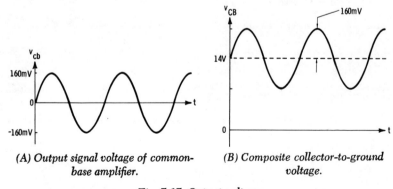

(A) Output signal voltage of common-base amplifier.

(B) Composite collector-to-ground voltage.

Fig. 7-17. Output voltages.

7-3 SUPERPOSITION OF DC & AC SIGNALS IN THE AMPLIFIER

shown in Fig. 7-17A, and the composite voltage across the collector and base v_{CB} is shown in Fig. 7-17B.

Example 7-7 Suppose that in a circuit like that of Fig. 7-8, $V_{EE} = -20$ V, $V_{CC} = 30$ V, $R_E = 40$ kΩ, $R_C = 36$ kΩ, $R_L = 9$ kΩ, and v_s is a sine-wave generator with a peak output of 4 mV and $r_e' \cong 50$ Ω. Find the values I_E, I_C, i_e, i_c, V_{CB}, v_{cb} and describe the waveforms of i_C and v_{CB}. What is the voltage gain A_e of this amplifier?

Answer 7-7 $I_E \cong I_C \cong 0.5$ mA, $i_{e(\text{peak})} \cong i_{c(\text{peak})} \cong 80$ μA, $V_{CB} \cong 12$ V, $v_{cb(\text{peak})} \cong 576$ mV. The waveform of i_C has a dc component value of 0.5 mA with an ac signal superimposed on it whose peak value is 80 μA, v_{CB} has a dc value of 12 V with an ac signal superimposed on it whose peak value is 576 mV. $A_e \cong 144$.—By Eqs. 6-1 and 6-2 we find that

$$I_E \cong \frac{V_{EE}}{R_E} = \frac{20 \text{ V}}{40 \text{ k}\Omega} = 0.5 \text{ mA} \cong I_C$$

Assuming that the reactances of the coupling capacitors are negligible, the source of signal voltage v_s is directly across the dynamic emitter resistance r_e'; see Fig. 7-10B. Since $r_e' \cong 50$ Ω in this case, by Ohm's law,

$$i_{e(\text{peak})} \cong \frac{v_{s(\text{peak})}}{r_e'} = \frac{4 \text{ mV}}{50 \text{ }\Omega} = 80 \text{ μA} \cong i_{c(\text{peak})}$$

With $I_C \cong 0.5$ mA, the dc voltage drop across R_C is $I_C R_C \cong (0.5 \text{ mA})(36 \text{ k}\Omega) = 18$ V. Therefore, the dc voltage across the collectors and base $V_{CB} = V_{CC} - I_C R_C \cong 30$ V $- 18$ V $= 12$ V.

The ac equivalent of the output is like the circuit of Fig. 7-15C. In this case the output of the constant-current source has a peak value of 80 μA. This $i_{c(\text{peak})} \cong 80$ μA flows through a total resistance r_L whose value in this case is, by Eq. 7-2,

$$r_L = \frac{36 \times 9}{45} = 7.2 \text{ k}\Omega$$

The output voltage $v_{cb} = i_c r_L$; thus

$$v_{cb(\text{peak})} = i_{c(\text{peak})} r_L \cong (80 \text{ μA})(7.2 \text{ k}\Omega) = 576 \text{ mV}$$

The voltage gain is

$$A_e = \frac{v_{cb}}{v_s} \cong \frac{576 \text{ mV}}{4 \text{ mv}} = 144$$

7-4 V-I CHARACTERISTICS AND DYNAMIC RESISTANCE OF THE EMITTER-BASE JUNCTION

Since the emitter-base diode junction is usually forward biased, it has forward biased characteristics that are very similar to the forward characteristics of ordinary diodes shown in Chap. 2, Figs. 2-4 and 2-5. We learned in the last section that if there is a small ac component current superimposed on the forward biasing dc current I_E, the ac component sees a resistance r_e', which may be called the dynamic input resistance or emitter resistance. The value of r_e' depends a lot on the magnitude of the forward bias dc current I_E. In fact, its value can be approximated by the equation

$$r_e' \cong \frac{25 \text{ mV}}{I_E} \qquad (7\text{-}3)^*$$

where 25 mV is a constant and I_E is the dc emitter current. Actually, this dynamic resistance r_e' is in the range

$$\frac{25 \text{ mV}}{I_E} \leqq r_e' \leqq \frac{50 \text{ mV}}{I_E}$$

Typically r_e' is in the lower rather than the upper portion of this range and therefore Eq. 7-3 is practical and used throughout the remainder of this text. See Appendix A-3.

Example 7-8 What is the approximate emitter resistance r_e' with each of the following emitter bias currents? (a) $I_E = 0.25$ mA, (b) $I_E = 0.5$ mA, (c) $I_E = 5$ mA.

Answer 7-8 (a) $r_e' \cong 100 \ \Omega$, (b) $r_e' \cong 50 \ \Omega$ (c) $r_e' \cong 5 \ \Omega$.— For part (a) we may show by Eq. 7-3 that

$$r_e' \cong \frac{25 \text{ mV}}{0.25 \text{ mA}} = 100 \ \Omega$$

for part (b)

$$r_e' \cong \frac{25 \text{ mV}}{0.5 \text{ mA}} = 50 \ \Omega$$

and for part (c)

$$r_e' \cong \frac{25 \text{ mV}}{5 \text{ mA}} = 5 \ \Omega$$

---

*See Appendix A-3.

V-I Characteristics & Dynamic Resistance of the Junction 149

As you can see, the emitter resistance r_e' decreases as the dc emitter current I_E increases. The reason why this is so may be understood by examining a V-I curve of the emitter-base junction like the one in Fig. 7-18. On this curve we will observe graphically how the emitter resistance r_e' is different with three different dc emitter currents, I_E, I_E' and I_E''.

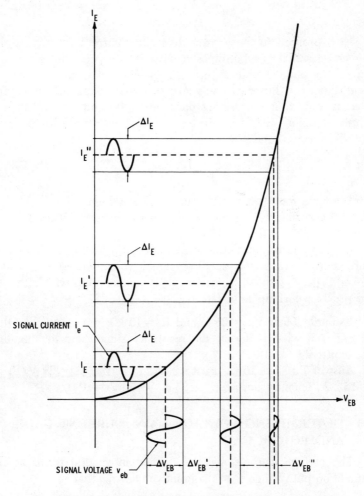

Fig. 7-18. V-I forward characteristics of emitter-base junction in common-base circuit.

Note first I_E, which has an ac signal i_e riding on it whose peak-to-peak amplitude is $\triangle I_E$. By projecting from the peaks of this signal to the right on to the curve and then down as shown, we get an idea of how the emitter-base voltage varies. In this case we have an emitter-base signal voltage v_{eb} whose peak-to-peak amplitude is $\triangle V_{EB}$. The dynamic emitter resistance thus by Ohm's law may be expressed as

$$r_e' \cong \frac{\triangle V_{EB}}{\triangle I_E} \qquad (7\text{-}4)$$

Now proceed to the greater dc emitter current I_E', which also has an ac signal i_e superimposed on it whose peak-to-peak amplitude is the same as before; that is, $\triangle I_E$. Note that this same amplitude signal i_e will not cause as large a signal voltage v_{eb} as before. That is, now that i_e is riding on a larger dc emitter current, the amplitude of v_{eb} is smaller. In this case v_{eb} has a peak-to-peak amplitude of $\triangle V_{EB}'$ and now

$$r_e' \cong \frac{\triangle V_{EB}'}{\triangle I_E} \qquad (7\text{-}5)$$

We have found that $\triangle V_{EB}$ in Eq. 7-4 is larger than $\triangle V_{EB}'$ in Eq. 7-5 and that $\triangle I_E$ is the same in both equations. Therefore, r_e' in Eq. 7-4 is larger than r_e' in Eq. 7-5.

In the same way note that the signal i_e on the largest dc emitter current I_E'' causes the smallest signal voltage drop v_{eb}, which means that r_e' with bias current I_E'' is smaller than with either of the previous currents I_E or I_E'.

Example 7-9 If $V_{EE} = 25$ V and $R_E = 10$ kΩ in a circuit like that of Fig. 7-8, what is it that the resistance signal sees looking into the emitter?

Answer 7-9 $r_e' \cong 10$ Ω.—First we find $I_E \cong V_{EE}/R_E = 25$ V/10 kΩ $= 2.5$ mA. Then by Eq. 7-3 $r_e' \cong 25$ mV/2.5 mA $= 10$ Ω.

7-5 DETERMINING VOLTAGE GAIN, CURRENT GAIN, AND POWER GAIN

The *voltage gain* A_e may be defined in general terms as the ratio of output voltage v_o to the input voltage v_{in}; that is,

$$\text{voltage gain } A_e = \frac{v_o}{v_{in}} \qquad (7\text{-}6)$$

7-5 DETERMINING VOLTAGE, CURRENT, & POWER GAINS

We noted in Sec. 7-3 that the common-base amplifier is capable of considerable voltage gain. Here we will learn how to determine the approximate voltage gain of the common-base amplifier by simple examination of the circuit components and, at most, a couple of calculations. In the common-base circuit like that of Fig. 7-8, the input voltage v_{in} is the signal voltage at the emitter-base junction and is called v_{eb}, while the output voltage v_o is the signal voltage at the collector-base junction and is called v_{cb}. Thus, for the common-base amplifier

$$\text{voltage gain } A_e = \frac{v_{cb}}{v_{eb}} \tag{7-6A}$$

To pursue the business of determining gain of a circuit like that of Fig. 7-8 we can draw its ac equivalents as in Fig. 7-19. Fig. 7-19A shows the circuit of Fig. 7-8 with components V_{EE}, V_{CC}, C_1, and C_2 replaced with short circuits. Fig. 7-19B shows also an ac equivalent showing that, as signal sees it, the input resistance to the transistor is r_e' and that the transistor output acts like a constant-current source i_c. Fig. 7-19C shows a simplified version of Fig. 7-19B. Since R_E is much larger than r_e', their total resistance is very nearly equal to r_e' as the source v_s sees it. Also since the transistor works into a load consisting of R_C and R_L in parallel, their total resistance is shown as r_L.

In Fig. 7-19C we can show by Ohm's law that the input voltage is

$$v_{eb} = i_e r_e' \tag{7-7}$$

and similarly the output voltage

$$v_{cb} = i_c r_L \tag{7-8}$$

Substituting Eqs. 7-7 and 7-8 into Eq. 7-6A, we get

$$A_e = \frac{v_{cb}}{v_{eb}} = \frac{i_c r_L}{i_e r_e'} \cong \frac{r_L}{r_e'} \tag{7-9}$$

Note that since $i_e \cong i_c$, these currents cancel, showing that the voltage gain of the common-base transistor is approximately equal to the ratio of load ac resistance r_L to the dynamic resistance r_e'.

Example 7-10 Referring back to Ex. 7-7, in the circuit described, that is, where $V_{EE} = -20$ V, $V_{CC} = 30$ V, $R_E = 40$ kΩ, $R_C = 36$ kΩ, $R_L = 9$ kΩ, and $v_{s(\text{peak})} = 4$ mV in Fig. 7-8, solve for voltage gain A_e using Eqs. 7-3 and 7-9.

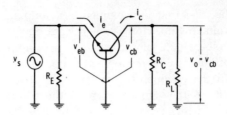

(A) V_{BB}, V_{CC}, C_1, and C_2 replaced with short circuits.

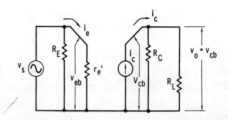

(B) With emitter-to-base resistance r_e' and constant-current source i_c.

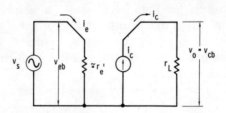

(C) More simplified version of (B) above.

Fig. 7-19. Ac equivalent circuits of the circuit in Fig. 7-8.

Answer 7-10 As shown in the answer to Ex. 7-7, $I_E \cong 20$ V/40 kΩ = 0.5 mA. By Eq. 7-3 we can show that $r_e' \cong 25$ mV/0.5 mA = 50 Ω, which verifies the r_e' value given in Ex. 7-7. Also in the answer to Ex. 7-7 we found that $r_L = 7.2$ kΩ. Thus, by Eq. 7-9 we solve

$$A_e \cong \frac{r_L}{r_e'} = \frac{7.2 \text{ k}\Omega}{50 \Omega} = 144$$

which agrees with our previous solution to Ex. 7-7.

Example 7-11 What is the voltage gain of a circuit like that of Fig. 7-8 if $V_{EE} = 15$ V, $V_{CC} = 20$ V, $R_E = 60$ kΩ, $R_C = 40$ kΩ, and $R_L = 10$ kΩ?

7-6 High-Resistance Signal Source vs Gain

Answer 7-11 $A_e \cong 80$.—The dc emitter current $I_E \cong 15 \text{ V}/60 \text{ k}\Omega$ $= 0.25$ mA. By Eq. 7-3 $r_e' \cong 25 \text{ mV}/0.25 \text{ mA} = 100 \ \Omega$. The signal from the collector i_c sees R_C and R_L in parallel; thus, it works into a total resistance $r_L = 40 \times 10/50 = 8 \text{ k}\Omega$. By Eq. 7-9 the voltage gain $A_e \cong 8000/100 = 80$.

The *current gain* A_i is the ratio of output current to the input current. In general

$$A_i = \frac{i_{\text{out}}}{i_{\text{in}}} \qquad (7\text{-}10)$$

Specifically for the common-base amplifier,

$$A_i = \frac{i_c}{i_e} \cong \alpha = h_{fb} \qquad (7\text{-}10\text{A})$$

Obviously, since α is almost equal to but less than 1, the current gain of the common-base amplifier is about 1.

The *power gain* A_p of an amplifier in general is the ratio of signal output power p_o to the signal input power p_{in}:

$$A_p = \frac{p_o}{p_{\text{in}}} \qquad (7\text{-}11)$$

It is also equal to the product of the voltage gain A_e and current gain A_i:

$$A_p = A_e A_i \qquad (7\text{-}12)$$

Since the current gain of the common-base amplifier is about 1, its power gain A_p is about equal to the voltage gain A_e.

Example 7-12 What is the power gain A_p of the circuit described in Ex. 7-11?

Answer 7-12 $A_p \cong 80$.—Since the circuit is a common-base amplifier, its $A_i = \alpha \cong 1$, and since $A_e \cong 80$, the power gain $A_p = A_e A_i \cong 1 \times 80$.

7-6 HIGH-RESISTANCE SIGNAL SOURCE VS GAIN

The relatively low input resistance of the common-base amplifier limits its applicability. Most signal sources have internal resistances larger than r_e' and therefore tend to be loaded down if worked into a common-base amplifier. (A signal source is *loaded down* if its load resistance is much smaller than its internal resistance). Loading down a signal source greatly reduces its output voltage and its efficiency and can cause distortion. For example, a signal generator may, with a normal load, have a sine-wave output voltage, but when it is loaded down the output voltage is much smaller in amplitude or may be a partially clipped or asymmetrical or an otherwise distorted wave. We will consider here the loss of amplitude caused by loading down a signal source and its effect on the overall gain of the common-base amplifier stage.

We can at this time review methods of determining internal resistance of a signal source and its effect on the output voltage. Fig. 7-20 shows a signal source with output terminals 1 and 2. The load is connected to the source by closing switch S. Components r_s and v_s in series represent the equivalent circuit inside the source. When the switch S is open, the voltage read across terminals 1 and 2 is v_s, that is, $e_o = v_s$ because the voltage drop across r_s is zero. When S is open, v_s is called the *open-circuit voltage* of the signal source. When the switch S is closed, r_s has a voltage drop e_s, causing the terminal e_o to be less than v_s.

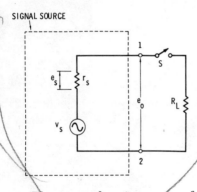

Fig. 7-20. Signal source is equivalent to voltage source v_s in series with internal resistance r_s.

The internal resistance r_s of the signal source in Fig. 7-20 can be determined with the equation

$$r_s = \frac{\triangle e_o}{\triangle i_o} \qquad (7\text{-}13)$$

where $\triangle e_o$ means *change in* output voltage and $\triangle i_o$ means *change in* output current. Thus, when a load on a power supply is changed, both the output voltage e_o and output current i_o will change too. The difference between the initial and final output voltages is di-

7-6 HIGH-RESISTANCE SIGNAL SOURCE VS GAIN

vided by the difference between the initial and final output currents when solving for a signal source internal resistance.

Example 7-13 Suppose that a source like that in Fig. 7-20 has an output voltage $e_o = 6$ V (rms) when the switch S is open. When S is closed, e_o drops to 4 V (rms). Load resistance $R_L = 1$ kΩ. Find the values of the internal resistance r_s and the open-circuit voltage v_s.

Answer 7-13 $r_s = 500$ Ω, $v_s = 6$ V.—Since the output voltage dropped from 6 V to 4 V, the change in output voltage is $\triangle e_o = 6 - 4 = 2$ V. With the switch S open the output current i_o is zero. With S closed the ouput current is $i_o = e_o/R_L = 4$ V/1 kΩ = 4 mA (rms) in this case. Therefore, the change in current is $\triangle i_o = 4 - 0 = 4$ mA. By Eq. 7-13

$$r_s = \frac{2 \text{ V}}{4 \text{ mA}} = 500 \text{ Ω}$$

Since v_s is the open-circuit voltage, that is, the voltage output e_o when the output terminals 1 and 2 are open (switch S open), it is 6 V in this case.

Example 7-14 If in Fig. 7-20, switch S is closed, $r_s = 500$ Ω, $v_s = 6$ V (rms) and R_L is reduced to 100 Ω, what is the output voltage e_o?

Answer 7-14 $e_o = 1$ V (rms).—In this case v_s sees a total resistance of $r_s + R_L = 500 + 100 = 600$ Ω. Therefore, by Ohm's law the output current $i_o = 6$ V/600 Ω = 10 mA (rms). The output voltage e_o is the voltage drop across R_L; thus $e_o = i_o R_L = (10 \text{ mA})(100 \text{ Ω}) = 1$ V (rms). Or

$$e_o = \frac{v_s R_L}{r_s + R_L} = \frac{6 \times 100}{600} = 1 \text{ V (rms)}$$

It is interesting to note in these two examples that the output voltage e_o drops when the load current is increased.

Fig. 7-21A shows a common-base amplifier driven by a signal source whose internal resistance is r_s. The signal source may be a signal generator, an ac power supply, or another amplifier. In any case, however, it may be shown to be equivalent to a single voltage source v_s in series with an internal resistance r_s. If r_s is not negligible, the voltage gain from v_s to the output v_o will be less than the

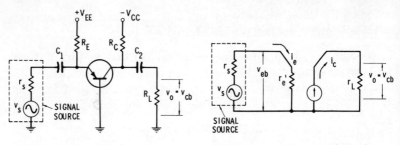

(A) Signal source has internal resistance r_s.

(B) Ac equivalent of amplifier shown in (A).

Fig. 7-21. Common-base amplifier driven by signal source.

voltage gain of the stage, which is the gain from v_{eb} to v_o. In other words, when $r_s > 0$, then

$$\frac{v_o}{v_{eb}} > \frac{v_o}{v_s}$$

The ac equivalent circuit of Fig. 7-21A is shown in Fig. 7-21B. The signal source works into a load whose resistance is approximately equal to r_e'. The emitter bias resistor R_E is not included because normally its resistance is very much greater than the emitter resistance r_e' and thus as signal from the source sees it, the total resistance of R_E and r_e' in parallel is approximately equal to r_e'. That is, the total load resistance as the signal source sees it in Fig. 7-21A is

$$\frac{R_E r_e'}{R_E + r_e'} \cong r_e' \quad \text{if} \quad R_E \gg r_e'$$

The signal current out of the signal source therefore is about equal to the emitter signal current i_e. Thus, by Ohm's law

$$i_e \cong \frac{v_s}{r_s + r_e'} \quad \text{or} \quad v_s \cong i_e(r_s + r_e')$$

We may now show that the gain from v_s to the output v_o is

$$\frac{v_o}{v_s} = \frac{v_{cb}}{v_s} = \frac{i_c r_L}{i_e(r_s + r_e')} \cong \frac{i_c r_L}{i_c(r_s + r_e')} = \frac{r_L}{r_s + r_e'} \quad (7\text{-}14)$$

Typically, because the emitter resistance r_e' is quite small—usually under 100 Ω—the internal resistance r_s is often much larger than r_e'. Therefore, Eq. 7-14 above can be simplified to $v_o/v_s \cong r_L/r_s$. In other words, if $r_s \gg r_e'$, then

7-7 SIGNAL AND LOAD LINES

$$\frac{v_o}{v_s} \cong \frac{r_L}{r_s} \qquad (7\text{-}15)$$

In general, if r_s is about ten times larger than $r_e{}'$ we can use Eq. 7-15 instead of Eq. 7-14.

Example 7-15 In a circuit like that of Fig. 7-21A, find the stage gain v_o/v_{eb} and the overall gain v_o/v_s, where $V_{EE} = 12$ V, $V_{CC} = 24$ V, $R_E = 48$ kΩ, $R_C = 50$ kΩ, $R_L = 200$ kΩ, and $r_s = 400$ Ω.

Answer 7-15 Stage gain $v_o/v_{cb} \cong 400$, overall gain $v_o/v_s \cong 80$.— First we find that the dc emitter bias current $I_E \cong V_{EE}/R_E = 12$ V/48 kΩ = 0.25 mA. Therefore $r_e{}' \cong 25$ mV/$I_E = 25$ mV/0.25 mA = 100 Ω. By Eq. 7-9 $v_o/v_{eb} \cong r_L/r_e{}' = 40{,}000/100 = 400$, where $r_L = R_C R_L/(R_C + R_L) = 40$ kΩ. Since r_s is not ten times greater than $r_e{}'$, we use Eq. 7-14 and find that $v_o/v_s \cong r_L/(r_s + r_e{}') = 40{,}000/500 = 80$.

Example 7-16 Referring to the circuit described in Ex. 7-15, suppose that it is driven by a signal source whose internal resistance $r_s = 5$ kΩ instead of 400 Ω. Find the stage gain v_o/v_{eb} and overall gain v_o/v_s.

Answer 7-16 $v_o/v_{eb} \cong 400$, $v_o/v_s \cong 8$.—The internal resistance of the signal source has no effect on the stage gain and it remains the same as before. In this case, since r_s is more than ten times larger than $r_e{}'$, we may use Eq. 7-15 to find the overall gain: $v_o/v_s \cong r_L/r_s = 40$ kΩ/5 kΩ = 8.

7-7 SIGNAL AND LOAD LINES

If we draw a load line and locate the operating point of a common-base amplifier, we can determine the maximum signal amplitude the amplifier can handle without causing serious distortion. We also get a better *picture* of what is happening to the voltages and currents in the amplifier when its input is driven with a signal source.

For example, the circuit in Fig. 7-22 is shown driven by a signal source whose open-circuit peak voltage $v_{s(\text{peak})}$ is 0.6 V. Since the value of r_s is relatively large, we can find the gain of this amplifier as we did in Ex. 7-16; that is, by Eq. 7-15

$$\frac{v_o}{v_s} \cong \frac{25{,}000 \ \Omega}{2{,}000 \ \Omega} = 12.5$$

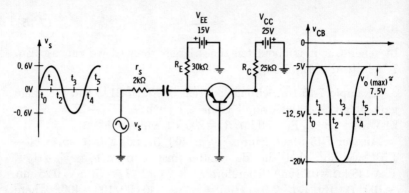

Fig. 7-22. Common-base amplifier with relatively large input and output signals.

Therefore

$$v_o \cong 12.5 v_s$$

or

$$v_{o(peak)} \cong 12.5 v_{s(peak)} = 12.5(0.6 \text{ V}) = 7.5 \text{ V}$$

This signal v_o is superimposed on the dc collector-to-base voltage V_{CB} as shown, where $V_{CB} = V_{CC} - I_C R_C$. In this case

$$I_C \cong I_E \cong \frac{V_{EE}}{R_L} = \frac{15 \text{ V}}{30 \text{ k}\Omega} = 0.5 \text{ mA}$$

Thus, $V_{CB} = 25 - (0.5 \text{ mA})(25 \text{ k}\Omega) = 12.5 \text{ V}$.

The variations of collector current and voltage may be shown on a load line. The load line for the circuit of Fig. 7-22 is drawn in Fig. 7-23. Since $I_C \cong I_E \cong 0.5$ mA, the operating point Q is on the load line directly to the right of $I_C = 0.5$ mA. Note that directly below the operating point Q we have $V_{CB} = 12.5$ V, verifying our calculation above.

Signal v_s works into a resistance $r_s + r_e'$, ignoring the shunting effect of R_E because its resistance is much larger than r_e'. In fact,

$$r_e' \cong \frac{25 \text{ mV}}{I_E} = \frac{25 \text{ mV}}{0.5 \text{ mA}} = 50 \text{ }\Omega$$

which is also much smaller than r_s. Thus, we may say that v_s sees a load resistance that is about equal to r_s alone. By Ohm's law, then, the signal emitter current is $i_e \cong v_s/r_s$. In this case, therefore,

7-7 SIGNAL AND LOAD LINES

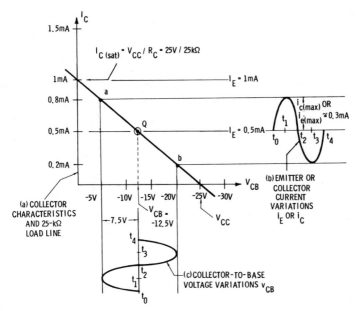

Fig. 7-23. Load line for the circuit of Fig. 7-22 with variations caused by the signal.

$$i_{e(\text{peak})} \cong \frac{v_{s(\text{peak})}}{r_s} = \frac{0.6 \text{ V}}{2 \text{ k}\Omega} = 0.3 \text{ mA}$$

which, of course, is also the approximate magnitude of the peak collector current $i_{c(\text{peak})}$. This signal current is shown in (b) of Fig. 7-23 and its peaks are projected to points a and b on the load line. Directly below points a and b are the minimum and maximum variations of the collector voltage (c). Note that the peak value of the collector voltage signal $v_{cb(\text{max})}$ is 7.5 V, which verifies the answer we obtained mathematically before.

In other words, when no signal is applied to the input of the circuit of Fig. 7-22, $v_s = 0$ and the collector current and voltage are steady at $I_C = 0.5$ mA and $V_{CB} = -12.5$ V, which are the coordinates of the operating point Q. When v_s is applied with a peak value of 0.6 V, the collector current i_C rises from 0.5 mA to a maximum of 0.8 mA at the positive peaks of the input voltage while the voltage v_{CB} changes from -12.5 V to -5 V at the same instants. On the peaks of the negative alternations, $i_C = 0.2$ mA and $v_{CB} = -20$ V.

An advantage of sketching a load line and locating an operating point of an amplifier is that it graphically shows us its limits of operation. If we were to drive the amplifier in Fig. 7-22 with a larger signal than the one shown, the collector current will rise to values larger than 0.8 mA and perhaps into saturation on the positive peaks. On the negative peaks the collector current will drop to values below 0.2 mA and perhaps into cutoff. In other words, large signals can drive the amplifier into saturation and cutoff during portions of each cycle. This causes *clipping* of the collector current and voltage waveforms as shown in Fig. 7-24. Note that even

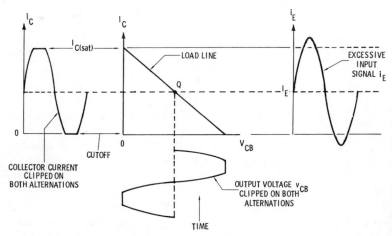

Fig. 7-24. Output voltage clipped on both alternations if input signal is excessive.

though the emitter current waveform i_E is sinusoidal, the collector current i_C and collector-to-base voltage v_{CB} are clipped sinusoidal waveforms. Thus, by driving the stage too hard we may cause distortion in the output voltage v_{CB} due to saturation on the positive alternations and cutoff on the negative alternations.

When designing an amplifier, it is usually desirable to place the operating point on the load line midway between the saturation and cutoff points as in Figs. 7-23 and 7-24. This enables the amplifier to handle larger signals without distortion than would be possible if the operating Q point were nearer to either the saturation or cutoff point. For example, Fig. 7-25 shows what might happen if the emitter bias current I_E is low, thus placing the oper-

7-7 Signal and Load Lines

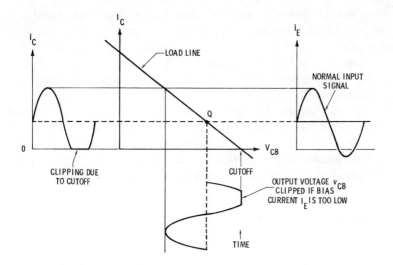

Fig. 7-25. If dc bias current I_E is too low, output is clipped on negative alternations even though input signal is not excessive.

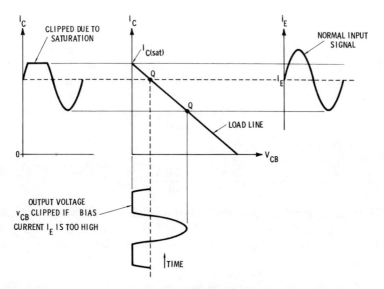

Fig. 7-26. Clipping caused by saturation when emitter bias current I_E is too large.

ating point Q lower on the load line. Though the input signal i_E is not excessively large, the amplifier is driven into cutoff on the negative peaks of the input cycle. Thus, the negative peaks of the output signal waveform are clipped off as shown, and the amplifier is not being utilized to give the maximum possible undistorted output.

We find similar difficulties if the operating point Q is too high on the load line as shown in Fig. 7-26, which may be caused by high bias current I_E. In this case the positive peaks of the output voltage waveform are clipped because the amplifier is driven into saturation during a portion of each input cycle.

Example 7-17 Sketch a load line and locate the operating point of the circuit of Fig. 7-27.

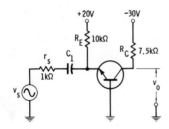

Fig. 7-27. Common-base amplifier circuit.

Answer 7-17 See (a) in Fig. 7-28.—The load line goes through the $I_{C(\text{sat})}$ on the vertical scale and voltage value equal to V_{CC} on the horizontal scale of the collector characteristic. In this case

$$I_{C(\text{sat})} \cong \frac{V_{CC}}{R_C} = \frac{30 \text{ V}}{7.5 \text{ k}\Omega} = 4 \text{ mA}$$

The emitter bias current is

$$I_E \cong I_C \cong \frac{V_{EE}}{R_E} = \frac{20 \text{ V}}{10 \text{ k}\Omega} = 2 \text{ mA}$$

Thus, the operating point, which is the point at which the $I_E = 2$ mA curve crosses the load line, is located to the right of $I_C = 2$ mA on the vertical scale and is halfway between saturation and cutoff points.

Example 7-18 Referring to the circuit of Fig. 7-27, sketch the output voltage waveforms and indicate their peak values when (a)

7-7 SIGNAL AND LOAD LINES

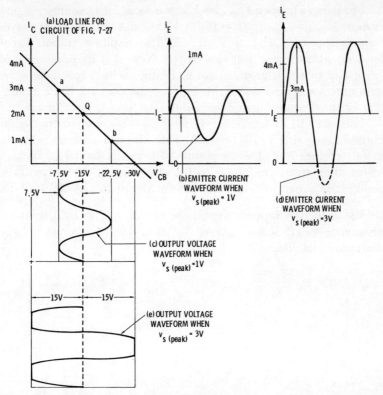

Fig. 7-28. Points a and b on the load line mark limits of excursions above and below operating point when $i_{e(\text{peak})} = 1$ mA.

v_s is a sine wave with a peak value of 1 V, and (b) v_s is sinusoidal with a peak of 3 V.

Answer 7-18 (a) v_o is a sine wave with a peak of 7.5 V, (b) v_o is a sine wave clipped on positive and negative alternations, each alternation having a peak of 15 V; see (c) and (e) in Fig. 7-28.—
(a) In this circuit (Fig. 7-27), bias current $I_E \cong 20$ V/10 kΩ = 2 mA. Thus, the emitter input resistance $r_e' \cong 25$ mV/2 mA = 12.5 Ω. Since both resistances r_s and R_E are much larger than r_e', we can approximate the emitter signal current with Ohm's law: $i_e \cong v_s/r_s$. Incidentally, by Eq. 7-15, the overall gain of this amplifier is

$$\frac{v_o}{v_s} \cong \frac{r_L}{r_s} = \frac{7.5 \text{ k}\Omega}{1 \text{ k}\Omega} = 7.5$$

For part (a), when $v_{s(peak)} = 1$ V, we solve for the emitter signal current $i_{e(peak)} \cong v_{s(peak)}/r_s = 1$ V/1 kΩ = 1 mA. This current waveform is shown in (b) of Fig. 7-28. The resulting output waveform v_o, which is v_{cb}, is shown in (c). Note that its peak value is 7.5 V and that it is greater than v_s by the factor 7.5, which is the gain of this amplifier. Since this input signal does not cause saturation or cutoff at any time, the output waveform is not clipped.

For part (b) of Ex. 7-18, when $v_{s(peak)} = 3$ V, the emitter signal current $i_{e(peak)} \cong 3$ V/1 kΩ = 3 mA. This current waveform is shown in (d) of Fig. 7-18. This amplitude of i_e causes saturation on half of the alternations and cutoff on the other half of the alternations, resulting in a clipped output waveform shown in (e) of Fig. 7-18.

Example 7-19 Suppose that in the circuit of Fig. 7-27, R_E is increased to 40 kΩ and $v_{s(peak)} = 3$ V. What does the output voltage waveform look like?

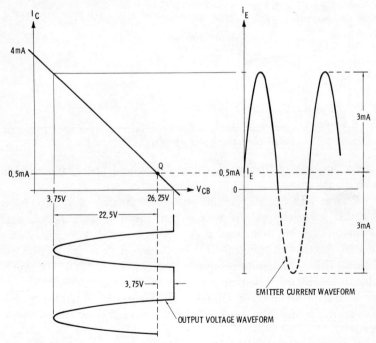

Fig. 7-29. Output voltage waveform for circuit of Fig. 7-27 with $R_E = 40$ kΩ and $v_{s(peak)} = 3$ V.

7-8 AC LOAD LINES

Answer 7-19 The output voltage waveform is clipped on half of the alternations. The unclipped alternations have a peak of 22.5 V and the clipped alternations have a 3.75 V peak; see Fig. 7-29.—In this case the emitter bias current is

$$I_E \cong \frac{V_{EE}}{R_E} = \frac{20 \text{ V}}{40 \text{ k}\Omega} = 0.5 \text{ mA}$$

Thus, the operating point Q is on the lower portion of the load line. The peak value of the emitter current is found as in the answer to Ex. 7-18 and is 3 mA. Thus, on portions of the negative alternations of i_E, the transistor is driven *hard* into cutoff, causing a substantial portion of these alternations to be clipped off of the output voltage waveform as shown.

7-8 AC LOAD LINES

As mentioned before, the total load in the collector circuit is not always a single resistor like R_C in Fig. 7-27. In fact, usually the load is more complex as in Fig. 7-30. The additional load R_L

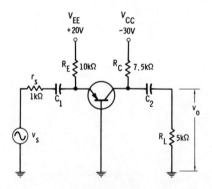

Fig. 7-30. Common-base pnp amplifier.

coupled to the collector through C_2 is not necessarily a simple resistor. That is, R_L may represent the input resistance to another amplifier stage, or it may be the input resistance to a speaker. The point is, R_L is the load to which we want to deliver the signal amplified by the circuit. The capacitor value is so chosen that its reactance is negligible at the operating frequency. It serves to block dc and thus prevents the V_{CC} supply from placing dc voltage across R_L, yet ac signal in the collector sees C_2 as a short.

Thus, as ac signal in the collector sees it, R_C and R_L are in parallel, with a total resistance

$$r_L = \frac{R_C R_L}{R_C + R_L} = 3 \text{ k}\Omega$$

So dc in the collector sees a load of 7.5 kΩ while ac sees much less, or 3 kΩ, in this circuit. These may be called *dc* and *ac loads* respectively, and load lines for each may be drawn on the collector characteristics. The load lines drawn up to now are all dc load lines. Now we will consider methods of drawing the ac load lines.

To analyze a circuit like the one in Fig. 7-30 with load lines, we must first draw the dc load line and the operating point, and then

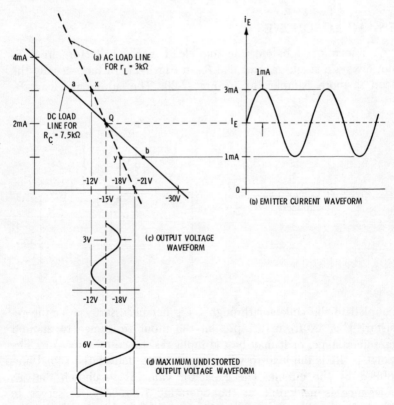

Fig. 7-31. Points x and y on the ac load mark limits of excursions above and below operating point when $i_{e(\text{peak})} = 1$ mA.

7-8 AC LOAD LINES

the ac load line is drawn through the operating point. The dc load line is drawn as in the previous examples; that is, R_L may be ignored because it does not affect the dc saturation point $I_{C(\text{sat})}$ and the dc cutoff point V_{CC}. Fig. 7-31 shows both load lines of the circuit of Fig. 7-30. Note that the dc load line is the same as for the circuit of Fig. 7-27 and that the ac load line is drawn through the operating point Q. If we can determine one more point that is on the ac load line, a straight line drawn through it and the operating point Q will give us the exact position of the ac load line. In other words, point x in Fig. 7-31 can be determined before the ac load line is drawn. A straight line through it and point Q is the ac load line.

As an example let's suppose v_s is a sine wave with a peak of 1 V as in (a) of Ex. 7-18. Thus, the emitter and collector signal currents have a peak of about 1 mA. Now this collector signal current i_c sees the ac load of 3 kΩ in the collector. The output voltage v_{cb} therefore by Ohm's law is $v_{cb} = i_c r_L$. In this case since we have $i_{c(\text{peak})} \cong 1$ mA, we find that

$$v_{cb(\text{peak})} = i_{c(\text{peak})} \times r_L \cong (1 \text{ mA})(3 \text{ k}\Omega) = 3 \text{ V}$$

This tells us that with an input signal $i_{e(\text{peak})} \cong 1$ mA the output signal voltage $v_{cb(\text{peak})}$ is 3 V, which causes the collector voltage v_{CB} to vary above and below its quiescent value of -15 V by 3 V. In other words, with an input emitter signal current like (b) in Fig. 7-31, the output voltage has a waveform (c) that varies from -15 V in the positive direction to -12 V and in the negative direction to -18 V. If we project a horizontal line to the left of the positive peak of i_E and a vertical line up from $v_{CB} = -12$ V, the two lines intersect at point x. A straight line through points x and Q is the ac load line. Similarly, we could have projected to the left of $i_E = 1$ mA and also up from $v_{CB} = -18$ V. These projections intersect at point y and a straight line through it and point Q is the ac load line.

Notice that the ac load line crosses the V_{CB} axis at -21 V. This tells us that signal cannot drive the collector-to-base voltage more negatively than -21 V. Thus, the maximum collector signal voltage that this amplifier (Fig. 7-30) can handle without serious distortion has a peak value of 6 V as shown in (d) of Fig. 7-31.

Example 7-20 Referring to the circuit of Fig. 7-30, what maximum amplitude of v_s will give us the maximum undistorted output voltage like (d) of Fig. 7-31 if $r_s = 2$ kΩ?

Answer 7-20 $v_{s(\text{peak})} \cong 4$ V.—In this circuit, $r_e' \cong 25$ mV/$I_E =$ 25 mV/2 mA = 12.5 Ω. Since $r_s \gg r_e'$, by Eq. 7-15 we find the overall gain

$$\frac{v_o}{v_s} \cong \frac{r_L}{r_s} = \frac{3 \text{ k}\Omega}{2 \text{ k}\Omega} = 1.5$$

Since $v_o/v_s \cong 1.5$, then $v_s \cong v_o/1.5$, or

$$v_{s(\text{peak})} \cong \frac{v_{o(\text{peak})}}{1.5} = \frac{6 \text{ V}}{1.5} = 4 \text{ V}$$

The value $v_{o(\text{peak})} = 6$ V is the peak value of the largest signal output voltage possible without clipping as shown in (d) of Fig. 7-31.

Example 7-21 If in circuit of Fig. 7-30 v_s is a sine wave voltage with a peak value of 5 V and $r_s = 2$ kΩ, what does the output voltage waveform look like?

Answer 7-21 The output voltage is a sine wave with negative alternations clipped, and having a peak amplitude of 6 V. The unclipped positive alternations have a peak of 7.5 V.—In the previous example we found that the overall gain $v_o/v_s \cong 1.5$. Since in this case $v_{s(\text{peak})} = 5$ V, then $v_{o(\text{peak})} \cong 1.5 v_{s(\text{peak})} = 1.5 \times 5 = 7.5$ V. Thus this input v_s tends to vary the output by 7.5 V alternately to each side of the quiescent value. That is, on positive alternations v_{CB} swings from -15 V to -7.5 V and on negative alternations it swings from -15 V to -22.5 V but it cannot go more negative

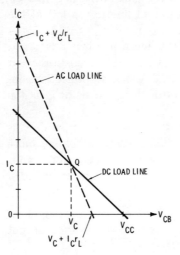

Fig. 7-32. Finding ac load-line cutoff and saturation points mathematically.

7-8 AC LOAD LINES

than -21 V, which is the cutoff voltage on the ac load line; see Fig. 7-31. Thus, clipping occurs at $v_{CB} = -21$ V, which gives the clipped alternations a 6 V peak value.

In Fig. 7-31, the ac cutoff point can be obtained graphically by noting where the ac load line crosses the horizontal V_{CB} axis. We may find this cutoff point, and also the saturation point, mathematically with more accuracy.

As shown in Fig. 7-32, the end points of the ac load line are

$$i_{C(\text{sat})} = I_C + \frac{V_C}{r_L} \qquad (7\text{-}16)^*$$

and

$$v_{C(\text{cutoff})} = V_C + I_C r_L \qquad (7\text{-}17)^*$$

where

$i_{C(\text{sat})}$ with the lower case i is saturation current on the ac load line,

$v_{C(\text{cutoff})}$ with the lower case v is the cutoff voltage on the ac load line,

V_C is the absolute value of collector-to-base voltage at the operating point Q,

I_C is the collector current at the operating point Q,

r_L is the ac resistance on the collector.

By the *absolute value* of V_C we mean that we express it as a positive value even though the V_{CC} source is negative with respect to ground. For example, in the circuit of Fig. 7-30, we found that the coordinates of the operating Q are $V_C = -15$ V and $I_C = 2$ mA and that the ac load was $r_L = 3$ kΩ. Thus, the saturation current, by Eq. 7-16, is

$$i_{C(\text{sat})} = 2 \text{ mA} + \frac{15 \text{ V}}{3 \text{ k}\Omega} = 2 \text{ mA} + 5 \text{ mA} = 7 \text{ mA}$$

Similarly, by plugging these knowns into Eq. 7-17 we find the cutoff voltage:

$$v_{C(\text{cutoff})} = 15 \text{ V} + (2 \text{ mA})(3 \text{ k}\Omega) = 21 \text{ V}$$

Note that this mathematical solution for $v_{C(\text{cutoff})}$ is consistent with the graphical one shown in Fig. 7-31.

---
*See Appendix A-4.

These formulas, Eqs. 7-16 and 7-17, offer an easy way of determining the exact position of the ac load line. After the dc load line and operating point are determined, either $i_{C(\text{sat})}$ or $v_{C(\text{cutoff})}$ is found. A straight line through either the saturation or cutoff point and the operating Q is the ac load line.

Example 7-22 In a circuit like that of 7-30, $V_{EE} = 15$ V, $V_{CC} = -24$ V, $R_E = 60$ kΩ, $R_C = 40$ kΩ, and $R_L = 10$ kΩ. Sketch the collector characteristics and both the dc and ac load lines.

Answer 7-22 See Fig. 7-33.—First the saturation cutoff points on the dc load line are found:

$$I_{C(\text{sat})} = \frac{V_{CC}}{R_C} = \frac{24 \text{ V}}{40 \text{ k}\Omega} = 0.6 \text{ mA}, \qquad V_{CB(\text{cutoff})} = V_{CC} = -24 \text{ V}$$

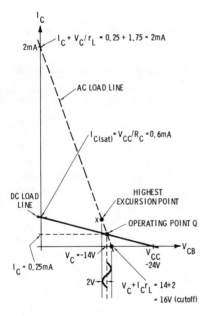

Fig. 7-33. Load lines on collector characteristics of common-base amplifier.

The dc load line is drawn through these two points. The quiescent collector current is

$$I_C \cong I_E \cong \frac{V_{EE}}{R_E} = \frac{15 \text{ V}}{60 \text{ k}\Omega} = 0.25 \text{ mA}$$

Projecting to the right of this I_C value to the dc load line marks the operating point. Projecting down from the operating point Q as shown in Fig. 7-33, we note that $V_c = -14$ V. We can also solve mathematically for the quiescent collector-to-base voltage (absolute value) as follows:

$$V_C = V_{CC} - I_C R_C = 24 - 10 = 14 \text{ V}$$

The ac load on the collector is $r_L = (40 \times 10)/50 = 8$ kΩ.

Now that we know the values of I_C, V_C, and r_L, we plug them into Eqs. 7-16 and 7-17 to find the end points of the ac load line:

$$i_{C(\text{sat})} = I_C + \frac{V_C}{r_L} = 0.25 \text{ mA} + \frac{14 \text{ V}}{8 \text{ k}\Omega} = 2 \text{ mA}$$

$$v_{C(\text{cutoff})} = V_C + I_C r_L = 14 + (0.25 \text{ mA})(8 \text{ k}\Omega) = 16 \text{ V}$$

Example 7-23 Referring to the circuit described in Ex. 7-22, what is the peak value of the maximum possible undistorted signal voltage output?

Answer 7-23 $v_{o(\text{max})} = 2$ V.—Note in Fig. 7-33 that there is only 2 V between the quiescent collector-to-base voltage V_C and cutoff point $V_C + I_C r_L$ on the ac load line. Thus, in order to avoid clipping on the negative alternations, the peak of the signal must be kept at 2 V or less.

7-9 THE OPTIMUM OPERATING POINT

In the last two example problems and on the load lines drawn for the circuit described in them (Fig. 7-33), we found the operating point Q at about the center of the dc load line; that is, almost halfway between ($I_C = 0.6$ mA, $V_{CB} = 0$) and ($I_C = 0$, $V_{CB} = -24$ V). However, this operating point Q is far from being centered on the ac load line; that is, it is relatively close to the cutoff point and far from the saturation point on the ac load line. As pointed out in Ex. 7-23, this operating point limits the output signal amplitude to a peak of 2 V or less if serious distortion is to be avoided. We can improve on our undistorted output signal capability somewhat if we properly increase the emitter bias current I_E. This will increase the quiescent value of I_C and will place the operating point Q higher and to the left on the dc load line. The ac load line drawn through this new operating point Q has a greater distance from point Q to the cutoff point. Thus, if the operating point and ac load line in Fig. 7-33 are moved to the left, we may obtain a new operating point Q that is centered on the ac load line as shown in Fig. 7-34.

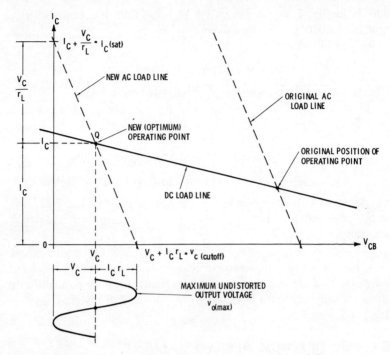

Fig. 7-34. Increasing undistorted output signal capability.

Note in Fig. 7-34 that the original and new load lines have the same slope (steepness), which must be the case if the ac load r_L is not changed. In other words, the slope of a load line is determined by the value of load resistance. Note also that the quiescent current I_C is centered between zero and $i_{C(sat)}$ on the vertical axis and that the quiescent voltage V_C is centered between zero and $v_{c(cutoff)}$ on the horizontal axis. Since $v_{c(cutoff)} = V_C + I_C r_L$ we can see that V_C and $I_C r_L$ are equal to each other on the horizontal axis in Fig. 7-34. That is,

$$V_C = I_C r_L \tag{7-18}$$

at the optimum operating point. By Kirchhoff's law we also know that V_C, which is the collector-to-base dc voltage, plus $I_C R_C$, which is the dc voltage drop across R_C, must equal the source voltage V_{CC}. Thus,

$$V_C = V_{CC} - I_C R_C \tag{7-19}$$

7-9 THE OPTIMUM OPERATING POINT

If we substitute the right side of Eq. 7-18 for V_C in Eq. 7-19, we get

$$I_C r_L = V_{CC} - I_C R_C \qquad (7\text{-}20)$$

By solving Eq. 7-18 for I_C we can show that at the optimum operating point

$$I_C = \frac{V_{CC}}{R_C + r_L} \qquad (7\text{-}21)$$

Example 7-24 Referring to the circuit of Fig. 7-30, suppose that $V_{EE} = 15$ V, $V_{CC} = -24$ V, $R_C = 40$ kΩ, and $R_L = 10$ kΩ as in Ex. 7-22. However, you are required to determine the value of R_E that will provide the optimum operating point. What is the largest undistorted peak signal output $v_{o(\max)}$ with your R_E value?

Answer 7-24 $R_E = 30$ kΩ, $v_{o(\max)} = 4$ V.—By Eq. 7-21, the quiescent collector current at the optimum operating point is

$$I_C = \frac{24 \text{ V}}{40 \text{ k}\Omega + 8 \text{ k}\Omega} = 0.5 \text{ mA}$$

Thus, the emitter bias current $I_E \cong 0.5$ mA too, and from Eq. 6-1 we have

$$R_E \cong \frac{V_{EE}}{I_E} = \frac{15 \text{ V}}{0.5 \text{ mA}} = 30 \text{ k}\Omega$$

In Fig. 7-34 it is shown that the peak value of the maximum undistorted output voltage is equal to V_C or $I_C r_L$, which, by Eq. 7-10, is equal to

$$V_C = 24 - 0.5 \times 40 = 4 \text{ V}$$

or, by Eq. 7-18,

$$I_C r_L = 0.5 \times 8 = 4 \text{ V}$$

Compare now $v_{o(\text{peak})} = 2$ V obtained in Ex. 7-23 and shown in Fig. 7-33, with $v_{o(\text{peak})} = 4$ V obtained in this case by using the optimum operating point shown in Fig. 7-34. In this case the undistorted output signal capability was doubled by using the proper bias current I_E.

Example 7-25 Referring to the circuit described in Ex. 7-22, what is the largest peak signal voltage $v_{s(\text{peak})}$ that can be used to drive the input if $r_s = 4$ kΩ?

Answer 7-25 $v_{s(\text{peak})} \cong 1$ V.—It may be helpful to refer to an ac equivalent circuit of Fig. 7-30 as shown in Fig. 7-21B. In this case, since $v_{o(\text{peak})} = 2$ V (*cf.* Ex. 7-23) and $r_L = 8$ kΩ, by Ohm's law

$$i_{c(\text{peak})} = \frac{v_{o(\text{peak})}}{r_L} = \frac{2 \text{ V}}{8 \text{ k}\Omega} = 0.25 \text{ mA} \cong i_{e(\text{peak})}$$

With $r_s \gg r_e'$, the signal source v_s sees a total load very nearly equal to r_s. Thus $v_{s(\text{peak})} \cong i_{e(\text{peak})} \times r_s = 0.25 \times 4 = 1$ V.

Example 7-26 Referring to the circuit described in Ex. 7-24 and using $R_E = 30$ kΩ to obtain the optimum operating point, what is the largest $v_{s(\text{peak})}$ that can be used to drive the input if $r_s = 4$ kΩ?

Answer 7-26 $v_{s(\text{peak})} = 2$ V.—Again using the ac equivalent (Fig. 7-21B), with $v_{o(\text{peak})} = 4$ V in this case, $i_{c(\text{peak})} = 4$ V/8 kΩ = 0.5 mA $\cong i_{e(\text{peak})}$. And therefore $v_{s(\text{peak})} \cong 0.5 \times 4 = 2$ V by Ohm's law.

Another approach to the solution to this problem is to find the overall gain of the stage first. Since $r_s \gg r_e'$, we find this gain with Eq. 7-15. Thus

$$\frac{v_o}{v_s} \cong \frac{r_L}{r_s} = \frac{8 \text{ k}\Omega}{4 \text{ k}\Omega} = 2$$

and therefore since $v_s \cong v_o/A_e$, then $v_{s(\text{peak})} \cong v_{o(\text{peak})}/2 = 4$ V/2 = 2 V.

7-10 OPERATING POINT AFFECTED BY LEAKAGE I_{CBO}

We have so far considered how the operating point of an amplifier is determined by the emitter current I_E, which is the only consideration necessary if the leakage current I_{CBO} or I_{CO} is negligible. With the assumption that $I_{CO} \cong 0$, we used Eq. 6-4, the equation for the dc alpha:

$$\alpha_{\text{dc}} \cong \frac{I_C}{I_E}$$

which, when rearranged, gives

$$I_C \cong \alpha_{\text{dc}} I_E$$

or

$$I_C \cong I_E$$

if $\alpha_{\text{dc}} = \alpha \cong 1$.

7-10 OPERATING POINT AFFECTED BY LEAKAGE I_{CBO}

However, if I_{CO} is not negligible, the collector current I_C is not solely dependent on I_E as shown in the equations above but instead the following is applicable:

$$I_C = \alpha I_E + I_{CO} \qquad (7\text{-}22)$$

Note that if I_{CO} is made zero in Eq. 7-22 above, that it simplifies to Eq. 6-4. Also, if I_E were made zero in Eq. 7-22, which can be accomplished in circuitry by disconnecting the emitter in the common-base amplifier and letting it hang open, the equation simplifies to

$$I_C = I_{CO} \qquad (7\text{-}23)$$

This last equation agrees with the definition, given in the last chapter, for leakage I_{CO}; that is $I_{CO} \cong I_{CBO}$ and is the current across the reverse biased collector-base junction with emitter lead open. (Review Sec. 6-1 and Fig. 6-2 as needed.)

This leakage I_{CO} may become a problem if the ambient temperature increases significantly above normal room temperature, about 25°C, especially with germanium transistors. As shown in Eq. 7-22, if the term I_{CO} increases, it causes an increase in I_C. This increase in I_C moves the operating point up the load line and may cause clipping of the output signal. That is, an increased leakage I_{CO} caused by an increased temperature causes all of the I_E curves to rise as shown in Fig. 7-35. The dashed lines indicate the posi-

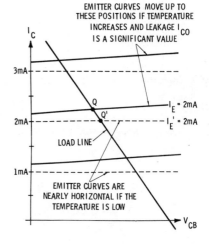

Fig. 7-35. Operating point rises from Q' to Q if leakage I_{CO} increases due to temperature increase.

tions of the curves before a temperature increase. The solid lines represent positions of the curves after a temperature increase. If the circuit emitter current is 2 mA, point Q' is the operating point at the low temperature. A higher temperature moves the curves and therefore the operating point up the load line to a new position Q. Actually, if the temperature increase is quite large, the operating point may move up to, or at least near to, the saturation point. This, of course, limits the maximum possible unclipped variations in the output signal.

Example 7-27 Suppose that the circuit of Fig. 7-30 is used in a hot environment. When the emitter lead of the transistor is disconnected and left open, a dc voltmeter across the collector and base leads reads 26.25 V. (a) What is the leakage I_{CO}? (b) What are I_C and V_{CB} when the emitter is reconnected?

Answer 7-27 (a) $I_{CO} = 0.5$ mA = 500 µA, (b) $I_C \cong 2.5$ mA, $V_{CB} \cong 11.25$ V.—(a) If the emitter is open, that is, $I_E \cong 0$, and if the full source voltage V_{CC} is not across the collector-base junction, leakage current exists and flows through R_C, causing a voltage drop across it. In this case, the voltage across R_C is found by solving Eq. 7-19 for $I_C R_C$ and substituting known values of V_{CC} and V_{CB} into it. Thus, $I_C R_C = V_{CC} - V_{CB} = 30 - 26.25 = 3.75$ V. Thus, since $I_C R_C = 3.75$ V, we can show that $I_C = 3.75$ V/7.5 kΩ = 0.5 mA = 500 mA. With the emitter open, I_C is I_{CO}; see Eq. 7-23. (b) With the emitter reconnected, $I_E \cong V_{EE}/R_E = 2$ mA. Assuming that $\alpha \cong 1$, by Eq. 7-22 $I_C = \alpha I_E + I_{CO} = 2$ mA + 0.5 mA = 2.5 mA. The collector-to-base voltage is $V_{CB} = V_{CC} - I_C R_C = 30 - 2.5 \times 7.5 = 30 - 18.75 = 11.25$ V.

PROBLEMS

7-1. In a circuit like that of Fig. 7-36, what is the collector saturation current and what value of R_E will cause the collector to saturate?

7-2. In a circuit like that of Fig. 7-37, what is the collector saturation current and what value of R_E will cause the collector to saturate?

7-3. Referring to the circuit of Fig. 7-36, what value of R_E will cause a collector-to-base voltage $V_{CB} = 35$ V?

7-4. Referring to the circuit of Fig. 7-37, what value of R_E will cause $V_{CB} = 4$ V?

PROBLEMS

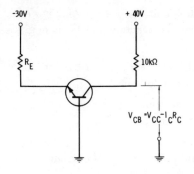

Fig. 7-36. Circuit for Probs. 7-1, 7-3, and 7-7.

Fig. 7-37. Circuit for Probs. 7-2, 7-4, and 7-8.

7-5. If the circuit of Fig. 7-5 is operating normally, what will happen to the readings on the meters M_1, M_2, and M_3 if the slide on the potentiometer R_2 is moved to the right?

7-6. What will happen to the readings on the meters in the circuit of Fig. 7-5 if the slide on the potentiometer R_1 is moved to the left?

7-7. Referring to the circuit in Fig. 7-36, what are the collector current I_C and voltage V_C values if $R_E = 20$ kΩ?

7-8. Referring to the circuit in Fig. 7-37, what are the collector current and voltage values if $R_E = 15$ kΩ?

7-9. What is the ac load r_L in the circuit of Fig. 7-38, assuming that the reactance of C_2 is negligible?

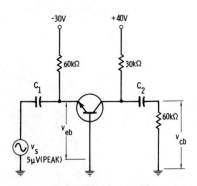

Fig. 7-38. Circuit for Probs. 7-9, 7-11, 7-13, 7-15, and 7-17.

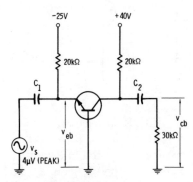

Fig. 7-39. Circuit for Probs. 7-10, 7-12, 7-14, 7-16, and 7-18.

7-10. If $R_C = 20$ kΩ and $R_L = 30$ kΩ in the circuit of Fig. 7-39, what is the ac load r_L, assuming that the reactance of C_2 is negligible?

7-11. What resistance r_{in} does the source v_s see in the circuit of Fig. 7-38, assuming C_1 has negligible reactance?

7-12. What resistance r_{in} does the source v_s see in the circuit of Fig. 7-39, assuming C_1 has negligible reactance?

7-13. Referring to the circuit of Fig. 7-38, what are the peak values of emitter signal current i_e and collector signal current i_c?

7-14. In the circuit of Fig. 7-39, what are the peak values of emitter signal current i_e and the collector signal current i_c?

7-15. What is the voltage gain v_{cb}/v_s of the circuit in Fig. 7-38 and what is the peak of its output v_{cb}?

7-16. What is the voltage gain v_{cb}/v_s of the circuit of Fig. 7-39 and what is the peak of its output v_{cb}?

7-17. What are the quiescent values of dc collector current and voltage in the circuit shown in Fig. 7-38?

7-18. What are the quiescent values of dc collector current and voltage in the circuit shown in Fig. 7-39?

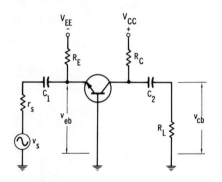

Fig. 7-40. Circuit for Probs. 7-25 through 7-30, 7-37, and 7-38.

7-19. If in a circuit like that of Fig. 7-8, $V_{EE} = -12$ V, $V_{CC} = 9$ V, $R_E = 9.6$ kΩ, $R_C = 3$ kΩ, and $R_L = 3$ kΩ, what are the approximate values of the following: (a) current gain A_i, (b) voltage gain v_{cb}/v_s, (c) power gain A_p, (d) quiescent dc collector current I_C, (e) quiescent dc collector-to-base voltage V_{CB}, and (f) the average dc collector power dissipation P_C?

Problems

7-20. If in a circuit like that of Fig. 7-8, $V_{EE} = -40$ V, $V_{CC} = 40$ V, $R_E = 50$ kΩ, $R_C = 20$ kΩ, and $R_L = 4$ kΩ, what are the approximate values of the following: (a) current gain A_i, (b) voltage gain v_{cb}/v_s, (c) power gain A_p, (d) quiescent dc collector current I_C, (e) quiescent dc collector-to-base voltage V_C, and (f) the average dc collector power dissipation P_C?

7-21. A certain signal source has an open-circuit output voltage of 6 mV rms. With a 6 kΩ load its output voltage drops 4.8 mV. What is the internal resistance r_s of this source?

7-22. A certain signal source has an open-circuit output voltage of 20 mV rms. With a 4 kΩ load, the output voltage drops to 16 mV. What is the internal resistance of this source?

7-23. If the signal source described in Prob. 7-21 is used to drive a 1 kΩ load, what is the rms voltage across the load?

7-24. If the signal source described in Prob. 7-22 is used to drive a 3-kΩ load, what is the rms voltage across the load?

7-25. Suppose that in a circuit like that of Fig. 7-40, $V_{EE} = -30$ V, $R_E = 60$ kΩ, $v_{s(\text{peak})} = 3$ mV, and $r_s = 100$ Ω. What are the approximate values: (a) dynamic emitter resistance r_e', and (b) peak of the emitter signal current $i_{e(\text{peak})}$?

7-26. Suppose that in a circuit like that of Fig. 7-40, $V_{EE} = -20$ V, $R_E = 20$ kΩ, $v_{s(\text{peak})} = 8$ mV, and $r_s = 75$ Ω. What are the approximate values of: (a) dynamic emitter resistance r_e', and (b) peak emitter signal current $i_{e(\text{peak})}$?

7-27. Referring to the circuit of Fig. 7-40 and using the data from Prob. 7-25 with $V_{CC} = 30$ V, $R_C = 30$ kΩ, and $R_L = 20$ kΩ, what is: (a) the peak of the output voltage $v_{cb(\text{peak})}$, (b) the voltage gain of the stage v_{cb}/v_{eb}, and (c) the overall gain of the circuit v_{cb}/v_s?

7-28. Referring to the circuit of Fig. 7-40 and using the data of Prob. 7-26 with $V_{CC} = 20$ V, $R_C = 10$ kΩ, and $R_L = 40$ kΩ, what is: (a) the peak of the output voltage $v_{cb(\text{peak})}$, (b) the voltage gain of the amplifier v_{cb}/v_{eb}, and (c) the overall gain of the circuit v_{cb}/v_s?

7-29. If in the circuit of Fig. 7-40 $V_{EE} = -12$ V, $V_{CC} = 12$ V, $R_E = 20$ kΩ, $R_C = R_L = 10$ kΩ, $r_s = 2$ kΩ, and $v_{s(\text{peak})} = 0.4$ V, what are the approximate values of the following: (a) the voltage gain of the amplifier v_{cb}/v_{eb}, (b) the overall voltage gain of the circuit, and (c) the peak of the output voltage $v_{cb(\text{peak})}$?

7-30. If in the circuit Fig. 7-40, $V_{EE} = -32$ V, $V_{CC} = 24$ V, $R_E = 20$ kΩ, $R_C = 10$ kΩ, $R_L = 30$ kΩ, $r_s = 2.5$ kΩ, and $v_{s(\text{peak})} = 0.5$ V, what are the approximate values of the following: (a) the voltage gain of the amplifier v_{cb}/v_{eb}, (b) the overall voltage gain of the circuit, and (c) the peak of the output voltage $v_{cb(\text{peak})}$?

7-31. Sketch the dc and ac load lines for the circuit described in Prob. 7-29. Indicate the values where these load lines cross the I_C and V_{CB} axes and show the coordinates (I_C and V_C quiescent values) of the operating point.

7-32. Sketch the dc and ac load lines for the circuit described in Prob. 7-30. Indicate the values where these load lines cross the I_C and the V_{CB} axes and show the coordinates (I_C and V_C quiescent values) of the operating point.

7-33. What is the peak value of the largest possible unclipped signal output $v_{cb(\text{peak})}$ in the circuit described in Prob. 7-29, assuming that v_s can be varied?

7-34. What is the peak value of the largest possible unclipped signal output $v_{cb(\text{peak})}$ in the circuit described in Prob. 7-30, assuming that v_s can be varied?

7-35. Suppose that we want to replace R_E to obtain the optimum operating point in the circuit described in Prob. 7-29. (a) What R_E value will you use? (b) With your R_E value, what is the peak value of the largest possible unclipped signal output $v_{cb(\text{peak})}$, assuming that v_s is a variable signal source?

7-36. Suppose that in the circuit described in Prob. 7-30, we want to replace R_E to obtain the optimum operating point. (a) What R_E value will you use? (b) With your R_E value, what is the peak value of the largest possible unclipped signal output $v_{cb(\text{peak})}$, assuming that v_s is a variable signal source?

7-37. Sketch the output voltage waveform v_o for a circuit like that of Fig. 7-40, where $V_{EE} = -40$ V, $V_{CC} = 60$ V, $R_E = 10$ kΩ, $R_C = 8$ kΩ, $R_L = 2$ kΩ, $r_s = .8$ kΩ and the output of the signal source is sinusoidal with $v_{s(\text{peak})} = 7$ V.

7-38. Sketch the output voltage waveform v_o for a circuit like that of Fig. 7-40, where $V_{EE} = -20$ V, $V_{CC} = 24$ V, $R_E = 10$ kΩ,

PROBLEMS

$R_C = 8$ kΩ, $R_L = 2$ kΩ, $r_s = 1.8$ kΩ and the output of the signal source is sinusoidal with $v_{s(peak)} = 4.5$ V.

7-39. Suppose that you are to modify the circuit described in Prob. 7-37 to obtain the optimum operating point. (a) What value of R_E will you use? (b) With your new R_E value, sketch v_{cb} waveform where $v_{s(peak)} = 7$ V. (c) With your R_E value, sketch v_{cb} waveform if $v_{s(peak)} = 10$ V.

7-40. Suppose that you are to modify the circuit described in Prob. 7-38 to obtain the optimum operating point. (a) What value of R_E will you use? (b) Using your R_E value, sketch v_{cb} waveform where $v_{s(peak)} = 4$ V. (c) Using your R_E value, sketch v_{cb} waveform if $v_{s(peak)} = 5$ V.

8

Junction Transistors in Common-Emitter Connection

Part I: DC Considerations

In the last chapter we learned that the common-base amplifier, though simple to build, has undesirable features. As we noted in some examples and problems in that chapter, the common-base amplifier input resistance r_e' is low and therefore it loads down any source whose internal resistance is high, that is used to drive the amplifier input. Since the common-base amplifier output acts like a constant-current source, that is, like a source with very large internal resistance, the output of one common-base amplifier cannot be connected to the input of another (connected in *cascade*) if reasonably good gain is to be expected. Also the inherent low current gain A_i of the common-base circuit limits its power gain A_p capability.

The common-emitter (or CE) amplifier is so called because usually its emitter lead is common with the input and output signals. For example, typically the emitter is grounded and the signal source v_s is connected across the base and ground, while the output signal is taken off the collector and ground. Thus, the emitter and ground are common to input and output. The common-emitter

8-1 I_C VS I_B (CUTOFF, SATURATION, AND LEAKAGE DEFINED)

amplifier has a much larger input resistance and is capable of much more power gain than the common-base circuit. These features make the common-emitter amplifier the most popular configuration. Because of its popularity, the discussion in this chapter is rather thorough and is divided into two parts. Part I concentrates on various methods of biasing CE circuits and other dc considerations. Common-emitter circuit action when ac (signal) is applied is covered in Part II.

8-1 I_C VS I_B (CUTOFF, SATURATION, AND LEAKAGE DEFINED)

Simple circuits connected in common emitter are shown in Fig. 8-1. Normally, as with the common-base connection, the collector-base junction is reverse biased while the base-emitter junction is forward biased. In this case a single V_{CC} source does the job. We may better see how in Fig. 8-2, which considers the pnp circuit. Note in Fig. 8-2A that the V_{CC} supply tends to cause electron flow

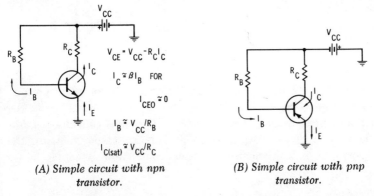

(A) Simple circuit with npn transistor.

(B) Simple circuit with pnp transistor.

Fig. 8-1. Basic common-emitter configurations.

down through R_C. However, the V_{CC} source sees the collector-base junction in reverse bias via ground. That is, the majority carriers (holes) in the collector material are drawn up toward the negative terminal of this supply while the majority carriers (electrons) in the thin base material are drawn down toward the positive terminal through ground. The majority carriers thus drawn away from the junction prevent recombinations of electrons and holes which support current across it, if there is no current in the base lead

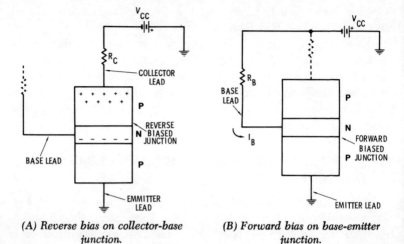

(A) Reverse bias on collector-base junction.

(B) Forward bias on base-emitter junction.

Fig. 8-2. Bias on pnp transistor junctions in common-emitter circuit.

and if the leakage current is negligible. Fig. 8-2B shows how the base-emitter junction is forward biased through R_B and ground. An analysis similar to that above may be made for the npn circuit.

With no current in the base lead, as would be the case if the base lead is open, as shown in Fig. 8-3A, the collector-base junction acts like any reverse biased diode junction and normally conducts very little current. Of course, if minority carriers are present, some leakage current I_{CEO} flows. In this symbol for leakage in the common-emitter circuit, the subscript CEO means *collector-to-emitter* leakage with the *base* lead open. For any given transistor the leakage I_{CEO} in the common-emitter connection is much larger

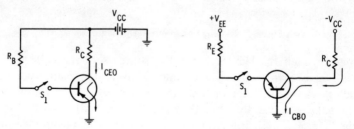

(A) With base lead open, I_{CEO} flows supported by minority carriers.

(B) With emitter lead open, I_{CBO} flows supported by minority carriers.

Fig. 8-3. Leakage currents and terminology.

8-1 I_C vs I_B (Cutoff, Saturation, and Leakage Defined)

than the leakage I_{CBO}, which is the collector-to-base leakage with emitter lead open. In fact their relationship is shown in the following equation:

$$I_{CEO} = (\beta + 1)I_{CBO} \qquad (8\text{-}1)$$

If the β of the transistor is greater than 30 or so, we may simplify Eq. 8-1 above to

$$I_{CEO} \cong \beta I_{CBO} \qquad (8\text{-}2)$$

In Chap. 6, Eq. 6-5, we learned that the dc beta is

$$\beta_{dc} = h_{FE} \cong \frac{I_C}{I_B}$$

There is also a term called *ac beta*, which may be defined as follows:

$$\beta = h_{fe} \cong \frac{i_c}{i_b} \qquad (8\text{-}3)$$

The symbol β without subscript is usually assumed to be the ac beta. Actually the ac and dc beta values for any given transistor are so nearly the same that usually in practice, and for the remainder of this book, you may assume that they are and that

$$h_{FE} \cong h_{fe} = \beta \qquad (8\text{-}4)$$

Example 8-1 Suppose that a certain transistor has these characteristics: maximum $I_{CBO} = 1$ μA, $h_{FE} = h_{fe} = 80$. What is the maximum current in the collector lead in circuit Fig. 8-3A?

Answer 8-1 With the base lead open, $I_C = I_{CEO} \cong \beta I_{CBO} = 80 \times (1 \text{ μA}) = 80 \text{ μA}$.

Example 8-2 Suppose that in a circuit like that of Fig. 8-3A you measure the current in R_C to be 0.5 mA. Using the same transistor, what is the current in R_C if the base lead is closed and the emitter lead opened as in Fig. 8-3B and if $\beta = 200$?

Answer 8-2 $I_{CBO} \cong 2.5$ μA.—Solving Eq. 8-2 for I_{CBO} we can show that

$$I_{CBO} \cong \frac{I_{CEO}}{\beta} = \frac{0.5 \text{ mA}}{200} = 2.5 \text{ μA}$$

It should be noted that the leakage I_{CBO} may be small and insignificant but I_{CEO} may not be because it is multiplied by the β of the transistor. Transistor manufacturers are continually improving the purity of their semiconductors. With fewer impurities there are fewer minority carriers to support leakage current, and therefore problems caused by leakage are now becoming less severe.

If the base circuit is closed as in Fig. 8-1, base current I_B flows. Referring to Fig. 8-2B, existence of I_B causes electron flow into the base material. In other words, the source V_{CC} injects electrons into the base via R_B in Fig. 8-2B. These injected electrons tend to refill the depletion region in the base with majority carriers which then support current across the reverse biased collector-base junction. In other words, the region depleted of its majority carriers at the reverse biased collector-base junction becomes smaller, more or less, depending on how many majority carriers are injected into the base. Reducing the depletion region increases the current through it. Thus, increasing the base current increases the number of majority current carriers injected into the base (which are electrons) in Fig. 8-2B. This causes a reduced depletion region allowing increased current across the collector-base junction that supports the collector current I_C. In fact, if the leakage current is negligible, base and collector currents are related as follows:

$$I_C \cong \beta I_B \qquad (8\text{-}5)$$

If leakage cannot be ignored, their relationship is more accurately shown as:

$$I_C = \beta I_B + I_{CEO} \qquad (8\text{-}6)$$

or, by substituting Eq. 8-2 into Eq. 8-6 above, we can show

$$I_C \cong \beta I_B + \beta I_{CBO} = \beta(I_B + I_{CBO}) \qquad (8\text{-}7)$$

Note that with Eq. 8-5, if $I_B = 0$, then $I_C \cong 0$, and that with Eq. 8-6, if $I_B = 0$, then $I_C = I_{CEO}$.

The approximate value of I_B can be found by Ohm's law in circuits as in Fig. 8-1A. Since the base-emitter junction is a forward biased diode junction, it has only a few tenths dc volt drop across it. Thus, most of the V_{CC} source voltage appears across R_B. Thus, by Ohm's law,

$$I_B \cong \frac{V_{CC}}{R_B} \qquad (8\text{-}8)$$

Example 8-3 Suppose that in the circuit of Fig. 8-1A, $V_{CC} = 24$ V, $R_B = 960$ kΩ, $R_C = 7$ kΩ, $\beta = 80$, and the leakage is negligible.

8-1 I_C vs I_B (Cutoff, Saturation, and Leakage Defined)

What are the values of I_B, I_C, and V_{CE}, where V_{CE} is the collector-to-emitter voltage?

Answer 8-3 $I_B \cong 25$ μA, $I_C \cong 2$ mA, $V_{CE} = 10$ V.—By Eq. 8-8, we first find that

$$I_B \cong \frac{V_{CC}}{R_B} = \frac{24 \text{ V}}{960 \text{ k}\Omega} = 25 \text{ μA}$$

By Eq. 8-5 we now solve for the collector current

$$I_C \cong \beta I_B = 80(25 \text{ μA}) = 2 \text{ mA}$$

By Kirchhoff's law $V_{CC} = V_{CE} + I_C R_C$; therefore

$$V_{CE} = V_{CC} - I_C R_C \qquad (8\text{-}9)$$

and in this case

$$V_{CE} = 24 \text{ V} - (2 \text{ mA})(7 \text{ k}\Omega) = 24 - 14 = 10 \text{ V}$$

Example 8-4 Suppose that in the circuit described in Ex. 8-3, the leakage current is not negligible but instead $I_{CBO} = 5$ μA. What are values I_B, I_C, and V_{CE} in this case?

Answer 8-4 $I_B \cong 25$ μA, $I_C \cong 2.4$ mA, $V_{CE} \cong 7.2$ V.—Now with $I_{CBO} = 5$ μA, by Eq. 8-2, leakage

$$I_C = \beta I_B + I_{CEO} = 2 + 0.4 = 2.4 \text{ mA}$$

Therefore by Eq. 8-6 we find that

$$I_C = \beta I_B + I_{CEO} = 2 + 0.4 = 2.4 \text{ mA}$$

where I_B is the same as in the previous problem; that is, $I_B \cong V_{CC}/R_B = 25$ μA. Thus, by Eq. 8-9 $V_{CE} = 24 - 2.4 \times 7 = 24 - 16.8 = 7.2$ V.

It is interesting to note here that if the leakage current is not insignificant, it may noticeably affect the collector current I_C and the voltage across collector and emitter, V_{CE}.

As with the common-base connection of the transistor, there is a maximum possible collector current $I_{C(\text{sat})}$ for a given circuit. And as in the last chapter, its value is found with Eq. 7-1. That is, for circuits like those in Fig. 8-1, the maximum current will flow through R_C when all of the source V_{CC} is dropped across it. Thus, by Ohm's law,

$$I_{C(\text{sat})} \cong \frac{V_{CC}}{R_C}$$

When the transistor is cut off (not conducting), the full source V_{CC} appears across the collector and emitter. Thus, the range over which I_C can be varied by varying I_B, is from 0 to $I_{C(\text{sat})}$ if leakage is negligible. If leakage is not negligible, I_C may be varied from I_{CEO} to $I_{C(\text{sat})}$. In all problems in this chapter, assume that the leakage is negligible unless otherwise specified.

Example 8-5 Suppose that in a circuit like that of Fig. 8-3A, $V_{CC} = 20$ V, $R_C = 4$ kΩ, $R_B = 500$ kΩ, and $ß = 50$. (a) With the switch S_1 open, what values of I_C and V_{CE} would you expect? (b) With S_1 closed, what are the values of I_C and V_{CE}?

Answer 8-5 (a) $I_C \cong 0$, $V_{CE} \cong 20$ V, (b) $I_C \cong 2$ mA, $V_{CE} \cong 12$ V. —With the switch S_1 open, the base current $I_B = 0$ and thus $I_C \cong ßI_B = 0$. With no current through R_C there is no voltage drop across it and the full source voltage is dropped across the collector and emitter. When S_1 is closed, we find that the base current is approximately 40 μA:

$$I_B \cong \frac{V_{CC}}{R_B} = \frac{20\text{ V}}{500\text{ kΩ}} = 40\text{ μA}$$

Therefore $I_C \cong ßI_B = 50(40 \text{ μA}) = 2$ mA. And in this case, then, $V_{CE} = V_{CC} - I_C R_C = 20 - (2\text{ mA})(4\text{ kΩ}) = 20 - 8 = 12$ V.

Example 8-6 Suppose that in the circuit described in Ex. 8-5, the transistor is replaced with one having $ß = 100$ that is suspected of having high leakage. With the switch S_1 open, you read 1 V across R_C. What are the leakage currents I_{CEO} and I_{CBO}?

Answer 8-6 $I_{CEO} = 0.25$ mA, $I_{CBO} = 2.5$ μA.—With the switch S_1 open, the voltage across R_C must be due to leakage I_{CEO}. Thus, by Ohm's law $I_{CEO} = 1$ V/4 kΩ $= 0.25$ mA. Solving Eq. 8-2 for I_{CBO} we get $I_{CBO} = I_{CEO}/ß = 0.25$ mA/100 $= 2.5$ μA.

Example 8-7 Referring to Ex. 8-6, what are values I_C and V_{CE} with this transistor when S_1 is closed, accounting for leakage current?

Answer 8-7 $I_C = 3$ mA, $V_{CE} = 3$ V.—As long as the base-emitter junction is forward biased, I_B value is determined by the values of V_{CC} and R_B. Thus, it is found to be 40 μA as in Ex. 8-5. Since I_{CEO} is not to be considered negligible, we can find the collector current

8-1 I_C vs I_B (Cutoff, Saturation, and Leakage Defined)

with Eq. 8-6. $I_C = \beta I_B + I_{CEO} = 100(40 \ \mu A) + 0.25 \ mA = 4 + 0.25 = 4.25 \ mA$. Thus, $V_{CE} = V_{CC} - I_C R_C = 20 - (4.25 \ mA)(4 \ k\Omega) = 20 - 17 = 3 \ V$.

Example 8-8 In a circuit like that of Fig. 8-3A, $R_B = 1.2 \ M\Omega$, $R_C = 6 \ k\Omega$, $V_{CC} = 12 \ V$, and the transistor $\beta = 50$. At room temperature, about 25°C, a voltmeter across R_C reads 0.3 V when switch S_1 is open. (a) What are the leakage currents I_{CEO} and I_{CBO}? (b) What is the collector current when the switch S_1 is closed? (c) If the temperature increases to 45°C and the transistor is a germanium type, what are the leakage currents I_{CEO} and I_{CBO}? (d) What is the collector current at 45°C when S_1 is closed? HINT: Leakage current in a reverse biased germanium pn junction roughly doubles with every 10°C increase in temperature; see Sec. 3-5.

Answer 8-8 (a) $I_{CEO} = 50 \ \mu A$, $I_{CBO} \cong 1 \ \mu A$, (b) $I_C = 550 \ \mu A$, (c) $I_{CEO} = 0.2 \ mA$ or $200 \ \mu A$, $I_{CBO} \cong 4 \ \mu A$, (d) $I_C = 0.7 \ mA$ or $700 \ \mu A$.—(a) At 25°C, the 0.3-V drop across R_C is caused by leakage current through this resistor. Its value by Ohm's law is $I_{CEO} = 0.3 \ V/6 \ k\Omega = 50 \ \mu A$. By Eq. 8-2, $I_{CBO} \cong I_{CEO}/\beta = 50 \ \mu A/50 = 1 \ \mu A$.

(b) When S_1 is closed, nearly the full 12 V is across R_B and therefore $I_B \cong 12 \ V/1.2 \ M\Omega = 10 \ \mu A$. Now by Eq. 8-6, $I_C = \beta I_B + I_{CEO} = 50(10 \ \mu A) + 50 \ \mu A = 550 \ \mu A$.

(c) At 45°C, a temperature increase of 20°C, the leakage is roughly doubled twice; that is, it is doubled once for each 10°C rise. Thus, we can multiply each of the previous leakage values by 2 twice and thus find that the new leakage currents are: $I_{CEO} = 50 \ \mu A \times 2 \times 2 = 200 \ \mu A$, and $I_{CBO} = 1 \ \mu A \times 2 \times 2 = 4 \ \mu A$.

(d) With S_1 closed at 45°C we find that the collector current increases to

$$I_C = \beta I_B + I_{CEO} = 50(10 \ \mu A) + 200 \ \mu A = 700 \ \mu A$$

The base current in this circuit does not change because, regardless of the temperature and amount of leakage, the 12 V source voltage remains essentially across R_B, ensuring a nearly constant current through it.

Example 8-9 If in a circuit like that of Fig. 8-1B, $V_{CC} = 21 \ V$, $R_C = 7 \ k\Omega$, and $\beta = 150$, (a) what is the largest value of R_B that will cause $I_{C(sat)}$ to flow in R_C? (b) What value of R_B will cause $I_C = 1.5 \ mA$?

Answer 8-9 (a) 1.05 MΩ, (b) 2.1 MΩ.—(a) In this case $I_{C(\text{sat})} = V_{CC}/R_C = 21$ V/7 kΩ = 3 mA. The maximum base current that will still have control of the collector can be found by solving Eq. 8-5 for I_B. That is, since $I_B \cong I_C/\beta$, then $I_{B(\text{sat})} \cong 3$ mA/150 = 20 μA. Base currents larger than this will not cause proportional increases in I_C. Solving Eq. 8-8 for R_B we can show that

$$R_B \cong \frac{V_{CC}}{I_B} \cong \frac{21 \text{ V}}{20 \text{ μA}} = 1.05 \text{ MΩ}$$

Smaller values of R_B than this will cause proportionally larger values of I_B but not I_C. (b) To obtain $I_C = 1.5$ mA, we need $I_B \cong I_C/\beta = 1.5$ mA/150 = 10 μA. Thus, we would use $R_B \cong V_{CC}/I_B = 21$ V/10 μA = 2.1 MΩ.

8-2 EMITTER FEEDBACK FOR STABILITY

The circuits shown in Fig. 8-1 may be called "base biased" circuits because their base bias current I_B is determined primarily by the values of the base resistor R_B and the source V_{CC}. See Eq. 8-8. A base biased circuit is simple and easy to build but has a potenitally serious disadvantage. Its quiescent collector current and therefore its operating point Q are subject to changes with temperature changes if the transistor leakage is not negligible, as we saw in Exs. 8-7 and 8-8. This may cause more or less clipping and distortion of the output signal with temperature changes when this base biased circuit is used as an amplifier. You should note in the preceding examples that while I_C and V_{CE} may change with temperature changes, the base current I_B does not. In different circuit designs you will see that I_B actually decreases if leakage increases.

One such different design is shown in Fig. 8-4. We may call this type of circuit *base biased with emitter feedback*. Use of an emitter resistor R_E tends to cause I_B to decrease if I_{CEO} increases, and conversely if I_{CEO} decreases, I_B increases. This action stabilizes, that is, tends to prevent changes in, the values of I_C and V_{CE} even though I_{CEO} varies with temperature variations. Why this is so may be explained as follows: In Fig. 8-4, the voltage across R_B *is not* the source voltage V_{CC}. Instead, the sum of the voltages across R_B and R_E is equal to V_{CC}, neglecting the small drop across the forward biased base-emitter junction. Thus, if a temperature increase in-

8-2 EMITTER FEEDBACK FOR STABILITY

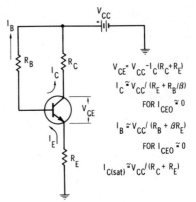

Fig. 8-4. Circuit using emitter feedback for temperature stability.

creases both leakage I_{CEO} and I_C, the current up through R_E increases, causing more voltage drop across R_E. This in turn causes less voltage drop across R_B and decreases I_B. The reduced I_B tends to limit the increase in I_C. That is, referring to the circuit in Fig. 8-4 and Eq. 8-6,

$$I_C = \beta I_B + I_{CEO}$$

we can observe that if I_{CEO} increases while I_B simultaneously decreases, these oppositely varying values on the right side tend to prevent large changes in I_C on the left side. You will observe this in examples later.

The following are equations applicable to circuits like that of Fig. 8-4:

$$I_C \cong \frac{V_{CC}}{R_E + R_B/\beta} + \left(\frac{R_B + R_E}{R_B + \beta R_E}\right) I_{CEO} \qquad (8\text{-}10)^*$$

Or if the leakage is negligible, that is, $I_{CEO} \cong 0$, Eq. 8-10 above simplifies to

$$I_C \cong \frac{V_{CC}}{R_E + R_B/\beta} \qquad (8\text{-}11)$$

Also

$$I_B \cong \frac{V_{CC}}{R_B + \beta R_E} - \left(\frac{R_E}{R_B + \beta R_E}\right) I_{CEO} \qquad (8\text{-}12)^*$$

But again if $I_{CEO} \cong 0$, Eq. 8-12 above simplifies to

---

*See Appendix A-5.

$$I_B \cong \frac{V_{CC}}{R_B + \beta R_E} \tag{8-13}$$

and by Kirchhoff's law the collector-to-emitter voltage is

$$V_{CE} = V_{CC} - I_C(R_C + R_E) \tag{8-14}$$

When selecting component values for a circuit like that in Fig. 8-4, usually the simpler Eqs. 8-11 and 8-13 can be used even though a significant I_{CEO} exists. Small compensations can be made as you will see.

The collector saturation current $I_{C(\text{sat})}$ is easily found by noting that since $V_{CE} \cong 0$ at saturation, the full source voltage V_{CC} is across R_C and R_E. Thus by Kirchhoff's and Ohm's laws

$$V_{CC} = R_C I_C + R_E I_E$$

and since $I_C \cong I_E$, then

$$V_{CC} \cong R_C I_C + R_E I_C = (R_C + R_E) I_C$$

Dividing both sides of the above equation by $R_C + R_E$ and rearranging terms we get

$$I_C \cong \frac{V_{CC}}{R_C + R_E}$$

at saturation. Therefore

$$I_{C(\text{sat})} \cong \frac{V_{CC}}{R_C + R_E} \tag{8-15}$$

where $I_{C(\text{sat})}$ denotes the collector saturation current.

Example 8-10 Suppose that in a circuit like that in 8-4, $R_B = 2$ MΩ, $R_C = 8$ kΩ, $R_E = 4$ kΩ, and $V_{CC} = 18$ V. The transistor $h_{FE} \cong h_{fe} \cong 100$. Neglecting leakage, find $I_{C(\text{sat})}$, I_C, I_B, and V_{CE}.

Answer 8-10 $I_{C(\text{sat})} \cong 1.5$ mA, $I_C \cong 0.75$ mA, $I_B \cong 7.5$ μA, $V_{CE} \cong 9$ V.—With Eq. 8-15 we find the saturation collector current

$$I_{C(\text{sat})} \cong \frac{V_{CC}}{R_C + R_E} = \frac{18 \text{ V}}{8 \text{ kΩ} + 4 \text{ kΩ}} = 1.5 \text{ mA}$$

Since leakage is negligible, we can find the quiescent values of collector and base currents with Eqs. 8-11 and 8-5 respectively as follows:

$$I_C \cong \frac{V_{CC}}{R_E + R_B/\beta} \cong \frac{18 \text{ V}}{4 \text{ kΩ} + 2 \text{ MΩ}/100} = \frac{18 \text{ V}}{24 \text{ kΩ}} = 0.75 \text{ mA}$$

8-2 EMITTER FEEDBACK FOR STABILITY

Solving Eq. 8-5 for I_B and substituting we have

$$I_B \cong \frac{I_C}{\beta} = \frac{750 \ \mu A}{100} = 7.5 \ \mu A$$

We could have solved for I_B first with Eq. 8-13:

$$I_B \cong \frac{V_{CC}}{R_B + \beta R_E} = \frac{18 \ V}{2 \ M\Omega + 100(4 \ k\Omega)} = \frac{18 \ V}{2.4 \ M\Omega} = 7.5 \ \mu A$$

And then by Eq. 8-5 we find

$$I_C \cong \beta I_B = 100(7.5 \ \mu A) = 0.75 \ mA$$

By Eq. 8-14 the collector-to-emitter voltage is

$$V_{CE} = V_{CC} - I_C(R_C + R_E) = 18 - 0.75(12) = 18 - 9 = 9 \ V$$

Example 8-11 If the circuit described in Ex. 8-10 has a leakage $I_{CEO} = 0.75$ mA at a higher temperature, what are the values I_C, I_B, and V_{CE} when $h_{FE} \cong h_{fe} \cong 100$?

Answer 8-11 $I_C \cong 1.375$ mA, $I_B \cong 6.25 \ \mu A$, $V_{CE} \cong 1.5$ V.—Since the leakage is significant, we use Eq. 8-10 to find the collector current:

$$I_C \cong \frac{V_{CC}}{R_E + R_B/\beta} + \left(\frac{R_B + R_E}{R_B + \beta R_E}\right) I_{CEO} = 0.75 + 0.625 = 1.375 \ mA$$

Once the value of I_C is known we can find the base current I_B by either Eq. 8-12 or Eq. 8-6 as follows:

$$I_B \cong \frac{V_{CC}}{R_B + \beta R_E} - \left(\frac{R_E}{R_B + \beta R_E}\right) I_{CEO} \cong 7.5 - 1.25 = 6.25 \ \mu A$$

or solving Eq. 8-6 for I_B we have

$$I_B \cong \frac{I_C - I_{CEO}}{\beta} = \frac{1.375 - 0.75}{100} = \frac{0.625 \ mA}{100} = 6.25 \ \mu A$$

By Eq. 8-14 the collector-to-emitter voltage is

$$V_{CE} = V_{CC} - I_C(R_C + R_E) = 18 - 16.5 = 1.5 \ V$$

Example 8-12 Referring back to the circuit of Fig. 8-1B, if $V_{CC} = -18$ V, $R_B = 2.4$ MΩ, $R_C = 12$ kΩ and the transistor $h_{FE} \cong h_{fe} \cong 100$, what are the values of I_C, I_B, and V_{CE}? First (a) assume that the leakage is negligible, then (b) use leakage $I_{CEO} = 0.75$ mA.

Answer 8-12 (a) $I_C \cong 0.75$ mA, $I_B \cong 7.5 \ \mu A$, $V_{CE} \cong 9$ V; (b) $I_C \cong 1.5$ mA, $I_B \cong 7.5 \ \mu A$, $V_{CE} \cong 0$.—Whether leakage is negligible or not the base current I_B stays constant for practical purposes in this circuit. This is true because the drop across the forward biased base-emitter junction remains negligible regardless of temperature and leakage current changes. Thus nearly the full V_{CC} voltage is always across R_B and the base current I_B is always determined with Eq. 8-8. Thus, in this case

$$I_B \cong \frac{V_{CC}}{R_B} = \frac{18 \text{ V}}{2.4 \text{ M}\Omega} = 7.5 \text{ }\mu\text{A}$$

If the leakage is negligible ($I_{CEO} \cong 0$), then by Eq. 8-5

$$I_C \cong \beta I_B = 100(7.5 \text{ }\mu\text{A}) = 0.75 \text{ mA}$$

and by Eq. 8-9 the collector-to-emitter voltage is

$$V_{CE} = V_{CC} - R_C I_C \cong 18 - 0.75(12) = 18 - 9 = 9 \text{ V}$$

On the other hand, if $I_{CEO} = 0.75$ mA, we find the collector current I_C by Eq. 8-6

$$I_C = \beta I_B + I_{CEO} = 0.75 + 0.75 = 1.5 \text{ mA}$$

Again applying Eq. 8-9 we find that $V_{CE} \cong 18 - 1.5(12) = 18 - 18 = 0$ V.

Example 8-13 Suppose that you are required to build a circuit like that of Fig. 8-4 and use values of $V_{CC} = 24$ V and $R_C = 10$ kΩ. You are also to provide an approximate quiescent collector current $I_C = 1$ mA and to prevent the collector current from ever exceeding 2 mA, which means that you can set $I_{C(\text{sat})} = 2$ mA. Select the values of R_B and R_E that will fulfill these requirements. The transistor $h_{FE} \cong h_{fe} \cong 80$.

Answer 8-13 $R_B \cong 1.76$ MΩ, $R_E = 2$ kΩ.—We can find R_E first by solving Eq. 8-15 for R_E. That is, since

$$I_{C(\text{sat})} \cong \frac{V_{CC}}{R_C + R_E}$$

then

$$R_C + R_E \cong \frac{V_{CC}}{I_{C(\text{sat})}}$$

and

$$R_E \cong \frac{V_{CC}}{I_{C(\text{sat})}} - R_C = \frac{24 \text{ V}}{2 \text{ mA}} - 10 \text{ k}\Omega = 2 \text{ k}\Omega$$

8-2 EMITTER FEEDBACK FOR STABILITY

Assuming that leakage is negligible, we can solve Eq. 8-11 for R_B. Thus, since

$$I_C \cong \frac{V_{CC}}{R_E + R_B/\beta}$$

then

$$R_E + R_B/\beta \cong \frac{V_{CC}}{I_C}$$

and then

$$R_B/\beta \cong \frac{V_{CC}}{I_C} - R_E$$

and finally

$$R_B \cong \beta \left(\frac{V_{CC}}{I_C} - R_E \right) = 80 \left(\frac{24 \text{ V}}{1 \text{ mA}} - 2 \text{ k}\Omega \right) \cong 1.76 \text{ M}\Omega$$

Note in Exs. 8-11 and 8-12 that a leakage I_{CEO} of 0.75 mA causes more or less change in I_C and V_{CE} depending on whether the emitter is connected directly to ground or has a resistance R_E in series with it. In both cases, however, I_C did change with a change in I_{CEO}. Thus, if you intend on using either circuit, with or without emitter feedback, in an environment whose temperature is expected to rise a little, use an I_C that is a little less than half of $I_{C(\text{sat})}$. In other words, though the calculated values of R_B and R_E, in the last example, will work, you may be better off by providing a little more room for I_C increases. That is, with a quiescent $I_C \cong 1$ mA and with $I_{C(\text{sat})} \cong 2$ mA, an increase in temperature might easily drive I_C up by 1 mA, thus causing saturation. If a smaller value of quiescent I_C is used initially, there will be a greater range over which it can increase before reaching saturation of 2 mA. By using an R_B value of 1.8 MΩ or 1.9 MΩ or even 2 MΩ instead of 1.76 MΩ, you can set the quiescent I_C a little below 1 mA when leakage is negligible. Then when I_{CEO} is significant at some higher temperature, the quiescent I_C might settle at about 1 mA. Thus, you can often design circuits using equations that assume $I_{CEO} \cong 0$. Then use a slightly larger than calculated base resistance R_B, which will help keep the quiescent I_C below $I_{C(\text{sat})}$ even though I_{CEO} increases.

Now we can note some differences between the simple base biased circuit (Fig. 8-1) and the circuit with base bias with emitter feedback (Fig. 8-4). Exs. 8-10 and 8-11 show that for a specific circuit with base bias with emitter feedback the collector current I_C increases from 0.75 mA to 1.375 mA, the collector-to-emitter voltage V_{CE} decreases from 9 V to 1.5 V and the base current I_B decreases from 7.5 μA to 6.25 μA when the leakage I_{CEO} increases from about 0 to 0.75 mA. However, in Ex. 8-12 we see that with the *base biased* circuit, the collector current I_C increases from 0.75 mA to 1.5 mA, and the collector-to-emitter voltage V_{CE} decreases from 9 V to 0 V and the base current I_B remains constant when the leakage I_{CEO} increases from about 0 to 0.75 mA. By comparison we see that the circuit with emitter feedback (Fig. 8-4) has less change in I_C and V_{CE} than does the simple *base biased* circuit even though the change in leakage I_{CEO} is the same in both circuits. Thus use of an emitter resistance tends to stabilize the I_C and V_{CE} values in the face of temperature and leakage changes. Actually, the larger the value of R_E, the better the stability. Large R_E values, however, bring up some other disadvantages as you will see when we consider load lines for emitter stabilized circuits later. Specifically, a large R_E tends to limit the operating range of the circuit when used as an amplifier. That is, it can reduce the maximum signal output capability of the circuit.

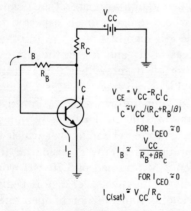

Fig. 8-5. Circuit using collector feedback for temperature stability.

8-3 COLLECTOR FEEDBACK FOR TEMPERATURE STABILITY

The base biased collector feedback circuit shown in Fig. 8-5 also has better stability of its I_C and V_{CE} values than the simple base biased circuit. Like the emitter feedback circuit covered in the previous section, the circuit with collector feedback has a decreased I_B

8-3 COLLECTOR FEEDBACK FOR TEMPERATURE STABILITY

when leakage I_{CEO} increases. This tends to hold I_C and V_{CE} relatively constant.

The action of stability may be explained as follows: The base bias current I_B value is determined by the voltage drop across R_B and its resistance value. That is, more or less voltage drop across R_B causes more or less I_B respectively. Note that by Kirchhoff's law, the voltage across R_B must be equal to the source voltage V_{CC} minus the voltage across R_C. An increased temperature and leakage I_{CEO} will cause some increase in collector current I_C. This increased I_C will increase the voltage drop across R_C and cause a decrease in voltage across R_B. With a decreased voltage across R_B the current I_B through it decreases too. Therefore, when I_{CEO} goes up, I_B goes down, and vice versa. By examining Eq. 8-6 you can see that changes of I_{CEO} and I_B in opposite directions, one increasing with the other decreasing, tend to prevent changes in I_C. Actually, I_C does increase a small amount with increases in I_{CEO}. In general, smaller values of R_B tend to make the circuit more stable; that is, I_C will change less with a given change in I_{CEO} if R_B is made smaller. A relatively small R_B, however, tends to lower the input impedance as seen by a signal source connected to the base. As you will see later, this can affect operation when the circuit is used as an amplifier.

The following are equations applicable to the circuit in Fig. 8-5:

$$I_C \cong \frac{V_{CC}}{R_C + R_B/\beta} + \left(\frac{R_B}{R_B + \beta R_C}\right) I_{CEO} \qquad (8\text{-}16)*$$

If the leakage is negligible ($I_{CEO} \cong 0$), the second term on the right side of Eq. 8-16 above drops out and we have

$$I_C \cong \frac{V_{CC}}{R_C + R_B/\beta} \qquad (8\text{-}17)$$

Also

$$I_B \cong \frac{V_{CC}}{R_B + \beta R_C} - \left(\frac{R_C}{R_B + \beta R_C}\right) I_{CEO} \qquad (8\text{-}18)*$$

And if $I_{CEO} \cong 0$, the second term of Eq. 8-18 drops out, simplifying it to

$$I_B \cong \frac{V_{CC}}{R_B + \beta R_C} \qquad (8\text{-}19)$$

---

*See Appendix A-6.

The collector-to-emitter voltage is

$$V_{CE} = V_{CC} - R_C I_C \qquad (8\text{-}20)$$

The collector saturates when the full source V_{CC} is dropped across R_C. Therefore,

$$I_{C(\text{sat})} \cong \frac{V_{CC}}{R_C} \qquad (8\text{-}21)$$

In most practical cases, we can use the simpler equations, Eqs. 8-17 and 8-19 as you will see later, when we select components for circuits like that of Fig. 8-5.

Example 8-14 If in Fig. 8-5, $V_{CC} = 24$ V, $R_B = 1.6$ MΩ, $R_C = 16$ kΩ, and the transistor $\beta = 100$, (a) find I_C, I_B, and V_{CE}, assuming that the leakage $I_{CEO} \cong 0$, and (b) find I_C, I_B, and V_{CE} where the leakage $I_{CEO} = 0.75$ mA.

Answer 8-14 (a) $I_C \cong 0.75$ mA, $I_B \cong 7.5$ μA, $V_{CE} \cong 12$ V, (b) $I_C \cong 1.125$ mA, $I_B \cong 3.75$ μA, $V_{CE} \cong 6$ V.—(a) Neglecting leakage I_{CEO}, we use Eq. 8-17 and find the collector current as

$$I_C \cong \frac{V_{CC}}{R_C + R_B/\beta} = \frac{24 \text{ V}}{16 \text{ kΩ} + 1600 \text{ kΩ}/100} = 0.75 \text{ mA}$$

We now determine base current I_B with either Eq. 8-5 solved for I_B or Eq. 8-19:

$$I_B \cong \frac{I_C}{\beta} = \frac{0.75 \text{ mA}}{100} = 7.5 \text{ μA}$$

or

$$I_B \cong \frac{V_{CC}}{R_B + \beta R_C} = \frac{24 \text{ V}}{1600 \text{ kΩ} + 100(16 \text{ kΩ})} = 7.5 \text{ μA}$$

With Eq. 8-20 we solve for the collector-to-emitter voltage

$$V_{CE} = V_{CC} - R_C I_C = 24 - 16(0.75) = 24 - 12 = 12 \text{ V}$$

(b) When leakage $I_{CEO} = 0.75$ mA, we use Eq. 8-16 and find the collector current

$$I_C \cong \frac{V_{CC}}{R_C + R_B/\beta} + \left(\frac{R_B}{R_B + \beta R_C}\right) I_{CEO} = 0.75 + 0.5(0.75) = 1.125 \text{ mA}$$

The base current I_B is determined with either Eq. 8-6 solved for I_B or Eq. 8-18.

8-3 COLLECTOR FEEDBACK FOR TEMPERATURE STABILITY

$$I_B = \frac{I_C - I_{CEO}}{\beta} \cong \frac{1.125 \text{ mA} - 0.75 \text{ mA}}{100} = 3.75 \text{ μA}$$

or

$$I_B \cong \frac{V_{CC}}{R_B + \beta R_C} - \left(\frac{R_C}{R_B + \beta R_C}\right) I_{CEO} = 7.5 - 3.75 = 3.75 \text{ μA}$$

Also,

$$V_{CE} = V_{CC} - I_C R_C \cong 24 - 1.125(16) = 24 - 18 = 6 \text{ V}$$

Example 8-15 Suppose that you are to build a circuit like that of Fig. 8-5 and use values $V_{CC} = 9$ V and $R_C = 3$ kΩ. (a) What value of R_B will you use to obtain the optimum operating point—a point where I_C is half of $I_{C(sat)}$? (b) What are the quiescent values of I_C and V_{CE} with your value of R_B in the circuit? (c) What is the collector power dissipation P_C with your R_B? The transistor $\beta \cong 60$.

Answer 8-15 (a) $R_B \cong 180$ kΩ, (b) $I_C \cong 1.5$ mA, $V_{CE} \cong 4.5$ V, (c) $P_C \cong 6.75$ mW.—Given $V_{CC} = 9$ V and $R_C = 3$ kΩ, we can find the saturation current with Eq. 8-21:

$$I_{C(sat)} \cong \frac{V_{CC}}{R_C} = \frac{9 \text{ V}}{3 \text{ kΩ}} = 3 \text{ mA}$$

Since the load is a single resistor R_C, that is, dc and ac loads are the same, we can reason that the optimum operating point current I_C should be halfway between cutoff (0 mA) and saturation (3 mA), which is to say $I_C \cong 1.5$ mA. Eq. 7-21 in the previous chapter can be used to find the optimum operating point in this circuit too. Thus in this case ac load r_L seen by the collector is $R_C = 3$ kΩ. And therefore with Eq. 7-21 we get

$$I_C = \frac{V_{CC}}{R_C + r_L} \cong \frac{9 \text{ V}}{3 \text{ kΩ} + 3 \text{ kΩ}} = 1.5 \text{ mA}$$

Since our desired quiescent $I_C \cong 1.5$ mA, we can use it with Eq. 8-17 to find the required R_B. Solving Eq. 8-17 for R_B we work as follows: Since

$$I_C \cong \frac{V_{CC}}{R_C + R_B/\beta} \tag{8-17}$$

then

$$R_C + R_B/\beta \cong \frac{V_{CC}}{I_C}$$

and

$$R_B/\beta \cong \frac{V_{CC}}{I_C} - R_C$$

and finally

$$R_B \cong \beta\left(\frac{V_{CC}}{I_C} - R_C\right) = 60\left(\frac{9\text{ V}}{1.5\text{ mA}} - 3\text{ k}\Omega\right) = 180\text{ k}\Omega$$

Now with Eq. 8-20

$$V_{CE} = V_{CC} - R_C I_C \cong 9 - 3(1.5) = 4.5\text{ V}$$

Referring back to Eq. 6-9 in Chap. 6, we find that the collector power is $P_C \cong V_{CE} I_C = (4.5\text{ V})(1.5\text{ mA}) = 6.75\text{ mW}$.

8-4 EMITTER FEEDBACK WITH TWO SUPPLIES FOR TEMPERATURE STABILITY

Fig. 8-6 shows a circuit using emitter feedback with two power supplies. This circuit has very good stability of its quiescent I_C

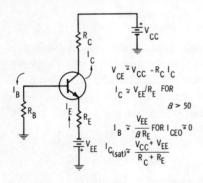

Fig. 8-6. Emitter feedback with two power supplies for temperature stability.

and V_{CE} values even though variations in leakage I_{CEO} are significant. An interesting feature of this circuit is that the voltage drop across R_E is nearly equal to the supply voltage V_{EE} if the β of the transistor is relatively high, that is if $\beta > 50$. Thus, in practice, the quiescent dc voltage at the emitter with respect to ground is about zero or just a few tenths of a volt.

The reason this circuit has very good temperature stability may be explained as follows: As with all other transistor circuits cov-

8-4 EMITTER FEEDBACK WITH TWO SUPPLIES

ered thus far, the base-emitter junction is forward biased, resulting in a few tenths of a volt across it. Since this is an npn-type transistor, the base is positive with respect to the emitter or we can say that the emitter is negative with respect to the base by a few tenths of a volt. This small potential difference will vary slightly with variations in base current I_B. Thus, slight increases or decreases in the forward voltage drop across the base-emitter junction means there are increases or decreases in I_B respectively. Now if leakage I_{CEO} increases, which increases I_C and I_E, the voltage across R_E increases. This makes the emitter to ground voltage more positive (less negative). As mentioned above, the emitter is normally negative with respect to the base. By making the emitter less negative to ground it also becomes less negative with respect to the base. Therefore, the forward drop across the base-emitter junction is reduced when the voltage across R_E is increased. This reduced base-to-emitter voltage must coincide with a reduced value of I_B. Thus, in this circuit, as I_{CEO} increases, I_B decreases, and vice versa. These oppositely varying values, I_{CEO} and I_B, oppose each other in Eq. 8-6 and tend to minimize changes in I_C.

A larger ratio of R_E to R_B gives better stability. Too large a value of R_E, however, reduces the unclipped output signal capability of the circuit when used as an amplifier, as you will see. If R_B is too small, on the other hand, it will tend to load down a signal source connected to the base. Actually, the specific value of R_B is not critical especially with high-β transistors. Typically, R_B values greater than about 10 kΩ but less than ten times R_E are used. Thus, typically

$$10 \text{ k}\Omega \leq R_B \leq 10 \, R_E \qquad (8\text{-}22)$$

For this circuit with emitter feedback and two supplies (Fig. 8-6), the equations for collector and base currents are as follows:

$$I_C \cong \frac{V_{EE}}{R_E + R_B/\beta} + \left(\frac{R_B + R_E}{R_B + \beta R_E}\right) I_{CEO} \qquad (8\text{-}23)^*$$

And if $I_{CEO} \cong 0$, Eq. 8-23 above simplifies to

$$I_C \cong \frac{V_{EE}}{R_E + R_B/\beta} \qquad (8\text{-}24)$$

With high-β transistors in this circuit, resistance R_E is much greater than the ratio R_B/β; that is, typically $R_E \gg R_B/\beta$ when $\beta > 50$. This allows us to further simplify Eq. 8-24 to

$$I_C \cong \frac{V_{EE}}{R_E} \tag{8-25}$$

Also

$$I_B \cong \frac{V_{EE}}{R_B + \beta R_E} - \left(\frac{R_E}{R_B + \beta R_E}\right) I_{CEO} \tag{8-26}*$$

And if $I_{CEO} \cong 0$, Eq. 8-26 simplifies to

$$I_B \cong \frac{V_{EE}}{R_B + \beta R_E} \tag{8-27}$$

Also if $\beta > 50$ in this circuit, Eq. 8-27 simplifies further to

$$I_B \cong \frac{V_{EE}}{\beta R_E} \tag{8-28}$$

By Kirchhoff's law for the circuit of Fig. 8-6,

$$V_{CC} + V_{EE} = R_C I_C + R_E I_E + V_{CE} \tag{8-29}$$

As mentioned before, the emitter resistor voltage drop $R_E I_E$ is nearly equal to the source V_{EE} if the transistor β is large. Thus, assuming that $V_{EE} \cong R_E I_E$, we can simplify Eq. 8-29 by subtracting V_{EE} from the left side and $R_E I_E$ from the right side to get

$$V_{CC} \cong R_C I_C + V_{CE}$$

This when rearranged shows that

$$V_{CE} \cong V_{CC} - R_C I_C \tag{8-30}$$

The collector saturation current $I_{C(\text{sat})}$, which is the maximum possible collector current, theoretically exists when the transistor acts like a short; that is, when $V_{CE} = 0$. In a circuit like that of Fig. 8-6, therefore, the sum of the two source voltages is dropped across both resistances R_C and R_E when the collector is saturated. Thus, by Kirchhoff's law $V_{CC} + V_{EE} = R_C I_{C(\text{sat})} + R_E I_{E(\text{sat})}$. Since $I_C \cong I_E$, which means that $I_{C(\text{sat})} \cong I_{E(\text{sat})}$, we can simplify and rearrange the above Kirchhoff's voltage equation as follows:

$$V_{CC} + V_{EE} \cong R_C I_{C(\text{sat})} + R_E I_{C(\text{sat})}$$

or

$$V_{CC} + V_{EE} \cong (R_C + R_E) I_{C(\text{sat})}$$

---

*See Appendix A-7.

8-4 EMITTER FEEDBACK WITH TWO SUPPLIES

and

$$I_{C(\text{sat})} \cong \frac{V_{CC}+V_{EE}}{R_C+R_E} \tag{8-31}$$

Example 8-16 If in a circuit like that of Fig. 8-6, $V_{CC}=30$ V, $V_{EE}=15$ V, $R_B=30$ kΩ, $R_C=20$ kΩ, $R_E=20$ kΩ, and the transistor $h_{FE} \cong h_{fe} \cong 100$, what are (a) the values I_C, I_B, V_{CE}, and P_C when leakage $I_{CEO}=0.75$ mA, and (b) what are these values when $I_{CEO} \cong 0$?

Answer 8-16 (a) $I_C \cong 0.75$ mA, $I_B \cong 0$, $V_{CE} \cong 14.8$ V, $P_C \cong 11.2$ mW. (b) $I_C \cong 0.75$ mA, $I_B \cong 7.5$ μA, $V_{CE} \cong 15$ V, $P_C \cong 11.25$ mW.—Since the leakage I_{CEO} is significant in part (a), we can use Eqs. 8-23 and 8-26 to find the collector and base currents. Thus,

$$I_C \cong \frac{V_{EE}}{R_E+R_B/\beta} + \left(\frac{R_B+R_E}{R_B+\beta R_E}\right) I_{CEO} = \frac{15 \text{ V}}{20 \text{ k}\Omega+30 \text{ k}\Omega/100}$$
$$+ \left(\frac{30 \text{ k}\Omega+20 \text{ k}\Omega}{30 \text{ k}\Omega+100(20 \text{ k}\Omega)}\right) 0.75 \text{ mA}$$

$$\cong 0.74 \text{ mA} + 0.0184 \text{ mA} \cong 0.758 \text{ mA}$$

And, from Eq. 8-26,

$$I_B \cong \frac{V_{EE}}{R_B+\beta R_E} - \left(\frac{R_E}{R_B+\beta R_E}\right) I_{CEO} = \frac{15 \text{ V}}{30 \text{ k}\Omega+100(20 \text{ k}\Omega)}$$
$$- \left(\frac{20 \text{ k}\Omega}{30 \text{ k}\Omega+100(20 \text{ k}\Omega)}\right) 0.75 \text{ mA}$$

$$\cong 7.4 \text{ μA} - 7.4 \text{ μA} \cong 0$$

Since the transistor β is large, we can use Eq. 8-30 to find the collector-to-emitter voltage:

$$V_{CE} \cong V_{CC} - R_C I_C \cong 30 - 20(0.758) \cong 30 - 15.2 = 14.8 \text{ V}$$

The collector power dissipation, by Eq. 6-9, is

$$P_C = V_{CE} I_C \cong (14.8 \text{ V})(0.758 \text{ mA}) \cong 11.2 \text{ mW}$$

In part (b), when the leakage I_{CEO} is insignificant, and since the value of R_E is much greater than the ratio R_B/β and the transistor β is relatively large, collector and base currents can be found with Eqs. 8-25 and 8-28. Thus,

$$I_C \cong \frac{V_{EE}}{R_E} = \frac{15 \text{ V}}{20 \text{ k}\Omega} = 0.75 \text{ mA}$$

and

$$I_B \cong \frac{V_{EE}}{\beta R_E} = \frac{15 \text{ V}}{100(20 \text{ k}\Omega)} = 7.5 \text{ }\mu\text{A}$$

The collector-to-emitter voltage V_{CE} and the collector dissipation P_C are found here as in part (a).

This last example points out some interesting facts about the circuit with emitter feedback with two supplies in Fig. 8-6. Note that the collector current I_C remained essentially constant even though the leakage I_{CEO} changed significantly. Also note that this was caused by the fact that as I_{CEO} increased, the base current I_B decreased.

Selecting component values for a circuit like that in Fig. 8-6 is easier than for previous circuits covered in this chapter. If you use a high-β transistor, and stay within the limits given in Eq. 8-22, you can use the simpler equations 9-25 and 9-30 for *all* practical purposes.

Example 8-17 Suppose that you are required to design a circuit with emitter feedback and two supplies. One supply is $+9$ V and the other is -9 V to ground. The quiescent values of collector current and collector-to-emitter voltage are to be 1 mA and 4.5 V respectively. The transistor β is high. Find the required values of R_B, R_C, and R_E. What is the saturation current $I_{C(\text{sat})}$, using your component values?

Answer 8-17 10 k$\Omega < R_B <$ 90 kΩ, $R_C \cong 4.5$ kΩ, $R_E \cong 9$ kΩ, $I_{C(\text{sat})} \cong 1.33$ mA.—First we solve Eq. 8-25 for R_E. Thus, since $I_C \cong V_{EE}/R_E$, then $R_E \cong V_{EE}/I_C$, and in this case $R_E \cong 9$ V/1 mA $= 9$ kΩ. Now solving Eq. 8-30 for R_C we can show that since $V_{CE} \cong V_{CC} - R_C I_C$, then $R_C I_C \cong V_{CC} - V_{CE}$ and

$$R_C \cong \frac{V_{CC} - V_{CE}}{I_C} = \frac{9 \text{ V} - 4.5 \text{ V}}{1 \text{ mA}} = 4.5 \text{ k}\Omega$$

The value of R_B is chosen arbitrarily. Its value need only to fit the limits of Eq. 8-22. Thus, since generally 10 k$\Omega < R_B < 10 R_E$, then more specifically in this case

$$10 \text{ k}\Omega < R_B < 90 \text{ k}\Omega$$

From this we may say that R_B can be any value from about 10 kΩ to about 90 kΩ. The larger values should be avoided unless you are

8-5 EMITTER FEEDBACK AND BASE BIAS

certain that the β of the transistor is quite high. If R_B is too low, on the other hand, it may cause excessive "loading" on the signal source that is eventually used to drive this circuit. If 40 kΩ is used in this case it is somewhat a "safe" median between the possible limits.

With Eq. 8-31 we find the collector saturation current

$$I_{C(\text{sat})} \cong \frac{V_{CC} + V_{EE}}{R_C + R_E} = \frac{9\text{ V} + 9\text{ V}}{4.5\text{ k}\Omega + 9\text{ k}\Omega} = \frac{18\text{ V}}{13.5\text{ k}\Omega} \cong 1.33\text{ mA}$$

8-5 EMITTER FEEDBACK AND BASE BIAS WITH A VOLTAGE DIVIDER

A popular circuit, which possesses very good stability of its quiescent I_C and V_{CE} values, even when subjected to large temperature changes, is shown in Fig. 8-7. As you can see, this circuit has an emitter feedback resistor R_E. It uses a voltage divider (R_1 and R_2) to obtain base bias, and, in contrast to the circuit of Fig. 8-6, it uses only one voltage source, V_{CC}.

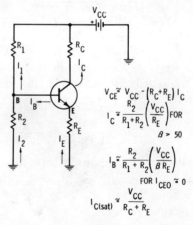

Fig. 8-7. Emitter feedback and base bias with voltage divider.

Typically the values of R_1 and R_2, in the circuit of Fig. 8-7, are values such that currents I_1 and I_2 are much larger than the base current I_B, which is to say that

$$I_1 \cong I_2 \gg I_B$$

Therefore if I_B varies, its variations have negligible effect on either I_1 or I_2 and we may assume that these currents are practically constant. This means that the constant current I_2 keeps the voltage from point B to ground constant regardless of what happens to currents I_B, I_C, and I_E. Normally, with an npn transistor the voltage at B is a few tenths of a volt more positive than the voltage at E with respect to ground. The potential difference at points B and E is the voltage drop across the forward biased base-emitter junction. Now if the leakage I_{CEO} increases, due to a temperature increase, both I_C and I_E increase. The larger I_E makes point E more positive with respect to ground, which reduces the potential between points B and E. This reduced V_{BE} reduces the base current I_B, which in turn limits the increase in the I_C and I_E values. The fact is, then, if I_{CEO} increases or decreases, the base current I_B decreases or increases respectively, and this tends to limit the changes in I_C. See Eq. 8-6.

For this circuit (Fig. 8-7), the collector current is

$$I_C \cong \frac{R_2}{R_1+R_2}\left(\frac{V_{CC}}{R_E}\right) + \left(\frac{R_1 R_2}{\beta R_E(R_1+R_2)}\right) I_{CEO} \quad (8\text{-}34)*$$

if the β of the transistor is high (over 50) and R_2 is not more than about ten times larger than R_E.

Typically the factor in the brackets of Eq. 8-34 is a very small value. This tends to make the second term on the right side of Eq. 8-34 insignificant compared with the value of the first term even if the leakage I_{CEO} is significant. Thus, since the second term in Eq. 8-34 is usually negligible, the following simpler version can be used:

$$I_C \cong \frac{R_2}{R_1+R_2}\left(\frac{V_{CC}}{R_E}\right) \quad (8\text{-}35)$$

With few exceptions, where I_{CEO} is large and R_E and β are unusually low, Eq. 8-35 is used with circuits like that of Fig. 8-7.

Since $I_C = \beta I_B + I_{CEO}$ by Eq. 8-6, substituting this into the above equation we get

$$\beta I_B + I_{CEO} \cong \frac{R_2}{R_1+R_2}\left(\frac{V_{CC}}{R_E}\right)$$

Rearranging terms gives

$$\beta I_B \cong \frac{R_2}{R_1+R_2}\left(\frac{V_{CC}}{R_E}\right) - I_{CEO}$$

Dividing both sides by β

8-5 EMITTER FEEDBACK AND BASE BIAS

$$I_B \cong \frac{R_2}{R_1+R_2}\left(\frac{V_{CC}}{\beta R_E}\right) - \frac{I_{CEO}}{\beta} \qquad (8\text{-}36)*$$

And if the leakage I_{CEO} is negligible, Eq. 8-36 simplifies to

$$I_B \cong \frac{R_2}{R_1+R_2}\left(\frac{V_{CC}}{\beta R_E}\right) \qquad (8\text{-}37)$$

An equation for the collector-to-emitter voltage V_{CE} may be found with a Kirchhoff's voltage equation for the loop containing V_{CC}, R_C, the transistor, and R_E. That is,

$$V_{CC} = R_C I_C + V_{CE} + R_E I_E \qquad (8\text{-}38)$$

Solving the above for V_{CE} gives us

$$V_{CE} = V_{CC} - R_C I_C - R_E I_E$$

Since $I_C \cong I_E$ the above may be rewritten as

$$V_{CE} \cong V_{CC} - R_C I_C - R_E I_C$$

or

$$V_{CE} \cong V_{CC} - (R_C + R_E)I_C \qquad (8\text{-}39)$$

Since the collector is saturated when the full V_{CC} voltage is dropped across R_C and R_E, which is to say when $V_{CE}=0$, Eq. 8-38 above becomes

$$V_{CC} \cong R_C I_{C(\text{sat})} + R_E I_{E(\text{sat})}$$

And since $I_{C(\text{sat})} \cong I_{E(\text{sat})}$, the above may be shown as

$$V_{CC} \cong R_C I_{C(\text{sat})} + R_E I_{C(\text{sat})}$$

or

$$V_{CC} \cong (R_C + R_E)I_{C(\text{sat})}$$

Solving for $I_{C(\text{sat})}$ we finally get

$$I_{C(\text{sat})} \cong \frac{V_{CC}}{R_C + R_E} \qquad (8\text{-}40)$$

In circuits like that of Fig. 8-7, larger values of R_E give better temperature stabilities. As with other circuits using emitter feedback, however, larger R_E values tend to reduce the maximum un-

---

*See Appendix A-8.

clipped output signal capability of this circuit when used as an amplifier. In general, good temperature stability and amplifier characteristics are usually achieved if the value of R_E is between the values of R_C and one-tenth of R_C, and if the value of R_2 is between the values of R_E and ten times R_E. In other words, the stability and amplifier characteristics are usually good when

$$\frac{R_C}{10} \leq R_E \leq R_C \tag{8-41}$$

and when

$$R_E \leq R_2 \leq 10 R_E \tag{8-42}$$

The values of resistances R_C and R_E are also determined by other design considerations, as you will see in later sections. So at this time we can limit ourselves to determining the values of R_1 and R_2 where the values of R_C and R_E are specified.

Example 8-18 Suppose that in a circuit like that of Fig. 8-7 you have to use $R_C = R_E = 10$ kΩ and $V_{CC} = 40$ V. What values of R_1 and R_2 will you use to obtain quiescent I_C and V_{CE} values that are the coordinates of the optimum operating point?

Answer 8-18 If the transistor β is high, acceptable voltage divider values are between $R_1 = 30$ kΩ, $R_2 = 10$ kΩ, and $R_1 = 300$ kΩ, $R_2 = 100$ kΩ provided that the ratio $R_2/(R_1 + R_2) \cong 0.25$.—With a single resistor R_C as the load in the collector (no difference between dc and ac load lines), the optimum operating point collector current I_C is halfway between zero and $I_{C(\text{sat})}$; see Sec. 7.7. First we determine the saturation current $I_{C(\text{sat})}$ with Eq. 8-40:

$$I_{C(\text{sat})} \cong \frac{V_{CC}}{R_C + R_E} = \frac{40 \text{ V}}{10 \text{ k}\Omega + 10 \text{ k}\Omega} = 2 \text{ mA}$$

Therefore at the optimum operating point, the quiescent $I_C = I_{C(\text{sat})}/2 = 1$ mA. Now we have all factors in Eq. 8-35 except the ratio $R_2/(R_1 + R_2)$. Thus, we can solve Eq. 8-35 for this ratio. If

$$I_C \cong \frac{R_2}{R_1 + R_2} \left(\frac{V_{CC}}{R_E} \right)$$

then by rearranging terms and substituting known values, we have

$$\frac{R_2}{R_1 + R_2} \cong \frac{R_E I_C}{V_{CC}} = \frac{(10 \text{ k}\Omega)(1 \text{ mA})}{40 \text{ V}} = 0.25$$

8-5 EMITTER FEEDBACK AND BASE BIAS

Since a desirable value of R_2 is approximately between the values of R_E and ten times R_E, we can select an R_2 with a resistance between 10 kΩ and 100 kΩ in this case. Then the necessary value of R_1 to be used in the voltage divider with our selected R_2 is found by solving the above equation for R_1. Thus, since

$$\frac{R_2}{R_1 + R_2} \cong 0.25$$

then

$$\frac{R_2}{0.25} \cong R_1 + R_2$$

and then rearranging terms and substituting values,

$$R_1 \cong \frac{R_2}{0.25} - R_2 = \frac{10 \text{ k}\Omega}{0.25} - 10 \text{ k}\Omega = 30 \text{ k}\Omega$$

when $R_2 = 10$ kΩ. Or

$$R_1 \cong \frac{100 \text{ k}\Omega}{0.25} - 100 \text{ k}\Omega = 300 \text{ k}\Omega$$

when $R_2 = 100$ kΩ.

In this case, R_1 and R_2 values of 60 kΩ and 20 kΩ, or 150 kΩ and 50 kΩ, or 240 kΩ and 80 kΩ, etc., would work too. However, the larger values of these resistances tend to reduce temperature stability and affect the accuracy of Eq. 8-35, especially if $β$ is not very high. Therefore, lower values of R_1 and R_2 are preferable, but if they are too low they will tend to load down a signal source that may be connected to the base when the circuit is used as an amplifier.

Example 8-19 If in a circuit like that of Fig. 8-7, $R_C = 10$ kΩ, $R_E = 2$ kΩ, and $V_{CC} = 24$ V, what values of R_1 and R_2 will provide the optimum operating point?

Answer 8-19 Assuming high $β$ transistors, the optimum operating point is nearly obtained with voltage divider values between $R_1 = 22$ kΩ, $R_2 = 2$ kΩ and $R_1 = 20$ kΩ, provided that the ratio $R_2/(R_1 + R_2) \cong 0.0833$.—With these component values

$$I_{C(\text{sat})} \cong \frac{V_{CC}}{R_C + R_E} = \frac{24 \text{ V}}{12 \text{ k}\Omega} = 2 \text{ mA}$$

The quiescent operating current is half of $I_{C(\text{sat})}$, or $I_C \cong 1$ mA in this case. As in Ex. 8-18 we solve for the ratio

$$\frac{R_2}{R_1+R_2} \cong \frac{R_E I_C}{V_{CC}} = \frac{(2\text{ k}\Omega)(1\text{ mA})}{24\text{ V}} = \frac{1}{12} \cong 0.0833$$

In this case since $R_E = 2$ kΩ, we know that R_2 should be between 2 kΩ and 20 kΩ by Eq. 8-42. Therefore when $R_2 = 2$ kΩ,

$$R_1 \cong \frac{R_2}{0.0833} - R_2 = \frac{2\text{ k}\Omega}{0.0833} - 2\text{ k}\Omega \cong 22\text{ k}\Omega$$

and when $R_2 = 20$ kΩ,

$$R_1 \cong \frac{20\text{ k}\Omega}{0.0833} - 20\text{ k}\Omega \cong 220\text{ k}\Omega$$

To avoid loading down a signal source connected to the base and to provide good stability, use of $R_1 = 110$ kΩ and $R_2 = 10$ kΩ would be practical. Note that in this problem the ratio $R_1/R_2 \cong 11:1$.

Example 8-20 Suppose a circuit like the one shown in Fig. 8-7 has component values $R_1 = 90$ kΩ, $R_2 = 10$ kΩ, $R_C = 8$ kΩ, $R_E = 2$ kΩ, $V_{CC} = 9$ V, and the transistor β is 100. (a) Neglecting leakage, find $I_{C(\text{sat})}$, I_C, I_B, V_{CE}, and P_C. (b) If $I_{CEO} = 0.4$ mA, find $I_{C(\text{sat})}$, I_C, I_B, V_{CE}, and P_C.

Answer 8-20 (a) $I_{C(\text{sat})} \cong 0.9$ mA, $I_C \cong 0.45$ mA, $I_B \cong 4.5$ μA, $V_{CE} \cong 4.5$ V, $P_C \cong 20.3$ mW, (b) $I_{C(\text{sat})} \cong 0.9$ mA, $I_C \cong 0.468$ mA, $I_B \cong 0.5$ μA, $V_{CE} \cong 4.32$ V, $P_C \cong 20.2$ mW.—In part (a), when leakage is negligible, we find the collector saturation current with Eq. 8-40:

$$I_{C(\text{sat})} \cong \frac{V_{CC}}{R_C + R_E} = \frac{9\text{ V}}{10\text{ k}\Omega} = 0.9\text{ mA}$$

The quiescent collector current is found with Eq. 8-35:

$$I_C \cong \frac{R_2}{R_1+R_2}\left(\frac{V_{CC}}{R_E}\right) = \frac{10\text{ k}\Omega}{90\text{ k}\Omega + 10\text{ k}\Omega}\left(\frac{9\text{ V}}{2\text{ k}\Omega}\right)$$
$$\cong 0.1(4.5\text{ mA}) = 0.45\text{ mA}$$

Use Eq. 8-37 or Eq. 8-5 to find the quiescent base current,

$$I_B \cong \frac{R_2}{R_1+R_2}\left(\frac{V_{CC}}{\beta R_E}\right) = 0.1(45\text{ }\mu\text{A}) = 4.5\text{ }\mu\text{A}$$

or

$$I_B \cong I_C/\beta = \frac{0.45\text{ mA}}{100} = 4.5\text{ }\mu\text{A}$$

8-6 COLLECTOR CHARACTERISTICS OF THE COMMON-EMITTER

Eq. 8-39 determines the collector-to-emitter voltage:

$$V_{CE} \cong V_{CC} - (R_C + R_E)I_C = 9 - (10\text{ k}\Omega)(0.45\text{ mA}) = 4.5\text{ V}$$

And by Eq. 6-10 we find the collector power dissipation:

$$P_C \cong V_{CE}I_C \cong (4.5\text{ V})(4.5\text{ mA}) \cong 20.3\text{ mW}$$

In part (b) where $I_{CEO} = 0.4$ mA, the collector saturation current is found exactly as in part (a), that is,

$$I_{C(\text{sat})} = 0.9\text{ mA}$$

The quiescent collector current can be determined with Eq. 8-34 if we intend to keep track of the leakage:

$$I_C \cong \frac{10}{90+10}\left(\frac{9\text{ V}}{2\text{ k}\Omega}\right) + \left(\frac{90(10)}{100(2)(90+10)}\right)0.4\text{ mA}$$

$$\cong 0.45\text{ mA} + (0.045)0.4\text{ mA}$$

$$\cong 0.45 + 0.018 = 0.468\text{ mA}$$

But notice in this case that because ß is large, the second term on the right is small, showing us that we could just as well have used the simpler Eq. 8-35 as in part (a), and simply state the fact that with or without leakage $I_C \cong 0.45$ mA.

The quiescent base current is found by using either Eq. 8-36 or Eq. 8-6 when solved for I_B:

$$I_B \cong \frac{R_2}{R_1+R_2}\left(\frac{V_{CC}}{\text{ß}R_E}\right) - \frac{I_{CEO}}{\text{ß}} = 4.5\text{ }\mu\text{A} - \frac{0.4\text{ mA}}{100} = 4.5 - 4 = 0.5\text{ }\mu\text{A}$$

or since $I_C = \text{ß}I_B + I_{CEO}$, then

$$I_B = \frac{I_C - I_{CEO}}{\text{ß}} = \frac{0.05\text{ mA}}{100} = 0.5\text{ }\mu\text{A}$$

As in part (a), Eqs. 8-39 and 6-10 are used to find that

$$V_{CE} \cong 9\text{ V} - (8\text{ k}\Omega + 2\text{ k}\Omega)(0.468\text{ mA}) = 4.32\text{ V}$$

and that

$$P_C \cong (4.32\text{ V})(0.468\text{ mA}) \cong 20.2\text{ mW}$$

8-6 COLLECTOR CHARACTERISTICS OF THE COMMON-EMITTER CIRCUIT

Typical collector characteristics of the common-emitter transistor circuit are shown in Fig. 8-8. They can be plotted with a circuit

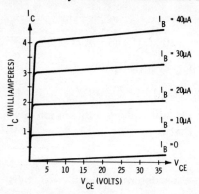

Fig. 8-8. Collector characteristics of the common-emitter circuit.

like that in Fig. 8-9 and they indicate the dependence of I_C on I_B and the near independence of I_C on V_{CE}. For example, suppose that the transistor shown in Fig. 8-9 has the characteristics shown in Fig. 8-8. If the potentiometer R_1 is adjusted so that $I_B = 10$ μA, as read on the microammeter M_1, and potentiometer R_2 is adjusted so that $V_{CE} = 20$ V, as read on voltmeter M_4, I_C will be about 1 mA, as read on milliammeter M_2. Note on the characteristics that if V_{CE} is varied above or below 20 V by adjustment of R_2, while I_B is held constant at 10 μA by adjustment of R_1 as needed, the value of I_C will increase or decrease very little. On the other hand, if R_1 is adjusted so that $I_B = 20$ μA, or 30 μA, or 40 μA, the I_C value will increase to about 2 mA, or 3 mA, or 4 mA respectively. Thus, the value of I_C is sensitive to changes in the value of I_B but is rather insensitive to changes in V_{CE}.

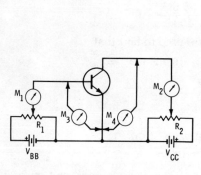

Fig. 8-9. Circuit that can be used to plot collector characteristics.

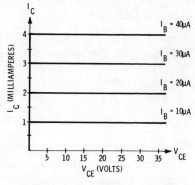

Fig. 8-10. Ideal collector characteristics of common-emitter circuit.

8-6 COLLECTOR CHARACTERISTICS OF THE COMMON-EMITTER

The relatively small increases in I_C due to increases in V_{CE} even though I_B is held constant is caused by increased leakage I_{CEO} that exists with increased voltage. Larger collector reverse bias voltages tend to tear more valence electrons from their parent atoms, thus producing more electron-hole pairs to support a larger leakage. If the leakage could be made zero, the I_B curves would become horizontal as shown in Fig. 8-10. These are called *ideal* collector characteristics. They show I_C's total independence of changes in V_{CE}. For most practical purposes we can assume that the transistors we use have ideal characteristics unless they have unusually high leakage. Such an assumption simplifies matters considerably; that is, as shown in the previous sections, the equations for I_C and I_B are much simpler if leakage $I_{CEO} \cong 0$.

The approximate values of β_{dc} and β, or in other words h_{FE} and h_{fe}, can be determined from the collector characteristics. Likewise if the values h_{FE} and h_{fe} are specified, you can estimate the collector characteristics. For example, in Fig. 8-8, when $V_{CE} = 20$ V and $I_B = 20$ μA, then $I_C \cong 2.1$ mA. Therefore by Eq. 6-5

$$\beta_{dc} = h_{FE} \cong \frac{I_C}{I_B} \cong \frac{2.1 \text{ mA}}{20 \text{ μA}} = 105$$

Actually, a more accurate approximation of β can be determined by solving Eq. 8-6 for β. That is, since $I_C = \beta I_B + I_{CEO}$, then

$$\beta = \frac{I_C - I_{CEO}}{I_B}$$

From Fig. 8-8, we can read the approximate leakage I_{CEO}. Remember that I_{CEO} is the leakage current across the collector and emitter leads through the transistor when the base lead is open ($I_B = 0$). Thus, if $V_{CE} = 20$ V and $I_B = 0$, the collector current I_C *is* the leakage I_{CEO}, which can be estimated at about 0.2 mA in this case. Therefore when $V_{CE} = 20$ V and $I_B = 20$ μA, then $I_C \cong 2.1$ mA as estimated off Fig. 8-8. And since $I_{CEO} \cong 0.2$ mA, then

$$\beta_{dc} \cong h_{FE} \cong \frac{2.1 \text{ mA} - 0.2 \text{ mA}}{20 \text{ μA}} = 95$$

Either method of estimating the value of β shows that in this case, in round numbers, $\beta = 100$. Since dc and ac values of β are approximately equal, we can say that

$$h_{FE} \cong h_{fe} \cong 100$$

for the transistor whose characteristics are shown in Fig. 8-8. Note that this agrees with the ideal characteristics in Fig. 8-9, which show that ratio of I_C to I_B is always exactly 100.

On the other hand, if the ß of a transistor is given, the ideal collector characteristics are easily estimated. A graph with horizontal I_B curves like Fig. 8-10 may be drawn; only the values of I_B on the I_B curves will differ, depending on the ß of the transistor. Fig. 8-11 shows estimated ideal collector characteristics for a transistor whose ß is 80. Note that the I_B curve extending horizontally from $I_C = 1$ mA is identified as $I_B = 12.5$ μA. This value was arrived at by solving Eq. 8-5 for I_B and substituting knowns into it. That is, since $I_C \cong ßI_B$ when leakage $I_{CEO} \cong 0$, then with $I_C = 1$ mA and ß = 80

$$I_B \cong \frac{I_C}{ß} = \frac{1 \text{ mA}}{80} = 12.5 \text{ μA}$$

Similarly the ß value is divided into 2 mA, 3 mA, etc., to find the I_B values 25 μA, 37.5 μA, etc., that are shown in Fig. 8-11.

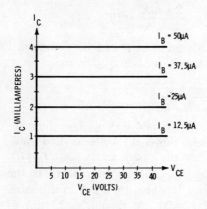

Fig. 8-11. Ideal collector characteristics for common-emitter circuit when ß = 80.

Example 8-20 If $h_{FE} \cong h_{fe} \cong 200$ for a certain type of transistor, sketch its ideal common-emitter collector characteristics.

Answer 8-20 Similar to those shown in Fig. 8-11 except that the I_B curves from bottom to top are 5 μA, 10 μA, 15 μA, and 20 μA instead of 12.5 μA, 25 μA, etc.

Example 8-21 Suppose that the collector characteristics of a certain transistor are like those in Fig. 8-8 but the I_B curves from bottom to top are identified as zero, 20 μA, 40 μA, 60 μA, and

8-7 Load Lines of the Various Common-Emitter Circuits

80 μA instead of zero, 10 μA, 20 μA, 30 μA, and 40 μA. What is the approximate β of the transistor?

Answer 8-21 $\beta \cong 50$.—In this case, when $I_B = 20$ μA, the collector current $I_C \cong 1$ mA. Therefore by Eqs. 6-5 and 8-4,

$$h_{FE} \cong \frac{I_C}{I_B} \cong \frac{1 \text{ mA}}{20 \text{ μA}} = 50$$

and

$$h_{FE} \cong h_{fe} \cong \beta \cong 50$$

8-7 LOAD LINES OF THE VARIOUS COMMON-EMITTER CIRCUITS

Dc load lines for CE circuits are sketched on common-emitter collector characteristics much like they are sketched for common-base circuits on common-base collector characteristics; see Sec. 7-2. Two points, the saturation point and the cutoff point, are found first. A straight line drawn through these two points *is* the dc load line. At the saturation point, we can assume that theoretically the transistor acts as a short circuit; therefore this point's coordinates are $I_{C(\text{sat})}$ and $V_{CE} = 0$. At the cutoff point, the transistor acts as an open, neglecting leakage, and the full dc source voltage appears across the collector and emitter leads. Thus, its coordinates theoretically are $I_C = 0$ and $V_{CE(\text{cutoff})}$. The value $V_{CE(\text{cutoff})}$ is equal to the V_{CC} source voltage if only one source is used or it is equal to $V_{CC} + V_{EE}$ if two voltage sources are used, as in Fig. 8-6.

Example 8-22 Sketch a load line and identify the operating point for a circuit such as that in Fig. 8-1B, where $V_{CC} = -18$ V, $R_B = 2.4$ MΩ, $R_C = 12$ kΩ, and the transistor $h_{FE} \cong h_{fe} \cong 100$.

Answer 8-22 See Fig. 8-12.—This is the same circuit described in Ex. 8-12. In its answer (a) we found the quiescent I_C, I_B, and V_{CE} to be 0.75 mA, 7.5 μA, and 9 V respectively. We can now find the saturation current with Eq. 7-1:

$$I_{C(\text{sat})} \cong \frac{V_{CC}}{R_C} = \frac{18 \text{ V}}{12 \text{ kΩ}} = 1.5 \text{ mA}$$

At the cutoff point, the full V_{CC} supply voltage is the maximum collector-to-emitter voltage $V_{CE(\text{cutoff})} \cong V_{CC} = 18$ V.

Example 8-23 Suppose that the transistor in the circuit de-

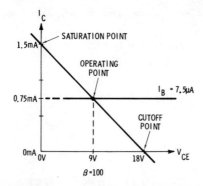

Fig. 8-12. Load line and operating point for the circuit described in Exs. 8-12 and 8-22.

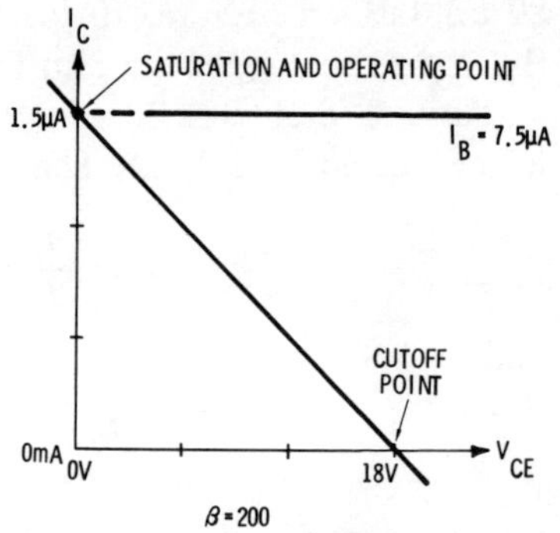

Fig. 8-13. Load line and operating point for the circuit described in Exs. 8-12 and 8-23.

scribed in the previous example is replaced with one having $β = 200$. Sketch the load line and identify the operating point.

Answer 8-23 See Fig. 8-13.—In this simple base biased circuit, because the voltage across the forward biased base-emitter junction is negligible, the voltage across and the current through resistor R_B are practically constant. Therefore a different $β$ will not noticeably affect the value of I_B. Thus, as in the last problem, $I_B \cong 7.5$ μA. But, in this case, by Eq. 8-5 we find that

$$I_C \cong βI_B = 200(7.5 \text{ μA}) = 1.5 \text{ mA}$$

And with Eq. 8-9,

$$V_{CE} = V_{CC} - R_C I_C \cong 18 - 12(1.5) = 0 \text{ V}$$

The saturation current is found as in Ex. 8-22; that is,

$$I_{C(\text{sat})} \cong 1.5 \text{ mA}$$

Notice in this case that the quiescent I_C is about equal to $I_{C(\text{sat})}$. Thus, in the simple base biased circuit, if the transistor is replaced with one with a larger $β$, saturation may easily occur. In other words, in simple base biased circuits like that in Fig. 8-1 it is apparent that if transistors of various types, including various $β$ values, are interchanged, the position of the operating point is significantly affected.

Example 8-24 Sketch a dc load line and identify the operating

8-7 LOAD LINES OF THE VARIOUS COMMON-EMITTER CIRCUITS

point for the circuit Fig. 8-4 where $R_B = 2$ MΩ, $R_C = 8$ kΩ, $R_E = 4$ kΩ, and $V_{CC} = 18$ V. The transistor $\beta \cong 100$ and its leakage is negligible.

Answer 8-24 See Fig. 8-14.—This circuit is the same as the one described in Ex. 8-10. In Ans. 8-10 we found $I_{C(sat)} \cong 1.5$ mA, $I_C \cong 0.75$ mA, $I_B \cong 7.5$ μA, and $V_{CE} \cong 9$ V. The load line is drawn through the $I_{C(sat)}$ current and the $V_{CE(cutoff)}$, which is 18 V in this case.

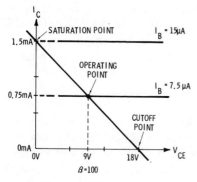

Fig. 8-14. Load line and operating point for the circuit described in Exs. 8-10 and 8-24.

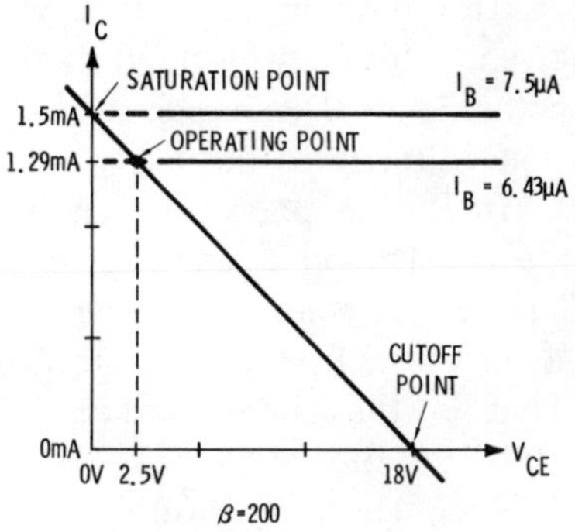

Fig. 8-15. Load line and operating point of the circuit described in Exs. 8-10 and 8-25.

Example 8-25 Suppose that in the circuit described in the previous example, the transistor is replaced with one having $\beta = 200$. Sketch the dc load line and identify the operating point.

Answer 8-25 See Fig. 8-15.—The values $I_{C(sat)}$ and $V_{CE(cutoff)}$ are found exactly as in Ex. 8-10. That is, $I_{C(sat)} \cong 1.5$ mA and $V_{CE(cutoff)} \cong 18$ V. By Eq. 8-11 we find that

$$I_C \cong \frac{V_{CC}}{R_E + R_B/\beta} = \frac{18 \text{ V}}{4 \text{ kΩ} + 2 \text{ MΩ}/200} \cong 1.29 \text{ mA}$$

And with Eq. 8-13,

$$I_B \cong \frac{V_{CC}}{R_B + \beta R_E} = \frac{18 \text{ V}}{2 \text{ MΩ} + 200(4 \text{ kΩ})} \cong 6.43 \text{ μA}$$

The last four examples illustrate the fact that the simple base bias circuit, which uses no feedback, is less stable than a circuit that uses emitter feedback. The use of resistance R_E in the emitter stabilizes the common-emitter circuit, not only under temperature changes, but also if its transistor is replaced with one with a different β.

It is interesting to note that when R_E is used in the circuit the base current I_B actually decreases with higher-β transistors, whereas I_B is practically constant in the simpler base biased circuit. If the value of R_E were larger in Ex. 8-25, the operating point would have moved up less and I_B would have decreased more. In other words, a larger R_E tends to make the circuit more stable.

Also note in Figs. 8-12 through 8-15 that the I_B curves of given values are located higher on the collector characteristics with higher-β transistors. That is, with high-β transistors, the I_C value directly to the left of any given I_B curve is relatively large.

In the common-base circuit, we noted that the I_E curves of the collector characteristics rise with temperature increases. Similarly in the common-emitter circuits, the I_B curve rises, even more so in the collector characteristic when the temperature increases. That is, the curve marked 7.5 μA in Fig. 8-14 may cross the center of the load line as shown at, perhaps, room temperature. At a higher temperature, however, this 7.5-μA curve may rise to the position shown in Fig. 8-15. An important point to note now is that the I_B curves are higher on the collector characteristics of higher-β transistors and that they are also higher at higher operating temperatures. Thus, circuits that are designed to have a more stable operating point with large changes in β are also more stable under conditions of large temperature changes.

In general, more feedback tends to make the circuit operating point less dependent on the β value of the transistor. So if you are required to build circuits using transistors whose β and leakage values are uncertain and where temperature changes are likely to be large, it is wise to use considerable feedback. That is, relatively large R_E values in the circuits in Figs. 8-4, 8-6, and 8-7 and relatively small R_B values in circuits like that in Fig. 8-5 tend to stabilize the operating point Q.

Example 8-26 Sketch a load line and identify the operating point for the circuit using collector feedback as in Fig. 8-5 where $V_{CC} = 24$ V, $R_B = 1.6$ MΩ, $R_C = 16$ kΩ, and the transistor $\beta = 100$.

Answer 8-26 See Fig. 8-16.—This circuit is the same as the one described in Ex. 8-14. Neglecting leakage in part (a) we found

8-7 LOAD LINES OF THE VARIOUS COMMON-EMITTER CIRCUITS

that $I_C \cong 0.75$ mA, $I_B \cong 7.5$ μA, and $V_{CE} \cong 12$ V. Now with Eq. 8-21 we find the saturation current

$$I_{C(\text{sat})} \cong \frac{V_{CC}}{R_C} = \frac{24 \text{ V}}{16 \text{ k}\Omega} = 1.5 \text{ mA}$$

At the lower end of the load line, when $I_C = 0$, the full 24 V source appears across the collector and emitter terminals; therefore $V_{CE(\text{cutoff})} = 24$ V.

Example 8-27 Sketch the load line and identify the operating point for the circuit described in the previous example but assume that the transistor is replaced with one having $\beta = 200$.

Answer 8-27 See Fig. 8-17.—By Eq. 8-17,

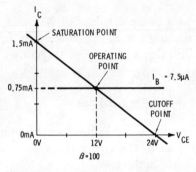

Fig. 8-16. Load line and operating point for the circuit described in Exs. 8-14 and 8-26.

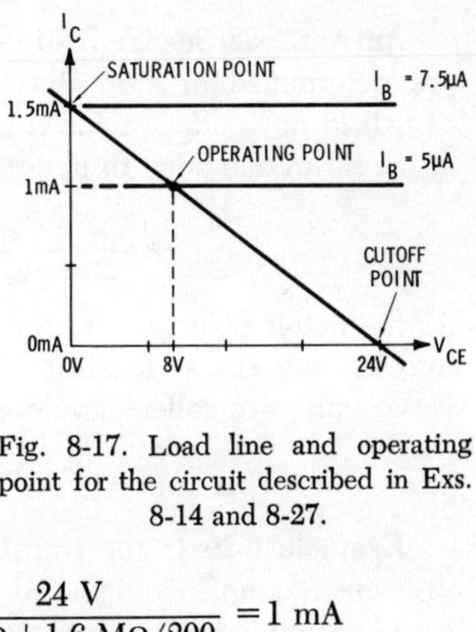

Fig. 8-17. Load line and operating point for the circuit described in Exs. 8-14 and 8-27.

$$I_C \cong \frac{V_{CC}}{R_C + R_B/\beta} = \frac{24 \text{ V}}{16 \text{ k}\Omega + 1.6 \text{ M}\Omega/200} = 1 \text{ mA}$$

With Eq. 8-19

$$I_B \cong \frac{V_{CC}}{R_B + \beta R_C} = \frac{24 \text{ V}}{1.6 \text{ M}\Omega + 200(16 \text{ k}\Omega)} = 5 \text{ μA}$$

And with Eq. 8-20

$$V_{CE} = V_{CC} - R_C I_C = 24 - 16(1) = 8 \text{ V}$$

Since the load R_C is unchanged, the saturation and cutoff points on the load line are unchanged too.

Example 8-28 If the circuit in Fig. 8-6 has component values

$V_{CC} = 30$ V, $V_{EE} = 15$ V, $R_B = 30$ kΩ, $R_E = 20$ kΩ, and the transistor $\beta = 100$, sketch its load line and identify its operating point.

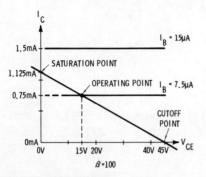

Fig. 8-18. Load line and operating point for the circuit described in Exs. 8-16 and 8-28.

Answer 8-28 See Fig. 8-18.—The quiescent values of I_C and I_B are determined in part (b) of the answer to Ex. 8-16; that is, $I_C \cong 0.75$ mA and $I_B \cong 7.5$ μA. With Eq. 8-31 we find the current at the saturation point to be

$$I_{C(sat)} = \frac{V_{CC} + V_{EE}}{R_C + R_E} = \frac{45 \text{ V}}{40 \text{ k}\Omega} = 1.125 \text{ mA}$$

At the cutoff point, where the transistor acts like an open looking into its collector and emitter, the sum of the two voltage sources is the emitter-to-collector voltage. Thus, in this case,

$$V_{CE(cutoff)} \cong V_{CC} + V_{EE} = 30 \text{ V} + 15 \text{ V} = 45 \text{ V}$$

Example 8-29 If the transistor in the circuit described in the previous example is replaced with one with $\beta = 200$, sketch the load line and indicate the operating point.

Answer 8-29 See Fig. 8-19A.—The saturation and cutoff points are the same here as in the last problem. Also the I_C is unchanged because β is not a factor in Eq. 8-25; therefore changes in β have nearly no effect on the value of I_C. As in the last example, then,

$$I_C \cong \frac{V_{EE}}{R_E} = \frac{15 \text{ V}}{20 \text{ k}\Omega} = 0.75 \text{ mA}$$

Since I_C is unchanged, the quiescent collector-to-emitter voltage V_{CE} is unchanged too. By Eq. 8-30,

$$V_{CE} \cong V_{CC} - R_C I_C = 30 - 20(0.75) = 15 \text{ V}$$

8-7 LOAD LINES OF THE VARIOUS COMMON-EMITTER CIRCUITS

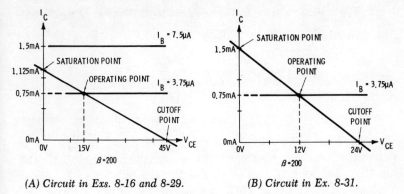

(A) Circuit in Exs. 8-16 and 8-29.   (B) Circuit in Ex. 8-31.

Fig. 8-19. Load lines and operating points.

The base current, however, decreases with larger β values. By Eq. 8-28,

$$I_B \cong \frac{V_{EE}}{\beta R_E} = \frac{15 \text{ V}}{200(20 \text{ k}\Omega)} = 3.75 \text{ }\mu\text{A}$$

It is apparent that the operating point is very stable in this circuit with emitter feedback and two supplies. Also if the β is doubled by changing the transistor, the base current I_B decreases to about half its previous value. This tends to hold I_C constant.

Example 8-30 In a circuit as shown in Fig. 8-7, $V_{CC} = 24$ V, $R_1 = 130$ kΩ, $R_2 = 30$ kΩ, $R_C = 10$ kΩ, $R_E = 6$ kΩ and the transistor $\beta = 100$. (a) Sketch the load line and identify the operating point. Assume $I_{CEO} \cong 0$. (b) If the transistor is replaced with one having $\beta = 200$, sketch the load line and identify the operating point.

Answer 8-30 (a) See Fig. 8-16. In this circuit, current I_C, as in the last example, is independent of β according to Eq. 8-35. In this case, then,

$$I_C \cong \frac{R_2}{R_1 + R_2} \left(\frac{V_{CC}}{R_E} \right) = \frac{30}{160} \left(\frac{24 \text{ V}}{6 \text{ k}\Omega} \right) = 0.75 \text{ mA}$$

With Eq. 8-37 we find that

$$I_B \cong \frac{R_2}{R_1 + R_2} \left(\frac{V_{CC}}{\beta R_E} \right) \cong 7.5 \text{ }\mu\text{A}$$

By Eq. 8-39 we find that the quiescent collector-to-emitter voltage is

$$V_{CE} \cong V_{CC} - (R_C + R_E)I_C \cong 24 - (16)0.75 = 24 - 12 = 12 \text{ V}$$

The saturation current by Eq. 8-40 is

$$I_{C(\text{sat})} \cong \frac{V_{CC}}{R_C + R_E} = \frac{24 \text{ V}}{16 \text{ k}\Omega} = 1.5 \text{ mA}$$

At the cutoff point

$$V_{CE(\text{cutoff})} \cong V_{CC} = 24 \text{ V}$$

(b) See Fig. 8-19B. Since current I_C and therefore V_{CE} are independent of the transistor β value according to Eq. 8-35, they are the same in this case as in the last example. Also $I_{C(\text{sat})}$ and $V_{CE(\text{cutoff})}$ are the same. Thus, in this case, as in the last one, $I_C \cong 0.75$ mA, $V_{CE} \cong 12$ V, $I_{C(\text{sat})} \cong 1.5$ mA, and $V_{CE(\text{cutoff})} \cong 24$ V. The base current, however, decreases and, in this case, by Eq. 8-37, is

$$I_B \cong \frac{R_2}{R_1 + R_2}\left(\frac{V_{CC}}{\beta R_E}\right) = \frac{30}{160}\left(\frac{24 \text{ V}}{200(6 \text{ k}\Omega)}\right) = 3.75 \text{ }\mu\text{A}$$

which shows us that in this circuit if a transistor with twice the previous β is used, I_B tends to decrease to half of its previous value just as in the circuit using emitter feedback with two supplies. Conversely, if a lower-β transistor is used instead, the current I_B increases.

PROBLEMS (PART I)

8-1. If $R_B = 800$ kΩ and the transistor $\beta = 80$ in the circuit of Fig. 8-20, find the quiescent values of I_C and V_{CE}, and determine the approximate collector power dissipation P_C. Assume that leakage is negligible.

8-2. If $R_B = 800$ kΩ and the transistor $\beta = 100$ in the circuit of Fig. 8-20, find the quiescent values of I_C and V_{CE}, and determine the approximate collector power dissipation P_C. Assume that leakage is negligible.

8-3. In the circuit of Fig. 8-20, when switch S is open, 0.5 V is read across the collector resistor. If the transistor $\beta = 80$, what are the leakage currents I_{CEO} and I_{CBO}?

8-4. In the circuit of Fig. 8-20, when switch S is open, 1 V is read across the collector resistor. If the transistor $\beta = 100$, what are the leakage currents I_{CEO} and I_{CBO}?

PROBLEMS (PART I)

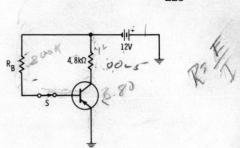

Fig. 8-20. Circuit for Probs. 8-1 through 8-8.

8-5. In the circuit of Fig. 8-20, if $R_B = 800$ kΩ and the transistor $\beta = 80$ and its $I_{CEO} = 150$ μA at 25°C, what are the approximate quiescent values of I_C, V_{CE}, and P_C at (a) 25°C and at (b) 55°C? The transistor is a germanium type.

8-6. In the circuit of Fig. 8-20, if $R_B = 800$ kΩ and the transistor $\beta = 100$ and its $I_{CEO} = 200$ μA at 25°C, what are the approximate quiescent values of I_C, V_{CE}, and P_C at (a) 25°C and at (b) 35°C? The transistor is a germanium type.

8-7. If the transistor $\beta = 80$ in the circuit of Fig. 8-20, what value of R_B would you use to obtain the optimum operating point? Assume $I_{CBO} \cong 0$.

8-8. If the transistor $\beta = 100$ in the circuit of Fig. 8-20, what value of R_B would you use to obtain the optimum operating point? Assume $I_{CBO} \cong 0$.

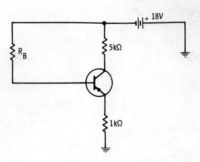

Fig. 8-21. Circuit for Probs. 8-9 through 8-12, 8-25, and 8-26.

8-9. If in the circuit of Fig. 8-21, $R_B = 1.7$ MΩ, the transistor $\beta = 100$ and its leakage is negligible, what are (a) the quiescent I_C and collector-to-emitter voltage V_{CE}, and (b) the end points on the load line $I_{C(sat)}$ and $V_{CE(cutoff)}$?

8-10. If in the circuit of Fig. 8-21, $R_B = 560$ kΩ, the transistor $\beta = 70$ and its leakage is negligible, what are (a) the quiescent I_C and collector-to-emitter voltage V_{CE}, and (b) the end points on the load line $I_{C(sat)}$ and $V_{CE(cutoff)}$?

8-11. If in Fig. 8-21 the transistor β is 100 and its leakage is negligible, what value of R_B will you use to obtain the optimum operating point, which is the point at which the quiescent collector current I_C is half the collector saturation current $I_{C(sat)}$?

8-12. If in Fig. 8-21 the transistor β is 70 and its leakage is negligible, what value of R_B will you use to obtain the optimum operating point?

8-13. In the circuit of Fig. 8-22, if $V_{CC} = -20$ V, $R_C = 5$ kΩ, $R_B = 900$ kΩ and the transistor $\beta = 60$, what are the values of (a) quiescent I_C and V_{CE}, and (b) the end points of the load line $I_{C(sat)}$ and $V_{CE(cutoff)}$? Assume $I_{CEO} \cong 0$.

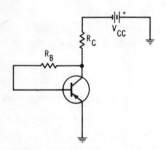

Fig. 8-22. Circuit for Probs. 8-13 through 8-16, 8-27, and 8-28.

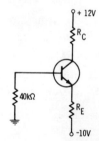

Fig. 8-23. Circuit for Probs. 8-17 through 8-20, 8-29 through 8-32.

8-14. In the circuit of Fig. 8-22, if $V_{CC} = -24$ V, $R_C = 10$ kΩ, $R_B = 700$ kΩ, and the transistor $\beta = 50$, what are the values of (a) the quiescent I_C and V_{CE}, and (b) the end points of the load line $I_{C(sat)}$ and $V_{CE(cutoff)}$? Assume $I_{CEO} \cong 0$.

8-15. In the circuit of Fig. 8-22, if $V_{CC} = -20$ V, $R_C = 5$ kΩ and the transistor $\beta = 60$ and $I_{CEO} \cong 0$, what value of R_B would you use to obtain the optimum operating point?

8-16. If in the circuit of Fig. 8-22, $V_{CC} = -24$ V, $R_C = 10$ kΩ, and the transistor $\beta = 50$, and $I_{CEO} \cong 0$, what value of R_B would you use to obtain the optimum operating point?

Problems (Part I)

8-17. If $R_C = 5$ kΩ, $R_E = 10$ kΩ, and the transistor $\beta = 80$ and its leakage is negligible in the circuit of Fig. 8-23, what are the values of (a) quiescent I_C and V_{CE}, and (b) the end points of the load line $I_{C(\text{sat})}$ and $V_{CE(\text{cutoff})}$?

8-18. If $R_C = 8$ kΩ, $R_E = 20$ kΩ, and the transistor $\beta = 60$ and its leakage is negligible in the circuit of Fig. 8-23, what are the values of (a) quiescent I_C and V_{CE}, and (b) the end points of the load line $I_{C(\text{sat})}$ and $V_{CE(\text{cutoff})}$?

8-19. If in the circuit of Fig. 8-23, $R_C = 5$ kΩ, what value of R_E would you select to obtain $V_{CE} \cong 6$ V, assuming that $I_{CEO} \cong 0$?

8-20. If in Fig. 8-23, $R_C = 8$ kΩ, what value of R_E would you select to obtain $V_{CE} \cong 7$ V, assuming that $I_{CEO} \cong 0$?

8-21. If in Fig. 8-24, $R_1 = 210$ kΩ, $R_2 = 40$ kΩ, $R_E = 4$ kΩ, and if the transistor $\beta = 100$ and its leakage is negligible, what are the values of (a) the quiescent I_C and V_{CE} and (b) the end points on the load line $I_{C(\text{sat})}$ and $V_{CE(\text{cutoff})}$?

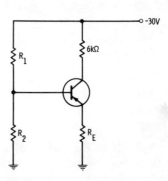

Fig. 8-24. Circuit for Probs. 8-21 through 8-24, 8-35, and 8-36.

8-22. If in the circuit of Fig. 8-24, $R_1 = 85$ kΩ, $R_2 = 15$ kΩ, $R_E = 1.5$ kΩ, and if the transistor $\beta = 60$, and its leakage is negligible, what are the values of (a) the quiescent I_C and V_{CE} and (b) the end points on the load line $I_{C(\text{sat})}$ and $V_{CE(\text{cutoff})}$?

8-23. If in the circuit of Fig. 8-24, $R_1 = 210$ kΩ, $R_2 = 40$ kΩ, and the transistor leakage is negligible, what value of R_E would you use to obtain a quiescent $I_C \cong 1$ mA?

8-24. If in the circuit of Fig. 8-24, $R_1 = 170$ kΩ, $R_2 = 30$ kΩ, and the transistor leakage is negligible, what value of R_E would you use to obtain a quiescent $I_C \cong 1$ mA?

Additional Problems

8-25. Referring back to the circuit of Fig. 8-21, if $R_B = 1.7$ MΩ the silicon transistor $\beta = 100$, and leakage $I_{CEO} = 0.2$ mA at 25°C, find the values of (a) the quiescent I_C at 25°C, and (b) the quiescent I_C at 37°C.

8-26. Referring back to Fig. 8-21, if $R_B = 560$ kΩ, the germanium transistor $\beta = 70$, and $I_{CEO} = 0.2$ mA at 25°C, find the value of (a) the quiescent I_C at 25°C, and (b) the quiescent I_C at 45°C.

8-27. In the circuit of Fig. 8-22, if $V_{CC} = -20$ V, $R_C = 5$ kΩ, $R_B = 700$ kΩ, the germanium transistor $\beta = 50$, and $I_{CEO} = 300$ μA at 25°C, what are the values of (a) the quiescent I_C at 25°C, and (b) the quiescent I_C at 45°C?

8-28. In the circuit of Fig. 8-22, if $V_{CC} = -24$ V, $R_C = 10$ kΩ, $R_B = 700$ kΩ the germanum transistor $\beta = 50$, and $I_{CEO} = 300$ μA at 25°C, what are the values of (a) the quiescent I_C at 25°C, and (b) the quiescent I_C at 45°C?

8-29. If $R_C = 5$ kΩ, $R_E = 10$ kΩ, the germanium transistor $\beta = 80$, and the $I_{CEO} = 0.5$ mA at 30°C in the circuit of Fig. 8-23, what are the values of (a) the quiescent I_C at 30°C, and (b) the quiescent I_C at 50°C?

8-30. If $R_C = 8$ kΩ, $R_E = 20$ kΩ, the germanium transistor $\beta = 60$, and $I_{CEO} = 310$ μA at 30°C in the circuit of Fig. 8-23, what are the values of (a) the quiescent I_C at 30°C, and (b) the quiescent I_C at 50°C?

8-31. If in the circuit of Fig. 8-23, $R_C = 5$ kΩ, $R_E = 10$ kΩ, $I_{CEO} \cong 0$, and if the transistor with $\beta = 80$ is replaced with one having $\beta = 160$, find (a) I_C when $\beta = 80$, (b) I_C when $\beta = 160$, and (c) the percent of increase in I_C that is caused by this replacement.

8-32. If in Fig. 8-23 $R_C = 8$ kΩ, $R_E = 20$ kΩ, $I_{CEO} \cong 0$, and if the transistor with a β of 60 is replaced with one having a β of 120 (β increased 100 percent), find (a) I_C when $\beta = 60$, (b) I_C when $\beta = 120$, and (c) the percent of increase in I_C that is caused by this replacement.

8-33. Referring back to the circuit described in Prob. 8-1, what is the quiescent I_C if the transistor is replaced with one having $\beta = 160$?

8-34. Referring back to the circuit described in Prob. 8-2, what is the quiescent I_C if the transistor $\beta = 200$?

8-8 V-I Characteristics and the Dynamic Resistance

8-35. If in the circuit of Fig. 8-24, $R_1 = 210$ kΩ, $R_2 = 40$ kΩ, $R_E = 4$ kΩ and if the transistor $\beta = 100$ and its leakage is $I_{CEO} = 1$ mA at some higher than room temperature, (a) what are the values of the quiescent I_C and V_{CE} and (b) how do these compare with the I_C and V_{CE} values found for the same circuit with negligible transistor leakage in Prob. 8-21?

8-36. If in the circuit of Fig. 8-24, $R_1 = 85$ kΩ, $R_2 = 15$ kΩ, $R_E = 1.5$ kΩ, and if the transistor $\beta = 60$ and its leakage is $I_{CEO} = 2$ mA at some higher than room temperature, (a) what are the quiescent values of I_C and V_{CE} and (b) how do these compare with the I_C and V_{CE} values found for the same circuit with negligible transistor leakage in Prob. 8-22?

Part II: AC Considerations

Once having established popular methods of biasing common-emitter type circuits, we can proceed to use such circuits as signal amplifiers. In this part of the chapter we will consider the effect a signal voltage, applied to the base, has on the voltage at the collector. Also by sketching the dc and ac load lines, we will see the importance of a proper operating point. After obtaining an understanding of how signal voltages and currents "ride on" the dc bias values, and how the gain factors and maximum output signal capabilities are affected by the circuit components, you will be in a good position to build your own amplifiers that possess whatever characteristics you need or desire, provided that the characteristics you seek are within obtainable limits.

8-8 V-I CHARACTERISTICS AND THE DYNAMIC RESISTANCE OF THE BASE-EMITTER JUNCTION

In all the versions of the common-emitter circuits that we have considered, the base-emitter junctions are forward biased. The typical V-I characteristics of these base-emitter diode junctions are shown in Fig. 8-25. Note their similarity to the characteristics of the ordinary diode; see Figs. 2-4 and 2-5, Chap. 2. They are also similar to the common-base circuit's forward V-I emitter-base characteristics shown in Fig. 7-18, Chap. 7.

If a signal source v_s is applied across the forward biased base-emitter junction through a coupling capacitor C_1 as shown in Fig.

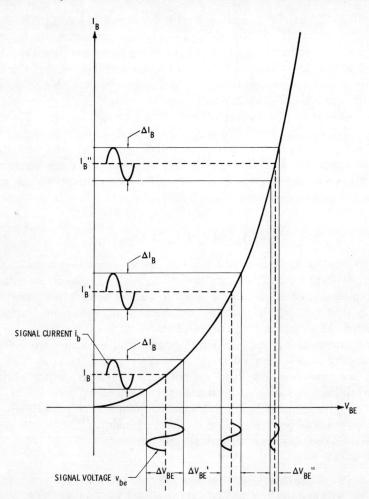

Fig. 8-25. V-I characteristics of the base-emitter junction in the common-emitter circuit.

8-26A, the source v_s causes a signal current to ride on the dc base current; that is, i_b is superimposed on I_B. Depending on the magnitude of I_B, the ac component i_b sees more or less resistance across the base-emitter junction. More specifically, the resistance seen by the source v_s looking into the base, call it r_b', decreases if the base bias current I_B is increased. Of course, then r_b' increases if I_B is decreased. Since base-emitter junction possesses ac resistance r_b', the

8-8 V-I CHARACTERISTICS AND THE DYNAMIC RESISTANCE

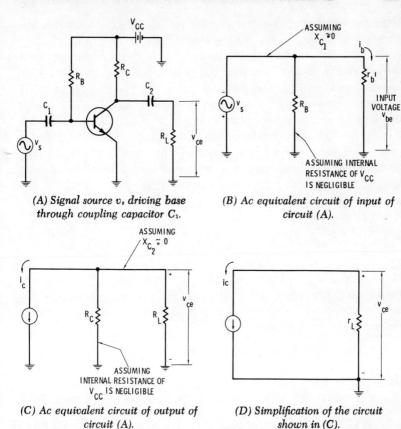

(A) Signal source v_s driving base through coupling capacitor C_1.

(B) Ac equivalent circuit of input of circuit (A).

(C) Ac equivalent circuit of output of circuit (A).

(D) Simplification of the circuit shown in (C).

Fig. 8-26. Ac considerations of the common-emitter circuit.

ac equivalent of the input of the circuit in Fig. 8-26A can be shown as in Fig. 8-26B. Assuming that the reactance of C_1 at the frequency of the source v_s and the internal resistance of the dc source V_{CC} are negligible, both C_1 and V_{CC} are replaced with short circuits in the ac equivalent.

If the amplitude of the signal base current i_b is small compared to the value of base bias current I_B, as is certainly the case with the signal components riding on bias currents I_B' and I_B'' in Fig. 8-25, the ac resistance seen looking into the base may be approximated with the equation

$$r_b' \cong \frac{25 \text{ mV}}{I_B} \tag{8-43}$$

Junction Transistors in Common-Emitter Connection

Note the similarity of this equation to Eq. 7-3 given in the last chapter. By examination, we see that 25 mV is a constant, and r_b' varies inversely with I_B. The reason for this inverse relationship is shown in Fig. 8-25. Note that the relatively low base current I_B has a signal, with a peak-to-peak value of $\triangle I_B$, riding on it. It causes a signal voltage, with a peak-to-peak value of $\triangle V_{BE}$, across the base-emitter junction. Now a larger base bias current like I_B', having the same amplitude of signal superimposed on it, causes a smaller signal voltage $\triangle V_{BE}'$ across the base-emitter junction. Similarly, if the signal current rides on an even larger bias current like I_B'', the resulting signal across the base-emitter junction has a still smaller amplitude $\triangle V_{BE}''$. Since r_b' is a dynamic resistance, it can be shown to be

$$r_b' \cong \frac{\triangle V_{BE}}{\triangle I_B}$$

It is obvious in this equation that if the numerator $\triangle V_{BE}$ is made smaller for any given value of $\triangle I_B$, the r_b' value must become smaller too. And as shown in Fig. 8-25, larger base bias currents do cause smaller $\triangle V_{BE}$ values.

Example 8-31 In a circuit like that in Fig. 8-26A, $V_{CC} = 9$ V, $v_{s(peak)} = 5$ mV, and the reactance of C_1 is negligible. Find the values r_b' and the peak value of the base signal current $i_{b(peak)}$ (a) when $R_B = 900$ kΩ, (b) when $R_B = 1.8$ MΩ, and (c) when $R_B = 3.6$ MΩ.

Answer 8-31 (a) $r_b' \cong 2500$ Ω, $i_{b(peak)} \cong 2$ μA, (b) $r_b' \cong 5000$ Ω, $i_{b(peak)} \cong 1$ μA, (c) $r_b' \cong 10$ kΩ, $i_{b(peak)} \cong 0.5$ μA. By Eq. 8-8 the dc bias current is

$$I_B \cong \frac{V_{CC}}{R_B} = \frac{9 \text{ V}}{900 \text{ k}\Omega} = 10 \text{ μA}$$

By Eq. 8-43 the dynamic resistance seen looking into the base is

$$r_b' \cong \frac{25 \text{ mV}}{I_B} = \frac{25 \text{ mV}}{10 \text{ μA}} = 2500 \text{ Ω}$$

The current in r_b' is the base signal current i_b. Therefore, since the signal voltage v_s is directly applied to r_b' in this case, by Ohm's law $i_b = v_s/r_b'$ or

$$i_{b(peak)} = \frac{v_{s(peak)}}{r_b'} \cong \frac{5 \text{ mV}}{2.5 \text{ k}\Omega} = 2 \text{ μA}$$

8-8 V-I CHARACTERISTICS AND THE DYNAMIC RESISTANCE

In part (b), then,

$$I_B \cong \frac{9 \text{ V}}{1.8 \text{ M}\Omega} = 5 \text{ }\mu\text{A}$$

and

$$r_b' \cong \frac{25 \text{ mV}}{5 \text{ }\mu\text{A}} = 5000 \text{ }\Omega$$

Therefore

$$i_{b(\text{peak})} \cong \frac{5 \text{ mV}}{5 \text{ k}\Omega} = 1 \text{ }\mu\text{A}$$

Similarly, in part (c)

$$I_B \cong \frac{9 \text{ V}}{3.6 \text{ M}\Omega} = 2.5 \text{ }\mu\text{A}$$

and

$$r_b' \cong \frac{25 \text{ mV}}{2.5 \text{ }\mu\text{A}} = 10 \text{ k}\Omega$$

Therefore

$$i_{b(\text{peak})} \cong \frac{5 \text{ mV}}{10 \text{ k}\Omega} = 0.5 \text{ }\mu\text{A}$$

The base-to-emitter resistance r_b' in the common-emitter circuit is larger, in fact β times larger, than the emitter-to-base resistance r_e' in the common-base circuit provided that the bias currents are the same in either configuration. For example, since $I_C \cong \beta I_B$, then by rearranging terms we can show that

$$I_B \cong \frac{I_C}{\beta} \cong \frac{I_E}{\beta}$$

Substituting this into Eq. 8-43 we find

$$r_b' \cong \frac{25 \text{ mV}}{I_B} \cong \frac{25 \text{ mV}}{I_E/\beta} \cong \beta \frac{25 \text{ mV}}{I_E}$$

Now substituting Eq. 7-3 into the above we can show that

$$r_b' \cong \beta r_e' \tag{8-44}$$

This last equation will be useful in later sections.

8-9 DETERMINING A_e, A_i, AND A_p OF THE SIMPLE BASE BIASED COMMON-EMITTER AMPLIFIER

As mentioned in the last chapter, the voltage gain A_e of a device is the ratio of its output voltage v_o to its input voltage v_{in}. In the common-emitter amplifier circuit, the output voltage is the signal voltage across the collector and emitter leads and is called v_{ce}. The input voltage is applied to the base and emitter leads and is called v_{be}. Specifically for the common-emitter amplifier, therefore,

$$A_e = \frac{v_{ce}}{v_{be}} \quad (8\text{-}45)$$

Referring to the circuit in Fig. 8-26A, the ac equivalent of its output is shown in Figs. 8-26C and 8-26D. These show that the output of the common-emitter circuit acts like a constant-current source with an output signal current i_c. The magnitude of this constant current i_c depends on the magnitude of base signal current i_b. That is, since $\beta \cong i_c/i_b$ by Eq. 8-3A, then

$$i_c \cong \beta i_b$$

In Fig. 8-26D we can see that by Ohm's law the output voltage is

$$v_{ce} = r_L i_c \quad (8\text{-}46)$$

where, of course, r_L is the total resistance of R_C and R_L in parallel. Similarly in Fig. 8-26B we can show by Ohm's law that

$$v_{be} = r_b{'} i_b \quad (8\text{-}47)$$

In this case, $v_s \cong v_{be}$, assuming negligible reactance of C_1. Eq. 8-47 above can be modified by substituting Eqs. 8-44 and 8-3A into it as follows:

$$v_{be} = r_b{'} i_b \cong (\beta r_e{'}) i_b \cong r_e{'}(\beta i_b) \cong r_e{'} i_c$$

This last equation, $v_{be} \cong r_e{'} i_c$, and Eq. 8-46 substituted into Eq. 8-45 give us a simple and practical equation for voltage gain A_e:

$$A_e = \frac{v_{ce}}{v_{be}} \cong \frac{r_L i_c}{r_e{'} i_c} = \frac{r_L}{r_e{'}}$$

Notice that this gain equation is identical to the one for the common-base circuit given in Sec. 7-5 (Eq. 7-9).

8-9 Determining A_e, A_i, and A_p of Amplifier

The current gain A_i of the common-emitter transistor circuit is the ratio of its output signal current i_c to its input signal current i_b. This ratio is approximately the $ß$ of the transistor; thus

$$A_i = ß \cong \frac{i_c}{i_b} \qquad (8\text{-}48)$$

As with the common-base amplifier, the power gain A_p is the product of the voltage and current gains; that is,

$$A_p = A_e A_i \qquad (8\text{-}49)$$

Example 8-32 In a circuit like that of Fig. 8-26A, $R_B = 1.2$ MΩ, $R_C = 5$ kΩ, $R_L = 20$ kΩ, $V_{CC} = 12$ V, $v_s = 5$ mV (rms), the transistor $ß = 100$, and the reactances of the coupling capacitors are negligible for all frequencies of v_s. Find the voltage, current, and power gains, and the signal output voltage v_{ce}.

Answer 8-32 $A_e \cong 160$, $A_i \cong 100$, $A_p \cong 16{,}000$, $V_{ce} \cong 0.8$ V.—
First we can find the base bias current with Eq. 8-8:

$$I_B \cong \frac{12 \text{ V}}{1.2 \text{ MΩ}} = 10 \text{ μA}$$

Next we determine the quiescent collector current with Eq. 8-5:

$$I_C \cong ß I_B = 100(10 \text{ μA}) = 1 \text{ mA}$$

Knowing that $I_C \cong I_E$ we can now solve for r_e' with Eq. 7-3:

$$r_e' \cong \frac{25 \text{ mV}}{1 \text{ mA}} = 25 \text{ Ω}$$

Since in this case $r_L = R_C R_L / (R_C + R_L) = 4$ kΩ, with Eq. 7-9 we find

$$A_e \cong \frac{r_L}{r_e'} = \frac{4000}{25} = 160$$

The transistor current gain A_i is its $ß$, which is 100 in this case. Thus, the power gain by Eq. 8-49 is

$$A_p = A_e A_i \cong 160(100) = 16{,}000$$

Knowing the voltage gain A_e and the signal voltage input v_s, which is v_{be} because C_1 reactance is negligible, we can solve for the output voltage v_{ce} with Eq. 8-45. That is, since $A_e \cong v_{ce}/v_{be}$ then

$$v_{ce} = A_e v_{be} \cong 160(5 \text{ mV}) = 0.8 \text{ V}$$

Voltage and current signals in the circuit described in Ex. 8-32 are shown in Fig. 8-27. Their explanations are as follows:

(I) The input signal v_s is the signal applied to the base-emitter junction as long as the reactance of C_1 is negligible, in which case its waveform is shown in Fig. 8-27A.

(II) The input signal v_{be} is superimposed on the forward biased dc voltage drop across the base-emitter junction as shown in Fig. 8-27B. The dc component is shown as 0.7 V, assuming that the transistor is a silicon type.

(III) The source v_s sees the base-to-emitter resistance r_b', which is about equal to $\beta r_e'$, causing a base signal current $i_b \cong v_{be}/\beta r_e' \cong 5 \text{ mV}/2.5 \text{ k}\Omega = 2 \text{ }\mu\text{A}$ in this case. See Fig. 8-27C.

(IV) This base signal current i_b rides on the 10-μA base bias current as shown in Fig. 8-27D.

(V) Also, i_b is amplified by the β of the transistor, resulting in a collector signal current $i_c \cong \beta i_b \cong 100(2 \text{ }\mu\text{A}) = 0.2 \text{ mA}$; see Fig. 8-27E.

(VI) This i_c rides on the quiescent collector current I_C as shown in Fig. 8-27F.

(VII) The voltage gain A_e in Ex. 8-32 was found to be 160. Thus, the output voltage $v_{ce} = A_e v_{be} \cong 160(5 \text{ mV}) = 0.8 \text{ V}$. Or, knowing the output current i_c and the ac load r_L, we can find the output voltage with Ohm's law: $v_{ce} = r_L i_c \cong (4 \text{ k}\Omega)(0.2 \text{ mA}) = 0.8 \text{ V}$. See Fig. 8-27G.

(VIII) The output signal voltage v_{ce} rides on the quiescent dc collector-to-emitter voltage V_{CE} of 7 V in this case as shown in Fig. 8-27H.

You will note that the input and output voltages are 180° out of phase in the common-emitter amplifier whereas they are in phase in the common-base amplifier; see Figs. 7-11 and 7-17. Referring back to Fig. 8-1, we can see that when I_B flows toward or into the transistor, as in Fig. 8-1B, then I_C also flows into the transistor. On the other hand, when I_B flows away, I_C also flows away from the transistor, as shown in Fig. 8-1A. The point is, the I_B and I_C both flow toward or both flow away from the transistor. The same is true with instantaneous directions of signal currents i_b and i_c. When v_s is negative with respect to ground at some instant, i_b flows into the base or r_b' at the same instant as shown in Fig. 8-26B. This causes the instantaneous i_c to flow toward the transistor and up

8-9 DETERMINING A_e, A_i, AND A_p OF AMPLIFIER

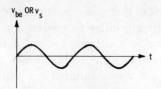

(A) Input signal voltage of 5 mV (rms).

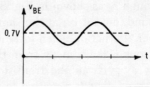

(B) 5 mV input signal riding on 0.7 V drop across base-emitter junction.

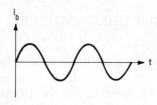

(C) Base signal current with amplitude $i_b \cong 2$ μA (rms).

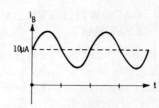

(D) Base signal current riding on dc base bias current I_B.

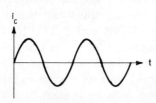

(E) Collector signal current $i_c \cong 0.2$ mA.

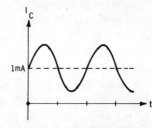

(F) Collector signal current riding on dc collector current I_C.

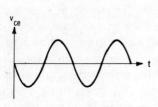

(G) Output signal voltage is $v_{ce} \cong 0.8$ V.

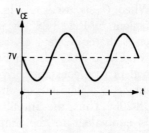

(H) Output voltage riding on quiescent V_{CE}.

Fig. 8-27. Signals in the circuit described in Ex. 8-32.

through r_L as shown in Fig. 8-26D. Note that when v_s is negative to ground, output v_{ce} at the same instant is positive to ground. On another alternation when instantaneous v_s is positive to ground, instantaneous i_b flows away from the transistor and r_b'. This causes i_c to flow away from the transistor too and down through r_L, producing a v_{ce} that is negative to ground at the same instant. Thus, v_{be} and v_{ce} are always out of phase with each other.

8-10 GAIN WITH THE AMPLIFIER USING EMITTER FEEDBACK

In Fig. 8-28A we have a base biased circuit using emitter feedback. It is working into a load R_L and has a signal voltage v_s applied to its input. The feedback resistance R_E can be bypassed with a capacitor C_3. If a bypass capacitor C_3 is used, its value is chosen so that its reactance is negligible at the lowest operating frequency of v_s. This causes signal to see C_3 as practically a short circuit which causes emitter signal current to flow around R_E. As far as signal is concerned, then, R_E is not even in the circuit when bypassed with C_3. When C_3 is used, the ac equivalent circuits of Fig. 8-28A may be shown as in Figs. 8-28B and 8-28C. Note in these equivalent circuits that all components whose reactances or resistances are negligible to signal currents have been replaced with short circuits. Particularly notice that R_E and its bypass capacitor C_3 have been replaced with a short in each equivalent.

On the other hand, if R_E is not bypassed, the ac equivalent circuits are like those in Figs. 8-28D and 8-28E. The voltage gain A_e of the circuit is significantly affected by whether or not R_E is bypassed, as you will see.

When C_3 is used in Fig. 8-28A, notice that the input of its equivalent in Fig. 8-28C is identical to the circuit in Fig. 8-26B, which is the equivalent of the input of the simple base biased circuit of Fig. 8-26A. Also in Fig. 8-28C, the output is shown to be equivalent to a constant-current source i_c working into a total load resistance r_L. This output circuit is identical to the equivalent in Fig. 8-26D. Thus, the amplifier in Fig. 8-28A when using a bypass C_3 is equivalent to the amplifier of Fig. 8-26A. Therefore gain equations that are applicable to one of these may also be used with the other. In other words, when R_E is bypassed in Fig. 8-28A, the voltage gain is, by Eq. 7-9,

8-10 Gain With the Amplifier Using Emitter Feedback

$$A_e \cong \frac{r_L}{r_e'}$$

The circuit in Fig. 8-28A acts quite differently when R_E is not bypassed. We will see that its voltage gain is reduced but is more stable; that is, the gain will not vary with temperature changes or with replacements of transistors of different types. Also the quality of amplification is better with an unbypassed R_E, which means that the output signal is less distorted.

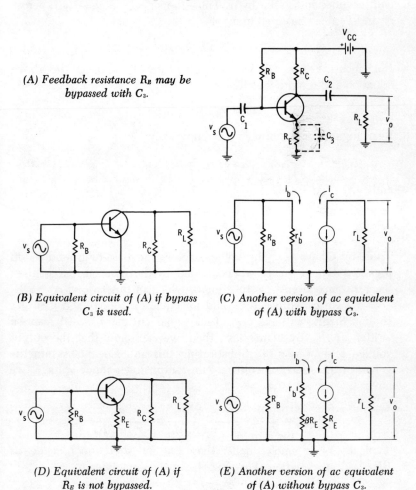

(A) Feedback resistance R_E may be bypassed with C_3.

(B) Equivalent circuit of (A) if bypass C_3 is used.

(C) Another version of ac equivalent of (A) with bypass C_3.

(D) Equivalent circuit of (A) if R_E is not bypassed.

(E) Another version of ac equivalent of (A) without bypass C_3.

Fig. 8-28. Common-emitter circuit with emitter feedback.

We see in Fig. 8-28E that base signal current i_b has to flow through resistances r_b' and βr_E in series. We would certainly expect that i_b has to flow through the dynamic base-to-emitter resistance r_b' but we might stop and wonder why i_b sees the emitter resistor R_E larger by a factor of β. The reason for this can be seen if a Kirchhoff's voltage equation is written for the loop containing components v_s, r_b', and R_E in Fig. 8-28D. Of course, r_b' represents the dynamic resistance of the forward biased base-emitter junction as seen by i_b. Thus, the signal voltage drops across r_b' and R_E must be equal to v_s, or

$$v_s = r_b' i_b + R_E i_e$$

But since $i_e \cong i_c \cong \beta i_b$, we can substitute βi_b for i_e in the above equation and rewrite it as follows:

$$v_s \cong r_b' i_b + R_E(\beta i_b)$$

Rearranging and factoring we can now show

$$v_s \cong (r_b' + \beta R_E) i_b \qquad (8\text{-}50\text{A})$$

and

$$\frac{v_s}{i_b} \cong r_b' + \beta R_E \qquad (8\text{-}50\text{B})$$

Thus, by Ohm's law, the voltage across the base-to-ground v_s, in Figs. 8-28D and 8-28E, when divided by the base current i_b gives the resistance from base to ground, which in this case is $r_b' + \beta R_E$.

Now we can proceed to find an equation for voltage gain A_e. In Fig. 8-28E we have an output signal current i_c which can be written as βi_b. By Ohm's law, then, we can show that the output voltage $v_o = r_L i_c \cong r_L \beta i_b$. Substituting this and Eq. 8-50A into the general expression for voltage gain, output over input voltage, we have

$$A_e = \frac{v_o}{v_s} \cong \frac{r_L \beta i_b}{(r_b' + \beta R_E) i_b} = \frac{\beta r_L}{r_b' + \beta R_E}$$

Typically r_b' is much smaller than βr_E and therefore $r_b' + \beta R_E \cong \beta R_E$, and the equation above may be further simplified to

$$A_e \cong \frac{\beta r_L}{\beta R_E} = \frac{r_L}{R_E} \qquad (8\text{-}51)$$

8-10 Gain With the Amplifier Using Emitter Feedback

The current and power gains are found as in the last section; that is, $A_i \cong \beta$ and $A_p = A_e A_i$.

Example 8-33 In the circuit shown in Fig. 8-28A, $V_{CC} = 24$ V, $R_B = 1.1$ MΩ, $R_C = 10$ kΩ, $R_E = 2$ kΩ, $R_L = 40$ kΩ, $v_s = 60$ μV (rms), and the transistor $\beta = 50$. Find (a) A_e, A_p, and v_o when R_E is by-passed, and (b) A_e, A_p, and v_o when R_E is not bypassed.

Answer 8-33 (a) $A_e \cong 320$, $A_p \cong 16{,}000$, $v_o \cong 19.2$ mV (rms), (b) $A_e \cong 4$, $A_p \cong 200$, $v_o \cong 240$ μV.—In part (a) when R_E is bypassed we can plan to use Eq. 7-9 to find A_e. However, first we must determine the quiescent I_E and r_e'. Because the capacitors act like open circuits to dc, the current of Fig. 8-4 is the dc equivalent of the circuit of Fig. 8-28A. Thus, the dc equations for the circuit of Fig. 8-4 are applicable here, and therefore by Eq. 8-11,

$$I_E \cong I_C \cong \frac{V_{CC}}{R_E + R_B/\beta} = \frac{24 \text{ V}}{2 \text{ kΩ} + 1.1 \text{ MΩ}/50} = 1 \text{ mA}$$

Now we find that

$$r_e' \cong \frac{25 \text{ mV}}{I_E} = \frac{25 \text{ mV}}{1 \text{ mA}} = 25 \text{ Ω}$$

Signal i_c sees R_C and R_L in parallel and therefore a total load resistance $r_L = 8$ kΩ. And now by Eq. 7-9 we can find the voltage gain:

$$A_e \cong \frac{r_L}{r_e'} = \frac{8000}{25} = 320$$

The current gain A_i is the β of the transistor, which is 50 in this case. So by Eq. 8-49

$$A_p = A_e A_i = 320(50) = 16{,}000$$

The output voltage is

$$v_o = A_e v_s \cong 320(60 \text{ μV}) = 19.2 \text{ mV}$$

In part (b) where R_E is not bypassed we use Eq. 8-51 to find voltage gain:

$$A_e \cong \frac{r_L}{R_E} = \frac{8000}{2000} = 4$$

Therefore the power gain is

$$A_p = A_e A_i = A_e \beta \cong 4(50) = 200$$

and the voltage output is

$$v_o \cong A_e v_s \cong 4(60 \ \mu V) = 240 \ \mu V$$

Obviously, the unbypassed R_E in the last example causes a large loss of voltage and power gain. However, some advantages of not using the bypass can be pointed out. In Eq. 8-51, the factors r_L and R_E are not subject to change even if the temperature changes or the transistor and its β value is changed. That is, r_L and R_E are practically constant; therefore, the voltage gain without a bypassed R_E, which is the ratio of r_L to R_E, is constant too. So without C_3, the gain is stable even though the ambient temperature varies and transistors of various types are interchanged in the circuit.

On the other hand, if R_E is bypassed, the voltage gain, though high, is unstable. As observed in previous sections, the quiescent values of I_C and I_E change when the temperature or the β is changed. Thus, a change in temperature or type of transistor will cause a change in I_E, which in turn causes a change in r_e', as can be seen by analyzing Eq. 7-3. If r_e' changes then gain A_e changes, too, by Eq. 7-9. In general, the voltage gain A_e tends to be unsteady and unpredictable when R_E is bypassed.

Frequently only a portion of the total resistance in the emitter is bypassed in a compromise to obtain good gain and good stability of the operating point too. In such cases the gain A_e is found with Eq. 8-51 in which R_E represents the unbypassed portion of resistance in the emitter lead.

Example 8-34 Suppose that the transistor in the circuit described in Ex. 8-33 is replaced with one having $\beta = 100$. Find the voltage gain A_e when (a) R_E is bypassed and (b) when R_E is not bypassed.

Answer 8-34 (a) $A_e \cong 593$, (b) $A_e \cong 4$.—In part (a), with this increased β we find the new quiescent emitter current with Eq. 8-11:

$$I_E \cong I_C \cong \frac{24 \ V}{2 \ k\Omega + 1100 \ k\Omega/100} \cong 1.85 \ mA$$

Therefore

8-10 GAIN WITH THE AMPLIFIER USING EMITTER FEEDBACK

$$r_e' \cong \frac{25 \text{ mV}}{1.85 \text{ mA}} \cong 13.5 \text{ }\Omega$$

By Eq. 7-9 the voltage gain is

$$A_e \cong \frac{8000}{13.5} \cong 593$$

In part (b), Eq. 8-51 is applicable because R_E is not bypassed. Neither of its variables, r_L or R_E, changes because the transistor has been changed. Thus, the gain and the method of finding it are the same as in the last example.

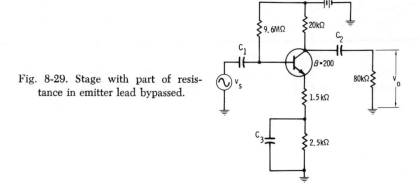

Fig. 8-29. Stage with part of resistance in emitter lead bypassed.

Example 8-35 What is the voltage gain of the circuit shown in Fig. 8-29?

Answer 8-35 $A_e \cong 10.7$.—There is a total resistance of 4 kΩ in the emitter as dc emitter current I_E sees it. As ac signal emitter current i_e sees it, however, there is only 1.5 kΩ if the reactance of C_3 is negligible. In this case then, from the signal's point of view, the total ac load in the collector r_L is equal to 20 kΩ||80 kΩ, or 16 kΩ and the resistance seen by the signal in the emitter lead is 1.5 kΩ. Thus, by Eq. 8-51

$$A_e \cong \frac{r_L}{R_E} = \frac{16{,}000}{1500} \cong 10.7$$

Note in this last example that the resistance seen in the emitter may be different, depending on how you look at it; that is, from either ac or dc point of view. In dc formulas like Eqs. 8-11, 8-13, 8-15, etc., R_E is the total resistance in the emitter lead, bypassed

242 JUNCTION TRANSISTORS IN COMMON-EMITTER CONNECTION

or not. On the other hand, R_E is only the unbypassed portion of the resistance in the emitter when ac formulas like Eq. 8-51 are used.

8-11 GAIN WITH THE COMMON-EMITTER AMPLIFIER USING COLLECTOR FEEDBACK

An amplifier using collector feedback is shown in Fig. 8-30A. Its ac equivalents are shown in Figs. 8-30B and 8-30C. Note in Fig. 8-30B as in Fig. 8-30A, R_B is connected across points B and C, that is, across base and collector leads. Signal source v_s puts out a current i_s that divides into two components i_b and i_p as shown in Fig. 8-30B. The base signal current i_b sees the base-to-emitter resistance which, as in previous circuits, is r_b' or $\beta r_e'$. Current i_p sees a resistance much smaller than the resistance R_B. The reason why can be shown by Kirchhoff's and Ohm's laws. Writing a Kirchhoff's voltage law equation for the loop aribitrarily starting, say, at ground, then proceeding up through v_s, across R_B to the right, down through r_L, and back to ground can be shown

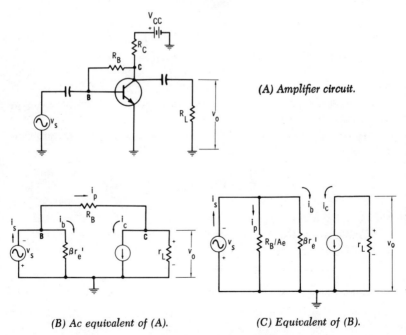

Fig. 8-30. Common-emitter amplifier using collector feedback.

8-11 Gain With the Amplifier Using Collector Feedback

$$v_s - R_B i_p + v_o = 0$$

But since the output voltage v_o is equal to gain A_e times the input voltage v_s, substituting this into the above gives

$$v_s - R_B i_p + A_e v_s = 0$$

And by rearranging and factoring we can show

$$v_s + A_e v_s - R_B i_p = (1 + A_e)v_s - R_B i_p = 0$$

Typically $A_e \gg 1$ and therefore $(1 + A_e) \cong A_e$, which may be substituted into the above as follows:

$$(1 + A_e)v_s - R_B i_p \cong A_e v_s - R_B i_p \cong 0$$

Rearranging the above we can show $A_e v_s \cong R_B i_p$, or

$$\frac{v_s}{i_p} \cong \frac{R_B}{A_e} \qquad (8\text{-}52)$$

or

$$i_p = \frac{v_s}{R_B/A_e}$$

Eq. 8-52 shows us by Ohm's law that an equivalent circuit may be drawn such that i_p is the current through a resistance R_B/A_e having a voltage source v_s across it; see Fig. 8-30C. Fig. 8-30C shows us that if the stage gain A_e is high, the resistance to i_p tends to be low—much lower than R_B alone. Thus, collector feedback may *load down* the source v_s if the latter has a high internal resistance.

As with previous circuits with no resistance in the emitter lead, the voltage gain of the amplifier is

$$A_e = \frac{v_o}{v_s} \cong \frac{r_L i_c}{\beta r_e' i_b} \cong \frac{r_L \beta i_b}{\beta r_e' i_b} = \frac{r_L}{r_e'}$$

which is Eq. 7-9 again.

The current and power gains may be found with Eqs. 8-48 and 8-49, which are applicable to all forms of common-emitter amplifiers covered in this chapter.

Example 8-36 In the circuit of Fig. 8-30A, $v_s = 20$ μV (rms), $R_B = 1.2$ MΩ, $R_C = 12$ kΩ, $R_L = 28$ kΩ, $V_{CC} = 24$ V, and the transistor $\beta = 100$. Find A_e, A_p, v_o, and the total resistance r_{in} as seen by the source v_s.

244   JUNCTION TRANSISTORS IN COMMON-EMITTER CONNECTION

Answer 8-36 $A_e \cong 336$, $A_p \cong 33{,}600$, $v_o \cong 6.72$ mV (rms), $r_{in} = 1.47$ kΩ.—If we are to find A_e with Eq. 7-9 we must first determine r_e' and therefore I_E. Since $I_E \cong I_C$, we use Eq. 8-17 to find

$$I_E \cong I_C \cong \frac{V_{CC}}{R_C + R_B/\beta} = \frac{24 \text{ V}}{12 \text{ k}\Omega + 1200 \text{ k}\Omega/100} = 1 \text{ mA}$$

Thus

$$r_e' \cong \frac{25 \text{ mV}}{I_E} \cong \frac{25 \text{ mV}}{1 \text{ mA}} = 25 \text{ }\Omega$$

In this case the ac load $r_L \cong R_C \| R_L = 12$ k$\Omega \| 28$ k$\Omega = 8.4$ kΩ, where $R_C \| R_L$ means that these resistances are in parallel, i.e., $R_C \| R_L = R_C R_L / (R_C + R_L)$. Therefore

$$A_e \cong \frac{r_L}{r_e'} = \frac{8400}{25} = 336$$

Since $A_i \cong \beta = 100$, then

$$A_p = A_e A_i \cong 336(100) = 33{,}600$$

Assuming that the reactance of C_1 is negligible, the input $v_s \cong v_{be} \cong 20$ μV, and by rearranging Eq. 8-45, we find the output voltage:

$$v_o = A_e v_{be} \cong 336(20 \text{ }\mu\text{V}) = 6.72 \text{ mV}$$

Note in the equivalent circuit of Fig. 8-30C that the source v_s works into resistances R_B/A_e and $\beta r_e'$ in parallel. Thus

$$r_{in} = \frac{R_B}{A_e} \;\Big\|\; \beta r_e' = \frac{1.2 \text{ M}\Omega}{336} \;\Big\|\; 100(25) \cong 3.57 \text{ k}\Omega \| 2.5 \text{ k}\Omega = 1.47 \text{ k}\Omega$$

Example 8-37 Suppose that we have a signal source with an open-circuit voltage of 50 μV and a high internal resistance r_i of 100 kΩ, as shown in Fig. 8-31A. What output voltage is across terminals 1-1 of this source (a) when it is working into an amplifier using emitter feedback as described in Ex. 8-33 but without the bypass C_3, and (b) when working into an amplifier using collector feedback as described in Ex. 8-36?

Answer 8-37 (a) 24 μV; see Fig. 8-26B, (b) 0.725 μV; see Fig. 8-31C.—In part (a) we are working with a circuit like that of Fig. 8-28A without the emitter bypass whose equivalent circuit therefore is shown in Fig. 8-28E. Thus, the source works into a load as shown in Fig. 8-31B, where $R_B = 1.1$ MΩ and

8-11 GAIN WITH THE AMPLIFIER USING COLLECTOR FEEDBACK

(A) Signal source with 50-μV open-circuit voltage and 100-kΩ internal resistance.

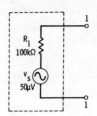

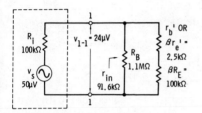

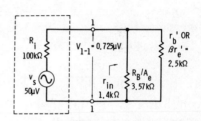

(B) High-resistance source working into circuit described in Ex. 8-33 without C_3.

(C) High-resistance source working into amplifier with collector feedback described in Ex. 8-36.

Fig. 8-31. Loading effect of emitter- and collector-feedback amplifiers.

$$r_b' + \beta R_E \cong \beta r_e' + \beta R_E = \beta(r_e' + R_E)$$

$$= 50(25 + 2000) \cong 100{,}000 \; \Omega$$

From the point of view of the source, these two resistances (R_B and $r_b' + (\beta R_E)$) are in parallel and have a total resistance of r_{in} of 1.1 MΩ‖100 kΩ ≅ 91.6 kΩ. The voltage across this r_{in} is the terminal voltage $v_{1\text{-}1}$ of the source when working into this amplifier. Since r_i and r_{in} are in series,

$$v_{1\text{-}1} = \frac{v_s r_{\text{in}}}{r_{\text{in}} + r_i} \cong \frac{(50 \; \mu\text{V})(91.6 \; \text{k}\Omega)}{91.6 \; \text{k}\Omega + 100 \; \text{k}\Omega} \cong 24 \; \mu\text{V}$$

In part (b), the amplifier input resistance $r_{\text{in}} \cong 1.47$ kΩ as was calculated in Ex. 8-36. In this case, then, the r_i of the source is in series with 1.47 kΩ, as shown in Fig. 8-31C. Therefore

$$v_{1\text{-}1} = \frac{v_s r_{\text{in}}}{r_{\text{in}} + r_i} \cong \frac{(50 \; \mu\text{V})(1.47 \; \text{k}\Omega)}{1.47 \; \text{k}\Omega + 100 \; \text{k}\Omega} \cong 0.725 \; \mu\text{V}$$

Since the output voltage $v_{1\text{-}1}$ of this high-resistance source is much smaller when working into the amplifier with collector feedback, we may say that collector feedback causes more loading of the signal source than does emitter feedback.

The amplifier with collector feedback does have advantages, though. It is capable of good voltage gain if driven by a signal source with low internal resistance. Its operating point is fairly stable with large changes in leakage I_{CEO}, as was pointed out in a previous section, and also with large changes in β. In Ex. 8-36 the circuit has a quiescent $I_C \cong 1$ mA with a transistor whose $\beta = 100$. Suppose the transistor is replaced with one having a $\beta = 200$, an increase of 100 percent, what is the new I_C? The new quiescent collector current is, by Eq. 8-17,

$$I_C \cong \frac{V_{CC}}{R_C + R_B/\beta} = \frac{24 \text{ V}}{12 \text{ k}\Omega + 1200 \text{ k}\Omega/200} \cong 1.33 \text{ mA}$$

which is a 33 percent increase. From this we can conclude that if the transistor in a circuit like that of Fig. 8-30A is replaced with one having a much larger or smaller β, the change in the operating point is not as large.

8-12 GAIN WITH OTHER COMMON-EMITTER AMPLIFIERS

The amplifier using an emitter resistor R_E and two dc power supplies is shown in Fig. 8-32A. If R_E is bypassed with C_3, whose reactance is negligible at the frequency of the signal source v_s, this amplifier ac equivalent circuit may be shown as in Fig. 8-32B. Without the bypass C_3 the circuit in Fig. 8-32C is this amplifier's ac equivalent.

Notice that both of these ac equivalents are exactly the same as those in Figs. 8-28C and 8-28E. Therefore the voltage gain equations that apply to the amplifier of Fig. 8-28A also apply to the one in Fig. 8-32A. That is,

$$A_e \cong \frac{r_L}{r_e{}'}$$

if R_E is bypassed, and

$$A_e \cong \frac{r_L}{R_E} \qquad (8\text{-}51)$$

if R_E is not bypassed.

8-12 Gain With Other Common-Emitter Amplifiers

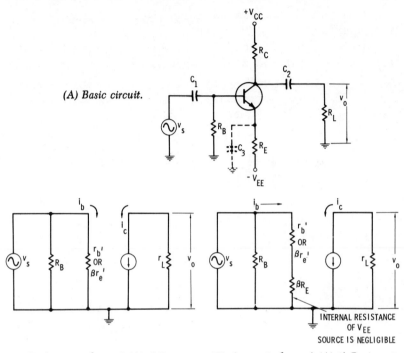

Fig. 8-32. Amplifier with emitter resistor and two dc voltage sources.

This amplifier (Fig. 8-32A) has excellent stability of its operating point not only with large changes in I_{CEO} but also with large changes in β. In general, use of smaller R_B and larger R_E improves the stability. However, if R_B is too small, it tends to load down the signal source driving the amplifier.

The circuit using an emitter resistor and obtaining base bias with a voltage divider, as shown in Fig. 8-33A, makes an amplifier with excellent stability. When R_E is bypassed with C_3, this circuit ac equivalent may be shown as in Fig. 8-33B. Without C_3, the ac equivalent is the circuit of Fig. 8-33C. Note that ac sees R_1 across the base lead and ground through the source V_{CC}, assuming that V_{CC}'s internal resistance is negligible. Source v_s therefore sees R_1 and R_2 in parallel and their total resistance may be treated as R_B in Figs. 8-28C and 8-28E. Thus, the voltage gain equations that applied to the circuits in Fig. 8-28 also apply to the circuits in

Fig. 8-33. Once again for voltage gain you can use Eq. 7-9 if R_E is bypassed and Eq. 8-51 if it is not.

In general, the circuit in Fig. 8-33A is made more stable by use of a smaller R_1 and R_2 and larger R_E but, as with the ampli-

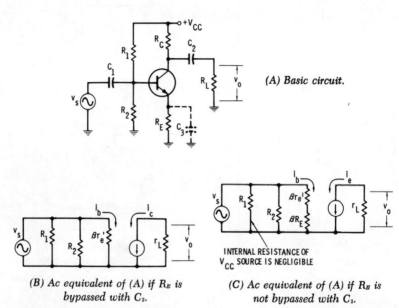

(A) Basic circuit.

(B) Ac equivalent of (A) if R_E is bypassed with C_3.

(C) Ac equivalent of (A) if R_E is not bypassed with C_3.

Fig. 8-33. Amplifier with emitter resistor and base bias with voltage divider.

fier with two supplies, if R_1 and R_2 are too small, which is like a small value of R_B in Fig. 8-32A, they tend to load down the signal source. If R_E is too large, the maximum unclipped output signal capability is reduced.

Example 8-38 If the circuit in Fig. 8-32A has $R_B = 30$ kΩ, $R_C = 20$ kΩ, $R_E = 10$ kΩ, $R_L = 80$ kΩ, $V_{CC} = 30$ V, $V_{EE} = 10$ V, the transistor $\beta = 100$, and $v_s = 100$ μV (rms), (a) find values A_e, A_p, and v_o when R_E is bypassed with C_3, and (b) find these values when R_E is not bypassed.

Answer 8-38 (a) $A_e \cong 640$, $A_p \cong 64{,}000$, $v_o \cong 64$ mV (rms), (b) $A_e \cong 1.6$, $A_p \cong 160$, $v_o \cong 160$ μV (rms).—In part (a), R_E is bypassed and Eq. 7-9 is applicable for gain. Thus, since $r_L \cong R_C \| R_L = 20$ kΩ $\|$ 80 kΩ $= 16$ kΩ, and by Eq. 8-25 the quiescent emitter current is $I_E \cong V_{EE}/R_E = 10$ V/10 kΩ $= 1$ mA. Then

8-12 GAIN WITH OTHER COMMON-EMITTER AMPLIFIERS

$$r_e' \cong \frac{25 \text{ mV}}{I_E} \cong \frac{25 \text{ mV}}{1 \text{ mA}} = 25 \text{ }\Omega$$

and

$$A_e \cong \frac{r_L}{r_e'} = \frac{16{,}000}{25} = 640$$

also

$$A_p = A_e A_i \cong A_e \beta \cong 640(100) = 64{,}000$$

and finally

$$v_o = A_e v_s \cong 640(100 \text{ }\mu\text{V}) = 64 \text{ mV (rms)}$$

In part (b), with R_E not bypassed, Eq. 8-51 is applicable and therefore

$$A_e \cong \frac{r_L}{R_E} = \frac{16 \text{ k}\Omega}{10 \text{ k}\Omega} = 1.6$$

and

$$A_p = A_e A_i = A_e \beta \cong 1.6(100) = 160$$

and finally

$$v_o = A_e v_s \cong 1.6(100 \text{ }\mu\text{V}) = 160 \text{ }\mu\text{V (rms)}$$

Example 8-39 If the circuit in Fig. 8-33A has $R_1 = 140$ kΩ, $R_2 = 10$ kΩ, $R_C = 18$ kΩ, $R_E = 2$ kΩ, $R_L = 72$ kΩ, $V_{CC} = 30$ V, $v_s = 80$ μV (rms) and the transistor $\beta = 100$, find the values A_e and v_o when (a) R_E is bypassed, and when (b) R_E is not bypassed.

Answer 8-39 (a) $A_e \cong 576$, $v_o \cong 46$ mV (rms), (b) $A_e \cong 7.2$, $v_o \cong 576$ μV (rms).—In this case $r_L \cong 18$ k$\Omega \| 72$ k$\Omega = 14.4$ kΩ. In this circuit, the quiescent collector current is found with Eq. 8-35:

$$I_C \cong \frac{R_2}{R_1 + R_2}\left(\frac{V_{CC}}{R_E}\right) = \frac{10}{150}\left(\frac{30 \text{ V}}{2 \text{ k}\Omega}\right) = 1 \text{ mA}$$

Thus,

$$r_e' \cong \frac{25 \text{ mV}}{1 \text{ mA}} = 25 \text{ }\Omega$$

And since Eq. 7-9 applies if R_E is bypassed,

$$A_e \cong \frac{14{,}400}{25} = 576$$

giving us an output

$$v_o \cong 576(80~\mu V) \cong 46~\text{mV (rms)}$$

(b) Without the bypass, Eq. 8-51 applies and

$$A_e \cong \frac{14.4~\text{k}\Omega}{2~\text{k}\Omega} = 7.2$$

and the output

$$v_o \cong 7.2(80~\mu V) = 576~\mu V~\text{(rms)}$$

While the gain factors obtained in the preceding examples were all different values, you should not come to any general conclusions that certain types of common-emitter bias arrangements are inherently better for gain than other common-emitter bias arrangements. The voltage gain A_e is largely dependent on the value of ac load r_L. In general, larger r_L's give larger voltage gains. The method of bias arrangement used depends on the required stability of the circuit and other related factors such as economy of construction, tolerances of components, operating environment, etc.

8-13 AC LOAD LINES

As with the common-base amplifier, the common-emitter amplifier ac load line is very useful in showing us the maximum output signal capabilities. When sketching the ac load line, the dc load line and the operating point Q are drawn first. The ac load line must intersect the operating point Q. If the ac and dc loads seen from the collector and emitter leads of the transistor are different, the slopes (steepnesses) of ac and dc loads are different. As in previous cases, the dc load line crosses the vertical and horizontal axes at $I_{C(\text{sat})}$ and $V_{CE(\text{cutoff})}$ respectively. The ac load line crosses these axes at points we can identify as $i_{C(\text{sat})}$ and $v_{CE(\text{cutoff})}$. See Fig. 8-34. The locations of these ac saturation and cutoff points are determined by the values of the ac load r_L, the quiescent I_C, and the quiescent collector-to-emitter voltage which is referred to as V_C. We can find these points with the following equations:

8-13 AC Load Lines

$$i_{C(\text{sat})} = I_C + \frac{V_C}{r_L} \quad (8\text{-}53)$$

$$v_{CE(\text{cutoff})} = V_C + r_L I_C \quad (8\text{-}54)$$

where
 I_C is the collector current at the operating point,
 V_C is the collector-to-emitter voltage at the operating point,
 r_L is the ac load resistance.

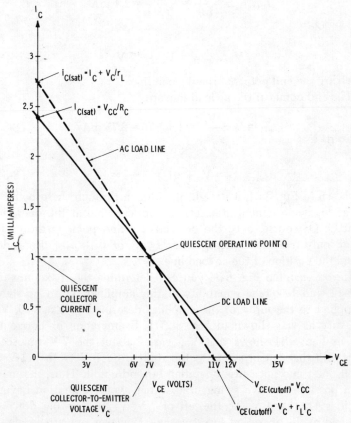

Fig. 8-34. Dc and ac load lines for the circuit described in Exs. 8-32 and 8-40.

Example 8-40 If the circuit shown in Fig. 8-26A has component values $R_B = 1.2$ MΩ, $R_C = 5$ kΩ, $R_L = 20$ kΩ, $V_{CC} = 12$ V, $v_s = 5$ mV (rms) and the transistor $\beta = 100$, sketch the dc and ac load lines and identify the operating point Q, and the saturation and cutoff points on both load lines. Assume that the reactances of both capacitors are negligible.

Answer 8-40 See Fig. 8-34: $I_C \cong 1$ mA, $V_C \cong 7$ V, $I_{C(\text{sat})} \cong 2.4$ μA, $i_{C(\text{sat})} \cong 2.75$ mA, $V_{CE(\text{cutoff})} \cong 12$ V, $v_{CE(\text{cutoff})} \cong 11$ V.—The work for I_C and r_L is shown in the answer to Ex. 8-32. At the operating point Q, $V_C = V_{CE} = V_{CC} - R_C I_C = 12 - 5(1) = 7$ V, by Eq. 8-20. Since the circuit is a simple base biased type, by Eq. 8-21,

$$I_{C(\text{sat})} \cong \frac{V_{CC}}{R_C} = \frac{12 \text{ V}}{5 \text{ k}\Omega} = 2.4 \text{ mA}$$

and DC

$$V_{CE(\text{cutoff})} = V_{CC} = 12 \text{ V}$$

which are the end points of the dc load line.

The end points of the ac load line are

$$i_{C(\text{sat})} = I_C + \frac{V_C}{r_L} = 1 + 1.75 = 2.75 \text{ mA} \tag{8-53}$$

and AC

$$v_{CE(\text{cutoff})} = V_C + r_L I_C = 7 + 4 = 11 \text{ V} \tag{8-54}$$

As shown in Fig. 8-34, a straight line drawn through these points is the ac load line which intersects the dc load line at the operating point Q. Of course, once the dc load line and point Q are established, only one more point, *either* $i_{C(\text{sat})}$ or $v_{CE(\text{cutoff})}$, is needed to find the position of the ac load line.

If you examine Fig. 8-34 you can determine the maximum unclipped signal voltage capability of this amplifier. If an ac signal is applied to the input of the amplifier described in Ex. 8-40, V_{CE} will vary as was shown in Fig. 8-27H. Examination of horizontal axis in Fig. 8-34 shows that V_{CE} varies about the 7 V quiescent value and that it can swing to the right from 7 V to the 11-V ac cutoff point and not cause operation beyond cutoff. Thus, the maximum possible positive swing of V_{CE} that avoids operation beyond cutoff is $11 - 7 = 4$ V. To the left of 7 V, V_{CE} can swing down to 0 V and avoid operation in saturation. However, when the circuit is amplifying symmetrical signals like sine waves, V_{CE} will swing almost equally to the right and then to the left of 7 V on every cycle of the input signal. Therefore, to avoid driving the collector into saturation, in this case, the output signal v_{ce} should not have a peak value greater than 4 V.

8-13 AC LOAD LINES

Example 8-41 Referring to the amplifier in Ex. 8-40, whose gain $A_e \cong 160$, what is the maximum peak value of the input v_s that will not cause clipping of the output signal? Also draw the output voltage waveforms if $v_{s(\text{peak})} = 40$ mV and if $v_{s(\text{peak})} = 50$ mV.

Answer 8-41 Maximum $v_{s(\text{peak})} \cong 25$ mV. See output waveforms (b) and (c) in Fig. 8-35 with inputs 40 mV(peak) and 50 mV(peak) respectively.—The largest unclipped output signal v_{ce} is shown as (a) in Fig. 8-35 with a peak value of 4 V. Since the stage gain is about 160,

$$v_s \cong \frac{v_{ce}}{A_e} = \frac{4 \text{ V}}{160} = 25 \text{ mV(peak)}$$

When $v_{s(\text{peak})} = 40$ mV, we expect the output signal voltage to have a peak 160 times larger; that is,

$$v_{ce(\text{peak})} \cong 160(40 \text{ mV}) = 6.4 \text{ V}$$

However, as shown as (b) in Fig. 8-35, this signal drives the circuit into cutoff for a portion of half the alternations, thus clipping them at a peak of 4 V.

When $v_{s(\text{peak})} = 50$ mV, we similarly expect 160 times this at the output; that is,

$$v_{ce(\text{peak})} \cong 160(50 \text{ mV}) = 8 \text{ V}$$

However, as shown as (c) in Fig. 8-35, this signal drives the circuit into cutoff for a portion of half the alternations and into saturation for a portion of the other alternations. Thus, the output signal, as viewed across R_L, has half of the alternations clipped at 7 V from a no-signal condition and the other half are clipped in the other direction at 4 V from a no-signal condition.

With resistance in the emitter lead as in Fig. 8-28A, the value of R_E affects the slope of the dc load line and of course the position of the operating point Q. The slope of the ac load line depends on whether R_E is bypassed or not. Methods of finding the load lines can be shown by referring back to some specific circuit, like the one described in Ex. 8-33; that is, with the circuit of Fig. 8-28A in which $V_{CC} = 24$ V, $R_B = 1.1$ MΩ, $R_C = 10$ kΩ, $R_E = 2$ kΩ, $R_L = 40$ kΩ, and the transistor $\beta = 50$. As previously, we solve for points on the dc load line first, and by Eq. 8-15 we find that

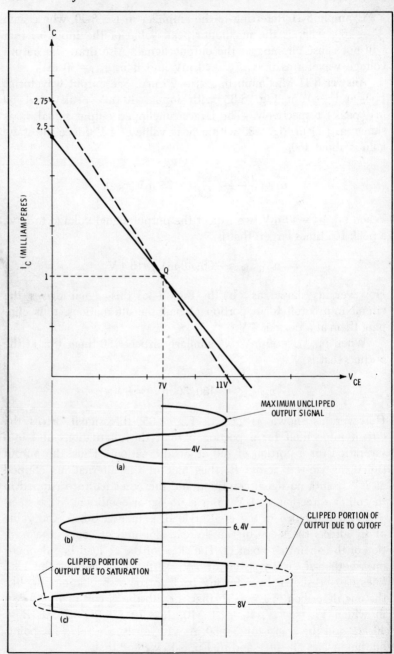

Fig. 8-35. Illustrating answers to Ex. 8-41.

8-13 AC LOAD LINES

$$I_{C(sat)} \cong \frac{V_{CC}}{R_C+R_E} = \frac{24\text{ V}}{10\text{ k}\Omega+2\text{ k}\Omega} = 2\text{ mA}$$

And since

$$V_{CC} = V_{CE(cutoff)} = 24\text{ V}$$

the dc load is drawn as shown in Fig. 8-36B. With Eq. 8-11 we find that at the operating point, $I_C \cong 1$ mA, as shown in Ex. 8-33.

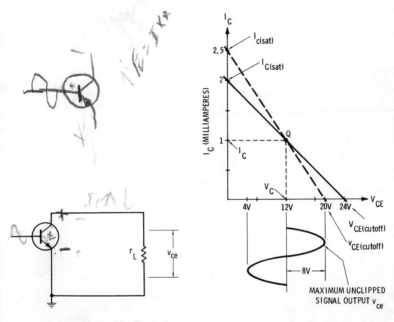

(A) *Transistor sees ac load r_L when R_E is bypassed.*

(B) *Dc and ac load lines for the circuit.*

Fig. 8-36. Dc and ac load lines for the circuit described in Ex. 8-33. with R_E bypassed.

Since point Q is centered on the load line we would expect to find V_C at 12 V, that is, centered between 0 V and 24 V on the V_{CE} axis. Mathematically, we can verify this with Eq. 8-14 and show that at point Q

$$V_C = V_{CE} = V_{CC} - I_C(R_C+R_E) = 24 - 1(12) = 12\text{ V}$$

The ac load line must go through point Q. Its exact slope can be found now by solving for one more point on the ac load line with either Eq. 8-53 or Eq. 8-54 and by drawing a straight line through it and point Q. For complete clarity, the points by both equations are solved here as follows if R_E is bypassed with C_3. Since the transistor sees an ac load $r_L = R_C \| R_L = 8$ kΩ, then

$$i_{C(\text{sat})} = I_C + \frac{V_C}{r_L} = 1 \text{ mA} + \frac{12 \text{ V}}{8 \text{ k}\Omega} = 2.5 \text{ mA}$$

and

$$v_{CE(\text{cutoff})} = V_C + r_L I_C = 12 + 8(1) = 20 \text{ V}$$

Note in Fig. 8-36 that since $V_C = 12$ V and $v_{CE(\text{sat})} = 20$ V, the maximum unclipped collector-to-emitter signal v_{ce} has a peak value of 8 V.

If R_E is not bypassed, it becomes part of the ac load as seen by the transistor. As shown in Fig. 8-32A, the transistor sees r_L and R_E in series, whose sum we can call r_L'. Thus, in this case, the transistor sees a total ac load of

$$r_L' = r_L + R_E = 8 \text{ k}\Omega + 2 \text{ k}\Omega = 10 \text{ k}\Omega$$

and Eq. 8-53 and 8-54 are modified to

$$i_{C(\text{sat})} = I_C + \frac{V_C}{r_L'} \qquad (8\text{-}55)$$

$$v_{CE(\text{cutoff})} = V_C + r_L' I_C \qquad (8\text{-}56)$$

Thus, in this case, the end points of the ac load line are

$$i_{C(\text{sat})} = 1 \text{ mA} + \frac{12 \text{ V}}{10 \text{ k}\Omega} = 2.2 \text{ mA}$$

$$v_{CE(\text{cutoff})} = 12 + (10)1 = 22 \text{ V}$$

as shown in Fig. 8-37. Note in this case that the collector-to-emitter signal v_{ce} can have as high as a 10-V peak value without clipping. The maximum unclipped output v_o, however, is considerably less than 10 V peak. As shown in Fig. 8-37A, v_{ce} is not v_o because some signal voltage is dropped across R_E. Since r_L and R_E are in series

$$v_o = \frac{v_{ce} r_L}{r_L + R_E}$$

Therefore, in this case, the maximum unclipped signal output is

$$v_o = \frac{(10 \text{ V})(8 \text{ k}\Omega)}{8 \text{ k}\Omega + 2 \text{ k}\Omega} = 8 \text{ V(peak)}$$

8-13 AC LOAD LINES

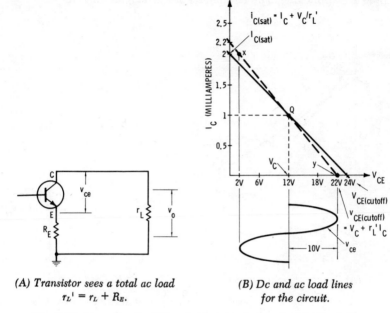

(A) Transistor sees a total ac load $r_L' = r_L + R_E$.

(B) Dc and ac load lines for the circuit.

Fig. 8-37. Dc and ac load lines for the circuit described in Ex. 8-33 without an emitter bypass.

Eq. 8-53 and 8-54 are also applicable to other common-emitter circuits that have bypassed R_E or no R_E at all. Similarly, Eqs. 8-55 and 8-56 are applicable with common-emitter circuits with unbypassed R_E. For instance, refer to the circuit described in Ex. 8-38. The following points are on the dc load line:

$$I_{C(\text{sat})} = \frac{V_{CC}+V_{EE}}{R_C+R_E} = \frac{40 \text{ V}}{30 \text{ k}\Omega} \cong 1.334 \text{ mA}$$

$$V_{CC}+V_{EE}=V_{CE(\text{cutoff})}=40 \text{ V}$$

As shown in Ex. 8-38, at the operating point $I_C \cong 1$ mA. Therefore, by Eq. 8-30,

$$V_C = V_{CE} \cong V_{CC} - R_C I_C = 30 - 20 = 10 \text{ V}$$

The dc load line is drawn through these saturation, operating, and cutoff points as shown in Fig. 8-38. The ac load line in this figure is drawn assuming that R_E is bypassed. Thus, the transistor works into a load r_L alone as shown in Fig. 8-38A and Eqs. 8-53 and 8-54 are applicable. The end points of the ac load line are

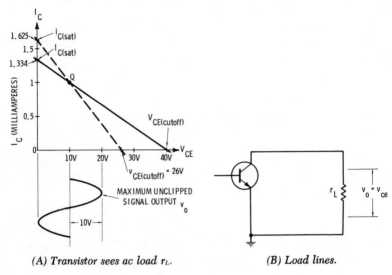

(A) Transistor sees ac load r_L. (B) Load lines.

Fig. 8-38. Dc and ac load lines for the circuit described in Ex. 8-38 with R_E bypassed.

$$i_{C(\text{sat})} \cong 1 \text{ mA} + \frac{10 \text{ V}}{16 \text{ k}\Omega} = 1 + 0.625 = 1.625 \text{ mA}$$

and

$$v_{ce(\text{cutoff})} \cong 10 \text{ V} + (16 \text{ k}\Omega)(1 \text{ mA}) = 26 \text{ V}$$

Note in this case that because the operating point Q is so high on the load line, the maximum unclipped output v_{ce} is limited by the saturation point instead of the cutoff point as in the previous example. The maximum unclipped output has a peak of 10 V as shown.

When R_E is not bypassed, the transistor sees a total ac load $r_L' = r_L + R_E$ as shown in Fig. 8-39A. In this case, then,

$$r_L' = 16 + 10 = 26 \text{ k}\Omega$$

and by Eqs. 8-55 and 8-56 the end points on the ac load line are

$$i_{C(\text{sat})} = I_C + \frac{V_C}{r_L'} = 1 + 0.385 = 1.385 \text{ mA}$$

8-13 AC LOAD LINES

and

$$v_{CE(\text{cutoff})} = V_C + r_L{}'I_C = 10 + 26 = 36 \text{ V}$$

Note that again the maximum unclipped signal voltage across the collector and emitter v_{ce} has a peak value of 10 V. However, due to the drop across R_E, the output v_o does not have a 10 V peak. Since the transistor sees r_L and R_E in series,

$$v_o = \frac{v_{ce}r_L}{r_L + R_E} = \frac{(10 \text{ V})(16 \text{ k}\Omega)}{16 \text{ k}\Omega + 10 \text{ k}\Omega} = 6.15 \text{ V}(\text{peak})$$

It is interesting to observe in this last equation that if R_E is increased, say to improve stability, the output v_o drops off.

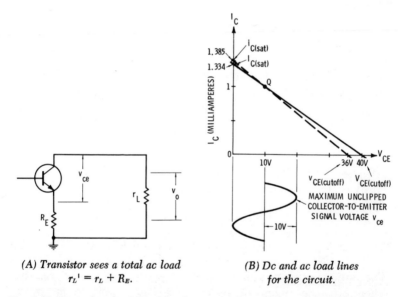

(A) Transistor sees a total ac load $r_L{}' = r_L + R_E$.

(B) Dc and ac load lines for the circuit.

Fig. 8-39. Dc and ac load lines for the circuit described in Exs. 8-38 and 8-42 without an emitter bypass.

Example 8-42 In Ex. 8-38, the circuit described (Fig. 8-32A) has component values $R_B = 30$ kΩ, $R_C = 20$ kΩ, $R_E = 10$ kΩ, $R_L = 80$ kΩ, $V_{CC} = 30$ V, $V_{EE} = 10$ V, and the transistor $\beta = 100$. What is the maximum amplitude of input signal $v_{s(\text{peak})}$ that can be applied to this circuit without clipping of the output v_o, if (a) R_E is bypassed and if (b) R_E is not bypassed.

Answer 8-42 (a) $v_{s(\text{peak})} \cong 15.6$ mV, (b) $v_{s(\text{peak})} \cong 6.25$ V.—As shown in Ex. 8-38, the voltage gain $A_e \cong 640$ with the bypass. And since the v_{ce} signal has a peak of 10 V as shown in Fig. 8-38, then

$$v_{s(\text{peak})} = \frac{v_{ce(\text{peak})}}{A_e} \cong \frac{10 \text{ V}}{640} \cong 15.6 \text{ mV}$$

Without the bypass, $A_e \cong 1.6$. Therefore, with the same $v_{ce} \cong 10$ V(peak) capability as shown in Fig. 8-39,

$$v_{s(\text{peak})} = \frac{v_{ce(\text{peak})}}{A_e} \cong \frac{10 \text{ V}}{1.6} = 6.25 \text{ V}$$

It is interesting to note here that if the stage gain is high, the input signal must be small if clipping of the output waveform is to be avoided.

Example 8-43 Referring to the circuit described in Ex. 8-39, which is the circuit of Fig. 8-33A, with the following component values: $R_1 = 140$ kΩ, $R_2 = 10$ kΩ, $R_C = 18$ kΩ, $R_E = 2$ kΩ, $R_L = 72$ kΩ, and $V_{CC} = 30$ V. Sketch the dc and ac load lines of this circuit, determine the peak value of the largest unclipped output $v_{o(\text{peak})}$, and determine the peak value of the largest input $v_{s(\text{peak})}$ that will not cause clipping, when (a) R_E is bypassed, and when (b) R_E is not bypassed.

Answer 8-43 (a) See Fig. 8-40A, $v_o = 10$ V (peak), $v_s \cong 17.35$ mV (peak), (b) See Fig. 8-40B, $v_o \cong 8.78$ V (peak), $v_s \cong 1.22$ V (peak).—The end points of the dc load line are

$$V_{CE(\text{cutoff})} = V_{CC} = 30 \text{ V}$$

$$I_{C(\text{sat})} \cong \frac{V_{CC}}{R_C + R_E} = \frac{30 \text{ V}}{20 \text{ kΩ}} = 1.5 \text{ mA}$$

The quiescent $I_C \cong 1$ mA, which was determined in Ex. 8-39. Therefore, the quiescent collector-to-emitter voltage is, by Eq. 8-39,

$$V_C = V_{CE} \cong 30 \text{ V} - (18 \text{ kΩ} + 2 \text{ kΩ})(1 \text{ mA}) = 10 \text{ V}$$

(a) With R_E bypassed the transistor works into a load $r_L = 14.4$ kΩ. Thus, the end points of the ac load line are, by Eqs. 8-54 and 8-53,

$$v_{CE(\text{cutoff})} = V_C + r_L I_C = 10 + 14.4(1) = 24.4 \text{ V}$$

and

8-13 AC LOAD LINES

$$i_{C(\text{sat})} = I_C + \frac{V_C}{r_L} = 1 \text{ mA} + \frac{10 \text{ V}}{14.4 \text{ k}\Omega} = 1.695 \text{ mA}$$

As shown in Fig. 8-40A, the V_{CE} voltage can swing to the left by as much as 10 V before the transistor goes into saturation. Therefor v_o, which is v_{ce} in this case, has an unclipped peak capability of 10 V. Since this circuit $A_e \cong 576$, (Ex. 8-39) and $A_e = v_o/v_s$, then

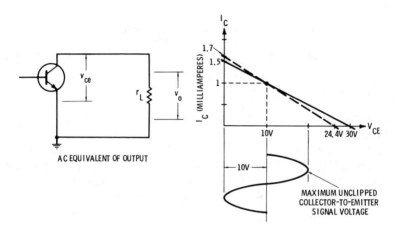

(A) For circuit in part (a) of Ex. 8-43.

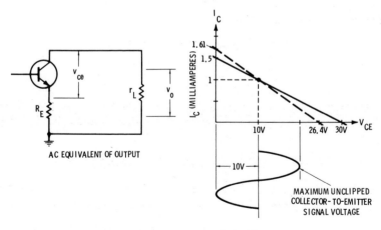

(B) For circuit in part (b) of Ex. 8-43.

Fig. 8-40. Dc and ac load lines.

$$v_s = \frac{v_o}{A_e} \cong \frac{10 \text{ V}}{576} \cong 17.35 \text{ mV (peak)}$$

(b) When R_E is not bypassed, the transistor works into a load $r_L' = r_L + R_E = 14.4 + 2 = 16.4$ kΩ; see the equivalent circuit in Fig. 8-40B. The end points on the ac load line in this case are

$$v_{CE(\text{cutoff})} = V_C + r_L' I_C = 10 + 16.4(1) = 26.4 \text{ V}$$

by Eq. 8-56, and, by Eq. 8-55,

$$i_{C(\text{sat})} = I_C + \frac{V_C}{r_L'} \cong 1 \text{ mA} + \frac{10 \text{ V}}{16.4 \text{ kΩ}} \cong 1.61 \text{ mA}$$

As in part (a), the maximum $v_{ce} = 10$ V (peak) if clipping is to be avoided. However, in this case v_{ce} is not the output v_o because some signal is dropped across R_E. Therefore

$$v_o = \frac{v_{ce} r_L}{r_L + R_E} \cong \frac{(10 \text{ V})(14.4)}{16.4} \cong 8.78 \text{ V(peak)}$$

As shown in Ex. 8-39, the voltage gain $A_e \cong 7.2$, which means that the maximum input is

$$v_s = \frac{v_o}{A_e} = \frac{8.78 \text{ V}}{7.2} = 1.22 \text{ V(peak)}$$

8-14 THE OPTIMUM OPERATING POINT

As in the common-base amplifier covered in Sec. 7-9, the maximum unclipped output signal capability of a common-emitter amplifier is achieved if the operating point is centered on the ac load line. If this is done, the amplifier is said to be biased at its *optimum operating point*. For common-emitter amplifiers whose emitters are directly grounded as in the simple base biased circuit of Fig. 8-26A, or as in the base biased circuit with collector feedback (Fig. 8-30A), the coordinates of the optimum operating point are found with Eqs. 7-21 and 7-18 given in the last chapter:

$$I_C = \frac{V_{CC}}{R_C + r_L}$$

and

$$V_C = I_C r_L$$

In common-emitter amplifiers that have resistors in the emitter lead, the presence of them affects the values of $I_{C(\text{sat})}$ and I_C. For

8-14 THE OPTIMUM OPERATING POINT

example, compare Eqs. 8-15 and 8-21. When no emitter resistance is present, by Eq. 8-21,

$$I_{C(sat)} \cong \frac{V_{CC}}{R_C}$$

And when emitter resistance R_E and only one source V_{CC} are used, Eq. 8-15 applies:

$$I_{C(sat)} \cong \frac{V_{CC}}{R_C + R_E}$$

Note that if R_E is added to the circuit, its value is simply added to the collector resistance R_C to give the total resistance that limits the dc current. Thus, in a similar way, Eq. 7-21 is modified by adding R_E to R_C. In other words, if an amplifier has R_E but it is bypassed, the coordinates of the optimum operating point are

$$I_C = \frac{V_{CC}}{R_C + R_E + r_L} \tag{8-57}$$

and, by Eq. 7-18,

$$V_C = r_L I_C$$

As previously shown, the transistor sees a total ac load $r_L' = r_L + R_E$ when R_E is not bypassed. Therefore, with the emitter not bypassed, the coordinates of the circuit optimum operating point can be found with Eqs. 8-57 and 7-18 above if r_L is replaced with r_L'. That is, if R_E is not bypassed, at the optimum operating point

$$I_C = \frac{V_{CC}}{R_C + R_E + r_L'} \tag{8-58}$$

and

$$V_C = r_L' I_C \tag{8-59}$$

Example 8-44 If in the simple base biased circuit of Fig. 8-26A, $R_C = 5$ kΩ, $R_L = 20$ kΩ, $V_{CC} = 12$ V, and the transistor $\beta = 100$, find the coordinates of the optimum operating point, the peak value of the maximum unclipped output signal v_{ce}, and the value of R_B that you would use to achieve this optimum point.

Answer 8-44 $I_C \cong 1.333$ mA, $V_C \cong 5.33$ V(peak), $R_B \cong 900$ kΩ. —Since the emitter is directly grounded in this case, the transistor works into an ac load $r_L = R_C \| R_L = 4$ kΩ and Eqs. 7-21 and 7-18

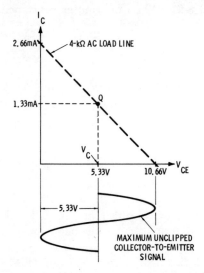

Fig. 8-41. Ac load line with optimum operating point Q.

are applicable. Thus, at the optimum operating point, Eq. 7-21 is used:

$$I_C = \frac{V_{CC}}{R_C + r_L} = \frac{12 \text{ V}}{5 \text{ k}\Omega + 4 \text{ k}\Omega} \cong 1.333 \text{ mA}$$

and, by Eq. 7-18,

$$V_C = r_L I_C \cong 4(1.333) \cong 5.33 \text{ V}$$

Note in Fig. 8-41 that since the optimum operating point Q is centered on the ac load line, V_C is centered between 0 V and, by Eq. 8-54,

$$v_{CE(\text{cutoff})} = V_C + r_L I_C \cong 5.33 + 4(1.333) \cong 10.66 \text{ V}$$

Thus, the signal can cause V_{CE} variations about the quiescent $V_C =$ 5.33 V by a peak of 5.33 V and not cause operation in cutoff or saturation.

Since we want $I_C \cong 1.333$ mA, and the transistor $\beta = 100$, we need a quiescent base current

$$I_B = \frac{I_C}{\beta} \cong 13.33 \text{ }\mu\text{A}$$

and therefore by Eq. 8-8

$$R_B \cong \frac{V_{CC}}{I_B} \cong \frac{12 \text{ V}}{13.33 \text{ }\mu\text{A}} \cong 900 \text{ k}\Omega$$

8-14 THE OPTIMUM OPERATING POINT

Example 8-45 If the circuit of Fig. 8-33A, $R_2 = 10$ kΩ, $R_C = 18$ kΩ, $R_E = 2$ kΩ, $R_L = 72$ kΩ, and $V_{CC} = 30$ V, as in the circuit described in Ex. 8-39, what are the coordinates of the optimum operating point and what is the peak value of the largest possible unclipped output signal for each of the following cases: (a) R_E is bypassed, (b) R_E is not bypassed.

Answer 8-45 (a) $I_C \cong 0.873$ mA, $V_C \cong 12.6$ V, $v_o = 12.6$ V (peak), (b) $I_C \cong 0.825$ mA, $V_C \cong 13.5$ V, $v_o \cong 11.85$ V(peak).— In part (a) with R_E bypassed, Eqs. 8-57 and 7-18 are applicable and at the optimum operating point Q

$$I_C \cong \frac{30 \text{ V}}{18 \text{ k}\Omega + 2 \text{ k}\Omega + 14.4 \text{ k}\Omega} \cong 0.873 \text{ mA}$$

by Eq. 8-57, and

$$V_C \cong 14.4(0.873) \cong 12.6 \text{ V}$$

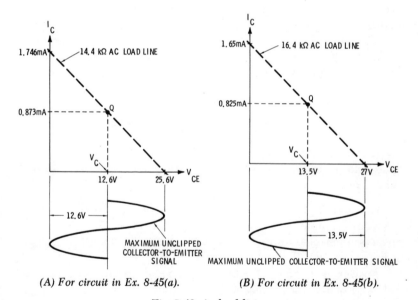

(A) For circuit in Ex. 8-45(a).   (B) For circuit in Ex. 8-45(b).

Fig. 8-42. Ac load lines.

by Eq. 7-18. This point is shown on a 14.4 kΩ load line in Fig. 8-42A. Notice that the optimum operating point Q is centered on the load line and that therefore V_{CE} can swing to either side of V_C by as much as 12.6 V and not cause clipping of the output signal. Compare this with the maximum peak output achieved with this circuit in Ex. 8-43 in which the operating point was not optimum.

In part (b) with R_E unbypassed, Eqs. 8-58 and 8-59 are applicable. Since the transistor sees an ac load $r_L' = r_L + R_E = 16.4$ kΩ in this case, at the optimum operating point Q

$$I_C \cong \frac{30 \text{ V}}{18 \text{ k}\Omega + 2 \text{ k}\Omega + 16.4 \text{ k}\Omega} \cong 0.825 \text{ mA}$$

and

$$V_C \cong 16.4(0.825) \cong 13.5 \text{ V}$$

This point is shown on a 16.4 kΩ load line in Fig. 8-42B. In this case the signal can vary about the quiescent $V_C = 13.5$ V value by a peak value of 13.5 V. Obviously, we can see that quiescent collector-to-emitter voltage value is equal to the peak value of the maximum unclipped collector-to-emitter signal voltage if the operating point is optimum.

Of course, due to signal voltage drop across R_E, voltage v_{ce} is not equal to v_o, where v_o is the output voltage across r_L. As we learned in a previous example, r_L and R_E are in series with v_{ce} applied to them. Therefore

$$v_o = \frac{v_{ce} r_L}{r_L + R_E} = \frac{13.5(14.4)}{16.4} \cong 11.85 \text{ V}$$

It may be interesting to verify the end points of the ac load lines shown in Figs. 8-42A and 8-42B with Eqs. 8-53 through 8-56. Try it.

Example 8-46 Again referring to the circuit described in the previous example, if you are to change the value of R_1 to achieve the optimum operating point, what value will you use if (a) R_E is bypassed, (b) R_E is not bypassed?

Answer 8-46 (a) $R_1 \cong 162$ kΩ, (b) $R_1 \cong 172$ kΩ.—Start by solving Eq. 8-35 for R_1. Since

$$I_C \cong \frac{R_2}{R_1 + R_2} \left(\frac{V_{CC}}{R_E} \right)$$

then

$$R_1 \cong R_2 \left(\frac{V_{CC}}{R_E I_C} - 1 \right)$$

8-15 USEFUL "RULE OF THUMB" DESIGN PROCEDURE

In part (a) we know $I_C \cong 0.873$ mA at the optimum operating point from the previous example. Therefore by substituting the knowns into the above equation we get

$$R_1 \cong 10 \text{ k}\Omega \left(\frac{30 \text{ V}}{2(0.873)} - 1 \right) \cong 162 \text{ k}\Omega$$

Similarly in part (b) we know that $I_C \cong 0.825$ mA at the optimum operating point and thus

$$R_1 \cong 10 \text{ k}\Omega \left(\frac{30 \text{ V}}{2(0.825)} - 1 \right) \cong 172 \text{ k}\Omega$$

8-15 USEFUL "RULE OF THUMB" DESIGN PROCEDURE

If you need an amplifier circuit, quick, with good stability and signal output capability, here's a very handy "rule of thumb" design procedure. Start with the circuit type (Fig. 8-33A) and do as follows:

(I) Pick a proper quiescent collector current that is within the transistor capability, typically about 1 mA.

(II) Choose an R_E that will drop *one-tenth* of the V_{CC} voltage.

(III) Choose an R_2 that is *ten times* larger than R_E.

(IV) Choose an R_1 that is *nine times* larger than R_2.

(V) Choose an R_C to drop *one-half* of the remaining source voltage; that is, one-half of the voltage left after one-tenth of V_{CC} is subtracted from V_{CC}.

Example 8-47 By "rule of thumb" procedure, design a common-emitter amplifier that must work from a 20 V dc source. Use a quiescent $I_C = 1$ mA.

Answer 8-47 $R_1 = 180$ kΩ, $R_2 = 20$ kΩ, $R_C = 9$ kΩ, $R_E = 2$ kΩ.— By "rule" (II) if R_E is to drop one-tenth of the dc source V_{CC}, and conduct 1 mA, its value is

$$R_E = \frac{2 \text{ V}}{1 \text{ mA}} = 2 \text{ k}\Omega$$

(III) Since R_2 is to be ten times R_E in this case

$$R_2 = 10 R_E = 10(2 \text{ k}\Omega) = 20 \text{ k}\Omega$$

(IV) And if R_1 is nine times R_2, we have

$$R_1 = 9 R_2 = 9(20 \text{ k}\Omega) = 180 \text{ k}\Omega$$

(V) If 2 V is dropped across R_E then 18 V must be across R_C and the transistor collector-emitter terminals. Resistance R_C is to drop one-half of this, or 9 V. Thus,

$$R_C = \frac{9 \text{ V}}{1 \text{ mA}} = 9 \text{ k}\Omega$$

Of course, if these exact resistance values are not available, you can use nearest available resistance values.

PROBLEMS (PART II)

8-37. Find the approximate values of I_E and dynamic resistance seen by the source v_s in the circuit of Fig. 8-43 with each of the following conditions: (a) $R=0$, (b) $R=1$ MΩ, (c) $R=3$ MΩ, and (d) $R=7$ MΩ. Assume that the reactance of C is negligible.

8-38. Find the approximate values of I_E and dynamic resistance seen by the source v_s in Fig. 8-43 for each of the following conditions: (a) $R=600$ kΩ, (b) $R=2$ MΩ, (c) $R=5.4$ MΩ, (d) $R=11.8$ MΩ. Assume that the reactance of C is negligible.

8-39. Source v_s is sinusoidal in the circuit of Fig. 8-44. If the transistor has $h_{FE} \cong h_{fe} \cong 80$, find the approximate (a) voltage gain, (b) current gain, (c) power gain, and (d) peak value of the signal output v_{ce} of this circuit.

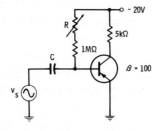

Fig. 8-43. Circuit for Probs. 8-37 and 8-38.

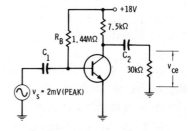

Fig. 8-44. Circuit for Probs. 8-39, 8-40, 8-65, and 8-67.

8-40. If the signal source v_s is sinusoidal in the circuit of Fig. 8-44 and the transistor $h_{FE} \cong h_{fe} \cong 120$, what are the approximate values of (a) voltage gain, (b) current gain, (c) power gain, and (d) the peak value of the output signal v_{ce} of this circuit?

PROBLEMS (PART II)

8-41. If switch S is closed in Fig. 8-45, what are (a) the approximate voltage gain A_e and (b) the value of $v_{ce(peak)}$, if $v_{s(peak)} = 1$ mV?

8-42. If switch S is closed in Fig. 8-46, what are (a) the approximate voltage gain A_e and (b) the value of $v_{ce(peak)}$, if $v_{s(peak)} = 1$ mV?

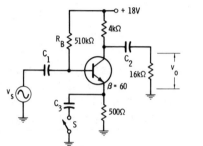

Fig. 8-45. Circuit for Probs. 8-41, 8-43, 8-66, and 8-68.

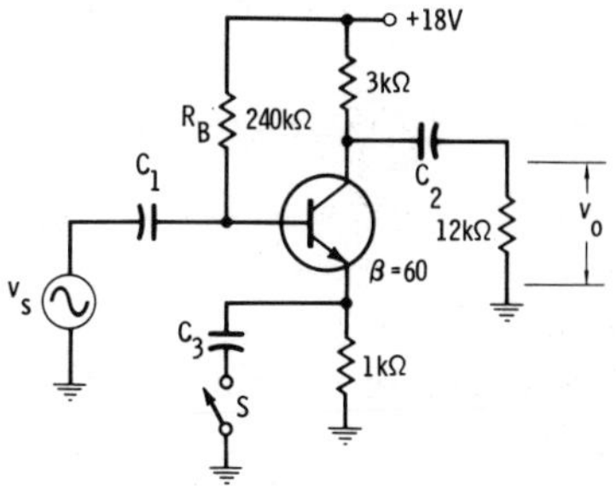

Fig. 8-46. Circuit for Probs. 8-42, 8-44, 8-51, 8-52, 8-69, and 8-71.

8-43. If switch S is open in Fig. 8-45, what are (a) the voltage gain A_e and (b) the value of $v_{ce(peak)}$, if $v_{s(peak)} = 200$ mV?

8-44. If switch S is open in Fig. 8-46, what are (a) the voltage gain and (b) the value of $v_{ce(peak)}$, if $v_{s(peak)} = 200$ mV?

8-45. Suppose that in the circuit of Fig. 8-47, $v_{s(peak)} = 400$ mV, $R_B = 6.6$ MΩ, $R_C = 25$ kΩ, $R_L = 100$ kΩ, $R_1 = 2$ kΩ, $R_2 = 3$ kΩ, and the transistor $\beta = 120$. What is the voltage gain of the circuit and the peak of the output signal $v_{o(peak)}$?

8-46. Suppose that in the circuit of Fig. 8-47, $v_{s(peak)} = 400$ mV, $R_B = 6$ MΩ, $R_C = 20$ kΩ, $R_L = 140$ kΩ, $R_1 = 1.5$ kΩ, $R_2 = 8.5$ kΩ, and the transistor $\beta = 120$. What is the voltage gain of the circuit and the peak of the output signal $v_{o(peak)}$?

8-47. If in the circuit of Fig. 8-48, $v_{s(peak)} = 2$ mV, $R_B = 720$ kΩ, $R_C = 6$ kΩ, $R_L = 24$ kΩ, $V_{CC} = -12$ V, and the transistor $h_{FE} \cong h_{fe} \cong 120$, what is the voltage gain and the peak of the output voltage $v_{o(peak)}$?

8-48. If in the circuit of Fig. 8-48, $v_{s(peak)} = 2$ mV, $R_B = 180$ kΩ, $R_C = 4.5$ kΩ, $R_L = 18$ kΩ, $V_{CC} = -9$ V, and the transistor $\beta = 120$, what is the voltage gain and the peak of the output signal $v_{o(peak)}$?

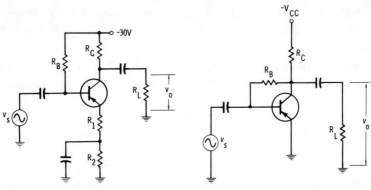

Fig. 8-47. Circuit for Probs. 8-45, 8-46, 8-55, 8-56, 8-70, and 8-72.

Fig. 8-48. Circuit for Probs. 8-48, 8-57, and 8-58.

8-49. In the circuit of Fig. 8-49, if $R_C = 5$ kΩ, $R_L = 20$ kΩ, $R_E = 10$ kΩ, $v_{s(peak)} = 1$ mV, and the transistor $\beta = 100$, what are the values of (a) voltage gain, (b) power gain, (c) dynamic resistance seen by the source v_s, and (d) the peak of the output signal $v_{o(peak)}$?

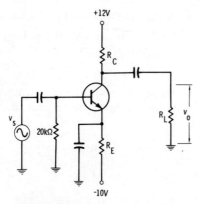

Fig. 8-49. Circuit for Probs. 8-49, 8-50, 8-59, 8-60, 8-73, and 8-74.

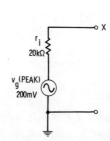

Fig. 8-50. Circuit for Probs. 8-51, 8-52, 8-55, 8-56, 8-57, and 8-58.

8-50. In the circuit of Fig. 8-49, if $R_C = 12$ kΩ, $R_L = 48$ kΩ, $R_E = 20$ kΩ, $v_{s(peak)} = 1$ mV, and the transistor $\beta = 80$, what are the values of (a) voltage gain, (b) power gain, (c) dynamic resistance seen by the source v_s, and (d) the peak of the output signal $v_{o(peak)}$?

PROBLEMS (PART II)

8-51. Suppose that you have a sine wave generator source whose open-circuit voltage has a peak of 200 mV and whose internal resistance is 20 kΩ, as shown in Fig. 8-50, and use it to drive the amplifier circuit in Fig. 8-45 in which switch S is closed; that is, r_i and v_g shown in Fig. 8-50 replace v_s shown in Fig. 8-45. What are the values of (a) the ac resistance r_{in} as seen by this generator to the right of its terminal X, (b) the peak value of signal at the base with respect to ground, and (c) the output signal $v_{o(peak)}$.

8-52. Suppose that the sine wave generator in Fig. 8-50 is used to drive the amplifier circuit of Fig. 8-46 in which switch S is closed. What are the values of (a) the ac resistance r_{in} as seen by this generator to the right of its terminal X, (b) the peak value of signal at the base with respect to ground, and (c) the output signal $v_{o(peak)}$.

8-53. Answer (a), (b), and (c) in Prob. 8-51 above assuming switch S is open.

8-54. Answer (a), (b), and (c) in Prob. 8-52 above assuming that switch S is open.

8-55. If the signal generator in Fig. 8-50 replaces the source v_s in the circuit of Fig. 8-47 in which $R_B = 6.6$ MΩ, $R_C = 25$ kΩ, $R_L = 100$ kΩ, $R_1 = 2$ kΩ, $R_2 = 3$ kΩ, and the transistor $\beta = 120$, what are the values of (a) the ac resistance r_{in} as seen by this generator to the right of its terminal X, (b) the peak value of signal at the base with respect to ground, and (c) the output signal $v_{o(peak)}$?

8-56. If the signal generator in Fig. 8-50 replaces the source v_s in the circuit of Fig. 8-47 in which $R_B = 6$ MΩ, $R_C = 20$ kΩ, $R_L = 140$ kΩ, $R_1 = 1.5$ kΩ, $R_2 = 8.5$ kΩ, and the transistor $\beta = 120$, what are the values of (a) ac resistance r_{in} as seen by this generator to the right of its terminal X, (b) the peak value of signal at the base with respect to ground, and (c) the output signal $v_{o(peak)}$?

8-57. What is the signal output $v_{o(peak)}$ of the circuit in Fig. 8-48 if the source v_s is replaced with the generator shown in Fig. 8-50 and if $R_B = 720$ kΩ, $R_C = 6$ kΩ, $R_L = 24$ kΩ, $V_{CC} = -12$ V, and the transistor $h_{FE} \cong h_{fe} \cong 120$?

8-58. What is the signal output $v_{o(peak)}$ of the circuit in Fig. 8-48 if its source v_s is replaced with the generator shown in Fig.

8-50 and if $R_B = 180$ kΩ, $R_C = 4.5$ kΩ, $R_L = 18$ kΩ, $V_{CC} = -9$ V, and the transistor $\beta = 120$?

8-59. What is the signal output $v_{o(peak)}$ of the circuit in Fig. 8-49 if its source v_s is replaced with the generator shown in Fig. 8-50 and if $R_C = 5$ kΩ, $R_L = 20$ kΩ, $R_E = 10$ kΩ, and the transistor $\beta = 90$?

8-60. What is the signal output $v_{o(peak)}$ of the circuit in Fig. 8-49 if its source v_s is replaced with the generator shown in Fig. 8-50 and if $R_C = 12$ kΩ, $R_L = 48$ kΩ, $R_E = 20$ kΩ, and the transistor $\beta = 90$?

8-61. If $r_i = 100$ kΩ and the transistor $\beta = 100$ in the circuit of Fig. 8-51, what is the output voltage $v_{o(peak)}$ if switch S is closed?

8-62. If $r_i = 250$ kΩ, the transistor $\beta = 80$, and switch S is closed in Fig. 8-51, what is the output voltage $v_{o(peak)}$?

8-63. If $r_i = 100$ kΩ, transistor $\beta = 100$, and switch S is open in Fig. 8-51, what is the output voltage $v_{o(peak)}$?

8-64. If $r_i = 250$ kΩ, switch S is open, and $\beta = 80$ in Fig. 8-51, what is the output voltage $v_{o(peak)}$?

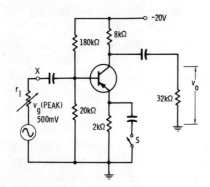

Fig. 8-51. Circuit for Probs. 8-61 through 8-64.

8-65. Referring back to 8-44, if $\beta = 80$, what are (a) the end points of the dc load line, (b) the quiescent values of I_C and V_{CE}, (c) the end points of the ac load line, and (d) the peak value of the largest possible unclipped output signal v_o? Sketch the load lines.

8-66. Referring back to Fig. 8-45, what are (a) the end points of the dc load line, (b) the quiescent values of I_C and V_{CE}, (c) the end points of the ac load line, and (d) the peak value of the largest possible unclipped output signal v_o? Assume that switch S is closed and sketch the load lines.

8-67. If we are to modify the circuit of Fig. 8-44 so that it will be biased at its optimum operating point, what are (a) the coordinates of the optimum operating point I_C and V_{CE}, (b) the value of base bias resistance R_B necessary to obtain this bias assuming that the other components remain unchanged and $\beta = 80$, and (c) the peak value of the largest possible unclipped output signal v_o? Sketch the load lines.

8-68. If we are to modify the circuit in Fig. 8-45 (S closed) so that it will be biased at its optimum operating point, what are (a) the coordinates of the optimum operating point I_C and V_{CE}, (b) the value of base bias resistance R_B needed to obtain this bias assuming that the other components remain unchanged, and (c) the peak value of the largest possible unclipped output signal v_o? Sketch the load lines.

8-69. Referring to the circuit of Fig. 8-46, if the switch S is open, what are (a) the end points of the dc load line, (b) the quiescent values of I_C and V_{CE}, (c) the end points of the ac load line, and (d) the peak value of the largest possible unclipped output signal v_o? Sketch the load lines.

8-70. Referring to Fig. 8-47, if $R_B = 6$ MΩ, $R_C = 20$ kΩ, $R_L = 140$ kΩ, $R_1 = 1.5$ kΩ, $R_2 = 8.5$ kΩ, and the transistor $\beta = 120$, what are (a) the end points of the dc load line, (b) the quiescent values of I_C and V_{CE}, (c) the end points of the ac load line, and (d) the peak value of the largest possible unclipped output signal v_o? Sketch the load lines.

8-71. If we are to modify the circuit of Fig. 8-46 so that it will be biased at the optimum operating point, if switch S is closed, what are (a) the coordinates of the optimum operating point I_C and V_{CE}, (b) the value of R_B needed to obtain this bias assuming that the other components are unchanged and (c) the peak value of the largest possible unclipped output signal v_o? Sketch the load lines.

8-72. If we are to modify the circuit as described in Prob. 8-70 so that it will be biased at the optimum operating point, what are (a) the coordinates of the optimum operating point I_C and V_{CE}, (b) the value of R_B needed to obtain this bias assuming that the other components are unchanged, and (c) the peak value of the largest possible unclipped output signal v_o? Sketch the load lines.

8-73. Referring to Fig. 8-49, if $R_C = 5$ kΩ, $R_L = 20$ kΩ, $R_E = 10$ kΩ, and the transistor $\beta = 90$, what are (a) the end points of the dc load line, (b) the quiescent values of I_C and V_{CE}, (c) the end points of the ac load line, (d) the peak value of the largest possible unclipped output signal v_o? Sketch the load lines.

8-74. Referring to Fig. 8-49, if $R_C = 12$ kΩ, $R_L = 48$ kΩ, $R_E = 20$ kΩ and the transistor $\beta = 90$, what are (a) the end points of the dc load line, (b) the quiescent values of I_C and V_{CE}, (c) the end points of the ac load line, (d) the peak value of the largest possible unclipped output signal v_o? Sketch the load lines.

8-75. Suppose that we are to modify the circuit described in Prob. 8-73 so that it will be biased at its optimum operating point. Therefore what are (a) the quiescent I_C and V_{CE} at the optimum operating point, (b) the value of R_E needed to obtain this operating point assuming that the other components are unchanged, and (c) the peak value of the largest possible unclipped output signal v_o? Sketch the load lines.

8-76. Suppose that we are to modify the circuit described in Prob. 8-74 so that it will be biased at its optimum operating point. Therefore what are (a) the quiescent I_C and V_{CE} at the optimum operating point, (b) the value of R_E needed to obtain this bias assuming that the other components are unchanged, and (c) the peak value of the largest possible unclipped output signal v_o? Sketch the load lines.

9

The Junction Transistor in the Common-Collector Connection

So far we have studied two methods of connecting transistors into amplifier circuits: the common-base (CB) and common-emitter (CE) connections. Here we will proceed with the common-collector (CC) connection, which is also called the *emitter follower* connection. Although the common-collector amplifier does provide some power gain A_p, its voltage gain A_e is less than 1. A useful application of the common-collector circuit is that it can be used to work a high-internal-resistance signal source into a low-resistance load without excessive loading of the source.

9-1 METHODS OF BIASING THE COMMON-COLLECTOR AMPLIFIER

As with the other connections, the emitter-base junction is forward biased while the collector-base junction is reverse biased. This is usually accomplished with circuitry similar to common-emitter amplifiers. For example, the common-collector circuit in Fig. 9-1 is very similar to the common-emitter circuit that uses emitter feedback and two supplies shown in Fig. 8-6 of the last

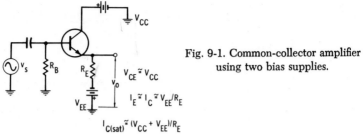

Fig. 9-1. Common-collector amplifier using two bias supplies.

chapter. Note in these two figures that the equations are similar, too. Since there is no resistance in the collector lead of the common-collector circuit in Fig. 9-1, that is, $R_C = 0$, Eq. 8-30 is modified as follows:

$$V_{CE} \cong V_{CC} - R_C I_C = V_{CC} - (0)I_C$$

so that

$$V_{CE} \cong V_{CC} \qquad (9\text{-}1)$$

Similarly, simplification of Eq. 8-31 can be shown:

$$I_{C(\text{sat})} \cong \frac{V_{CC} + V_{EE}}{R_C + R_E} = \frac{V_{CC} + V_{EE}}{0 + R_E}$$

and so

$$I_{C(\text{sat})} \cong \frac{V_{CC} + V_{EE}}{R_E} \qquad (9\text{-}2)$$

Eq. 8-25 can be used for the circuit of Fig. 9-1 as well as the circuit of Fig. 8-6 because R_C is not a factor. Thus the quiescent collector current is

$$I_C \cong I_E \cong \frac{V_{EE}}{R_E} \qquad (9\text{-}3)$$

As shown in Fig. 9-1, the output signal voltage v_o is taken off the emitter with respect to ground in the common-collector circuit. Assuming that the internal resistance of V_{CC} is negligible, the collector is at ac ground potential. This makes the collector common with the input voltage v_s and the output voltage v_o as shown in Fig. 9-2. This is more obvious in Fig. 9-2B, where we can see that the input voltage v_s is applied to the base and collector (ground) while the output v_o is taken off the emitter and collector (ground).

9-1 METHODS OF BIASING THE COMMON-COLLECTOR AMPLIFIER

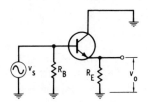

(A) Ac equivalent of common-collector amplifier.

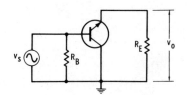

(B) Rearranged ac equivalent to common-collector amplifier.

Fig. 9-2. Reasons for the term "common collector."

Hence the circuit is appropriately called a "common-collector" or "grounded-collector" amplifier.

Another common-collector circuit connection is shown in Fig. 9-3. Note the similarity of this circuit and the one in Fig. 8-7. Since the only difference between the common-collector circuit in Fig. 9-3 and the common-base circuit in Fig. 8-7 is that the former has no resistance R_C in the collector lead, all equations applicable to the common-emitter circuit of Fig. 8-7 are also applicable to circuit of Fig. 9-3 with a small modification of those equations that contain a factor or term R_C. That is, in equations like Eqs. 8-39 and 8-40, R_C is replaced with zero, giving us the equations

$$V_{CE} \cong V_{CC} - R_E I_C \cong V_{CC} - R_E I_E \tag{9-4}$$

and

$$I_{E(\text{sat})} \cong I_{C(\text{sat})} \cong \frac{V_{CC}}{R_E} \tag{9-5}$$

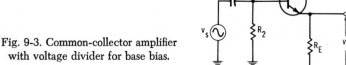

Fig. 9-3. Common-collector amplifier with voltage divider for base bias.

$I_{C(\text{sat})} \cong V_{CC}/R_E \qquad V_{CE} \cong V_{CC} - R_E I_E$

$I_C \cong I_E \cong \dfrac{R_2}{(R_1 + R_2)} (V_{CC}/R_E)$ FOR $\beta > 50$

Also

$$I_E \cong I_C \cong \frac{R_2}{R_1+R_2}\left(\frac{V_{CC}}{R_E}\right) \quad \text{if } \beta \text{ is large} \quad (9\text{-}6)$$

The circuits in Fig. 9-2 are ac equivalents of the circuit shown in Fig. 9-3 as well as of the circuit in Fig. 9-1. Resistance R_B is then the total resistance of R_1 and R_2 in parallel; that is, $R_B = R_1 \| R_2$.

Example 9-1 Suppose that the circuit in Fig. 9-1 has $R_B = 20$ kΩ, $R_E = 10$ kΩ, $V_{CC} = 30$ V, $V_{EE} = 20$ V, and a transistor whose β is 80. Find the quiescent V_{CE}, I_E, I_B, and the saturation currents $I_{C(\text{sat})}$ and $I_{E(\text{sat})}$.

Answer 9-1 $V_{CE} = 30$ V, $I_E \cong 2$ mA, $I_B \cong 25$ μA, $I_{C(\text{sat})} \cong I_{E(\text{sat})} \cong 5$ mA.—By Eq. 9-1 V_{CE} is the source V_{CC} voltage given as 30 V in this case. By Eq. 9-3

$$I_E \cong \frac{V_{EE}}{R_E} = \frac{20 \text{ V}}{10 \text{ k}\Omega} = 2 \text{ mA}$$

And since $I_E \cong I_C \cong \beta I_B$, then $I_B \cong I_E/\beta \cong 2$ mA/80 = 25 μA. Saturation occurs when the transistor $V_{CE} \cong 0$ and the sum of the source voltages appears across R_E. Thus, by Eq. 9-2,

$$I_{E(\text{sat})} \cong I_{C(\text{sat})} \cong \frac{V_{CC}+V_{EE}}{R_E} = \frac{50 \text{ V}}{10 \text{ k}\Omega} = 5 \text{ mA}$$

Example 9-2 In Fig. 9-3, $R_1 = 56$ kΩ, $R_2 = 40$ kΩ, $R_E = 10$ kΩ, and $V_{CC} = 24$ V. Assuming that the transistor β is high, find I_C and the quiescent V_{CE} and determine $I_{E(\text{sat})}$.

Answer 9-2 $I_C \cong 1$ mA, $V_{CE} \cong 14$ V, $I_{E(\text{sat})} \cong 2.4$ mA.—Using Eqs. 9-6, 9-4, and 9-5 respectively, we find

$$I_E \cong I_C \cong \frac{40}{56+40}\left(\frac{24 \text{ V}}{10 \text{ k}\Omega}\right) = 1 \text{ mA}$$

$$V_{CE} \cong 24 \text{ V} - (10 \text{ k}\Omega)(1 \text{ mA}) = 14 \text{ V}$$

$$I_{C(\text{sat})} \cong I_{E(\text{sat})} \cong \frac{24 \text{ V}}{10 \text{ k}\Omega} = 2.4 \text{ mA}$$

9-2 DETERMINING GAINS A_e, A_i, AND A_p FOR THE COMMON-COLLECTOR AMPLIFIER

The ac equivalent circuits of the common-collector amplifier, given in Fig. 9-2, can be modified further as shown in Fig. 9-4A.

9-2 DETERMINING GAINS FOR THE COMMON-COLLECTOR

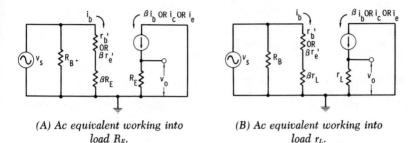

(A) Ac equivalent working into load R_E.

(B) Ac equivalent working into load r_L.

Fig. 9-4. Common-collector amplifier ac equivalent circuits.

Note the similarity of this circuit and the circuit in Fig. 8-28E, which is the ac equivalent of the common-emitter amplifier with unbypassed emitter resistance R_E. The differences are that R_C is a short circuit and the output is taken off R_E instead of R_C, in the ac equivalent of the common-collector amplifier. Both equivalents are similar in that the source v_s sees a resistance $r_b' + \beta R_E$ looking into the base. The reason why v_s sees the emitter resistance R_E larger by a factor of β was explained in Sec. 8-10 of the last chapter. A review of it may be useful at this time.

Note in Fig. 9-4A that the output voltage is $v_o = R_E i_e$. Also the input voltage v_s is across the series resistances r_b' and βR_E. Therefore $v_s = (r_b' + \beta R_E) i_b$. Since the voltage gain A_e is equal to the ratio of output to input voltages, we can show that

$$A_e = \frac{v_o}{v_s} = \frac{R_E i_e}{(r_b' + \beta R_E) i_b}$$

But since $i_e \cong \beta i_b$ and since $r_b' \cong \beta r_e'$, the above equation may be modified as follows:

$$A_e \cong \frac{R_E \beta i_b}{(\beta r_e' + \beta R_E) i_b} = \frac{R_E \beta i_b}{\beta(r_e' + R_E) i_b} = \frac{R_E}{r_e' + R_E} \quad (9\text{-}7)$$

The last equality shows us that the voltage gain A_e of the common-collector amplifier is always less than 1.

Example 9-3 Find the voltage and current and power gains of (a) the circuit described in Ex. 9-1, and (b) the circuit described in Ex. 9-2. Assume $\beta = 80$ in both cases.

Answer 9-3 (a) $A_e \cong 1$, $A_i \cong 80$, $A_p \cong 80$, (b) $A_e \cong 1$, $A_i \cong 80$, $A_p \cong 80$.—(a) In Ex. 9-1 we did determine that $I_E \cong 2$ mA. Therefore

$$r_e' \cong \frac{25 \text{ mV}}{I_E} = 12.5 \text{ }\Omega$$

Thus by Eq. 9-7 the voltage gain is

$$A_e \cong \frac{R_E}{r_e' + R_E} = \frac{10 \text{ k}\Omega}{12.5 \text{ }\Omega + 10 \text{ k}\Omega} \cong 1$$

The current gain is the ratio of output to input current. In this case, then,

$$A_i = \frac{i_e}{i_b} \cong \frac{i_c}{i_b} \cong \beta = 80$$

And since the power gain is the product of voltage and current gains,

$$A_p = A_e A_i \cong (1)80 = 80$$

(b) In Ex. 9-2 we found that $I_C \cong 1$ mA and therefore $r_e' \cong 25$ Ω. This, however, has little effect in Eq. 9-7 when R_E is large and again we find that $A_e \cong 1$. Thus, the values A_i and A_p are found exactly as above.

It is interesting to note in the last example that if the load resistance, which is R_E in this case, is very large compared to r_e', we may completely ignore r_e'. The value of r_e' may become significant when the load in the emitter lead is about 1 kΩ or less.

9-3 AC LOADS AND LOAD LINES

More often than not, R_E is not the only load in the emitter lead. Typically the common-collector amplifier is loaded with additional

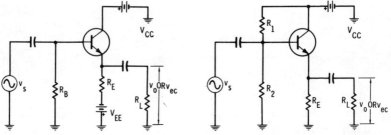

(A) *Circuit with two bias supplies.*   (B) *Circuit with voltage divider bias.*

Fig. 9-5. Common collector circuits capacitively coupled to load R_L, resulting in total emitter ac load $r_L = R_E \parallel R_L$.

9-3 AC Loads and Load Lines

resistance R_L as shown in Fig. 9-5. Resistance R_L is not necessarily a resistor but instead may represent the resistance of some practical lead like a speaker or the input of another amplifier, etc.

With a negligible reactance of the coupling capacitor, looking down from the emitter in either Fig. 9-5A or Fig. 9-5B, ac sees R_E and R_L in parallel, or a total ac load $r_L = R_E \| R_L$. Since R_E alone is no longer the load and r_L is instead, Eq. 9-7 is modified to

$$A_e \cong \frac{r_L}{r_e' + r_L} \qquad (9\text{-}8)$$

R_L is not in any dc current path and therefore it does not affect the dc formulas, Eqs. 9-1 through 9-6. The ac equivalent circuit is also modified as shown in Fig. 9-4B.

The typical load lines for the circuit of Fig. 9-5A are shown in Fig. 9-6A, and the total lines for the circuit of Fig. 9-5B are shown in Fig. 9-6B. For the circuit with two bias supplies in Fig. 9-5A, the upper end-point of the dc load line is found by noting that when the transistor is saturated and $V_{CE} = 0$, the total bias voltage $V_{CC} + V_{EE}$ is appears across R_E. Therefore as shown before, maximum possible emitter current is, by Eq. 9-2,

$$I_{E(\text{sat})} \cong \frac{V_{CC} + V_{EE}}{R_E}$$

The lower end-point of the dc load line is at the dc cutoff point. At the cutoff point the emitter current is about zero, causing zero voltage drop across R_E. The total bias voltage $V_{CC} + V_{EE}$ therefore appears across collector and emitter. That is,

$$V_{CE(\text{cutoff})} = V_{CC} + V_{EE} \qquad (9\text{-}9)$$

By similar reasoning, the end points of the dc load line, of the common-collector amplifier with the voltage divider, are determined. At saturation the source V_{CC} voltage is across R_E. Therefore,

$$I_{E(\text{sat})} \cong \frac{V_{CC}}{R_E} \qquad (9\text{-}10)$$

When the transistor is cut off, the source V_{CC} voltage appears across the transistor collector and emitter leads. Thus

$$V_{CE(\text{cutoff})} = V_{CC} \qquad (9\text{-}11)$$

As shown in Fig. 9-6, the end points of the ac load line are at points

$$i_{E(\text{sat})} = I_E + \frac{V_{CE}}{r_L} \tag{9-12}$$

and

$$v_{CE(\text{cutoff})} = V_{CE} + r_L I_E \tag{9-13}$$

where
I_E = quiescent dc emitter current,
V_{CE} = quiescent collector-to-emitter dc voltage,
r_L = ac load in the emitter circuit.

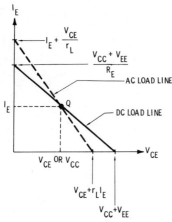

(A) For circuit in Fig. 9-5A.   (B) For circuit in Fig. 9-5B.

Fig. 9-6. Common-collector circuit load lines.

Example 9-4 If in Fig. 9-5A, $R_B = 20$ kΩ, $R_E = 12$ kΩ, $R_L = 48$ kΩ, $V_{CC} = 30$ V, $V_{EE} = 24$ V, and the transistor β is 100, plot dc and ac load lines and find I_E, V_{CE}, the peak value of the maximum unclipped output signal v_o, and the resistance r_{in} seen by signal looking into the base.

Answer 9-4 See Fig. 9-7, $I_E \cong 2$ mA, $V_{CE} \cong 30$ V, $v_o \cong 19.2$ V (peak), $r_{\text{in}} \cong 960$ kΩ.—The end points of the dc load line are, by Eqs. 9-2 and 9-11,

$$I_{E(\text{sat})} \cong \frac{V_{CC} + V_{EE}}{R_E} = \frac{54 \text{ V}}{12 \text{ kΩ}} = 4.5 \text{ mA}$$

9-3 AC Loads and Load Lines

and

$$V_{CE(cutoff)} = V_{CC} + V_{EE} = 54 \text{ V}$$

As the signal current i_e in the emitter sees it, the load is R_E and R_L in parallel, or

$$r_L = 12 \text{ k}\Omega \| 48 \text{ k}\Omega = 9.6 \text{ k}\Omega$$

and the quiescent values are, by Eqs. 9-3 and 9-1,

$$I_E \cong \frac{V_{EE}}{R_E} = \frac{24 \text{ V}}{12 \text{ k}\Omega} = 2 \text{ mA}$$

and

$$V_{CE} \cong V_{CC} = 30 \text{ V}$$

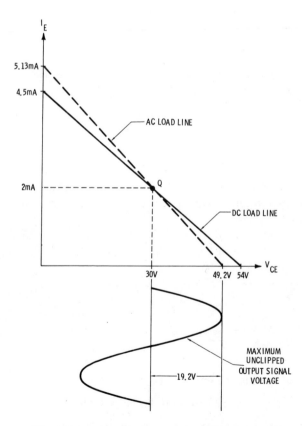

Fig. 9-7. Load lines for circuit described in Ex. 9-3.

Now, from Eqs. 9-12 and 9-13, we can find the end points of the ac load line:

$$i_{E(\text{sat})} = I_E + \frac{V_{CE}}{r_L} \cong 2 + 3.13 = 5.13 \text{ mA}$$

and

$$v_{CE(\text{cutoff})} = V_{CE} + r_L I_E \cong 30 + 19.2 = 49.2 \text{ V}$$

Thus, as shown in Fig. 9-7, output signal can swing the collector-to-emitter voltage from the quiescent value 30 V to cutoff at 49.2 V, by the amount 19.2 V, without clipping.

As shown in the ac equivalent circuit in Fig. 9-4B, signal i_b looking into the base sees a total resistance $r_{\text{in}} = \beta r_e' + \beta r_L$. In this case,

$$r_e' \cong \frac{25 \text{ mV}}{I_E} \cong 12.5 \text{ }\Omega$$

and

$$\beta r_L = 100(9.6 \text{ k}\Omega) = 960 \text{ k}\Omega$$

Needless to say, $\beta r_e'$ is insignificant compared to βr_L and therefore $r_{\text{in}} \cong 960$ kΩ.

Example 9-5 If in Fig. 9-5B, $R_1 = 300$ kΩ, $R_2 = 100$ kΩ, $R_E = 10$ kΩ, $R_L = 40$ kΩ, $V_{CC} = 20$ V, and the transistor $\beta = 80$, plot the dc and ac load lines and find I_E, V_{CE}, the peak value of the maximum unclipped output signal v_o, and the resistance seen by signal looking into the base r_{in}.

Answer 9-5 See Fig. 9-8, $I_E \cong 0.5$ mA, $V_{CE} \cong 15$ V, $v_o \cong 4$ V (peak), $r_{\text{in}} \cong 640$ kΩ.—In this case, the end points of the dc load line are, by Eqs. 9-10 and 9-11,

$$I_{E(\text{sat})} \cong \frac{V_{CC}}{R_E} \cong \frac{20 \text{ V}}{10 \text{ k}\Omega} = 2 \text{ mA}$$

and

$$V_{CE(\text{cutoff})} \cong V_{CC} = 20 \text{ V}$$

The total ac load in the emitter is

$$r_L = R_E \| R_L = 10 \text{ k}\Omega \| 40 \text{ k}\Omega = 8 \text{ k}\Omega$$

and at the operating point

9-3 AC Loads and Load Lines

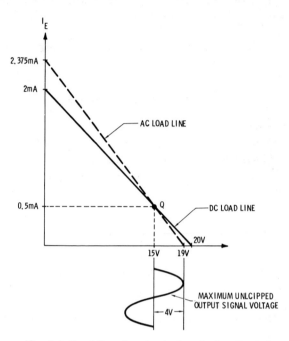

Fig. 9-8. Load lines for circuit described in Ex. 9-4.

$$I_E \cong \frac{100}{300+100}\left(\frac{20\text{ V}}{10\text{ k}\Omega}\right) = 0.5\text{ mA}$$

and

$$V_{CE} \cong 20\text{ V} - (10\text{ k}\Omega)(0.5\text{ mA}) = 15\text{ V}$$

by Eqs. 9-6 and 9-4. The end points of the ac load, therefore, are

$$i_{E(\text{sat})} = I_E + \frac{V_{CE}}{r_L} \cong 0.5 + 1.875 = 2.375\text{ mA}$$

and

$$v_{CE(\text{cutoff})} = V_{CE} + r_L I_E = 15 + 4 = 19\text{ V}$$

by Eqs. 9-12 and 9-13. The output signal voltage therefore can swing by a peak of 4 V as shown in Fig. 9-8. Looking into the base, the signal sees an input resistance

$$r_{\text{in}} = ßr_e' + ßr_L = 80(50\text{ }\Omega) + 80(8\text{ k}\Omega) \cong 640\text{ k}\Omega$$

286 JUNCTION TRANSISTOR IN THE COMMON-COLLECTOR CONNECTION

It is interesting to note in these last two examples that the input resistance of the common-collector amplifier is high, especially if compared to the common-base amplifier. This enables us to use a common-collector amplifier to match more efficiently a high-resistance source to a relatively low resistance load.

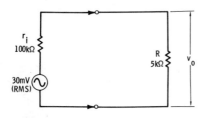

Fig. 9-9. Signal generator in series circuit.

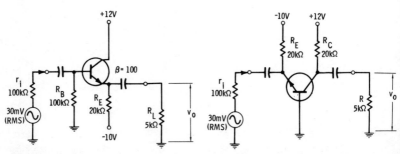

Fig. 9-10. Signal generator in CC circuit with two bias supplies.

Fig. 9-11. Signal generator in CB amplifier circuit.

Example 9-6 Solve for the voltage across R, shown as v_o, in each of the circuits shown in the following figures: (a) Fig. 9-9, (b) Fig. 9-10, and (c) Fig. 9-11.

Answer 9-6 (a) $v_o \cong 1.43$ mV, (b) $v_o \cong 13.3$ mV, (c) $v_o \cong 1.2$ mV.—Note that the same signal generator is used in each of the three circuits. In part (a), the internal resistance of the source r_i and the load R are simply in series, and therefore the voltage across R is

$$v_o = \frac{(30 \text{ mV})(5 \text{ k}\Omega)}{100 \text{ k}\Omega + 5 \text{ k}\Omega} = 1.43 \text{ mV}$$

In part (b), the signal source sees R_B in parallel with the input resistance of the transistor, which is $\beta r_e' + \beta r_L$, as shown in Fig. 9-12A. In this case

9-3 AC LOADS AND LOAD LINES

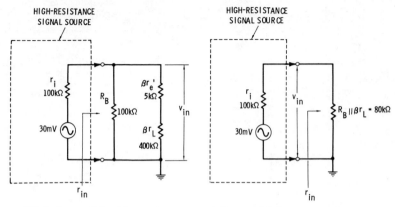

(A) R_B in parallel with transistor input impedance.

(B) Simplified version of the circuit shown in (A).

Fig. 9-12. Ac equivalents of the input of the circuit in Fig. 9-10.

$$\beta r_L = 100(R_E \| R_L) = 100(20 \text{ k}\Omega \| 5 \text{ k}\Omega) = 100(4 \text{ k}\Omega) = 400 \text{ k}\Omega$$

and, since $I_E \cong 10 \text{ V}/20 \text{ k}\Omega = 0.5 \text{ mA}$,

$$\beta r_e' \cong \beta \left(\frac{25 \text{ mV}}{I_E} \right) = 5 \text{ k}\Omega$$

which is insignificant compared to 400 kΩ. So for practical purposes, the signal source sees R_B in parallel with βr_L for a total input resistance

$$r_{in} \cong R_B \| \beta r_L = 100 \text{ k}\Omega \| 400 \text{ k}\Omega = 80 \text{ k}\Omega$$

as shown in Fig. 9-12B. The voltage across r_{in} is the input voltage v_{in} to the transistor; that is, it is the voltage at the base with respect to ground. In this case, then,

$$v_{in} = \frac{(30 \text{ mV})(80 \text{ k}\Omega)}{100 \text{ k}\Omega + 80 \text{ k}\Omega} \cong 13.3 \text{ mV}$$

The gain of this common-collector circuit is about 1; that is, by Eq. 9-8,

$$A_e \cong \frac{r_L}{r_e' + r_L} = \frac{4000}{50 + 4000} \cong 1$$

Therefore,

$$v_o = A_e v_{in} \cong 13.3 \text{ mV}$$

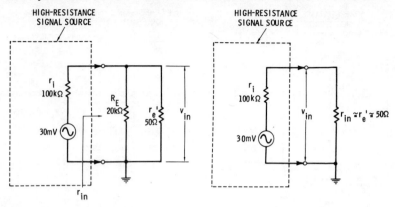

(A) R_E in parallel with r_e'. (B) Simplified version of (A).

Fig. 9-13. Ac equivalents of the input of the circuit in Fig. 9-11.

In part (c), we have a common-base amplifier that has a resistance r_e' looking into the emitter. Therefore, as the source sees it, R_E and r_e' are in parallel as shown in Fig. 9-13A. Since $r_e' = 25$ mV/I_E and $I_E = 10$ V/20 kΩ = 0.5 mA, the source sees a total input resistance

$$r_{in} = R_E \| r_e' \cong 20 \text{ k}\Omega \| 50 \text{ }\Omega \cong 50 \text{ }\Omega$$

as shown in Fig. 9-13B. Therefore the input voltage

$$v_{in} \cong \frac{(30 \text{ mV})(50 \text{ }\Omega)}{100 \text{ k}\Omega + 50 \text{ }\Omega} \cong 0.015 \text{ mV} = 15 \text{ }\mu\text{V}$$

The voltage gain of this common-base amplifier is much greater than 1. More specifically, by Eq. 7-9,

$$A_e \cong \frac{r_L}{r_e'} = \frac{4000}{50} = 80$$

Thus the output voltage is $v_o = A_e v_{in} \cong 80(15 \text{ }\mu\text{V}) = 1.2$ mV.

It is interesting to note in this last example that even though the voltage gain of the common-collector amplifier is less than 1, with it we obtained the largest voltage across R. Even the high voltage gain A_e of the common-base amplifier could not do better. The common-collector amplifier loaded the high-resistance source less than did the load R_L alone and certainly less than the common-base amplifier. We may conclude that the common-collector amplifier is useful as a device to match a high-resistance source to a low-resistance load while the common-base amplifier is practically worthless to amplify signals from high-resistance sources.

REVIEW QUESTIONS

9-1. What is the approximate voltage gain of the common-collector amplifier?

9-2. What can the common-collector amplifier be used for?

9-3. By what other term is the common-collector amplifier referred to?

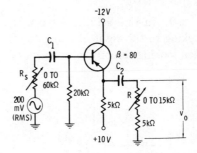

Fig. 9-14. Problem circuit with two bias supplies.

9-4. What is meant by the term "loading down the signal source?"

9-5. Is the common-collector capable of power gain?

9-6. What is the approximate input resistance, looking into the base, of the common-collector amplifier?

PROBLEMS

9-1. If variable resistance R, in Fig. 9-14, is set at 0 Ω, (a) what is the ac resistance seen looking directly into the base? (b) What is the ac resistance seen looking to the right from capacitor C_1?

9-2. If the variable resistance $R = 15$ kΩ in Fig. 9-14, what are the answers to questions (a) and (b) in Prob. 9-1 above?

9-3. If $R_s = R = 0$ Ω in Fig. 9-14, what is the output voltage v_o?

9-4. If $r_s = 0$ Ω and $R = 15$ kΩ in Fig. 9-14, what is the output voltage?

9-5. In Fig. 9-14, if $r_s = 60$ kΩ and $R = 0$ Ω, what is the output voltage?

9-6. If $r_s = 60$ kΩ and $R = 15$ kΩ in Fig. 9-14, what is the output voltage?

9-7. By sketching an ac load line for the circuit in Fig. 9-14, what is the peak value of maximum unclipped signal output v_o if $R = 15$ kΩ?

9-8. By sketching an ac load line for the circuit in Fig. 9-14, what is the peak value of the maximum unclipped signal output v_o if $R = 0$ Ω?

9-9. In Fig. 9-15, if $R = 0$ Ω, (a) what is the ac resistance seen looking to the right of capacitor C_1?, (b) what is the ac resistance seen looking directly into the base?

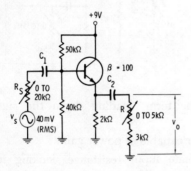

Fig. 9-15. Problem circuit with voltage-divider biasing.

9-10. If R is set at 5 kΩ in Fig. 9-15, what are the answers to questions (a) and (b) in Prob. 9-9 above?

9-11. If $r_s = 20$ kΩ, $v_s = 40$ mV, and $R = 0$ Ω in Fig. 9-15, what is the output voltage v_o?

9-12. If $r_s = 20$ kΩ and $R = 5$ kΩ in Fig. 9-15, what is the output voltage v_o?

9-13. By sketching an ac load line find the peak value of the maximum unclipped signal output v_o of the circuit in Fig. 9-15 if $R = 0$ Ω.

9-14. By sketching an ac load line find the peak value of the maximum unclipped output v_o of the circuit in Fig. 9-15 if $R = 5$ kΩ.

Problems

9-15. (a) What is the approximate dc voltage at the emitter with respect to ground in the circuit of Fig. 9-14? (b) What dc emitter-to-ground voltage V_E would you expect in the circuit of Fig. 9-14 if the 20 kΩ resistor becomes open due to a broken connection.

9-16. (a) What is the approximate dc emitter-to-ground voltage in the circuit of Fig. 9-15? (b) What dc emitter-to-ground voltage would you expect in the circuit of Fig. 9-15 if the 2 kΩ resistor burned open?

9-17. What is the average collector power dissipation P_C in the circuit of Fig. 9-14?

9-18. What is the average collector power dissipation P_C in the circuit of Fig. 9-15?

10

The Silicon Controlled Rectifier

The term *thyristor* defines a broad range of semiconductor devices used mainly as electronically controlled switches. Their internal construction is different from that of transistors. We learned that the transistor has two junctions, which means it is a three-layered device: either pnp or npn type. The thyristor, on the other hand, typically has four layers of semiconductor materials, and thus is called a *pnpn* device.

Thyristors are popular in civilian, industrial, and military equipment: battery chargers, welders, light flashers, voltage regulators, dc and ac motor controls, circuit breakers, battery operated vehicles, and light dimmers, to name only a few.

10-1 GENERAL CHARACTERISTICS

Among the most popular members of the thyristor family is the *reverse blocking triode thyristor,* commonly called the silicon controlled rectifier or SCR; see Fig. 10-1. It is similar to the diode in some ways. Like the diode, the SCR is nonconducting when reverse biased. Unlike the diode, however, its conduction can be controlled when forward biased. That is, when forward bias is placed across the anode and cathode leads, conduction does not occur unless a current is allowed to pass through the *gate* lead.

10-1 GENERAL CHARACTERISTICS

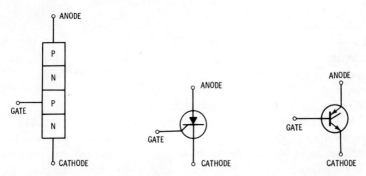

(A) The SCR a pnpn semiconductor device. (B) Schematic symbol of the SCR in popular use. (C) Schematic symbol proposed by IEEE.

Fig. 10-1. Diagrams of the silicon controlled rectifier.

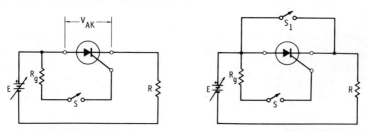

Fig. 10-2. Turning the SCR on.   Fig. 10-3. Turning the SCR off.

For example, suppose that in the circuit of Fig. 10-2, the variable dc source voltage E is increased from 0 V to a few volts while the switch S is open. The SCR is forward biased; however, it will not conduct because the gate lead is open and no gate current can flow. This causes the SCR to act like an open and the anode-to-cathode voltage V_{AK} is equal to the source voltage E. If the switch S is closed, the gate is made *positive* with respect to the cathode and gate current flows through R_g, causing the SCR to turn on. This means that the SCR starts to conduct; current flows through the load R and V_{AK} drops down to about 1 V or so. An interesting feature of the SCR is that after turn-on, reopening the switch S will not turn it off. Thus, the gate loses control after the SCR starts to conduct. The gate can regain control if the anode current I_A is momentarily interrupted, causing the SCR to turn off. Conduction of I_A can be stopped by reducing the source voltage

E to zero. Also a switch S_1 across the SCR, as shown in Fig. 10-3, if closed momentarily, will cause all of the circuit current to bypass the SCR. This causes the SCR to turn off because it requires a minimum anode current through it to hold it in conduction. This required minimum current is called the *holding current* I_H and is usually defined as the minimum current needed to hold the SCR in its conducting state while the gate lead is open.

A general understanding of why the gate loses control after turn-on may be obtained by looking at the SCR as an equivalent to a *complementary pair* of transistors. A complementary pair consists of a connection of one npn and one pnp type transistor in a way such that the collector of each is connected to the base of the other. For example, in Fig. 10-4, the pnpn structure shown in Fig. 10-4A may be split into two structures as shown in Fig. 10-4B yet

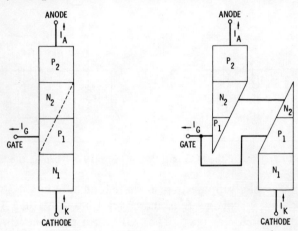

(A) The pnpn structure.   *(B) The pnpn when split.*

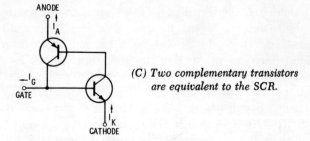

(C) Two complementary transistors are equivalent to the SCR.

Fig. 10-4. The SCR and its equivalents.

10-1 GENERAL CHARACTERISTICS

remain equivalent. Note in Fig. 10-4B that one structure forms an npn transistor while the other forms a pnp type. Therefore the circuit in Fig. 10-4C is equivalent to an SCR.

To review the facts, we can refer to the circuit in Fig. 10-5A. With the switch S open, the current in R remains essentially zero even after forward bias voltage E is applied. When S is closed, gate current flows which turns on the SCR, and anode current I_A flows through the load R. As long as the anode current I_A is greater than the holding current I_H, the SCR continues to conduct even if S is reopened. Now if we replace the SCR in this circuit with its equivalent, we have the circuit in Fig. 10-5B.

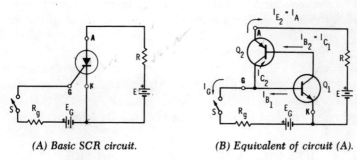

(A) Basic SCR circuit. *(B) Equivalent of circuit (A).*

Fig. 10-5. SCR circuit and its equivalent circuit.

Suppose that S is initially open when E is applied in Fig. 10-5B. Neither transistor conducts because each of their base currents is zero. That is, initially transistor Q_1 is cut off and its collector does not supply current to the base of Q_2, which keeps the latter cut off. As long as Q_2 is cut off, its collector does not supply current to the base of Q_1, thus holding Q_1 in cutoff. With both transistors cut off, the current from K to A through them is zero, neglecting leakage.

Now if the switch S is closed, a forward bias is placed across the base-emitter junction of Q_1, causing base current I_{B1} to flow. This causes Q_1 to conduct collector current I_{C1}, which is the base current I_{B2} of transistor Q_2. Current I_{B2} causes Q_2 to conduct and its emitter current I_{E2} is the anode current I_A.

With Q_2 in a conducting state, I_{B1} can flow up through it and therefore the existence of I_{B1} no longer depends on S being closed. That is, even with S reopened, I_{B1} flows up through conducting

Q_2, which keeps Q_1 conducting. Q_1 thus conducts base current I_{B2} and keeps Q_2 conducting. This process of each transistor keeping the other conducting is called "feedback" and is continuous unless voltage E is reduced or the anode current is reduced to a value below the holding current I_H.

So in either circuit, Fig. 10-2 or Fig. 10-5, momentarily closing the switch S turns on the SCR. The SCR then conducts anode current I_A through the load R. This current continues to flow, even though S is reopened, as long as I_A remains larger than I_H. Actually, I_A must be noticeably larger than I_H if we expect the SCR to turn on and stay on after it is triggered with a momentary gate current. That is, the SCR will not stay on after S is reopened unless I_A is somewhat larger than I_H. The minimum current required to keep the SCR conducting after S is reopened is called the *latching current* I_L. So if the load resistance R is too large or if the source voltage E is too small, the magnitude of I_A might be smaller than I_L and the SCR won't *latch* on and continue to conduct after a pulse of gate current flows through a momentarily closed switch S. This is rarely a problem in practice because the SCR is typically expected to conduct currents far larger than its latching current.

10-2 ANODE CHARACTERISTICS OF THE SCR

Fig. 10-6 shows typical characteristics of an SCR with the gate lead open. Current I_F and voltage V_F are the forward anode current and anode-to-cathode voltage respectively. Current I_R and voltage V_R are the reverse anode current and anode-to-cathode voltage. Obviously, the reverse bias characteristics of the SCR are similar to those of a diode. That is, I_R is a relatively small (leakage) value over a wide range of V_R values. At a maximum reverse voltage V_{ROM}, "avalanche conduction" occurs which is supported by an avalanche of minority carriers caused by V_{ROM}. This action is similar to zener conduction in the zener diode covered in Chap. 5. The letters in the subscript of the symbol V_{ROM} are from the words Maximum Reverse voltage with gate Open. In typical applications of the SCR, operation in the reverse avalanche region is avoided.

The forward characteristics show that conduction of I_F with V_F applied is small amounting to a usually negligible leakage up to a *forward breakover voltage* V_{FOM}. (The subscripts in V_{FOM} come from the words in the description: Maximum Forward voltage

10-2 ANODE CHARACTERISTICS OF THE SCR

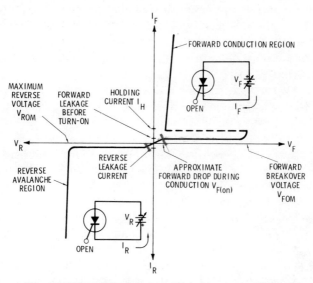

Fig. 10-6. *V-I* characteristics of the SCR with gate open.

with gate Open.) At voltage V_{FOM} or greater, the SCR *breaks over* into forward conduction and I_F suddenly increases, limited mainly by the external resistance in the anode circuit. The anode-to-cathode voltage then drops to a low value of $V_{F(on)}$—about a volt or so, depending on the magnitude of I_F and the size of the SCR. We may assume, for practical purposes, that the forward and reverse leakages are negligible.

The spec (specifications) sheets supplied by the manufacturers usually specify *repetitive* and *nonrepetitive* with the maximum voltage ratings. For example:

$V_{ROM(repetitive)}$—With the gate open, the maximum allowable reverse anode-to-cathode voltage (repetitive as in ac or continuous dc).

$V_{ROM(nonrepetitive)}$—With the gate open, the maximum allowable reverse anode-to-cathode transient voltage which is a non-repeating pulse of voltage that exists for only a small fraction of a second.

Similar descriptions are given for forward voltages $V_{FOM(repetitive)}$ and $V_{FOM(nonrepetitive)}$. Typically nonrepetitive values are about 20% to 50% higher than repetitive ones for any given SCR.

Example 10-1 If the circuit in Fig. 10-5A has a load resistance $R = 10$ Ω and the following specs for the SCR: $V_{ROM(rep)} = V_{FOM(rep)} = 100$ V, $V_{ROM(nonrep)} = V_{FOM(nonrep)} = 150$ V, $V_{F(on)} = 1$ V, $I_H = 20$ mA, find the anode current I_A at each of these applied voltages, assuming that the switch S is left open: (a) $E = 80$ V, (b) $E = 140$ V, (c) $E = 200$ V. Also after E is adjusted to 200 V, it is slowly reduced. (d) Find the minimum applied E that will keep the SCR turned on.

Answer 10-1 (a) $I_A \cong 0$, (b) $I_A \cong 13.9$ A, (c) $I_A \cong 19.9$ A, (d) $E_{(min)} \cong 1.2$ V.—Because the applied E forward biases the SCR and is increased from one dc value to another, the forward repetitive breakover voltage $V_{FOM(rep)}$ applies.

(a) The applied $E = 80$ V is less than $V_{FOM(rep)} = 100$ V and therefore does not turn on the SCR. Since the gate is open, the SCR remains nonconducting, except for a negligible leakage; thus $I_A \cong 0$ A.

(b) The applied voltage $E = 140$ V exceeds $V_{FOM(rep)}$, causing the SCR to turn even though the gate is open. When the SCR is conducting, the anode-to-cathode voltage $V_{F(on)} \cong 1$ V. Therefore the remaining 139 V is across R. The current through R is the anode current. Thus in this case $I_A \cong 139$ V/10 Ω = 13.9 A.

(c) The voltage $E = 200$ V exceeds $V_{FOM(rep)}$, causing the SCR to turn on. Since $V_{F(on)} \cong 1$ V, the remaining 199 V is across R and therefore $I_F = 199$ V/10 Ω = 19.9 A.

(d) After turn-on, the SCR continues to conduct until its anode current is reduced below the holding current I_H, which is 20 mA in this case. Thus the minimum current through R is 20 mA and its minimum voltage drop is, by Ohm's law, $R(I_H) = 10(20$ mA$) = 0.2$ V. With about 1 V dropped across the SCR, the minimum applied $E \cong 1.0 + 0.2 = 1.2$ V.

Example 10-2 Suppose that in the previous example, source E is adjusted from zero to 80 V and then the gate switch S is closed and reopened. (a) What is the value of I_A? (b) If the load resistance R is increased to 10 kΩ and again S is closed and reopened after E is adjusted up to 80 V, what is the value of I_A?

Answer 10-2 (a) $I_A \cong 7.9$ A, (b) $I_A \cong 0$.—Closing switch S places a positive voltage with respect to the cathode on the gate, admitting gate current which turns on the SCR if it has sufficient magnitude.

(a) When S is closed and the SCR turns on, about 1 V is dropped across the anode and cathode ($V_{F(on)} \cong 1$ V), and the remaining 79 V appears across R. Since $R = 10$ Ω, $I_A \cong 79$ V/10 Ω = 7.9 A. After turn-on, the gate loses control and an anode current $I_A \cong 7.9$ A continues even after S is reopened.

(b) In this case, with a load $R = 10$ kΩ, if the SCR conducts with $V_{F(on)} \cong 1$ V across it, the anode current $I_A \cong 79$ V/10 kΩ = 7.9 mA. This is below the 20-mA holding current value and the SCR will certainly turn off after S is reopened.

Example 10-3 If the polarity of the source E is reversed in the circuit described in Ex. 10-1, to what maximum value can the magnitude of E be adjusted while avoiding avalanche conduction?

Answer 10-3 About 100 V.—Since the specified $V_{ROM(rep)} = 100$ V, the safe maximum continuous dc reverse bias is 100 V. This same SCR can safely take a transient (reverse voltage pulse of very short time) up to 150 V because $V_{ROM(nonrep)} = 150$ V is specified.

10-3 GATE TRIGGER CHARACTERISTICS OF THE SCR

Referring back to Fig. 10-1A and looking into the gate and cathode leads, we see a pn diode junction. When anode current $I_A = 0$, the gate-to-cathode V-I characteristics are similar to those of an ordiary diode as shown in Fig. 10-7, curve No. 1. The minimum forward gate current I_{FG} that will turn on (trigger) an SCR is called the *dc gate trigger current* I_{GT}. At I_{GT} the voltage across the gate and cathode is the *dc gate trigger voltage* V_{GT}. The specific values of I_{GT} and V_{GT} vary with individual SCR's and are also affected by the temperature of the gate-cathode junction T_j, the anode load resistance R, and the anode source voltage E. Thus, manufacturers often provide general gate drive requirements in graphical form, as you will see. Note in Fig. 10-7 that there is a maximum reverse gate voltage V_{GRM} which, if exceeded, will cause avalanche conduction which can damage the SCR.

The gate V-I characteristics change after turn-on as shown in Fig. 10-7, curve No. 2. The position of V-I characteristic curve No. 2 depends a lot of the value of anode current I_A. During conduction, I_A causes a voltage drop across the gate-cathode junction, the gate being positive with respect to the cathode. That is, after the SCR is turned on, electrons flow up across the cathode-gate junction (Fig. 10-1A). This junction has some resistance and the large electron flow through it causes a voltage drop. This voltage drop is positive at the top (at the gate) with respect to the bottom (the cathode). This effect is shown on curve No. 2 in Fig. 10-7. Note

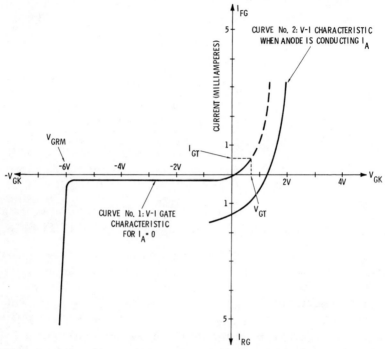

Fig. 10-7. Gate-cathode V-I characteristics of typical SCR.

that if $I_{FG} = 0$, as would be the case if the gate is left open after turn-on, then the gate-to-cathode voltage is $V_{GK} \cong 1.3$ V, as shown at the point where curve No. 2 crosses the V_{GK} axis. These changes in the gate-cathode V-I characteristics may be ignored in most applications, such as the circuits covered in this chapter.

General forward V-I gate *trigger* characteristics as shown in Fig. 10-8 are usually available. The actual gate to cathode V-I characteristics for all SCR's, for which these trigger characteristics apply, are between curves A and B. That is, due to the variety of possible operating conditions and inherent differences in individual SCR's of even the same type, the V-I characteristics differ. A specific unit (SCR) may under certain conditions have relatively

10-3 GATE TRIGGER CHARACTERISTICS OF THE SCR

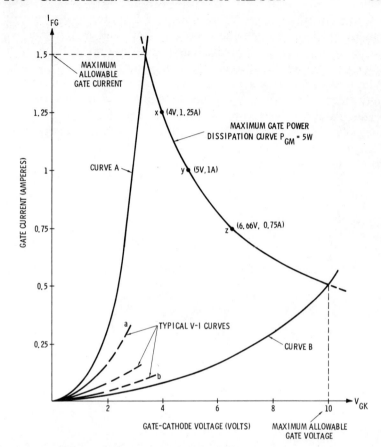

Fig. 10-8. Typical SCR gate trigger characteristics.

low dynamic resistance $\Delta V_{GK}/\Delta I_{FG}$ and V-I characteristics like curve A or somewhat below it like curve a. Another unit of the same type but in a different environment, may have higher dynamic resistance and V-I characteristics like curve B or somewhat above it like curve b. The point is that we can expect the SCR's V-I characteristics to lie on or between curves A and B.

The instantaneous maximum gate power dissipation P_{GM} curve shown in Fig. 10-8 as $P_{GM} = 5$ W means that the product of the peak gate-cathode voltage and peak gate current, $V_{GK}I_{FG}$, should never exceed 5 W. All points on this P_{GM} curve have coordinates (V_{GK}, I_{FG}) whose product is 5 W. For example, at point x, $V_{GK} =$

4 V and $I_{FG}=1.25$ A and their product is 5 W. Similarly, you can find that the product of the coordinates at each of the points y and z are likewise 5 W. Similar curves can be drawn if the maximum gate power dissipation P_{GM} is larger or smaller than 5 W. With smaller P_{GM} values, the curves are closer to the origin, while, conversely, with larger P_{GM} values the curves are farther from the origin. To avoid damage to the SCR, the gate must not be driven above its P_{GM} curve. For SCR's with trigger characteristics as in Fig. 10-8, gate current I_{FG} should never exceed 1.5 A and V_{GK} should never exceed 10 V.

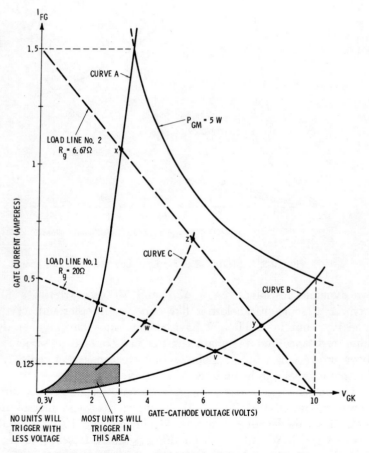

Fig. 10-9. SCR gate trigger characteristics with load lines.

10-3 GATE TRIGGER CHARACTERISTICS OF THE SCR

The *average gate power* $P_{G(av)}$ rating is less than the instantaneous maximum gate power P_{GM}, because $P_{G(av)}$ is the product of the average gate-cathode voltage and average gate current. Typically if $P_{GM} = 5$ W, then $P_{G(av)}$ is rated at one-tenth of it, or 0.5 W. We must avoid exceeding P_{GM} or $P_{G(av)}$ depending on how the SCR is triggered. If, for example, the switch S in Fig. 10-5A is closed for only a moment—only long enough to turn on the SCR—then reopened, gate voltage and current are applied for only an instant and their product should not exceed the instantaneous maximum power P_{GM}. On the other hand, if S is closed and left closed for long periods, then the product of gate voltage and current for this period of time should not exceed the $P_{G(av)}$ rating. As you will see, the gate drive is typically obtained from an ac source, which means that V_{GK} and I_{FG} are pulsating. Seldom is continuous dc used to turn on the SCR, and therefore the P_{GM} rating usually dictates the values of the gate drive circuit components.

Most gate trigger characteristics indicate the I_{FG} and V_{GK} values that will trigger (turn on) most SCR's for which the characteristics apply. For example, in Fig. 10-9, an $I_{FG} = 0.125$ A or a $V_{GK} = 3$ V will trigger most units, and even some will trigger with smaller values that are coordinates within the shaded area. As shown, no units will trigger with $V_{GK} = 0.3$ V or less, which is a typical minimum for SCR's. To trigger *all* units positively (definitely), the I_{FG} and V_{GK} should be well outside the shaded area. In practice, a good gate drive circuit is one that turns the SCR *on hard*, which means that it provides as large as possible I_{FG} and V_{GK} values momentarily that do not exceed the gate power rating P_{GM}.

Load lines can be drawn on the gate trigger characteristics to help us select the proper trigger circuit components and to get a more complete picture of the triggering action. For example, if $R_g = 20$ Ω, and $E_g = 10$ V and the switch S is momentarily closed in the circuit of Fig. 10-5A, the gate circuit load line is line No. 1 shown in Fig. 10-9. Its end points are found as they were for dc load lines in diode circuits (Sec. 3.4). That is, the left end point is on the I_{FG} axis at $I_{FG(max)}$, which is the gate current should the gate-to-cathode junction act like a short causing V_{GK} to equal zero and source voltage E_G to appear across R_g. Thus

$$I_{FG(max)} = \frac{E_G}{R_g} \qquad (10\text{-}1)$$

and in this case

$$I_{FG(\max)} = \frac{10 \text{ V}}{20 \text{ }\Omega} = 0.5 \text{ A}$$

The other end point is on the V_{GK} axis at the minimum possible gate-to-cathode voltage $V_{GK(\max)}$. If the gate-cathode junction acts like an open, the full source voltage E_G appears across it and therefore

$$V_{GK(\max)} = E_G \qquad (10\text{-}2)$$

In this case, then,

$$V_{GK(\max)} = 10 \text{ V}$$

Thus, if the gate-to-cathode characteristics of the SCR are on curve A, the gate voltage and current at the instant the switch S is closed are $V_{GK} \cong 2.3$ V and $I_{FG} \cong 0.38$ A (coordinates of point u). On the other extreme, if the characteristics are on curve B, then $V_{GK} \cong 6.4$ V and $I_{FG} \cong 0.18$ A (coordinates of point v). More likely the actual characteristics are on curve C, in which case $V_{GK} \cong 4$ V and $I_{FG} \cong 0.3$ A (coordinates at point w) at the instant S is closed. Note that regardless of the actual characteristics, the applied V_{GK} and I_{FG} are well above the minimum values required to trigger most units that have the values in the shaded area. In this case, then, a gate source with $E_G = 10$ V and $R_g = 20$ Ω will very likely trigger *all* units when S is closed.

For reasons covered later, it is recommended that SCR's be turned on as hard as possible, which means that the gate circuit load line should be as near as possible to the P_{GM} curve. For example, if in Fig. 10-5A, the gate circuit $E_G = 10$ V, but $R_g = 6.67$ Ω instead of 20 Ω, we would expect a larger gate current and voltage, which is in fact the case. With $R_g = 6.67$ Ω, we have load line No. 2 in Fig. 10-9. Note that this load line intersects curves A, B, and C at points x, y, and z, each of which has coordinates V_{GK} and I_{FG} that are now farther from the minimum trigger requirements in the shaded area and thus the SCR is turned on harder than before. We should understand, however, that while we want the SCR turned on hard with values V_{GK} and I_{FG} whose product is nearly equal to P_{GM}, the actual V_{GK} and I_{FG} should be applied for as short a time as possible while nevertheless allowing the SCR to still be turned on.

10-3 Gate Trigger Characteristics of the SCR

If an SCR is operated with an anode current near the maximum recommended value and is to be turned on from a high forward anode-to-cathode voltage V_F, the gate current should be significantly larger than the minimum required gate trigger current I_{GT}. Otherwise, destructive *hot spots* may occur within the pnpn structure. These hot spots occur because the turn-on process is not instantaneous and takes time to spread within the SCR. This action is shown in Fig. 10-10. Note that the pnpn structure is shown equivalent to several sections *a*, *b*, *c*, *d*, and *e*. A gate trigger current I_{GT}, in effect, turns on section *a* only. Thereafter the turn-on process spreads from section *a* to *b* to *c*, etc. However, before the spreading action takes place and section *a* only is conducting, the full anode current I_A flows through it. The depletion region at the reverse biased P_1N_2 junctions in nonconducting sections is wide, forcing all current across junction P_1N_2 of section *a*, causing a hot

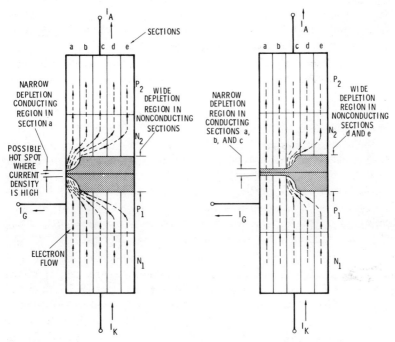

Fig. 10-10. Conduction paths in the SCR at the instant it is triggered by a relatively small gate current.

Fig. 10-11. Conduction paths in the SCR at the instant it is triggered by a relatively large gate current.

spot at this junction and possible damage to it. If section *a* is not damaged and the turn-on action spreads to the other sections, the conduction of I_A is distributed among all sections and no high-density current flows anywhere to cause a hot spot. The danger therefore exists at the instant the SCR is triggered.

If a large gate current is used to trigger the SCR, it has the tendency to turn on more than one section at the instant of its application, as shown in Fig. 10-11. Thus, if sections *a*, *b*, and *c* turn on together, the initial I_A is not concentrated in a relatively small area and hot spots are not likely to occur and damage the SCR.

The problem of hot spots does not exist if the anode current I_A, at the instant of turn-on, is not large or near the maximum value recommended by the manufacturer. It doesn't hurt, however, to drive the gate hard anyway, at least hard enough to work well outside of the minimum gate requirements (shaded area in Fig. 10-9).

Example 10-4 Suppose that in Fig. 10-5A, the SCR's trigger characteristics are given in Fig. 10-9 and its maximum recommended anode current* is $I_f = 20$ A rms. (Assume that an rms value is equivalent to the same value of continuous dc.) The anode load R is 12 Ω and source E is 120 V. The gate drive circuit R_g is 4 Ω and E_G is 2 V. (a) Will the SCR turn on when S is closed? (b) If the SCR does turn on, is its maximum current rating exceeded?

Answer 10-4 (a) Maybe, (b) no.—If a gate circuit load line is drawn on the characteristics in Fig. 10-9, it intersects 0.5 A on the I_{FG} axis and 2 V on the V_{GK} axis. It is therefore drawn through the shaded area and the SCR may or may not turn on. If the SCR does turn on, the forward voltage drop across it becomes negligible and nearly the full 120-V source appears across R. Thus

$$I_A \cong \frac{120 \text{ V}}{12 \text{ Ω}} = 10 \text{ A}$$

which obviously is less than the recommended maximum anode current I_f.

---

*In this chapter, the symbol I_A is used to represent anode current in general. The symbol I_f refers specifically to the maximum anode current recommended by the manufacturer of the SCR.

10-3 GATE TRIGGER CHARACTERISTICS OF THE SCR

Example 10-5 If we may change the value of E_G in the circuit described in Ex. 10-4, to either 3 V, 6 V, or 10 V, which would you choose, assuming everything else stays the same?

Answer 10-5 Use $E_G = 6$ V.—Sketch the load lines for all three cases. Since R_g is not changed, their slopes are the same; that is, the three load lines are parallel with the following end-points:

E_G	Left End Point	Right End Point
3 V	0.75 A	3 V
6 V	1.5 A	6 V
10 V	2.5 A	10 V

The load line when $E_G = 3$ V goes through the shaded area, which means that the SCR may or may not turn on, which makes it less than desirable.

The load line when $E_G = 6$ V is between the shaded area and P_{GM} curve, which means the SCR turns on fairly hard yet avoiding P_{GM}, indicating a desirable gate drive.

The load line when $E_G = 10$ V intersects the P_{GM} curve, which means that the maximum gate power dissipation capability might be exceeded, indicating a very undesirable gate drive.

Example 10-6 What is the minimum load resistance R that can be used in the circuit described in Ex. 10-4?

Answer 10-6 Minimum $R \cong 6$ Ω.—If the SCR is to be turned on for fairly long periods, by momentarily closing S, its anode current I_A should not exceed the maximum specified rms anode current I_f. Since this maximum current is 20 A and the load voltage is about 120 V, by Ohm's law the smallest load resistance is $R \cong 120$ V/ 20 A = 6 Ω.

If the graphical gate trigger characteristics, as in Fig. 10-9, are not made available, then the values of P_{GM} and the typical and maximum I_{GT} and V_{GT} usually are. For example, the manufacturer may provide gate information as follows:

	Typical	Maximum
I_{GT}	30 mA	80 mA
V_{GT}	1 V	3.5 V

$P_{GM} = 0.5$ W

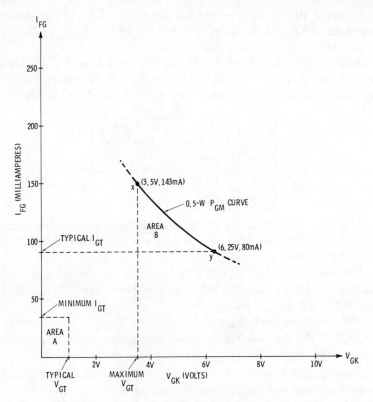

Fig. 10-12. Estimated graph of gate trigger characteristics.

This means that the SCR's for which this information applies will usually trigger with as little as 30 mA gate current and 1 V gate voltage. And all units will trigger with gate current and voltage as high as 80 mA and 3.5 V respectively. This information can be plotted on graph paper as shown in Fig. 10-12. The horizontal broken lines are drawn from 30 mA and 80 mA. The vertical broken lines are drawn from 1 V and 3.5 V. The 0.5 W P_{GM} curve is estimated by finding at least two points on it like x and y shown. If $V_{GK} = 3.5$ V, then the current I_{FG} required to cause 0.5 W power dissipation is

$$I_{FG} = 0.5 \text{ W}/3.5 \text{ V} \cong 143 \text{ mA}$$

Thus, point x on the P_{GM} curve has coordinates 3.5 V and 143 mA. Similarly, the coordinates or another point y may be found by assuming one coordinate like 80 mA; the other must therefore be

10-4 APPLICATIONS OF THE SCR

$$V_{GK} = \frac{0.5 \text{ W}}{80 \text{ mA}} = 6.25 \text{ V}$$

A smooth curve drawn through points x and y as shown in Fig. 10-12 is a close estimate of the actual P_{GM} curve. One or two more points can easily be added, such as one above $V_{GK} = 10$ V (coordinates 10 V, 50 mA) or one to the right of $I_{FG} = 200$ mA (coordinates 2.5 V, 200 mA), to give a better clue on how to sketch the P_{GM} curve.

Note that these broken lines are boundaries for areas A and B. The *typical* SCR, with these characteristics, will trigger if its gate circuit load line is just outside area A. Some units, however, may not and might require a gate circuit load line that goes through area B but, of course, is below the P_{GM} curve.

Example 10-7 If given these specs for a certain type SCR,

	Typical	Maximum
I_{GT}	10 mA	40 mA
V_{GK}	0.8 V	2.5 V

$P_{GM} = 0.5$ W

sketch (a) a graph of the gate trigger characteristics including the P_{GM} curve and areas A and B, and referring to Fig. 10-5A, sketch on your characteristics, the (b) load line if $E_G = 10$ V and $R_g = 2$ kΩ, (c) load line if $E_G = 10$ V and $R_g = 50$ Ω, (d) load line if $E_G = 12.5$ V and $R_g = 250$ Ω. (e) Which of the above three gate circuits would you recommend if all units are expected to trigger?

Answer 10-7 (a), (b), (c), and (d): see Fig. 10-13, (e) use $E_G = 10$ V, $R_g = 50$ Ω.—Load line (b) goes through area A, which means that the SCR may or may not trigger. Load line (c) goes through area B, which means that all units will trigger; thus, the gate circuit with this load line can be recommended. Load line (d) is outside of area A but does not go through area B; thus, the gate circuit with this load line will trigger *most* but probably not all units and it does not turn on the SCR as hard as it should to avoid hot spots.

10-4 APPLICATIONS OF THE SCR

SCR's are usually used to control the power delivered to a load where the source of power is 60 Hz ac. To start with a simple

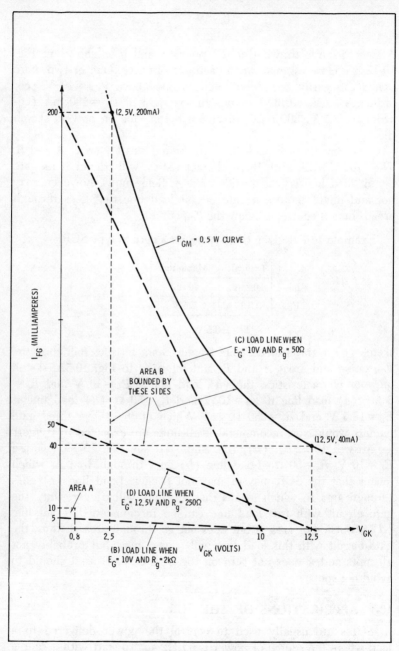

Fig. 10-13. Answers to Ex. 10-7.

10-4 APPLICATIONS OF THE SCR

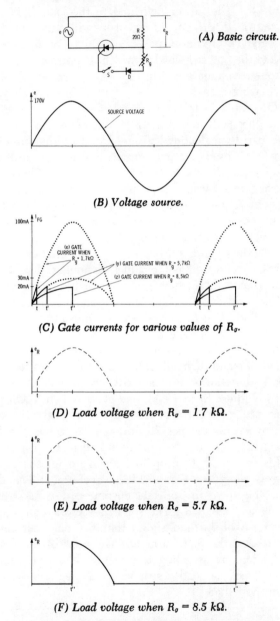

(A) Basic circuit.

(B) Voltage source.

(C) Gate currents for various values of R_g.

(D) Load voltage when $R_g = 1.7$ kΩ.

(E) Load voltage when $R_g = 5.7$ kΩ.

(F) Load voltage when $R_g = 8.5$ kΩ.

Fig. 10-14. Half-wave phase control circuit with SCR.

case, we have an application in the half-wave phase control circuit in Fig. 10-14A. The power delivered to the load R in this circuit may be controlled to a limited extent by adjustment of the variable resistance R_g. The required power capability of R_g need be small compared to the power in load R, which it controls.

The source voltage e in Fig. 10-14A is a standard 120 V, 60 Hz sine wave whose peak value is about 170 V as shown in Fig. 10-14B. Of course, while the switch S is open, no gate current flows and the SCR remains off (nonconducting), causing 0 V across the load R, neglecting leakage, for all alternations of e.

When S is closed, the gate circuit is complete and on positive alternations of e, the diode D is forward biased. Thus, current i_{FG} flows across the cathode-gate junction limited mainly by the resistances of R_g and R. This will turn on the SCR if i_{FG} exceeds the gate trigger current I_{GT}. If the SCR turns on, a voltage e_R (Fig. 10-14A) will appear across the load R.

On the other hand, during negative alternations, the SCR is reverse biased and will not conduct, assuming that its V_{ROM} rating is not exceeded. No gate current will flow either because the diode D is reverse biased and acts essentially like an open. The purpose of diode D is to prevent excessive negative gate-to-cathode voltages. The load voltage e_R, therefore, is zero for all negative alternations of e. Thus, the SCR can conduct only during positive alternations of e and then only if the gate current exceeds I_{GT}.

Suppose that the SCR's gate trigger current $I_{GT} = 20$ mA and that now we adjust R_g to 1.7 kΩ and close switch S. On positive alternations, the gate current will tend to rise as voltages e rises. However, when i_{FG} reaches 20 mA, the I_{GT} value, the SCR turns on, causing the voltage across the SCR and R_g to drop to about 1 V or so. This causes i_{FG} to drop to about zero for the remainder of the cycle as shown in Fig. 10-14C, waveform x. The dotted line represents the waveform of gate current that would exist if voltage e remained across R_g. The solid line in waveform x is the actual waveform of i_{FG} and shows that the gate current drops to about zero after the SCR turns on. The peak of the dotted waveform x *is* found by assuming that voltage e is continuously across R_g. Thus, since the load resistance R is negligible compared to R_g we can show that

$$i_{FG(\text{peak})} \cong \frac{e_{\text{peak}}}{R_g} \qquad (10\text{-}3)$$

10-4 Applications of the SCR

In this case, then,

$$i_{FG(\text{peak})} \cong \frac{170 \text{ V}}{1.7 \text{ k}\Omega} = 100 \text{ mA}$$

Note that since this waveform x tends to reach 100 mA in the first 90° of the cycle, it reaches 20 mA at time t, causing the SCR to turn on at this instant. The fact that time t occurs early in the cycle causes nearly the full source voltage e to appear across R; see the load voltage waveform in Fig. 10-14D.

Now suppose that we increase the gate resistance so that $R_g = 5.7$ kΩ. This will cause the gate current i_{FG} to rise at the rate shown in waveform y, Fig. 10-14C. The dotted line shows that if the source e would be continuously across R_g, the gate current would rise to

$$i_{FG(\text{peak})} \cong \frac{170 \text{ V}}{5.7 \text{ k}\Omega} \cong 30 \text{ mA}$$

However, because $I_{GT} = 20$ mA, the gate current reaches 20 mA at time t', causing the SCR to turn on and the gate current to become zero at this instant, t'; see the solid line waveform y. Thus, the source voltage e appears across the load R after t', resulting in a load voltage e_R as shown in Fig. 10-14E. Obviously, in this case, voltage is applied to the load at a later time. With voltage across R for a shorter period during each cycle, the average power delivered to R is decreased.

If we consider increasing R_g even more, say to 8.5 kΩ, the amplitude of i_{FG} is further decreased and its peak is

$$i_{FG(\text{peak})} \cong \frac{170 \text{ V}}{8.5 \text{ k}\Omega} = 20 \text{ mA}$$

as shown in waveform z in Fig. 10-14D. In this case i_{FG} reaches an instantaneous value of 20 mA at time t''. Thus, the SCR turns on at this instant t'' and the load R has the voltage waveform in Fig. 10-14F.

Use of an R_g larger than 8.5 kΩ in this circuit will reduce i_{FG} so that even its peak never reaches the 20-mA gate trigger current value and the SCR never turns on. This causes the load voltage e_R to remain zero for all alternations of the input voltage e.

To summarize, we can say that if R_g, in the circuit of Fig. 10-14A, is increased from small to larger resistance values, progressively more of the load voltage e_R is clipped off in the first 90° of the positive alternations. Attempts to clip off more than 90° by further increasing R_g cause a sudden loss of all load voltage.

Example 10-7 If the SCR shown in Fig. 10-14A definitely turns on with a gate current of 80 mA, what value of R_g will cause turn on at about the instant the source voltage e is at its positive peak value? The source e is sinusoidal with a peak of 170 V.

Answer 10-7 $R_g \cong 2.12$ kΩ.—Turn-on of the SCR will occur at the positive peak of e if the gate current reaches 80 mA at that instant. Thus, when the source e is at its positive peak of 170 V, the gate current should be about 80 mA or a little more. Since the gate current is limited mainly by R_g, we can rearrange Eq. 10-3 to find that

$$R_g \cong \frac{e_{\text{peak}}}{i_{FG(\text{peak})}} = \frac{170 \text{ V}}{80 \text{ mA}} \cong 2.12 \text{ k}\Omega$$

Example 10-8 Suppose that the SCR in the circuit of Fig. 10-14A has $I_{GT} = 10$ mA. If the applied voltage e has a peak of 40 V:

(a) What is the approximate value of R_g when the load voltage e_R has a waveform as in Fig. 10-15?

(b) What does the load voltage e_R look like when R_g is adjusted to 3 kΩ?

(c) What approximate value of R_g will cause the SCR to turn on near the peak of the positive alternations?

Answer 10-8 (a) $R_g \cong 1$ kΩ, (b) see Fig. 10-16, (c) $R_g \cong 4$ kΩ.—(a) Since the SCR turns on when the source reaches 10 V, the voltage across R_g is 10 V and the gate current is 10 mA at that instant. Thus, by Ohm's law,

$$R_g \cong \frac{10 \text{ V}}{10 \text{ mA}} = 1 \text{ k}\Omega$$

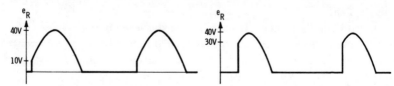

Fig. 10-15. Load voltage for the circuit in Fig. 10-14A.

Fig. 10-16. Load voltage for half-wave phase control circuit with $R_g = 3$ kΩ.

10-4 APPLICATIONS OF THE SCR

neglecting the drop across the gate-cathode junction.

(b) Because this SCR triggers when its gate current is 10 mA, the voltage across the 3-kΩ gate resistor is, by Ohm's law,

$$(3 \text{ k}\Omega)(10 \text{ mA}) = 30 \text{ V}$$

at the instant of turn-on. Thus if we still assume that $V_{GK} \cong 0$, then triggering occurs when the source e reaches 30 V on its positive alternations. See Fig. 10-16.

(c) If we want the SCR to turn on near the positive peaks of the source e, that is when about 40 V is across R_g, then by Ohm's law

$$R_g \cong \frac{40 \text{ V}}{10 \text{ mA}} = 4 \text{ k}\Omega$$

where, of course, 10 mA is the gate trigger current.

The ability to control turn-on of the SCR only during the first 90° of each positive alternation, as in the phase control circuit of

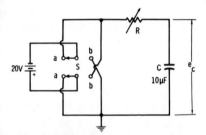

(A) With S as shown, C can change to −20 V with respect to ground.

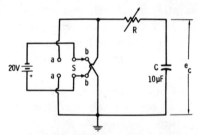

(B) With S as shown, C can charge to +20 V with respect to ground.

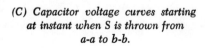

(C) Capacitor voltage curves starting at instant when S is thrown from a-a to b-b.

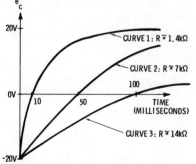

Fig. 10-17. Action of RC circuit.

Fig. 10-14, is often inadequate. Control over the full 180° of the conducting alternation is better and can be accomplished by addition of two more components: a capacitor C and diode D_2, as shown in Fig. 10-18. But before discussing how the circuit of Fig. 10-18 works, it may be useful to review the action of an RC circuit.

For example, in Fig. 10-17A, with the switch S in the position a-a as shown, 20 V is applied to the resistance R and capacitor C in series. The capacitor C will charge to the full 20 V, top plate negative with respect to ground, if S is left in position a-a long enough. If after C is fully charged, S is thrown into position b-b as shown in Fig. 10-17B, 20 V is again applied to R and C in series but this time it is of opposite polarity. This causes the capacitor to discharge and recharge again to 20 V but now the top plate is positive with respect to ground. After S is thrown from position a-a to b-b, the discharge and recharge action will take more or less time depending on the values of R and C. If the variable resistance R is adjusted to 0 Ω, then the discharge-recharge action is instantaneous and $+20$ V will appear at the top plate of C instantly after S is thrown into position b-b.

If the resistance R is greater than zero, it will definitely take time for the discharge-recharge action to occur. Curves of how voltage e_C varies with time with three different values of R are shown in Fig. 10-17C.

Curve 1 shows e_C vs t when $R \cong 1.4$ kΩ and that it takes about 10 ms for e_C to change from -20 V to 0 V after S is thrown into position b-b. Curve 2 shows that it takes about 50 ms for e_C to rise to 0 V when $R \cong 7$ kΩ. And about 100 ms is required for e_C to reach 0 V when $R \cong 14$ kΩ as shown on curve 3.

The point is, that with a larger resistance R in series with the capacitor, it takes longer for the voltage across the capacitor to change after the applied voltage polarity is reversed.

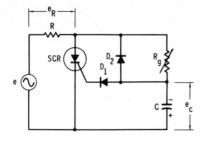

Fig. 10-18. Simple half-wave 0° to 180° phase control circuit.

10-4 APPLICATIONS OF THE SCR

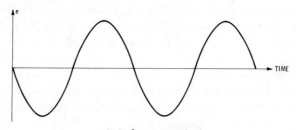

(A) Voltage e applied.

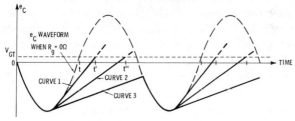

(B) Approximate curves of e_C vs time.

(C) Load voltage showing when e_C reaches V_{GT} at time t on each positive alternation.

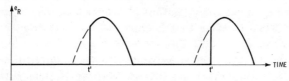

(D) Load voltage showing when e_C reaches V_{GT} at time t' on each positive alternation.

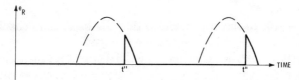

(E) Load voltage showing when e_C reaches V_{GT} at time t'' on each positive alternation.

Fig. 10-19. Waveforms for half-wave 0° to 180° phase control circuit.

Returning to the circuit in Fig. 10-18, the load is the resistance R and the power delivered to it is controlled by adjusting R_g. The applied voltage e is sinusoidal as shown in Fig. 10-19A. The capacitor C charges to the peak of the first negative alternation through the forward biased diode D_2. Negative gate current is prevented at this time by the reverse biased diode D_1. If R_g is greater than 0 Ω, the capacitor C cannot discharge and recharge to the peak of the next positive alternation because on positive alternations D_2 is reverse biased. Thus, capacitor C has to discharge through R_g and does so more or less quickly depending on the value of R_g. We can see this in Fig. 10-19B. Curves 1, 2, and 3 are approximate e_C vs time characteristics for three different finite values of R_g. The actual discharge curves are not straight lines, as we would expect after examining the bowed discharge curves between -20 V and 0 V in Fig. 10-17, but for ease of understanding the principles of the e_C vs time action, we may assume that the curves are not bowed but straight.

If we start with $R_g = 0$ Ω, the capacitor will charge as quickly on positive alternations as it did on negative ones and e_C will follow the applied voltage waveform shown as a broken line in Fig. 10-9B. Thus at the start of the positive alternation C starts to charge positively (top plate positive), and D_1 is forward biased, causing e_C to be equal to the gate-cathode voltage V_{GK}. When e_C reaches the gate trigger voltage V_{GT}, the SCR turns on at time t, causing a load voltage e_R waveform shown in Fig. 10-19C.

If R_g is increased, the e_C vs time characteristics may be on curve 1 in Fig. 10-19B showing that in this case the gate trigger voltage V_{GT} is reached at time t'. Thus, the SCR turns on at time t', resulting in load voltage e_R waveform shown in Fig. 10-19D.

If R_g is further increased, the e_C vs time characteristics may now be on curve 2. In this case V_{GT} is reached at time t'', causing the SCR to turn on at this time and an e_R waveform shown in Fig. 10-19E.

If R_g is still further increased so that the e_C vs time characteristics are on curve 3, the gate trigger voltage V_{GT} is never reached and the SCR does not turn on. This, of course, results in no output voltage e_R.

Thus, by varying R_g in the circuit of Fig. 10-18, we can cause the SCR to turn on at nearly any time during the positive alternations.

10-4 APPLICATIONS OF THE SCR

In the circuit of Fig. 10-18, the value of C and the maximum value of R_g should have a product $R_g \times C$ (time constant) of about 10 ms if the line frequency is 60 Hz and if we want control of turn on from about 0° to 180°. If the resistance R_g is not large enough, we may not have ability to cause SCR triggering in the 90° to 180° portion of the conducting alternation. For example, if $C = 0.25$ μF, then since

$$R_g C \cong 10 \text{ ms} \qquad (10\text{-}4)$$

we can solve for the gate resistance that will prevent turn-on of the SCR by simply rearranging Eq. 10-4. Thus, in this case, when

$$R_g \cong \frac{10 \text{ ms}}{C} = \frac{10 \text{ ms}}{0.25 \text{ }\mu\text{F}} = 40 \text{ k}\Omega$$

the capacitor will discharge slowly so that the e_C vs time characteristics are like curve 3 in Fig. 10-19B and the trigger voltage V_{GT} is never reached before the positive alternation is over. Of course if R_g is reduced gradually, the SCR will turn on progressively earlier in the positive alternations.

Example 10-9 Suppose that in the circuit of Fig. 10-18, the sinusoidal source $e = 120$ V (rms). What are the minimum required values of: (a) the V_{FOM} and V_{ROM} ratings of the SCR, (b) the voltage rating of the capacitor C, (c) the PIV rating of diode D_1, and (d) the PIV rating of diode D_2?

Answer 10-9 (a) $V_{FOM} \cong V_{ROM} \cong 170$ V, (b) the capacitor C must be capable of at least 170 V dc, (c) the diode D_1 must be able able to take reverse voltages of 170 V or more, (d) the diode D_2 must be able to take reverse voltages of 340 V or more.—(a) The source e in this case has a peak of about 170 V. The SCR therefore must be capable of withstanding 170-V forward and reverse breakover voltages.

(b) The capacitor C charges to 170 V, through D_2, at the peaks of the negative alternations of the source e.

(c) The capacitor voltage is across reverse biased D_1 and gate-to-cathode junction. While actually the capacitor maximum voltage, or 170 V, is distributed across both the diode D_1 and gate-cathode junction, it is safer to rate the diode PIV to 170 V. That is, assume that the gate-to-cathode voltage is negligible even though it is not. As an added precaution to protect the gate from excessive reverse voltages, a resistance is sometimes shunted across the gate and cathode leads. If this shunt resistance is in the order of 1 kΩ to 5 kΩ, its value is small compared to the reverse resistance of D_1 and therefore D_1 gets the brunt of the capacitor reverse biasing voltage.

(d) When R_g is set to its maximum value, the charge in capacitor C may have little time to discharge before the peak of the positive alternation comes along. Thus, it is possible to have about 170 V across C, in the polarity shown in Fig. 10-18, *plus* the 170 V positive peak of the source voltage, across the reverse biased D_2.

Example 10-10 If a 20 kΩ potentiometer is to be used for R_g in the circuit of Fig. 10-18, what value of capacitance C would be suitable if the source voltage e frequency is 60 Hz?

Answer 10-10 About 0.5 µF.—At 60 Hz, a time constant of 10 ms gives good control in the 90° to 180° portion of the positive alternations. Thus, by Eq. 10-4,

$$C \cong \frac{10 \text{ ms}}{R_g} = 0.5 \ \mu\text{F}$$

Another interesting circuit with an SCR is shown in Fig. 10-20. The brilliance of the lamp L in it can be varied from minimum to nearly maximum by adjusting R_1 from maximum to minimum resistance. Maximum brilliance can be achieved by closing switch S, which bypasses the SCR and its gate control circuitry. This circuit operation is similar to that of the circuit in Fig. 10-18 in that the time it takes the capacitor C to charge and discharge determines the instant when the SCR turns on during positive alternations of the source voltage e.

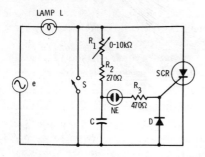

Fig. 10-20. Half-wave lamp dimmer.

10-4 APPLICATIONS OF THE SCR

Before proceeding with an analysis of the circuit in Fig. 10-20, we should have some understanding of the V-I characteristics of the neon bulb NE in it. Such characteristics are shown in Fig. 10-21B. The neon bulb consists of two electrodes and neon gas within a glass envelope. With small voltages across the neon bulb, zero or insignificant current flows through it. However, as the voltage is increased, it eventually becomes large enough to ionize the neon gas between the electrodes. When ionization occurs, the neon bulb conducts current, whose magnitude is determined by the external resistance and applied voltage. When conduction starts, the voltage across the bulb drops to a value that is smaller than the voltage initiating the ionization. For example, in Fig. 10-21, as the voltage

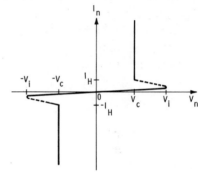

(A) Schematic diagram symbol for neon lamp.

(B) Approximate V-I characteristics of neon lamp.

Fig. 10-21. Neon lamp data.

across the bulb V_n is increased, the current I_n through it is negligible until the ionization potential V_i is reached. At V_i the gas in the bulb ionizes, causing the voltage to drop to a conduction value V_c and current to increase significantly. A minimum holding current I_H is required to keep the neon gas ionized. If the bulb current is reduced below I_H or if the voltage across it is reduced below V_c, deionization occurs and the bulb acts like an open. As shown in Fig. 10-21, regardless of the polarity of the bulb voltage V_n, its V-I characteristics are similar.

The function of the neon bulb NE in the circuit of 10-20 is to provide a relatively low resistance discharge path, after ionization, for the capacitor C through the gate-cathode junction of the SCR. For example, suppose that the potentiometer R_1 is adjusted

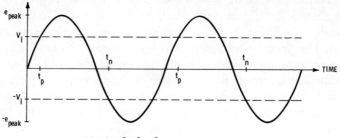

(A) Applied voltage waveform.

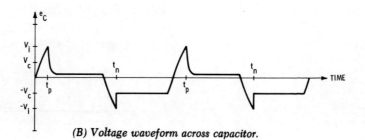

(B) Voltage waveform across capacitor.

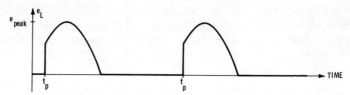

(C) Current waveform in capacitor.

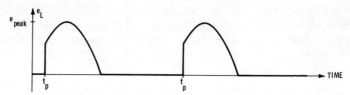

(D) Load voltage waveform.

Fig. 10-22. Waveforms for half-wave lamp dimmer of Fig. 10-20 when R_1 is adjusted to minimum resistance.

10-4 APPLICATIONS OF THE SCR

to minimum resistance. At the beginning of the positive alternation of source e, the voltage across the capacitor C will initially increase at about the same rate the source e increases because R_2 is a low-resistance charge path. This capacitor voltage e_C is across the neon lamp NE neglecting any drop across the forward biased gate-cathode junction. The bulb NE acts like an open until e_C reaches the bulb ionization voltage V_i. After ionization the effective resistance of the bulb decreases suddenly, allowing C to discharge (electrons flow counterclockwise) through the gate-cathode junction and R_3. The capacitor discharge current i_C triggers the SCR, which conducts current through the lamp L for the remainder of the positive alternation. We can see this shown graphically in Fig. 10-22. Note in Fig. 10-22B that e_C reaches V_i at time t_p, triggering the SCR at this time on each positive alternation. After time t_p, the capacitor voltage e_C drops quickly to almost zero and stays there until the source e starts its negative alternation. Note in Fig. 10-22C at time t_p, the capacitor discharge current is a sharp pulse existing for only a short time. Such a current pulse is excellent for triggering an SCR because it turns it on hard but causes very little power dissipation in the gate-cathode junction.

At the beginning of the negative alternation, the capacitor voltage e_C again increases, negatively this time, as source e increases. As shown in Fig. 10-22B, the capacitor voltage e_C reaches the bulb ionization potential V_i at time t_n. Thus, at time t_n ionization occurs, discharging C (electron flow clockwise) through NE, the forward biased diode D, and R_3. After time t_n, the capacitor voltage again decreases quickly to about either the bulb's conduction voltage V_c or zero, depending on whether the neon blub does not or does deionize. Neither alternative affects the circuit operation, however. As shown in Fig. 10-22C, the capacitor discharges quickly at time t_n in negative alternations. Note that this current flows through the diode D and not in the gate of the SCR. The diode D, in other words, provides a short across the gate-cathode junction during negative alternations, thus protecting the gate from negative gate voltages.

Fig. 10-22D shows us that since the SCR is triggered in the beginning of the positive alternations, the voltage across the lamp is nearly a half-wave rectified waveform. Actually, though not shown in Fig. 10-22D, there may be voltage across the lamp L during negative alternations, partly due to the conduction path

The Silicon Controlled Rectifier

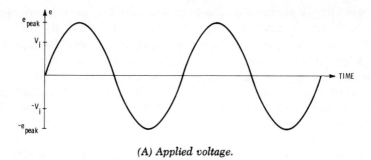

(A) Applied voltage.

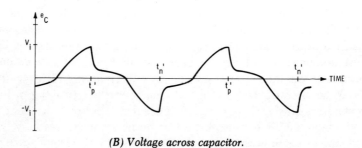

(B) Voltage across capacitor.

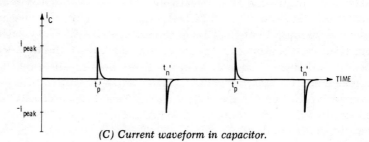

(C) Current waveform in capacitor.

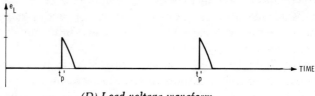

(D) Load voltage waveform.

Fig. 10-23. Waveforms for half-wave lamp dimmer of Fig. 10-20 when R_l is adjusted to maximum resistance.

10-4 APPLICATIONS OF THE SCR

present through R_1, R_2, R_3, the diode, and NE if it remains ionized. This would be more noticeable with low-wattage (high-resistance) lamps. The waveform in Fig. 10-22D is a good approximation, however, and does show us generally how the circuit works.

If we now consider adjusting R_1, in Fig. 10-20, to a large resistance value, the capacitor C will charge more slowly through R_1 and R_2. Therefore on both positive and negative alternations, it will take more time for the capacitor voltage e_C to reach the neon bulb ionization potential V_i. This causes triggering of the SCR in the latter portions of the positive alternations. The waveforms in this case are shown in Fig. 10-23. Note in Fig. 10-23B that e_C rises slowly and reaches V_i at time t_p' in positive alternations. Similarly, e_C reaches $-V_i$ at time t_n' in negative alternations. The positive current pulses in Fig. 10-23C are through the gate-cathode junction while the negative ones are through diode D. Since triggering occurs later at time t_p', the lamp voltage e_L waveform is as shown in Fig. 10-23D.

If R_1 is adjusted to maximum resistance 10 kΩ, the SCR will very likely never turn on. The capacitor voltage e_C will rise too slowly and will probably never reach the ionization voltage V_i. The term "probably" is used because neon bulbs of different types have different ionizing potentials. In some types V_i may be below 50 V while in others it may be larger than 100 V. Typically though, if $R_1 = 10$ kΩ and $C = 1$ μF, the SCR will not be triggered.

Example 10-11 Suppose that in the circuit of Fig. 10-20, the neon bulb NE fires (its gas ionizes) at 70 V and has a conducting voltage of 50 V. The source e is about 110 V to 120 V rms and the lamp L is rated 120 V, 200 W. You are to choose an SCR for this circuit. (a) What are the maximum forward and reverse anode-to-cathode voltages that are applied to this SCR? (b) What maximum reverse gate voltage $-V_{GK}$ and forward gate current i_{FG} would you expect? (c) What maximum anode current should the SCR be capable of?

Answer 10-11 (a) Ac peaks as high as 170 V may be across the anode and cathode in both forward and reverse bias, (b) $-V_{GK} \cong 0.7$ V, $i_{FG(\text{peak})} \cong 42.5$ mA, (c) $I_{A(\text{max})} \cong 0.833$ A rms.—Always during negative alternations and when the gate is not triggered on positive alternations, the SCR acts like an open. Thus the peak of the source voltage e is applied to the SCR which may be as large as

$$e_{\text{peak}} = 120 \times 1.414 \cong 170 \text{ V}$$

(b) No more than the forward drop of diode D can be across the gate-cathode junction during negative alternations, which is about 0.3 V or about 0.7 V, depending on whether the diode is germanium or silicon type respectively.

The gate current is a pulse whose peak value may be estimated as follows: With R_1 set at maximum resistance, the capacitor voltage e_C increases to the neon bulb firing voltage V_i, which is 70 V in this case. This 70 V is across the bulb just before its gas ionizes. After ionization, the bulb terminal voltage drops suddenly to 50 V, the conduction voltage V_c. At the instant of ionization, the voltage across the capacitor is still at 70 V. The difference between the capacitor and bulb voltages appears across R_3, neglecting the forward drop across the gate-cathode junction. Thus, in this case, for an instant of time, the voltage across R_3 is

$$e_C - V_c \cong 70 - 50 = 20 \text{ V}$$

and the current through R_3, which is the gate current, is

$$i_{FG(\text{peak})} = \frac{e_C - V_c}{R_3} \quad (10\text{-}5)$$

$$\cong \frac{20 \text{ V}}{470 \text{ }\Omega} = 42.5 \text{ mA}$$

This current is the peak value of the pulses shown in Fig. 10-23C because the capacitor discharge current *is* the gate current, at least during positive alternations.

(c) The SCR's maximum rms current can be closely approximated by first solving for the lamp current assuming that the full 120 V rms is applied to it. Since power $P = EI$, then solving for current we have $I = P/E$. In this case, then,

$$I = \frac{200 \text{ W}}{120 \text{ V}} \cong 1.67 \text{ A rms}$$

This 1.67 A is the rms current through the lamp when the full source voltage e (both alternations) is applied to the lamp. The SCR, however, can conduct current for a little less than half the alternations at best. That is, with R_1 adjusted to minimum resist-

10-4 APPLICATIONS OF THE SCR

ance the voltage across the lamp L may have a waveform somewhat like that in Fig. 10-22D. Thus, since the SCR conducts for about half of the alternations, its current is about half the lamp current that flows with both alternations neglecting the change in the filament resistance. In this case, then,

$$I_A \cong \frac{1.67 \text{ A}}{2} \cong 0.833 \text{ A rms}$$

The next larger size, say a 1-A rating, could be used for added safety factor.

Example 10-12 Referring to the previous example, if the SCR chosen has gate trigger characteristics shown in Fig. 10-12, will it turn on when the capacitor C discharges through the gate and the bulb NE?

Answer 10-12 Probably yes.—Since the circuit in the last example develops a gate current pulse of 42.5 mA, it is probably enough to turn on the SCR, judging from the trigger characteristics. The minimum I_{GT} in Fig. 10-12 can be estimated at about 30 mA and our 42.5 mA pulse exceeds this. Actually, when a neon bulb is used as part of the triggering circuitry, an SCR should be used whose gate trigger current I_{GT} is 10 mA or less. Thus, the capacitor discharge current more likely has an adequate amplitude to turn on the SCR. SCR's with large anode current ratings usually have larger gate current requirements too. In such cases we must use gate trigger circuitry that will provide larger gate current pulses. Examples of such circuitry are shown later.

With the circuit shown in Fig. 10-20, we learned that the brilliance of the lamp L may be gradually increased from minimum to about half of full brilliance by adjusting R_1 and that if

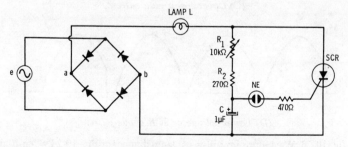

Fig. 10-24. Full-wave lamp dimmer.

we want full brilliance, switch S may be closed. Obviously, the change from about half to full brilliance is abrupt and smooth control is not possible. By a few modifications, we can have a circuit, shown in Fig. 10-24, that gives us smooth control from minimum to full brilliance with adjustments of R_1.

Note in Fig. 10-24 that a bridge rectifier is used to full-wave rectify the source voltage e. This provides forward bias on the anode-to-cathode junction of the SCR for *all* alternations of the source e. The voltage across points a and b has a waveform shown

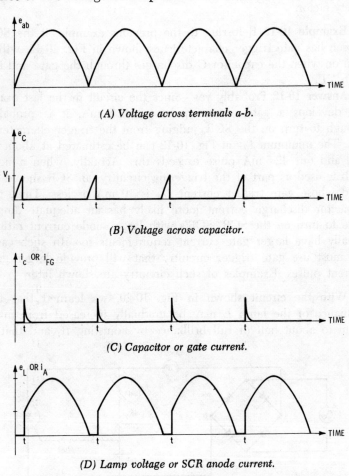

(A) *Voltage across terminals a-b.*

(B) *Voltage across capacitor.*

(C) *Capacitor or gate current.*

(D) *Lamp voltage or SCR anode current.*

Fig. 10-25. Waveforms for full-wave lamp dimmer of Fig. 10-24 when R_1 is adjusted to minimum resistance.

10-4 APPLICATIONS OF THE SCR

in Fig. 10-25A, which is the voltage applied to the lamp L and the SCR with its gate control circuitry. The capacitor C charges positively, with respect to the SCR's cathode, to the firing voltage V_i of the neon bulb on every alternation, provided R_1 is not adjusted to a large resistance value.

If R_1 is at minimum resistance, the capacitor will charge to V_i quickly, as shown in Fig. 10-25B. At time t of each alternation, the bulb NE fires and conducts a discharge current pulse, with the waveform shown in Fig. 10-25C, through the gate. The SCR,

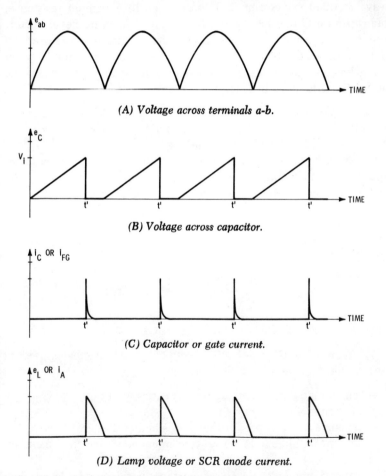

(A) Voltage across terminals a-b.

(B) Voltage across capacitor.

(C) Capacitor or gate current.

(D) Lamp voltage or SCR anode current.

Fig. 10-26. Waveforms for full-wave lamp dimmer of Fig. 10-24 when R_1 is adjusted to a large resistance.

therefore, is triggered at time t and the resulting lamp voltage e_L is shown in Fig. 20-25D.

When R_1 is adjusted to a large resistance value, it takes longer for the capacitor to charge to V_i, as shown in Fig. 10-26B. This causes the neon bulb NE to fire later—at time t' in this case. Thus, the gate current pulses occur, and the SCR turns on, at t' as shown in Figs. 10-26C and 10-26D. Further increases in the resistance of R_1, to a point, tend to cause even later triggering which causes the lamp voltage e_L to become a still smaller portion of the full-wave rectified waveform. If R_1 is adjusted to maximum resistance, the capacitor C never quite reaches V_i and triggering never occurs, resulting in no voltage to the lamp L. Sometimes it is necessary to put a resistance, about 1 kΩ to 5 kΩ, from gate to cathode when using low-current SCR's. This prevents accidental triggering of the SCR before the neon bulb ionizes.

Example 10-13 In the circuit of Fig. 10-24, if the source e is 110 V to 120 V rms and the lamp L is rated 120 V, 150 W, what current must the SCR be capable of conducting?

Answer 10-13 About 1.25 A rms.—With R_1 set at minimum resistance, the SCR and therefore the lamp L conduct for nearly the entire portion of each alternation as shown in Fig. 10-25D. This voltage waveform has approximately the same rms value as the source e has. Therefore, the SCR must be able to conduct a current about equal to the current the lamp L conducts when operated at its rated 120 V. In this case it is

$$I_A \cong \frac{P}{E} = \frac{150 \text{ W}}{120 \text{ V}} = 1.25 \text{ A rms}$$

10-5 GLOSSARY OF LETTER SYMBOLS

Here is a summary of letter symbols which we have used in describing the operation of SCR circuits.

e_C—The voltage across a capacitor that varies with time.
I_A—Any value of anode current.
i_C—The current in a capacitor that varies in amplitude with time.
I_f—The maximum anode current recommended by the manufacturer.
I_F—Forward anode current; the current in the anode lead when the anode is positive with respect to the cathode.

10-5 GLOSSARY OF LETTER SYMBOLS

i_{FG}—Forward gate current; not a constant dc.

I_{FG}—Forward gate current; the gate current that flows when the gate is positive with respect to the cathode.

I_{GT}—The forward gate current required to turn on the SCR.

I_H—Holding current; the minimum anode current required to hold the SCR in a conducting state.

I_L—Latching current; the minimum anode current required to keep the SCR conducting (latched on) after the gate is triggered with a pulse of current.

I_n—The current that flows through a neon bulb.

I_R—Reverse anode current; the current in the anode lead when the anode is negative with respect to the cathode.

$P_{G(av)}$—The recommended maximum average power that can be dissipated in the gate-cathode junction if gate current flows continuously.

P_{GM}—The recommended maximum instantaneous power that can be dissipated in the gate-cathode junction.

V_{AK}—Anode-to-cathode voltage.

V_c—The voltage across the neon bulb after it ionizes and conducts.

V_F—Forward anode-to-cathode voltage; a voltage positive on the anode with respect to the cathode.

$V_{F(on)}$—The forward voltage across the anode and cathode when the SCR is conducting (turned on).

V_{FOM}—A general term used to represent the maximum forward voltage across the anode and cathode which if exceeded causes the SCR to break over into forward conduction even though the gate is open.

$V_{FOM(nonrep)}$—Nonrepetitive maximum forward voltage; the maximum forward anode-to-cathode transient voltage that will not cause avalanche conduction.

$V_{FOM(rep)}$—Repetitive maximum forward breakover voltage; the maximum forward anode-to-cathode voltage that can be applied as a repetitive ac or continuous dc without causing the SCR to go into conduction.

V_{GK}—A general term for gate-to-cathode voltage.

V_{GRM}—The maximum reverse gate-to-cathode voltage that can be applied without causing avalanche conduction across the gate-cathode diode junction.

V_{GT}—The gate-to-cathode voltage when I_{GT} is flowing but before the SCR turns on.

V_i—The ionization potential of a neon bulb.

V_n—The voltage applied across a neon bulb.

V_R—Reverse anode-to-cathode voltage; a voltage negative on the anode with respect to the cathode.

V_{ROM}—A general term used to represent the maximum reverse voltage across the anode and cathode which if exceeded causes avalanche conduction.

$V_{ROM(nonrep)}$—Nonrepetitive maximum reverse voltage; the maximum reverse anode-to-cathode transient voltage that will not cause avalanche conduction.

$V_{ROM(rep)}$—Repetitive maximum reverse voltage; the maximum reverse anode-to-cathode voltage that can be applied as a repetitive ac or continuous dc without causing avalanche conduction.

PROBLEMS

10-1. While switch S is open in the circuit of Fig. 10-27, the source voltage is increased from 0 V to 40 V. (a) What voltage would you expect to read across the SCR? (b) What voltage should be across the load? Assume that $V_{FOM} \gg 40$ V.

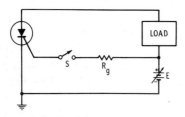

Fig. 10-27. SCR control circuit with variable dc source.

10-2. While S is open in the circuit of Fig. 10-27, the source voltage is increased from 0 V to 25 V. What voltage would you expect to read across (a) the SCR, and (b) the load? Assume that $V_{FOM} \gg 25$ V.

10-3. In a circuit as shown in Fig. 10-27, $R_g = 2$ kΩ, the load resistance is 10 Ω and the source voltage is increased from 0 V to 40 V while the switch S is open. The switch S is then closed momentarily. What are the approximate values of (a) gate current i_{FG} while S is closed and (b) anode current I_A, (c) voltage across the SCR, and (d) voltage across the load after S is reopened?

Problems

10-4. In a circuit as shown in Fig. 10-27, $R_g = 3$ kΩ, the load resistance is 24 Ω and the source voltage is increased from 0 V to 120 V while S is open. The switch S is then closed momentarily. What are the approximate values of (a) gate current i_{FG}, while S is closed, and (b) anode current I_A, (c) voltage across the SCR, and (d) voltage across the load after S is reopened?

10-5. If in the circuit of Fig. 10-28 the source e has the waveform of Fig. 10-29 in which $e_{peak} = 170$ V, $R_g = 5$ kΩ, the SCR's $I_{GT} = 30$ mA, and the V_{ROM} and V_{FOM} of the SCR are not exceeded, at approximately what time of the input cycle (Fig. 10-29) does the SCR turn on?

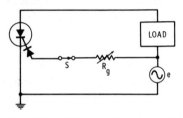

Fig. 10-28. SCR control circuit with ac source.

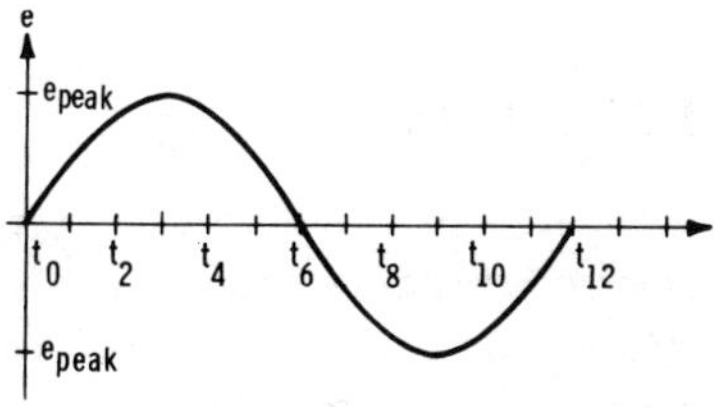

Fig. 10-29. Reference input waveform of SCR control circuit.

10-6. If in the circuit in Fig 10-28 the source e has the waveform of Fig. 10-29 with $e_{peak} = 40$ V, $R_g = 2$ kΩ, the SCR $I_{GT} = 10$ mA, and its V_{ROM} and V_{FOM} ratings are not exceeded, at approximately what time of the input cycle (Fig. 10-29) does the SCR turn on?

10-7. If R_g is adjusted to 10 kΩ in the circuit described in Prob. 10-5 above, at approximately what time of the input cycle does the SCR turn on? Assume that the load resistance is much less than 10 kΩ.

10-8. If R_g is adjusted to 5 kΩ in the circuit described in Prob. 10-6 above, at approximately what time of the input cycle does the SCR turn on? Assume that the load resistance is much less than 5 kΩ.

10-9. Referring again to the circuit described in Prob. 10-5, if R_g can be reduced down to nearly 0 Ω and the load consumes 200 W, what rms current must the SCR be capable of conducting?

10-10. Referring again to the circuit described in Prob. 10-6, if R_g can be reduced down to nearly 0 Ω and the load consumes 200 W, what rms current must the SCR be capable of conducting?

10-11. If the SCR in Fig. 10-27 has the following ratings; $V_{ROM(rep)} = V_{FOM(rep)} = 100$ V, $V_{ROM(nonrep)} = V_{FOM(nonrep)} = 130$ V, and the source E is adjusted gradually from 0 V to 150 V, what are the voltages across the load and SCR for each of the following source voltages: (a) 30 V, (b) 60 V, (c) 90 V, (d) 120 V, (e) 150 V? Assume that switch S is open.

10-12. If the SCR in Fig. 10-27 has the following ratings: $V_{ROM(rep)} = V_{FOM(rep)} = 200$ V, $V_{ROM(nonrep)} = V_{FOM(nonrep)} = 250$ V, and the source E is adjusted gradually from 0 V to 250 V, what are the voltages across the load and SCR at each of the following source voltages: (a) 70 V, (b) 140 V, (c) 210 V, (d) 250 V? Assume that switch S is open.

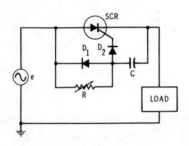

Fig. 10-30. SCR control circuit with two diodes.

10-13. Suppose that in the circuit described in Prob. 10-11, we use the same components but reverse the connections on the source E. If E is again increased gradually from 0 V to 150 V, what are the load and SCR voltages at each of the source E voltages given?

10-14. If the voltage waveform of Fig. 10-31B is across the load in circuit Fig. 10-30, to which of the following relative resistance values do you think R is adjusted: low, intermediate, high.?

10-15. If the voltage waveform Fig. 10-32A is across the load in the circuit of Fig. 10-30, to which of the following relative resistance values is R adjusted to: low, intermediate, high?

10-16. If R can be adjusted from about 500 Ω to 10 kΩ, and $C = 1$ μF, in Fig. 10-30, what approximate value of R will cause the SCR to turn on at time t_9 in the input cycle (Fig. 10-29)?

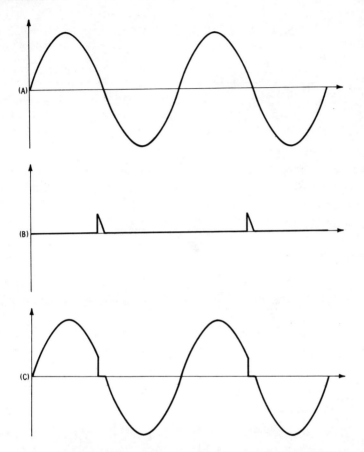

Fig. 10-31. Possible waveforms in the SCR circuit of Fig. 10-30.

10-17. By varying R in Fig. 10-30, we can cause the SCR to turn on within approximately what time interval of the input cycle (Fig. 10-29)?

10-18. What is the purpose of diode D_1 in the circuit of Fig. 10-30?

10-19. What is the purpose of diode D_2 in the circuit of Fig. 10-30?

10-20. With R_1 adjusted so that the lamp in Fig. 10-33 is almost out, what voltage waveform, in Fig. 10-34, would you expect to find across the lamp?

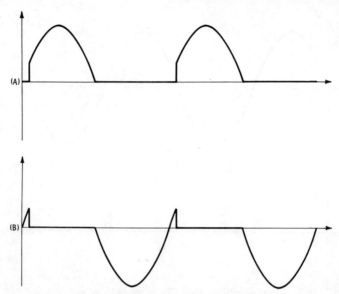

Fig. 10-32. Possible waveforms in the circuit of Fig. 10-30.

10-21. With R_1 adjusted so that the lamp in Fig. 10-33 is at full brilliance, what voltage waveform, in Fig. 10-34, would you expect to find across R_3?

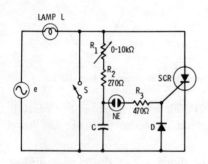

Fig. 10-33. Lamp dimmer circuit.

10-22. If R_1 is adjusted to 10 kΩ in Fig. 10-33, which of the voltage waveforms in Fig. 10-34 would you expect to find across (a) the lamp, and (b) the SCR?

10-23. If R_1 in Fig. 10-33, is adjusted so that the SCR triggers early in each positive alternation, which of the waveforms in Fig. 10-34 represents the *current* waveform in the gate lead of the SCR?

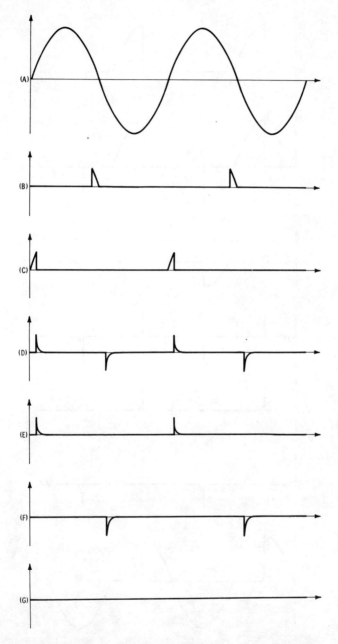

Fig. 10-34. Possible component waveforms for lamp dimmer circuit.

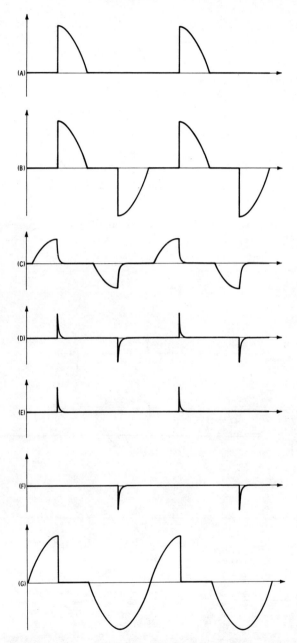

Fig. 10-35. Possible component waveforms for SCR circuit of Fig. 10-33.

PROBLEMS 339

10-24. If the source e is sinusoidal and the voltage waveform across the lamp in Fig. 10-33 has waveform A in Fig. 10-35, what does the voltage across the capacitor C look like? Select from Fig. 10-35.

10-25. If the source e is sinusoidal and the voltage waveform across the SCR in Fig. 10-33 has waveform G in Fig. 10-35, what does the voltage across the lamp look like? Select from Fig. 10-35.

10-26. If voltage waveform D, in Fig. 10-35, is viewed with a scope, across R_3 in the circuit of Fig. 10-33, what *current* waveform in Fig. 10-35 would you expect in the diode D?

10-27. Referring to the previous problem, what *current* waveform would you expect in the gate lead of the SCR?

10-28. Suppose that the neon bulb in Fig. 10-33 ionizes when the potential across it reaches 80 V, and has about 50-V drop after ionization. What is the peak value of the forward gate current? Assume that the drop across the gate-cathode junction is negligible and that R_1 is adjusted so that the SCR is being triggered on each cycle.

10-29. What is the answer to Prob. 10-28 above if the neon bulb is replaced with another one whose characteristics are $V_i = 60$ V and $V_c = 40$ V?

10-30. If the lamp in Fig. 10-33 is rated 200 W, 110 V, what maximum rms current is the SCR expected to conduct if the source $e = 110$ V rms?

10-31. If the lamp in the circuit of Fig. 10-33 is rated 60 W, 120 V, what maximum rms current is the SCR expected to conduct if the source $e = 120$ V rms?

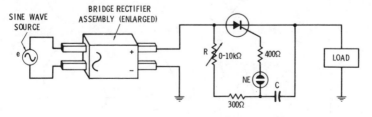

Fig. 10-36. Full-wave SCR control circuit.

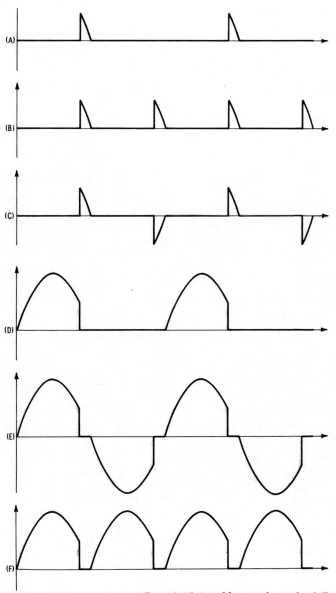

Fig. 10-37. Possible waveforms for full-wave

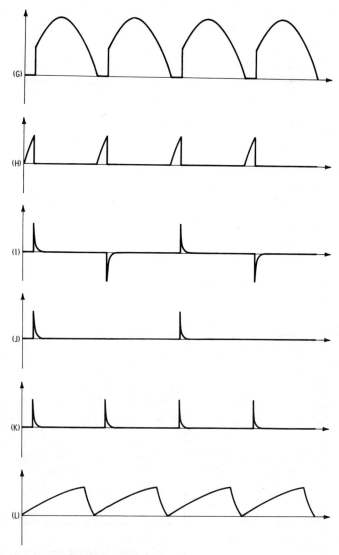

SCR control circuit of Fig. 10-36.

10-32. With a sine wave source $e = 120$ V rms applied to the circuit in Fig. 10-33, what voltage waveforms would you expect to find across (a) the lamp and (b) the SCR if the switch S is closed? Choose answers from Figs. 10-34 and 10-35.

10-33. If the lamp never lights in the circuit of Fig. 10-33, regardless of the adjustments made on R_1, and a complete sine wave is viewed across the capacitor C, which components are most likely defective?

10-34. If R in the circuit of Fig. 10-36 is adjusted to nearly its maximum value, what are the most likely waveforms across (a) the load and (b) the SCR? Select answer from Fig. 10-37.

10-35. If R in the circuit of Fig. 10-36 is adjusted to its minimum value, what are the most likely waveforms across (a) the load, and (b) the SCR? Pick from Fig. 10-37.

10-36. With R adjusted to its minimum value in the circuit of Fig. 10-36, which of the waveforms in Fig. 10-37 best represents the gate current waveform?

10-37. With R adjusted to its minimum value in the circuit of Fig. 10-36, which of the waveforms in Fig. 10-37 best represents the voltage across the capacitor C?

10-38. If $e = 100$ V rms and the load resistance is 20 Ω, what is the approximate required rms current rating of the SCR in the circuit of Fig. 10-36?

10-39. If $e = 110$ V rms and the load resistance is 10 Ω, what is the approximate required rms current rating of the SCR in the circuit of Fig. 10-36?

10-40. What is the approximate peak gate current $i_{FG(\text{peak})}$, in the circuit of Fig. 10-36, assuming that the voltage drop across the gate-cathode junction is negligible and if the neon bulb characteristics are $V_i = 70$ V and $V_c = 40$ V?

10-41. What is the approximate peak gate current $i_{FG(\text{peak})}$, in the circuit of Fig. 10-36, assuming that the voltage drop across the gate-cathode junction is negligible and the neon bulb characteristics are $V_i = 78$ V and $V_c = 32$ V?

10-42. Referring to the circuit described in Prob. 10-38, what are the minimum required $V_{RQM(\text{rep})}$ and $V_{FOM(\text{rep})}$ ratings of the SCR?

10-43. Referring to the circuit described in Prob. 10-39, what are the minimum required $V_{ROM(\text{rep})}$ and $V_{FOM(\text{rep})}$ ratings of the SCR?

11

Other Members of the Thyristor Family

The SCR, though popular, is not the only thyristor of interest to us. We should become familiar with some of the other thyristors, especially those that are used along with SCR's. Some are used in sources of gate current pulses for SCR's. Others replace SCR's in certain applications. These other thyristors are, for the most part, quite specialized in their possible applications due to their peculiar characteristics. In this chapter, we shall see a few thyristor family members, their characteristics, and some applications.

11-1 THE UNIJUNCTION TRANSISTOR

The *unijunction transistors* (UJT) is a three-terminal device like a junction transistor. Unlike the transistor, however, the UJT is not useful in signal amplifiers. Instead, it is usually used as a voltage controlled switch. In other words, the UJT may be used where we need a sudden decrease in resistance across two terminals and where this decrease can be triggered by a specific voltage applied to these terminals. Unlike the SCR, which is triggered by a gate *current*, the UJT is triggered with a *voltage*. As you will see, the UJT is used to generate sawtooth waveforms and current

11-1 THE UNIJUNCTION TRANSISTOR

pulses. Current pulses, as we learned in the last chapter, are especially handy for turning SCR's on hard while causing little power dissipation in the gate.

As the name implies, the *unijunction* transistor has only one junction of p- and n-type semiconductor materials. As shown in Fig. 11-1A, the UJT consists of a bar or block of n-type material to which a small pellet of p-type material is fused. Leads connected to n-type bar are called *base 1* (B_1) and *base 2* (B_2). Typically, base 2 is biased positively with respect to base 1 as shown in Fig. 11-1B. The lead connected to the pellet is called the *emitter* (E).

The bar of n-type material is lightly doped, which means that it has relatively few charge carriers available to support current through it while the emitter E lead is open. Thus, the resistance

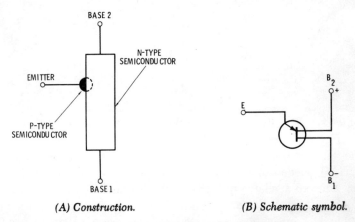

(A) *Construction.*   (B) *Schematic symbol.*

Fig. 11-1. The unijunction transistor.

looking into B_1 and B_2 is quite high if lead E is open. It is convenient to look at the bar as equivalent to two resistances as shown in Fig. 11-2. The diode in the emitter lead represents the diode formed by the pn junction.

In Fig. 11-3, the source voltage V_{BB} is distributed across R_{B1} and R_{B2} as in any voltage divider. The ratio of voltage across R_{B1}, call it V_{B1}, to the source voltage V_{BB} is called the *intrinsic standoff ratio* and is referred to with the Greek letter η (eta). With lead E open, resistances R_{B1} and R_{B2} are in series and the individual voltage drops across them are proportional to their resistance values. Therefore, the intrinsic standoff ratio η is also equal to the ratio of

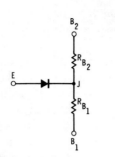

Fig. 11-2. Equivalent circuit of the UJT.

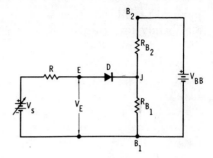

Fig. 11-3. Typical biasing polarities of the UJT.

R_{B1} to the total resistance of the bar ($R_{B1} + R_{B2}$). In equation form we have

$$\eta = \frac{V_{B1}}{V_{BB}} \quad (11\text{-}1)$$

or

$$\eta = \frac{R_{B1}}{R_{B1} + R_{B2}} \quad (11\text{-}2)$$

Manufacturers specify η values for the UJT's they make. Typically, the specified η values vary from 0.5 to 0.85.

The value of η is important in that it tells us the amount of V_E, in Fig. 11-3, needed to cause R_{B1} to change from a low-current conducting state to a relatively high-current conducting state. That is, with low values of V_E, current up through R_{B1} is very small or negligible. If V_E is increased to a value slightly larger than V_{B1}, the character of R_{B1} changes suddenly, allowing large current to flow through it and in the emitter lead. After this sudden change, the UJT is said to be *turned on* or *triggered*. The specific value of emitter voltage V_E that causes this change in character is called the *peak voltage* V_p and is dependent on the η value of the UJT and the source voltage V_{BB} in the following way:

$$V_p = V_F + \eta V_{BB} \quad (11\text{-}3)$$

where V_F is the forward drop across the diode junction. For example, if the UJT in the circuit of Fig. 11-4A has $\eta = 0.8$, it will trigger when the emitter E to B_1 voltage V_E reaches

$$V_p = V_F + \eta V_{BB} \cong 0.7 + 0.8(12) = 0.7 + 9.6 = 10.3 \text{ V}$$

11-1 THE UNIJUNCTION TRANSISTOR

assuming 0.7 V drop across the diode junction. The voltage across R_{B1} is the second term on the right side of the above equation, or it may be found by rearranging Eq. 11-1 and solving as follows: $V_{B1} = \eta V_{BB} = 0.8(12) = 9.6$ V.

We can get a better picture of the UJT's action by first noting that V_s and V_{BB} in Fig. 11-4A are bucking each other. When V_E is less than V_{B1}, the pn junction, represented by the diode D in the equivalent circuit (Fig. 11-4B), is reverse biased. Thus, at best, only a small, usually negligible, leakage emitter current I_E flows. If V_s is increased so that V_E reaches 9.6 V, zero voltage is across

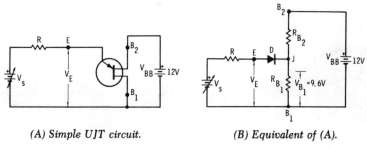

(A) Simple UJT circuit.   (B) Equivalent of (A).

Fig. 11-4. UJT with $\eta = 0.8$ and B_1 to B_2 bias voltage of 12V.

the diode D and it is on the verge of going into conduction. Increasing V_s and V_E by about another 0.7 V, the approximate forward drop across the diode D, causes the diode junction to become forward biased and capable of conducting a large I_E. Once the junction is forward biased, I_E flows across the diode junction and the resistance R_{B1} suddenly decreases to the point that the magnitude of I_E is limited mainly by the resistance R, which is in the emitter lead external to the UJT.

The reason why the resistance of R_{B1} decreases can be explained as follows: The n-type semiconductor forming the bar is only lightly doped and therefore normally has few charge carriers and therefore a high resistance. However, when the pn junction becomes forward biased and conducts I_E, additional charge carriers (holes) are injected into the bar. These injected holes are swept toward the B_1 lead because it is negative with respect to the emitter lead E. These holes recombine with electrons flowing out of the source V_s into B_1. A continuous recombination supports I_E. Of course, for every hole injected into the bar at J, an electron flows out of the bar at J and into the positive terminal of V_s through R. This action takes place in an avalanche fashion. That is, as I_E begins to flow and holes are injected into the bar, its conductivity increases, which further increases I_E. This injects even more holes and I_E further increases, etc. Thus, a short time after I_E is triggered by V_p, its magnitude greatly increases, limited mainly by the resistance R in the emitter circuit.

Although the resistance R_{B2} also decreases after turn-on, by far the greater reduction of resistance takes place in the lower section R_{B1} because it receives most of the injected holes. While I_E through the bar increases during the decrease in R_{B1}, the voltage V_{B1} de-

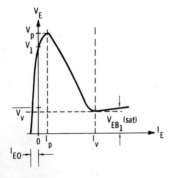

Fig. 11-5. V-I characteristic of emitter to base 1 terminals of typical UJT.

creases. This action is shown graphically in Fig. 11-5. When $V_E = 0$ V, the emitter to base 1 V-I characteristics show that a negative emitter current $-I_{EO}$ flows which is called the *emitter to base 2 leakage*. This is a clockwise current (electron flow) from left to right through R, pushed by the voltage source V_{BB} as shown in Fig. 11-6A. As V_E is increased, it starts to buck V_{BB} and the clockwise current $-I_E$ decreases (Fig. 11-6B). When V_E reaches V_1 in Fig. 11-5, the effect of V_{BB} is completely bucked out and $I_E = 0$. Further increases in V_E now cause a counterclockwise current as shown in Fig. 11-6C and the UJT is on its way to an avalanche. When V_E reaches V_p, the conduction of I_E increases from I_p to a valley current I_v and the voltage V_E drops to a valley voltage V_v. If I_E is increased beyond I_v, voltage V_E starts to increase again but only slightly. Voltage $V_{EB1(sat)}$ shown in Fig. 11-5 is the approximate voltage from the emitter to base 1 that you can expect if current I_E is larger than I_v. The major path of I_E after the UJT is

11-1 THE UNIJUNCTION TRANSISTOR

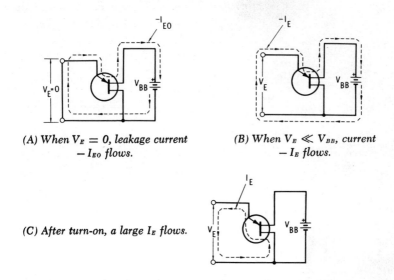

(A) When $V_E = 0$, leakage current $-I_{EO}$ flows.

(B) When $V_E \ll V_{BB}$, current $-I_E$ flows.

(C) After turn-on, a large I_E flows.

Fig. 11-6. Current paths in the UJT circuit.

triggered is shown in Fig. 11-6C. It should be understood that before triggering—that is, before V_E reaches V_p—the emitter current, whose path is shown in Figs. 11-6A and 11-6B, is small and may be considered negligible for most practical purposes. After the UJT is triggered, the emitter current, with the path shown in Fig. 11-6C, is relatively large.

We may get a better appreciation of the UJT by analyzing its operation in a practical circuit. The circuit with the UJT in Fig. 11-7 is called a *relaxation oscillator*. Generally, it works as follows: Referring to Figs. 11-5 and 11-7, when switch S is closed, capacitor C starts to charge through R_1 and R_2. The voltage e_C across the

Fig. 11-7. UJT relaxation oscillator.

capacitor is initially less than emitter peak voltage V_p, and the UJT is in a nonconducting state; that is, the emitter E to base 1 resistance (R_{B1}) is large initially. With time e_c increases and eventually reaches V_p and triggers the UJT and R_{B1} quickly decreases, providing a discharge path for C across the emitter-to-base 1 junction through R_4. Thus, the voltage e_C quickly decreases to the valley voltage V_v, typically about 2 V, causing the UJT to become nonconducting again and the capacitor to recharge. This process of discharge and recharge continually repeats itself, producing a sawtooth voltage waveform between E and ground as shown in Fig. 11-8A. Because the discharge is through a relatively small resistance, mainly being R_4, it occurs quickly and the discharge current is a pulse shown as i_{R4} in Fig. 11-8B. The voltage across R_4, as shown as e_{R4}, is therefore a pulse, too, during discharge.

If R_1 is adjusted from smaller to larger resistance, the capacitor charges more slowly. Thus, it takes more time for e_C to reach V_p, and the voltage at E to ground has the waveform shown in Fig.

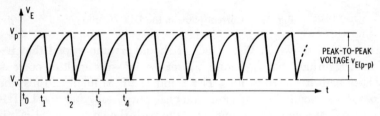

(A) *Voltage between emitter and ground.*

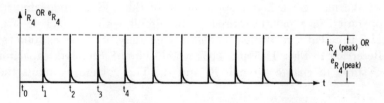

(B) *Voltage between B_1 and ground, or current through R_4.*

Fig. 11-8. Waveforms of UJT relaxation oscillator of Fig. 11-7 for relatively low R_1.

11-9A. We may conclude that the frequency of the sawtooth waveform at E and the spiked waveform at B_1 to ground is varied by varying R_1.

11-1 THE UNIJUNCTION TRANSISTOR

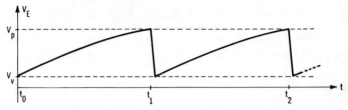

(A) Voltage between emitter and ground.

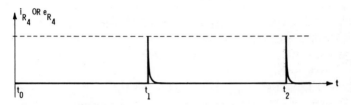

(B) Voltage between B_1 and ground, or current through R_4.

Fig. 11-9. Waveforms of UJT relaxation oscillator of Fig. 11-7 for increased R_1.

The total time of the full cycle of the capacitor charge and discharge, shown as the time from t_0 to t_1, or t_1 to t_2, etc., in Figs. 11-8 and 11-9, is called the *period of oscillation, T*. In other words, one sawtooth occurs in time T. Time T can be determined with either of the equations

$$T \cong RC \ln \frac{1}{1-\eta} \qquad (11\text{-}4)$$

or

$$T \cong 2.3 \, RC \log \frac{1}{1-\eta} \qquad (11\text{-}5)$$

where
  T is the period in seconds,
  R is the total resistance in the charging path of C, which is $R_1 + R_2$ in Fig. 11-7,
  C is the capacitance in farads,
  ln is the logarithm to the base e,
  log is the logarithm to the base 10,
  η is the intrinsic standoff ratio.

If approximations permit, assuming $\eta \cong 0.63$ simplifies the above equations to

$$T \cong RC \tag{11-6}$$

The number of periods that occur in one second is called the *frequency (f)* of the waveform and is found by the equation

$$f = \frac{1}{T} \tag{11-7}$$

where T is the period of the waveform.

Example 11-1 If R_1 is adjusted to 27 kΩ in the circuit of Fig. 11-7, what is the approximate period of each sawtooth and frequency of the sawtooth waveform?

Answer 11-1 $T \cong 15$ ms, $f \cong 66.6$ Hz.—Since an approximation is requested, we may assume that the intrinsic standoff ratio $\eta \cong 0.63$ and Eq. 11-6 is applicable. Thus, in this case, we find that the total resistance in the charging path of the capacitor C is 30 kΩ and

$$T \cong RC = (30 \text{ k}\Omega)(0.5 \text{ }\mu\text{F}) = 15 \text{ ms}$$

Substituting this answer into Eq. 11-7, we get

$$f = \frac{1}{T} \cong \frac{1}{15 \text{ ms}} \cong 66.6 \text{ Hz}$$

Example 11-2 What is the peak-to-peak value of the sawtooth waveform at point E to ground in Fig. 11-7 if $\eta \cong 0.63$?

Answer 11-2 $V_{E(p-p)} \cong 6.26$ V.—The upper peak occurs at the peak voltage V_p of the UJT which can be found by Eq. 11-3. Assuming that the diode formed at the junction of the emitter p-type material and bar n-type material has about 0.7 V forward drop, and assuming $\eta \cong 0.63$,

$$V_p = V_F + \eta V_{BB} \cong 0.7 + 0.63(12) = 8.26 \text{ V}$$

Since the valley voltage V_v is typically about 2 V, our sawtooth wave has a peak-to-peak amplitude that is the difference between V_p and V_v. In this case, then, $v_{E(p-p)} = V_p - V_v \cong 8.26 - 2 = 6.26$ V.

For reasons that are covered later, we should know the amplitudes of the discharge current i_{R_4} and the voltage e_{R_4}. These amplitudes may be affected by the *switching time* of the UJT, which is the time it takes the resistance between emitter E and base 1 (R_{B1}) to decrease from its maximum to minimum resistance value after triggering occurs. The switching time t_{on} of a given type

11-2 APPLICATIONS OF THE UJT

UJT can usually be obtained from the manufacturer's literature. Its effects on the circuit operation, however, are negligible if its value is one-tenth or less than the time constant of the discharge circuit consisting of capacitor C and R_4. That is, since capacitor C discharges through R_4, we may ignore the switching time t_{on} if it is about equal or less than one-tenth the product CR_4. For most practical purposes and therefore in this text, t_{on} of the UJT is ignored. Therefore we may assume that at the instant the UJT is triggered, the capacitor is effectively placed in shunt with R_4 because R_{B1} decreases to a negligible value. This tells us that as the capacitor voltage e_C increases during charge, it triggers the UJT at voltage V_p. Thus, the capacitor voltage of a value V_p less the valley voltage V_v appears across R_4 at that instant. The current through R_4 is the discharge current and the amplitude of its peak value at the instant of turn-on can be determined with the equation

$$i_{R4(peak)} = \frac{e_{R4(peak)}}{R_4} \cong \frac{V_p - V_v}{R_4}$$

Example 11-3 What are the instantaneous peak values of the voltage across and the current through R_4 in Fig. 11-7 assuming $\eta = 0.63$?

Answer 11-3 $e_{R4(peak)} \cong 6.26$ V, $i_{R4(peak)} \cong 125$ mA.—As shown in the last example, with $\eta = 0.63$, the UJT triggers at $V_p \cong 8.26$ V. Thus, a voltage pulse of about 8.26 V $- V_v \cong 2$ V appears across R_4 at the instant of turn-on. By Ohm's law, then, the current pulse through R_4 is found to be

$$i_{R4} = \frac{e_{R4}}{R_4} \cong \frac{6.26 \text{ V}}{50 \text{ }\Omega} \cong 125 \text{ mA}$$

11-2 APPLICATIONS OF THE UJT

We have already seen one application of the UJT which is in the relaxation oscillator of Fig. 11-7. Another worth noting is a circuit in which the UJT is used to trigger an SCR as shown in Fig. 11-10. This circuit is similar to the full-wave lamp dimmer circuit shown in Fig. 10-24 of the previous chapter with the exception that instead of the neon bulb we are now using a UJT as a part of the triggering mechanism.

With 120 V ac rms applied, a full-wave rectified voltage with a peak of about 170 V appears across a and b in the circuit of Fig.

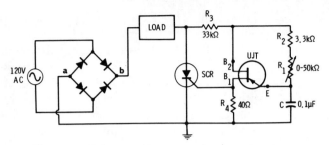

Fig. 11-10. Full-wave phase control with SCR and UJT.

11-10. Thus, the SCR is forward biased and capable of conducting on each alternation. Of course, it does so only if its gate is triggered. Generally, the circuit works as follows: Capacitor C starts to charge as the rectified voltage e_{ab}, shown in Fig. 11-11A, increases from zero. The capacitor voltage is at the UJT's emitter E with respect to base 1. As V_E increases, it eventually reaches V_p, which turns on the UJT. At the instant of turn-on the UJT allows the capacitor C to discharge through the cathode-gate junction of the SCR, thus triggering it. Once triggered, the SCR conducts load current for the remainder of the alternation.

When R_1 is adjusted to a relatively large resistance value, the capacitor C charges slowly, causing V_E to reach V_p late in the alternation. That is, with a large R_1, V_E may not reach V_p until time t_1 shown in the first two alternations of Fig. 11-11B. Thus, in this case, the UJT and the SCR turn on at time t_1, which is nearly at the end of each alternation, causing a load voltage e_L that is only a small portion of the rectified waveform, as shown in Fig. 11-11D. Of course, the average power dissipation in this load therefore is small, too.

If the resistance R_1 is decreased, capacitor C charges faster and V_E reaches V_p sooner, as shown in the third and fourth alternations in Fig. 11-11B. Thus, the UJT and SCR turn on at time t_2, resulting in a load voltage e_L as shown in Fig. 11-11D, indicating that now e_L is about half of the voltage e_{ab}. The power dissipation in the load therefore increases.

If R_1 is decreased even more, the capacitor charges quickly and V_E reaches V_p early in each alternation, causing turn-on of the UJT and SCR at time t_3 as shown in the last two alternations of Fig. 11-11B. Note that now the load voltage e_L has nearly the same waveform as the rectified voltage e_{ab}. Of course the largest power

11-2 APPLICATIONS OF THE UJT

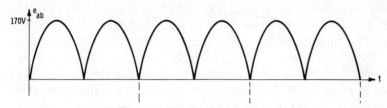

(A) Full-wave rectified voltage across a and b.

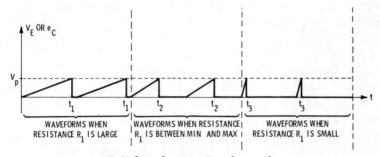

(B) Voltage between E and ground.

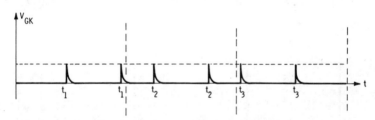

(C) Voltage from B_1 to ground, which is V_{GK} of SCR.

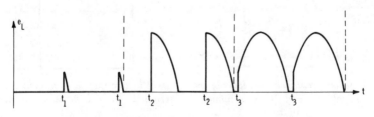

(D) Voltage across the load.

Fig. 11-11. Waveforms of full-wave phase control of Fig. 11-10.

is delivered to the load in this case. If R_1 is adjusted to its maximum resistance value, the capacitor C will likely never charge to voltage V_p and both the UJT and the SCR will never turn on.

The purpose of R_3 in the circuit of Fig. 11-10 is to keep the voltage across the B_1 and B_2 terminals (across resistance $R_{B1} + R_{B2}$) of the UJT within limits specified by the manufacturer. The resistance R_4 in the B_1 lead must be large enough to shunt most of the discharge current of capacitor C up through the cathode-gate junction of the SCR. That is, if R_4 is too small, most of the discharge current may flow through it, resulting in insufficient current in the gate lead and failure to trigger the SCR. Resistance R_4, however, cannot be too large. If it is, it may cause uncontrollable triggering of the SCR even though the capacitor never charges to V_p. The reason for this is that the UJT does not have infinite resistance before it turns on. Typically the resistance of the bar ($R_{B1} + R_{B2}$) is in the range of 5 kΩ to 10 kΩ. Since ideally the UJT is supposed to be nonconducting like an open switch before turn-on, we have the worse case when $R_{B1} + R_{B2} \cong 5$ kΩ. The leakage current through $R_{B1} + R_{B2}$ of the UJT causes voltage drops across R_3 and R_4 before turn-on. If R_4 is too large, the voltage drop across it may be larger than the gate trigger voltage V_{GT} of the SCR and thus may trigger

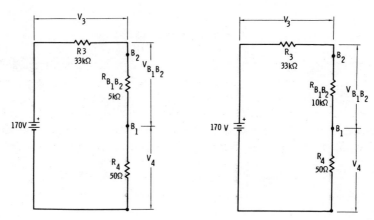

(A) When resistance seen looking into terminals B_1 and B_2 is a minimum of 5 kΩ.

(B) When resistance seen looking into terminals B_1 and B_2 is a maximum of 10 kΩ.

Fig. 11-12. Equivalent circuit of full-wave phase control of Fig. 11-10 before UJT turns on, looking to the right of the SCR.

11-2 APPLICATIONS OF THE UJT

it though the UJT never turns on. For example, when e_{ab} in Fig. 11-10 is at its peak, 170 V is distributed across R_3, the UJT, and R_4. Therefore, before the UJT and SCR turn on, we may show equivalent circuits as in Fig. 11-12. In these equivalents, the bar resistance $R_{B1} + R_{B2}$, shown as R_{B1B2}, may be between 5 kΩ and 10 kΩ as indicated in Figs. 11-12A and 11-12B respectively. In series circuit fashion we can find the voltage across R_4 in Fig. 11-12A as follows:

$$V_4 = \frac{(170 \text{ V}) R_4}{R_3 + R_{B1B2} + R_4} \cong \frac{(170 \text{ V})(50 \text{ }\Omega)}{38 \text{ k}\Omega} = 224 \text{ mV}$$

Similarly, in Fig. 11-12B we find that

$$V_4 \cong \frac{(170 \text{ V})(50 \text{ }\Omega)}{43 \text{ k}\Omega} \cong 198 \text{ mV}$$

In either case, V_4 is too small to be troublesome. However, if a much larger R_4 were used, we could have uncontrollable triggering. For example, if $R_4 = 1$ kΩ is used and if $R_{B1B2} \cong 5$ kΩ,

$$V_4 \cong \frac{(170 \text{ V})(1 \text{ k}\Omega)}{39 \text{ k}\Omega} \cong 4.36 \text{ V}$$

and if $R_{B1B2} \cong 10$ kΩ, then

$$V_4 \cong \frac{(170 \text{ V})(1 \text{ k}\Omega)}{44 \text{ k}\Omega} \cong 3.86 \text{ V}$$

In either case, V_4 now exceeds the V_{GT} value of most SCR's, which means that voltage V_4 may trigger the SCR before turn-on of the UJT does. Since this is an uncontrolled triggering, it is undesirable. Therefore, when choosing a resistance R_4 for the B_1 lead, it should have a voltage drop less than the gate trigger voltage V_{GT} of the SCR. This in equation form may be shown that since we must have

$$V_4 < V_{GT}$$

then

$$\frac{V_{s(\max)} R_4}{R_3 + R_{B1B2} + R_4} < V_{GT}$$

where R_{B1B2} is the lowest possible resistance looking into the B_1 and B_2 leads while the emitter lead E is open, R_3 and R_4 are the resistances in series with the B_2 and B_1 leads respectively and $V_{s(max)}$ is the instantaneous maximum voltage across resistances R_3, R_{B1B2}, and R_4. By solving the above equation for R_4 we can show that

$$R_4 < \frac{V_{GT}(R_3 + R_{B1B2})}{V_{s(max)} - V_{GT}} \quad (11\text{-}8)^*$$

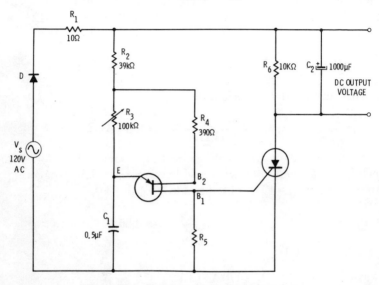

Fig. 11-13. Transformerless dc power supply.

Example 11-4 In the circuit of Fig. 11-13, what value of R_5 would you choose to prevent uncontrolled triggering if the SCR has the following trigger characteristics as supplied by the manufacturer:

Voltage	Minimum	Maximum
V_{GT}	0.5 V	2.5 V

Assume that R_{B1B2} is in the range of 5 kΩ to 10 kΩ; that is, 5 kΩ $\leq$ $R_{B1B2} \leq$ 10 kΩ.

Answer 11-4 About 130 Ω or less.—The maximum voltage occurs across R_5, which is the UJT gate-to-cathode voltage V_{GK}, when the 120-V-rms source is at its positive peak of about 170 V; thus $V_{s(max)}$ $\cong$ 170 V. If R_3 is set at near its maximum 100-kΩ value, R_1, R_2,

---

*See Appendix A-9.

11-2 APPLICATIONS OF THE UJT

and R_4 are effectively in series with the B_2 lead. Also if we ignore the SCR's gate current i_{FG} before triggering, R_5 is in series with the B_1 lead. Thus, from the point of view of V_s and before triggering occurs, the circuit of Fig. 11-13 can be shown as the equivalent in Fig. 11-14. As far as Eq. 11-8 is concerned, then, for its

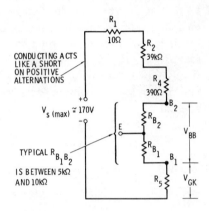

Fig. 11-14. Equivalent circuit of transformerless dc power supply as seen from V_s, with negligible i_C and i_{FG}.

R_3 we substitute $R_1 + R_2 + R_4$, or about 39.4 kΩ. Likewise for R_4 in Eq. 11-8 we substitute R_5, whose value we are to determine. Since we are to avoid accidental or uncontrolled triggering, we are concerned with the minimum specified V_{GT}, which in this case is 0.5 V. In other words, according to the manufacturer's specifications, some SCR's of this type could trigger with as little as 0.5 V gate-to-cathode voltage. We therefore want to keep the voltage across R_5 less than 0.5 V. The minimum R_{B1B2} or 5 kΩ is used in the equivalent circuit of Fig. 11-14 because the largest V_{GT} occurs with the smallest R_{B1B2}. To select safe R_5 we should do so considering the worst possible case, which is if R_{B1B2} is minimum. Thus, by Eq. 11-8 we find that

$$R_5 < \frac{(0.5 \text{ V})(39.4 \text{ k}\Omega + 5 \text{ k}\Omega)}{170 \text{ V} - 0.5 \text{ V}} \cong 132 \text{ }\Omega$$

Use of resistance larger than 132 Ω in series with the B_1 lead may cause uncontrolled triggering. To avoid such triggering, a smaller available resistor should be used; but not too small. Use of much less than 50 Ω in the B_1 lead may cause most of the capacitor's discharge current to flow through it instead of the gate lead, which could cause the SCR not to turn on when we want it to.

Example 11-5 Referring to the circuit of Fig. 11-13, if $R_5 = 100\ \Omega$ and the UJT $\eta = 0.6$ and its R_{B1B2} is in the range of 5 kΩ to 10 kΩ before turn-on, what is the approximate maximum gate-to-cathode voltage V_{GK} that could exist across the gate and cathode of the SCR before the UJT is triggered?

Answer 11-5 $V_{GK(\text{max})} \cong 382$ mV.—The voltage across R_5, call it V_5, is the gate-to-cathode voltage V_{GK}. If we assume that the gate current i_{FG} of the SCR, is negligible before it turns on, and R_3 is much larger than R_{B1B2}, then for practical purposes the source V_s sees R_1, R_2, R_4, R_{B1B2}, and R_5 in series. Since in this case, 5 kΩ $\leqq R_{B1B2} \leqq$ 10 kΩ, and $V_{s(\text{peak})} = 170$ V, in series circuit fashion we can find the range maximum gate-to-cathode voltage as follows: In general

$$V_{GK(\text{max})} \cong V_{5(\text{max})} \cong \frac{V_{s(\text{peak})} R_5}{R_t} \qquad (11\text{-}9)$$

where $R_t = R_1 + R_2 + R_4 + R_{B1B2} + R_5$. Thus, if, $R_{B1B2} = 5$ kΩ, then

$$V_{GK(\text{max})} \cong \frac{(170\ \text{V})(100\ \text{k}\Omega)}{44.5\ \text{k}\Omega} = 382\ \text{mV}$$

or if $R_{B1B2} = 10$ kΩ, then

$$V_{GK(\text{max})} \cong \frac{(170\ \text{V})(100\ \Omega)}{49.5\ \text{k}\Omega} = 344\ \text{mV}$$

Thus, in this case, you would choose an SCR whose gate trigger voltage V_{GT} is greater than 382 mV. Otherwise, the SCR may turn on at or near the positive peaks of the input voltage V_s even though R_3 is adjusted to maximum resistance.

As indicated, the circuit of Fig. 11-13 is a dc power supply. Without a transformer it converts the relatively high 120 V ac line voltage to a low dc in the range of about 8 V to 15 V, depending on the load resistance and adjustment of R_3. The dc output voltage increases if R_3 is decreased. This power supply has little ripple in the output voltage with load currents up to about 100 mA. It can easily supply up to 500 mA but with significant increase of ripple in the output.

The action of this circuit is similar to those previously covered that use a UJT. Diode D allows conduction only on positive alternations in the half-wave rectifier fashion. Capacitor C_1 is charged on positive alternations at a rate depending on the amount of

11-3 THE BIDIRECTIONAL TRIODE THYRISTOR (TRIAC)

resistance in series with it, which is mainly R_2 and R_3. (We may ignore the effect R_1 has on the time constant because it is so small. Its main function is to keep current in diode D and the SCR down to safe values while C_2 is initially charging.) When resistance R_3 is minimum, C_1 charges fairly quickly to V_p and turns on the UJT, which in turn triggers the SCR early in the alternation. This causes a relatively large average charging current to flow through the SCR, and C_2 and tends to provide larger dc output voltages. Adjustment of R_3 to a larger resistance value, on the other hand, causes C_1 to charge more slowly and therefore the SCR triggers later in the positive alternations. This reduces the average charging current, which in turn tends to reduce the dc output.

11-3 THE BIDIRECTIONAL TRIODE THYRISTOR (TRIAC)

The fact that the SCR can be turned on to conduct current only in one direction is often a disadvantage, especially when we need to control power in ac circuits. For example, due to the unilateral characteristics of the SCR in the circuit of Fig. 10-24 it was necessary to full-wave rectify the source voltage e in order to furnish load current on both alternations of each ac input cycle. Use of a *bidirectional triode thyristor*, commonly called the *Triac*, eliminates the need of the bridge rectifier and this reduces the number of circuit components, as you will see. The symbol for the Triac and its *V-I* characteristics are shown in Fig. 11-15.

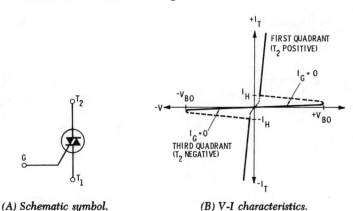

(A) Schematic symbol.   *(B) V-I characteristics.*

Fig. 11-15. Bidirectional triode thyristor, or Triac.

The Triac is capable of conducting current through its terminals T_1 and T_2 in either direction. However, it is nonconducting until a current larger than the gate trigger current I_{GT} passes through the gate lead or the voltage across T_1 and T_2 exceeds the Triac's *breakover voltage* V_{BO}. Both I_{GT} and V_{BO} are specified by the manufacturer. As with the SCR, the gate loses control after the Triac turns on. As the characteristic in Fig. 11-15B shows, the Triac conducts negligible current with voltages across T_1 and T_2 less than V_{BO} if the gate current I_G is zero. After being triggered by either a gate current I_G equal or greater than the specified I_{GT}, or by a terminal voltage greater than V_{BO}, the Triac turns on and conducts current well, and its terminal voltage drops to about one volt or so. The current magnitude is limited mainly by the resistance in series with the Triac. Once conducting, the Triac remains turned on until the current through it drops below its holding current I_H. An interesting feature is that the Triac can be turned on by either a positive or negative gate current, i.e., by current out of or current into the gate

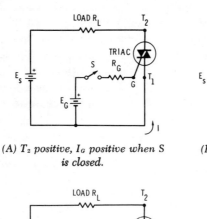

(A) T_2 positive, I_G positive when S is closed.

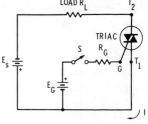

(B) T_2 negative, I_G positive when S is closed.

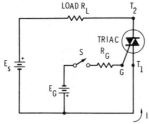

(C) T_2 positive, I_G negative when S is closed.

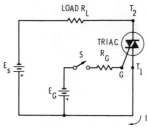

(D) T_2 negative, I_G negative when S is closed.

Fig. 11-16. Triggering modes for Triac.

11-3 THE BIDIRECTIONAL TRIODE THYRISTOR (TRIAC)

lead. This means that the Triac can be triggered by a voltage source that is either positive or negative with respect to T_1, which is arbitrarily used as the reference terminal. For example, the Triac in each of the circuits in Fig. 11-16 can be turned on by momentarily closing the switch S. When E_s is initially applied while switch S is open, the Triac is nonconducting, or off, provided that E_s is less than V_{BO}. The full E_s voltage appears across terminals T_1 and T_2. Closing S triggers the Triac, causing nearly the full E_s voltage to appear across the load R_L. That is, we may assume that the Triac acts like a short after turn-on if the current I is larger than the holding current I_H. By Ohm's law, the circuit current is

$$I \cong \frac{E_s}{R_L}$$

Once conducting current, the Triac cannot be turned off by reopening S. Removing or reducing E_s so that current I drops below I_H or momentarily placing a short across the Triac causes it to turn off.

The fact that the Triac conducts in either direction makes it useful as an electronically controlled ac circuit switch. A Triac used in such an application is shown in Fig. 11-17A. In this circuit, closing switch S turns on the Triac, making it conductive on positive and negative alternations, resulting in circuit waveforms as shown in Figs. 11-17B through 11-17D. When S is reopened, the Triac turns off as the alternation passes through zero, causing the circuit current to drop below I_H. The Triac thus turns off and remains off until the switch S is closed again. The gate current I_G required to trigger the Triac can be much smaller than the load current being controlled. This enables us to use a switch S that has relatively small current capacity to control a much larger current in a circuit like that of Fig. 11-17A.

Example 11-6 If we have a circuit like the one in Fig. 11-17A, where e is the 120-V rms ac line voltage, the load R_L is a lamp rated 300 W, 120 V, the gate source $E_G = 24$ V, and the Triac has the following characteristics:

maximum required gate trigger: $I_{GT(max)} = 100$ mA,
maximum terminal current: $I_{T(max)} = 4$ A rms, at ambient temperatures up to 50°C,
breakover voltage: $V_{BO} = 200$ V.

364　Other Members of the Thyristor Family

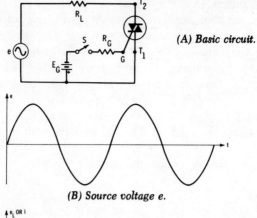

(A) Basic circuit.

(B) Source voltage e.

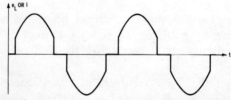

(C) Voltage across R_L or current through R_L after S is closed if peak current is not much larger than I_H.

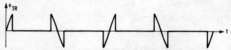

(D) Voltage across Triac after S is closed if peak current is not much larger than I_H.

(E) Voltage across R_L or current through R_L when S is open.

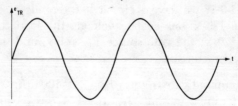

(F) Voltage across Triac when S is open.

Fig. 11-17. Triac as ac circuit switch.

11-3 THE BIDIRECTIONAL TRIODE THYRISTOR (TRIAC)

(a) what value of R_G would you select, and is the Triac capable of conducting the circuit current? (Assume the circuit environment is at room temperature.) (b) What instantaneous maximum voltage will we have across terminals T_1 and T_2, and does it exceed the breakover voltage?

Answer 11-6 (a) R_G should be about 240 Ω or a little less; yes, the Triac is capable of carrying the load current. (b) Peak T_2 to T_1 voltage is about 170 V; no, the V_{BO} voltage is not exceeded.—
(a) We may as a "rule of thumb" choose R_G by assuming that the voltage drop across the gate and T_1 leads is negligible. Therefore, after S is closed, the full E_G voltage appears across R_G. Since the current in R_G is the gate current, the maximum value of R_G may be determined by Ohm's law:

$$R_G \cong \frac{E_G}{I_{GT(max)}} = \frac{24 \text{ V}}{100 \text{ mA}} = 240 \text{ Ω}$$

This "rule of thumb" works because $I_{GT(max)}$ is the maximum gate current definitely required to turn on all Triacs of this type. We can therefore assume that most units will turn on with gate currents less than $I_{GT(max)}$. So even if there is a significant drop across the G and T_1 leads, which in this case will make I_G less than 100 mA when $R_G = 240$ Ω, this R_G value will likely work well anyway. A slightly smaller R_G, such as 220 Ω or 200 Ω, may be safe to use if we want to be very certain that the Triac will trigger, but we must be cautious because a too small R_G may cause excessive gate power dissipation and the Triac may be ruined.

The lamp rated 300 W, 120 V, means that it dissipates 300 W when 120 V rms is applied to it. When the switch S is closed in the circuit of Fig. 11-17A, nearly the entire sine wave is applied to the load, as shown in Fig. 11-17C. The lamp thus for all practical purposes has the total 120 V rms applied to it when the Triac is on. The current the load draws can be found by solving the power equation $P = EI$ for I and substituting the known P and E values into it. That is, since $P = EI$, then

$$I = \frac{P}{E}$$

In this case

$$I = \frac{300 \text{ W}}{120 \text{ W}} = 2.5 \text{ A rms}$$

This 2.5 A is the maximum current the load and the Triac will conduct in this circuit. The 4-A rating of the Triac means it is adequate for the job.

When the switch S is not closed, the Triac is nonconducting, or off, and the full source voltage e appears across its terminals T_1 and T_2. Since $e = 120$ V rms and is a sine wave, its peak value is $120 \times (1.414) \cong 170$ V. This 170 V appears across the Triac at the instances of positive and negative peaks when S is open, and it obviously does not exceed the 200 V breakover rating.

A simple phase controlled Triac is shown in Fig. 11-18. It is similar to the SCR half-wave phase controlled circuit of Fig. 10-20 in its theory of operation. In place of the SCR, use of a Triac pro-

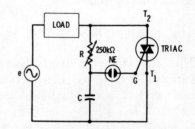

Fig. 11-18. Triac phase control circuit.

vides capability for load current on all alternations. The neon bulb NE, as before, serves to change the resistance suddenly in the gate circuit at a desired instant of the alternation and thus discharges the capacitor through the gate lead, which triggers the Triac. That is, on positive alternations, the Triac is initially nonconducting (off) and the capacitor C charges through R to the ionization potential V_i of the bulb. At V_i, the gas in the neon bulb ionizes and the bulb suddenly changes from a nonconducting to a conducting state. The capacitor thus discharges through NE and across the T_1 and gate G leads, triggering the Triac with a positive gate current pulse; see Fig. 11-19A.

On negative alternations, the capacitor C again charges, but now in the opposite polarity, through R. When capacitor C charges to V_i, the bulb ionizes as before, discharging capacitor C through the gate, triggering the Triac with a negative gate current pulse as shown in Fig. 11-19B.

As in previous circuits, capacitor C charges more or less quickly to V_i depending on the value of R. Setting R to a small resistance causes C to charge to V_i and thus the Triac triggers early in each

11-3 THE BIDIRECTIONAL TRIODE THYRISTOR (TRIAC)

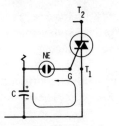

(A) Capacitor discharge path on positive alternations.

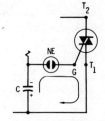

(B) Capacitor discharge path on negative alternations.

Fig. 11-19. Gate current paths for Triac phase control of Fig. 11-18.

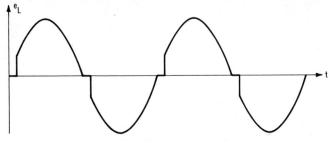

(A) When R has low resistance.

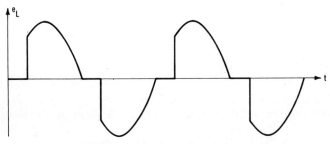

(B) When R has intermediate resistance.

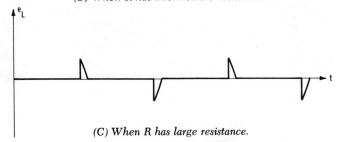

(C) When R has large resistance.

Fig. 11-20. Load voltage waveforms of Triac phase control circuit of Fig. 11-18.

alternation. The resulting load voltage waveform e_L is shown in Fig. 11-20A. As the resistance of R is progressively increased, triggering occurs later in each alternation, producing waveforms e_L shown in Figs. 11-20B and 11-20C.

The obvious advantage of the Triac, such as controlled conduction in either direction and triggering accomplished with either

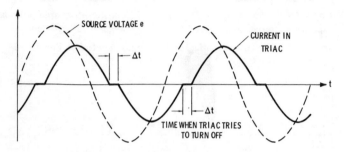

(A) Current in Triac lags source voltage e.

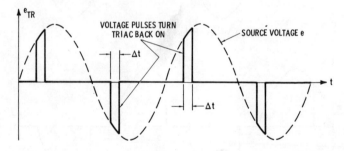

(B) Voltage across Triac terminals T_1 and T_2 when Triac tries to turn off.

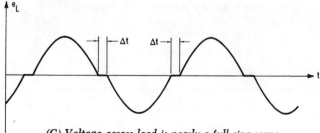

(C) Voltage across load is nearly a full sine wave.

Fig. 11-21. Waveforms in Triac phase control circuit when load is inductive, causing the gate to lose control.

11-3 THE BIDIRECTIONAL TRIODE THYRISTOR (TRIAC)

positive or negative pulses, could make us wonder why the ordinary SCR is used at all. Actually, the Triac has some serious disadvantages that limit its applications. Beside being unavailable at the large current capacities of some SCR's, the Triac often displays difficulty in turning off when the load is inductive. Turn-off with applied voltages at frequencies greater than 60 Hz is sometimes a problem, too. Thus once triggered, the Triac may not turn off again even though the current through it goes through zero twice each cycle of the input voltage and the gate is open. That is, the gate may not regain control at the end of each alternation. In other words, if the load is inductive in the circuit of Fig. 11-18, the Triac *tries* to turn off at the instant of zero current but it turns back on as the current passes through zero, even in absence of gate current. The Triac thus conducts most of the time and causes nearly the full source voltage waveform e to appear across the load. The load voltage e_L, therefore, always looks much like Fig. 11-21C regardless of the adjustment on R. Since most loads are somewhat inductive, such as lamp filaments, motor windings, heater coils, etc., the Triac could be unsatisfactory as a switch to control power to them.

The Triac's inability to turn off reliably when the load is inductive is caused by the fact that current in the Triac lags the source voltage e as shown in Fig. 11-21A. As the current goes through zero, the Triac tries to turn off and actually acts like an open for the short time intervals Δt shown. At these time intervals Δt, the source voltage e appears across the Triac terminals T_1 and T_2 as shown in Fig. 11-21B. The source e thus applied so suddenly on T_1 and T_2 acts as a fairly large amplitude voltage pulse. Such a fast rising voltage pulse has the ability to cause an avalanche of charge carriers within the Triac which turns it back on at the end of each interval Δt. The resulting load voltage e_L is as shown in Fig. 11-21C.

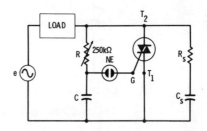

Fig. 11-22. Phase control with R_s and C_s added for more reliable turn-off of Triac when load is inductive.

The Triac may be made to turn off more reliably if a capacitance C_s is placed in shunt with it as shown in Fig. 11-22. This shunt capacitance C_s tends to round off the leading edges of the voltage pulses shown in Fig. 11-21B. That is, C_s accepts charge during time intervals Δt, causing voltage e_{TR} to increase more slowly than shown in Fig. 11-21B at the beginning of each interval Δt. This reduced rate of voltage increase across terminals T_1 and T_2 is much less likely to cause an avalanche of charge carriers and to prevent turn-off. The resistance R_s is used to prevent possible oscillations that can occur due to C_s and the load inductance forming a resonant circuit. For most practical purposes C_s of about 0.1 μF adequately rounds off the voltage pulses, and R_s of about 100 Ω prevents oscillations.

11-4 THE BIDIRECTIONAL DIODE THYRISTOR (DIAC)

We have so far learned that the neon bulb and UJT are useful devices in pulse generating circuitry and that the pulses produced by such circuitry are used to trigger the gates of SCR's and Triacs. Another device frequently used in the gate trigger circuitry is the *bidirectional diode thyristor,* often referred to as the Diac. Its symbol and V-I characteristics are shown in Fig. 11-23. As shown in its V-I characteristics, the Diac is nonconducting in either direction unitl the voltage across it reaches its breakover

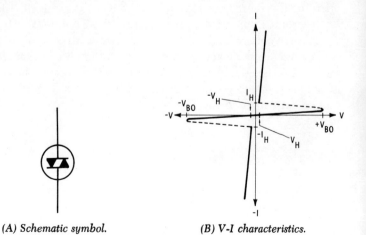

(A) *Schematic symbol.*   (B) *V-I characteristics.*

Fig. 11-23. Bidirectional diode thyristor, or Diac.

11-4 THE BIDIRECTIONAL DIODE THYRISTOR (DIAC)

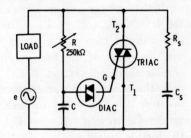

Fig. 11-24. Diac used as part of gate circuit to trigger Triac.

voltage, V_{BO}. When the terminal voltage reaches V_{BO}, the Diac suddenly becomes conductive and the terminal voltage drops to a much smaller value. After the Diac starts to conduct (is turned on), if the current through it is reduced below its holding current I_H, or the voltage across it drops below the holding voltage V_H, it turns off (stops conducting). The Diac has V-I characteristics very similar to the Triac's with the gate lead open; see Fig. 11-15B. Its characteristics are also comparable to those of the neon bulb (Fig. 10-21B). That is, both the neon bulb and Diac are two-terminal devices that are nonconducting (have high resistance)

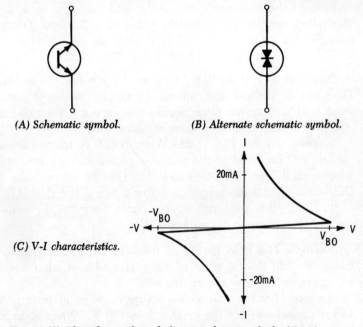

(A) Schematic symbol.

(B) Alternate schematic symbol.

(C) V-I characteristics.

Fig. 11-25. Three-layer silicon bidirectional trigger diode (Diac).

until the voltage across them exceeds the ionization voltage V_i or the breakover voltage V_{BO}. Thereafter the resistance of each suddenly drops to a much lower value. Therefore, the Diac can be used in place of the neon bulb as shown in Fig. 11-24, which is similar to the circuit in Fig. 11-22. In other words, in either circuit, the average power consumed by the load is increased or decreased by decreasing or increasing the resistance of R respectively. Components R_s and C_s serve to provide reliable control if the load is inductive.

Similar but not exactly equivalent to the bidirectional diode thyristor is the *three-layer silicon bidirectional trigger diode,* which due to its similarity is also called a Diac. The symbols for this device and its V-I characteristics are shown in Fig. 11-25. It too is nonconducting until its terminal voltage reaches its breakover voltage V_{BO}. Thereafter it turns on and the current suddenly increases more or less, depending on the amount of resistance in the circuit. As the V-I characteristics in Fig. 11-25C indicate that if the circuit current after turn-on is greater than about 20 mA, the terminal voltage decrease is significant. In effect, then, the two versions of the Diac are quite similar and both are used in pulse generating circuits that drive the gates of SCR's and Triacs. For example, the circuit in Fig. 11-24 is often shown with the symbol of Fig. 11-25A instead of the symbol in Fig. 11-23A.

Example 11-7 If in Fig. 11-24, source $e = 120$ V rms and the Diac has a forward and reverse breakover voltage $V_{BO} = 30$ V, what load voltage waveform would you expect if the resistance R is adjusted to its minimum value of about 0 Ω?

Answer 11-7 See Fig. 11-26.—When $R \cong 0$ Ω, the voltage across capacitor C increases as the source e passes through zero on each alternation. When V_C reaches 30 V, the Diac breakover voltage, the Diac turns on and discharges capacitor C across the T_1 and G leads of the Triac, thus triggering it. The Triac, therefore, becomes conductive after the source e reaches 30 V.

Example 11-8 If in Fig. 11-24 the load has 24 Ω resistance and source $e = 120$ V rms, what effective current and what breakover voltage must the Triac be capable of?

Answer 11-8 The current capability must be at least 5 A rms and the breakover voltage greater than 170 V.—When resistance R is adjusted to 0 Ω, the load voltage e_L is nearly equal to the source

11-4 THE BIDIRECTIONAL DIODE THYRISTOR (DIAC)

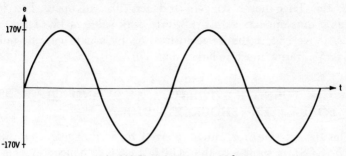

(A) 120 V rms sine wave voltage.

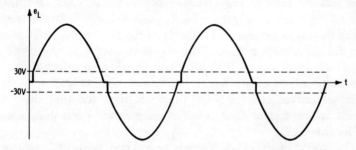

(B) Load voltage if Triac turns on when source e reaches instantaneous value of 30 V.

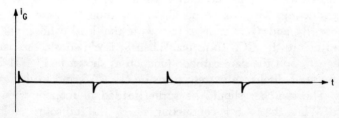

(C) Gate current waveform produced by capacitor C discharging through Diac.

Fig. 11-26. Waveforms of circuit described in Ex. 11-17.

e as was shown in Fig. 11-26. Thus, their effective values are nearly equal and the maximum current that the Triac must be capable of conducting is determined by Ohm's law:

$$I = \frac{E}{R} = \frac{120 \text{ V}}{24 \text{ }\Omega} = 5 \text{ A rms}$$

Of course, the Triac conducts less than this when R is adjusted to have more than zero resistance. If R is set at maximum resistance, the Triac never triggers and acts like an open. The full voltage e thus appears across it whose peak value is $120(1.414) \cong 170$ V. If we expect the Triac to turn on by a gate current pulse only, its V_{BO} rating must be larger than 170 V.

11-5 THE TURNOFF THYRISTOR (GCS) AND THE FOUR-LAYER (SHOCKLEY) DIODE

The *turnoff thyristor,* often referred to as the *gate controlled switch* (GCS), is similar to the SCR in that is is nonconducting in reverse bias and can be turned on by a positive gate current pulse when forward biased. As its name implies, the GCS can be turned back off by a negative gate pulse if operated within limits specified by the manufacturers. The ability of the GCS to turn off makes it useful as a switch in dc circuits such as the one shown in Fig. 11-27. In this circuit, if you close switch S_1 momentarily, the GCS turns on and conducts current through the load. Close switch S_2 for an instant and the GCS turns off, provided that the load current is not greater than the maximum anode current I_A specified for this GCS.

The circuit of Fig. 11-27A works as follows: With both switches S_1 and S_2 open and the source voltage E is applied, capacitor C_1 charges through R_1 to the source voltage E because the GCS is nonconducting. The charge on C_2 at this time is essentially zero because R_2 and C_2 are in shunt with the load whose voltage is zero before the GCS turns on. When S_1 is closed, C_1 discharges through R_a and the gate-cathode junction as shown in Fig. 11-27B. The pulse of discharge current turns on the GCS and load current flows. The switches should be spring loaded to reopen when released. While the GCS is conducting, nearly the full source voltage E appears across the load. Since R_2 and C_2 are in shunt with it, C_2 charges to voltage E. If we now close S_2, capacitor C_2 discharges through R_b and through the gate-cathode junction as shown in Fig. 11-27C. In this case, the discharge is such that the gate is negative with respect to the cathode. The GCS turns off and the circuit returns to its initial condition. The component values in this circuit are not critical. Resistors R_1 and R_2 should be at least 10 times larger than R_a and R_b to ensure that the capacitors discharge mainly through R_a and R_b. Resistors R_1 and R_2 should not be too large, however, to avoid a long charge time of the capacitors. Use

11-5 THE TURNOFF THYRISTOR AND FOUR-LAYER DIODE 375

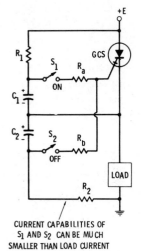

(A) Complete circuit.

CURRENT CAPABILITIES OF S_1 AND S_2 CAN BE MUCH SMALLER THAN LOAD CURRENT

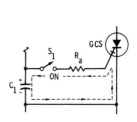

(B) Gate current path that causes turn-on.

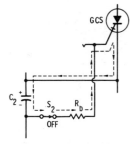

(C) Gate current path that causes turn-off.

Fig. 11-27. GCS used as switch in dc circuit.

of $C_1 = C_2 = 1$ μF and $R_1 = R_2 = 10$ kΩ works out well for most practical purposes. Resistors R_a and R_b should be large enough to prevent excessive instantaneous gate currents. For example, suppose that $E = 120$ V in Fig. 11-27A and the GCS has a specified maximum gate current of 200 mA. This means that the capacitors will charge to as much as 120 V. Neglecting the voltage drop across the gate-cathode junction, R_a and R_b should be large enough to keep current down to 200 mA with 120 V applied. That is, when the switch is closed, the 120 V of the capacitor appears across R_a or R_b, depending on whether the circuit is being turned on or off. Thus by Ohm's law we can use

$$R_a = R_b \cong \frac{120 \text{ V}}{200 \text{ mA}} = 600 \text{ }\Omega$$

Another interesting application of the GCS is shown in Fig. 11-28. This circuit is "sort of" an electronic fuse that can be designed to protect the load from excessive currents. The off switch S_2 in this case is shunted with a four-layer diode which is sometimes referred to as the *Shockley diode*. A few of the various sym-

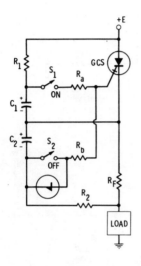

Fig. 11-28. GCS and four-layer diode used in circuit that serves as electronic fuse (circuit breaker).

bols used to represent, and the V-I characteristics of, the four-layer diode are shown in Figs. 11-29A and 11-29B respectively. As its V-I characteristics indicate, the four-layer diode is nonconducting with forward bias voltages less than a forward breakover V_{FBO}. Beyond V_{FBO} this device suddenly becomes conductive (turns on), much like the Diac, which is turned on by a breakover terminal voltage V_{BO}. Unlike the Diac, however, the reverse bias characteristics of the four-layer diode are similar to those of an ordinary diode. That is, as shown in Fig. 11-29B, if the reverse voltage is increased beyond the maximum specified value V_{RM}, the four-layer diode goes into avalanche conduction, which is a condition to be avoided. Thus, the four-layer diode is useful as a voltage operated switch in the forward direction only. However, it is available in rather low V_{FBO} ratings, and this feature makes it useful in applications like the circuit of Fig. 11-28 for reasons to be discussed here.

11-5 THE TURNOFF THYRISTOR AND FOUR-LAYER DIODE

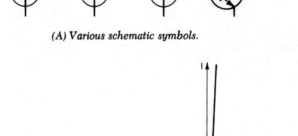

(A) Various schematic symbols.

(B) V-I characteristics.

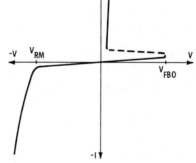

Fig. 11-29. Four-layer or Shockley diode.

In Fig. 11-28, when source voltage E is initially applied and switches S_1 and S_2 are open, the GCS is nonconducting. The full source voltage E appears across it, and since R_1 and C_1 are in shunt with the GCS, the capacitor C_1 charges to voltage E. The voltage across the load and R_F at this time is zero and there is no charge on C_2 because it is in shunt with R_F. Now if we close switch S_1 momentarily, C_1 discharges through the gate-cathode junction of the GCS and through R_a. The discharge current triggers the GCS and load current proceeds to flow. The load current causes a voltage drop across R_F and this voltage appears across C_2. The voltage across C_2 is also across the open switch S_2 and the four-layer diode. As long as this voltage is not greater than the diode's forward breakover rating V_{FBO}, C_2 has no discharge path. Note that if either S_2 is closed or the four-layer diode turns on (becomes conductive), the capacitor C_2 is discharged through R_b and the gate-cathode junction—either of which turns off the GCS. As long as the voltage drop across R_F does not exceed the V_{FBO} rating of the four-layer diode, it remains nonconducting. Now the load current becomes excessive in this circuit, the drop across R_F exceeds the V_{FBO} rating of the four-layer diode, causing it to turn on and discharge capacitor C_2 through the gate, thus turning off the GCS. The fact that the four-layer diode is available in low V_{FBO} ratings (under 10 V) means that the resistance of R_F may be very small compared to that of the load; that is, R_F drops only a small portion of the source E, allowing most of the source to appear across the load after the GCS turns on. As in the circuit of Fig. 11-27, components $C_1 = C_2 = 1$ μF and $R_1 = R_2 = 10$ kΩ work out well in most parctical cases. The other component values are determined by the maximum allowable load current, the GCS maximum gate current, and the V_{FBO} rating of the four-layer diode.

Example 11-9 Suppose that in the circuit shown in Fig. 11-27, the source $E = 40$ V, the GCS has characteristics shown in Fig. 11-30 and it can safely conduct up to 10 A anode current, and the load consumes 400 W at 40 V. (a) What component values would you use? (b) Can the GCS safely handle the load current?

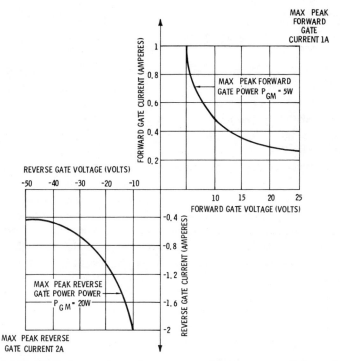

Fig. 11-30. Typical gate trigger characteristics of the gate controlled switch.

11-5 THE TURNOFF THYRISTOR AND FOUR-LAYER DIODE

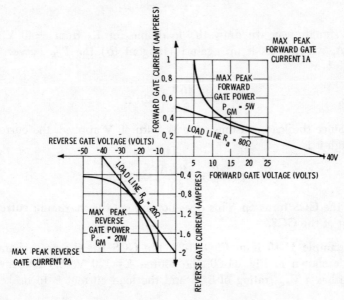

Fig. 11-31. Answer to Ex. 11-9.

Answer 11-9 (a) As mentioned previously, $R_1 = R_2 = 10$ kΩ and $C_1 = C_2 = 1$ μF will work in most cases; use $R_a = 80$ Ω, $R_b = 20$ Ω. (b) Yes, the GCS can handle the load current.—(a) The values of R_a and R_b can be found by plotting load lines on the gate characteristics of the GCS as shown in Fig. 11-31. Since C_1 charges to 40 V when the GCS is nonconducting and C_2 charges to 40 V when the GCS is turned on, the load lines for R_a and R_b must intersect the voltage axis at +40 V and −40 V respectively. Just as with the SCR, the GCS must be turned on hard to avoid hot spots; thus, the load line for R_a is drawn as close to the *maximum gate power dissipation curve as possible*. For similar reasons, the GCS must be turned off *hard,* which means that we should drive the gate as hard as possible during the turn-off process. Thus, the load line for R_b must likewise be drawn as close as possible to the P_{GM} curve. As shown in Fig. 11-31, the load line for R_a is drawn from an estimated +40 V on the forward gate voltage axis to 0.5 A on the forward gate current axis. This puts the load line reasonably close to the P_{GM} curve. Using the end points of the load line in Ohm's law we find that

$$R_a = \frac{40 \text{ V}}{0.5 \text{ A}} = 80 \text{ }\Omega$$

In a similar way we draw the load line for R_b from -40 V to -2 A, which puts it up against (tangent to) the P_{GM} curve. By Ohm's law

$$R_b = \frac{40 \text{ V}}{2 \text{ A}} = 20 \text{ }\Omega$$

(b) Since the load consumes 400 W with 40 V applied, the current through it is

$$I = \frac{P}{E} = \frac{400 \text{ W}}{40 \text{ V}} = 10 \text{ A}$$

after the GCS turns on. This does not exceed the maximum current rating of the GCS.

Example 11-10 If in Fig. 11-28, the GCS has the gate characteristic shown in Fig. 11-30, the source $E = 120$ V, the four-layer diode has a V_{FBO} rating of 5 V, and the load current is to be kept

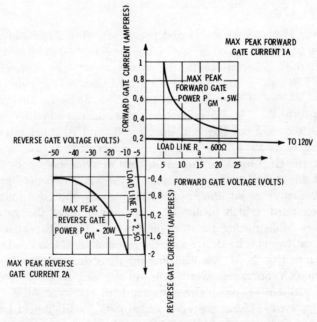

Fig. 11-32. Answer to Ex. 11-10.

11-5 The Turnoff Thyristor and Four-Layer Diode

under 10 A, what component values would you use that will turn off the GCS if the load current rises to 10 A or more?

Answer 11-10 $R_1 = R_2 = 10$ kΩ, $C_1 = C_2 = 1$ μF, $R_a \cong 600$ Ω, $R_b \cong 2.5$ Ω, $R_F \cong 0.5$ Ω.—As before, these values of R_1, R_2, C_1, and C_2 work out well for most practical purposes. If the GCS does not turn off when S_2 is closed, a slightly larger than 1 μF capacitor C_2 should be used. If the GCS turns off too slowly after the overload current occurs, a smaller R_2 can be used. If the GCS continually turns off with harmless transients in the load, a larger R_2 can be used.

The values of R_a and R_b are found by plotting load lines as shown in Fig. 11-32. Keeping the load line for R_a under the P_{GM} curve to prevent exceeding the maximum gate power dissipation, causes it to cross the forward gate current axis at about 0.2 A. The right end-point must be $+120$ V because that is the voltage across the capacitor before the GCS turns on. Thus,

$$R_a = \frac{120 \text{ V}}{0.2 \text{ A}} = 600 \text{ } \Omega$$

by Ohm's law.

Since the four-layer diode V_{FBO} rating is 5 V, we cannot have more than 5 V across the capacitor C_2 because the diode breaks over and discharges C_2 through R_b and the gate-cathode junction. The load line for R_b therefore is drawn from 5 V to 2 A, where 2 A is the largest peak gate current allowed according to the gate characteristics. In this case, then,

$$R_b = \frac{5 \text{ V}}{2 \text{ A}} = 2.5 \text{ } \Omega$$

assuming that the voltage across the diode is negligible after breakover. While this load line for R_b is not very near the P_{GM} curve, it should be satisfactory for most GCS's to which these gate characteristics apply. Use of a four-layer diode with a larger V_{FBO} rating will bring the load line nearer the P_{GM} curve, but this would require a larger voltage drop across R_F, thus reducing the load voltage.

Because the four-layer diode $V_{FBO} = 5$ V in this case, the GCS turn-off process occurs when the voltage on C_2 reaches 5 V, as was mentioned before. Thus, since C_2 and R_2 are in shunt with R_F, the GCS turns off when the voltage across R_F reaches 5 V or more.

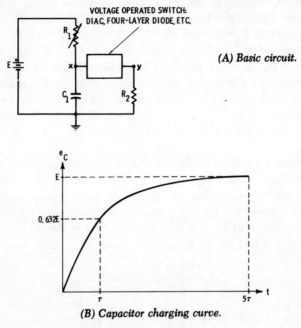

Fig. 11-33. Pulse generator and test circuit.

Since we want the GCS to *break* the circuit (stop load current) when the load current reaches 10 A or more, we want the 5-V breakover voltage across R_F when the current through it reaches the critical 10-A value. Then

$$R_F = \frac{5\text{ V}}{10\text{ A}} = 0.5\ \Omega$$

in this case.

If we ever need a sawtooth or pulse voltage generator, or need to check whether a certain neon bulb, Diac, or four-layer diode is working, we can use the simple circuit shown in Fig. 11-33A. In this circuit, if the source voltage E is larger than the voltage required to actuate the voltage operated switch, which is V_i in the case of the neon bulb, V_{BO} in the case of the Diac, or V_{FBO} in the case of a four-layer diode, the circuit will produce a sawtooth voltage at x and a pulse (spiked) voltage waveform at y with respect to ground. The frequency of the sawtooth or spiked

11-5 THE TURNOFF THYRISTOR AND FOUR-LAYER DIODE

waveform is determined by the time constant of the charging path and the distance between the voltage applied and the voltage required to trigger the voltage operated switch.

The time constant τ of the charging path is the product of the capacitance and the resistance in series with it. Thus, in circuit Fig. 11-33A,

$$\tau = R_1 C_1$$

where
  τ is the time constant in seconds,
  R_1 is the charging path resistance in ohms,
  C_1 is the capacitance being charged in farads.

If the voltage operated switch in this circuit did not interfere with the charging process, the voltage across capacitor C_1 would rise to about 63.2% of the applied voltage E in time τ as shown in Fig. 11-33B. Also note that the capacitor is fully charged in approximately the time 5τ. This means that if we have 10 kΩ in series with 1 μF capacitance and 40 V is applied, the capacitance will charge to 63.2% of 40 V, or about 25.3 V, in

$$\tau = (10 \text{ k}\Omega)(1 \text{ μF}) = 10 \text{ ms}$$

and it will take about $5\tau = 50$ ms to charge to the full 40 V applied.

If the resistance is reduced to 5 kΩ, but the applied E is unchanged, the same capacitance will charge to 25.3 V in

$$\tau = (5 \text{ k}\Omega)(1 \text{ μF}) = 5 \text{ ms}$$

and it will charge to 40 V in about $5\tau = 25$ ms.

Thus, if the voltage operated switch in the circuit of Fig. 11-33A is a Diac with a breakover voltage $V_{BO} = 30$ V and the applied voltage $E = 40$ V, the capacitance C_1 will charge to 30 V more or less slowly, depending on the value of R. At $e_C = 30$ V, the Diac turns on and discharges C_1 through R_2 quickly if R_2 is a much smaller resistance than R_1. When the discharge current of C_1 drops below the Diac's holding current, the Diac turns off and C_1 starts to charge again and the process repeats itself as shown in Fig. 11-34A. The natural charge characteristic of C_1 is shown by the broken line. Note that in this case each sawtooth takes a little more than time τ. Since time τ can be varied by varying R_1, the frequency of the sawtooth wave is increased or decreased by decreasing or increasing R_1 respectively. Of course, the spiked wave

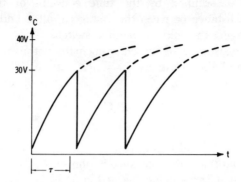

(A) For E = 40 V and voltage operated switch actuating at 30 V.

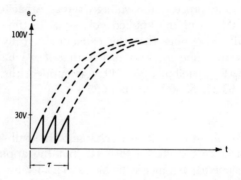

(B) For E = 100 V and voltage operated switch actuating at 30 V.

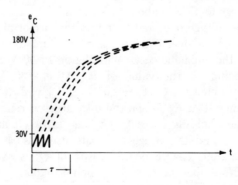

(C) For E = 180 V and voltage operated switch actuating at 30 V.

Fig. 11-34. Waveforms at x in pulse generator circuit of Fig. 11-33A with various applied voltages E.

11-5 THE TURNOFF THYRISTOR AND FOUR-LAYER DIODE

at y has the same frequency as the sawtooth at point x because it represents the discharge current waveform through R_2. We can *roughly* estimate the frequency of the sawtooth and spiked waveforms in this circuit with the following equations:

$$\frac{T}{\tau} \cong \frac{V_{BO}}{E} \tag{11-10}$$

with

$$T = \frac{1}{f} \tag{11-11}$$

where

T is the period in seconds of each sawtooth,
f is the number of sawtooth repetitions per second,
τ is the charging path time constant $R_1 C_1$,
V_{BO} is the breakover voltage of the Diac or the voltage required to actuate the voltage controlled switch,
E is the dc source voltage.

To get even rough estimates of f, the source E must be more than 25% larger than V_{BO}. Generally, the accuracy of Eq. 11-10 tends to improve as source E is made larger than V_{BO}. If E is increased but V_{BO} is not, the frequency of the sawtooth increases as shown in Figs. 11-34B and 11-34C. Note that the general shape of the charging curve (broken line) is the same as before but that now it rises to a larger applied voltage E. This causes C_1 to charge to V_{BO} and trigger the Diac sooner, even though the time constant τ is unchanged.

Example 11-11 If in the Fig. 11-33A, $E = 120$ V, $R_1 = 6$ kΩ, $C_1 = 0.5$ μF, and the voltage operated switch is a Diac with $V_{BO} = 20$ V, what is, roughly, the frequency of the sawtooth voltage at point x?

Answer 11-11 $f \cong 2$ kHz.—Solving Eq. 11-10 for T and substituting all knowns into it we work as follows: Since

$$\frac{T}{\tau} \cong \frac{V_{BO}}{E}$$

then

$$T \cong \tau \frac{V_{BO}}{E}$$

where $\tau = R_1C_1 = (6\text{ k}\Omega)(0.5\text{ μF}) = 3$ ms. Therefore

$$T \cong (3\text{ ms})\frac{20\text{ V}}{120\text{ V}} = 0.5\text{ ms}$$

Now by rearranging Eq. 11-11 we find

$$f = \frac{1}{T} \cong \frac{1}{0.5\text{ ms}} = 2\text{ kHz}$$

is the frequency of the sawtooth.

Example 11-12 (a) If the same Diac appearing in the circuit of the previous example is disconnected and then reconnected with its leads interchanged, what effect will it have on the circuit operation? (b) How should a four-layer diode be connected in the circuit of Fig. 11-33A?

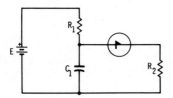

Fig. 11-35. Connecting four-layer diode in pulse generator circuit.

Answer 11-12 (a) It will have no effect, (b) see Fig. 11-35.—(a) Since the Diac has a breakover voltage V_{BO} in either polarity (see Fig. 11-23B), it may be connected either way in this circuit. Typically $-V_{BO}$ and V_{BO} are within 10% of each other numerically. (b) In this case the polarity of the switch is important. The four-layer diode acts like a voltage operated switch only when forward biased as shown in Fig. 11-35. If reverse biased heavily, it may go into avalanche conduction, causing its destruction.

REVIEW QUESTIONS

11-1. What does the term *intrinsic standoff ratio* mean, and to what kind of device does the term apply?

11-2. What is the meaning of the term *peak voltage* V_p when referring to UJT's?

11-3. What do the terms *valley voltage* and *valley current* mean, and to what kind of device do these terms apply?

11-4. Describe the construction of a relaxation oscillator.

11-5. How does a Triac differ from an SCR?

11-6. How can an SCR or a Triac be turned off after it starts conducting current?

11-7. When a Triac is used in an ac circuit, will it definitely turn off when the ac current goes through zero?

11-8. What is the purpose of capacitance that is sometimes used in shunt with Triacs?

11-9. What is a Diac, and how does it differ from a Triac?

11-10. What is a four-layer diode (Shockley diode) and how does it differ from the Diac?

PROBLEMS

11-1. If the B_2 lead is made 20 V positive with respect to the B_1 lead with a certain UJT whose intrinsic standoff ratio is 0.8, what emitter E to B_1 voltage will cause the device to turn on?

11-2. Given a UJT with an $\eta = 0.7$ and $V_{BB} = 12$ V applied, what is its peak voltage V_p?

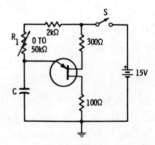

Fig. 11-36. UJT problem circuit.

11-3. If the UJT in Fig. 11-36 has an $\eta = 0.82$, what is the peak-to-peak voltage at the emitter E with respect to ground? Assume that the valley voltage is about 2 V.

11-4. If the UJT in Fig. 11-36 has an $\eta = 0.63$, what is the peak-to-peak voltage at the emitter with respect to ground? Assume that $V_v = 2$ V.

11-5. If R_1 is adjusted to 50 kΩ, $C = 1$ μF and $\eta = 0.63$ in Fig. 11-36, what is the frequency of the sawtooth output voltage?

11-6. If $C = 1$ μF and $\eta = 0.63$ in Fig. 11-36, what are the minimum and maximum output frequencies obtainable by adjustments of R_1?

11-7. If the UJT in Fig. 11-10 has a maximum $R_{B1B2} = 10$ kΩ, what is the maximum possible instantaneous voltage across the base 1 and base 2 terminals of the UJT?

11-8. If the UJT in the circuit of Fig. 11-10 has a maximum $R_{B1B2} = 12$ kΩ, what is the maximum possible instantaneous voltage across the B_1 and B_2 terminals of the UJT?

11-9. Suppose that you have a circuit as shown in Fig. 11-10 whose component values are as shown except for R_4. You are to select the largest usable R_4 that will not cause uncontrolled triggering. The UJT has $\eta = 0.75$, $V_{GT(min)} = 0.5$ V, and 5 kΩ $\leq R_{B1B2} \leq$ 10kΩ. Assume that the gate current of the SCR and the emitter current of the UJT are negligible before turn-on occurs.

11-10. Suppose that you have a circuit as shown in Fig. 11-10 whose component values are as shown except R_4. You are to select the largest usable R_4 that will not cause uncontrolled triggering. The UJT has $\eta = 0.8$, $V_{GT(min)} = 0.25$ V, and 8 kΩ $\leq R_{B1B2} \leq 12$ kΩ. Assume that the gate current of the SCR and the emitter current of the UJT are negligible before turn-on occurs.

11-11. Referring to Fig. 11-13, if $R_5 = 80$ Ω, and the UJT $\eta = 0.6$, and its R_{B1B2} is in the range of 5 kΩ to 10 kΩ before triggering, what is the maximum gate-to-cathode voltage V_{GK} before the UJT turns on?

11-12. Referring to Fig. 11-13. if $R_5 = 100$ Ω, and the UJT $\eta = 0.8$, and its R_{B1B2} is in the range of 8 kΩ to 12 kΩ before triggering, what is the maximum possible gate-to-cathode voltage V_{GK} before the UJT turns on?

11-13. If the source $e = 40$ V (rms) and the load resistance is 10 Ω, in the circuit of Fig. 11-22, what rms current must the Triac be capable of conducting?

11-14. If the source $e = 120$ V (rms) and the load's resistance is 40 Ω, in the circuit of Fig. 11-22, what rms current must the Triac be capable of conducting?

11-15. Suppose that the voltage shown in Fig. 11-37A is applied to the circuit in Fig. 11-24. If R is adjusted so that the voltage waveform of Fig. 11-37B is across the load, what voltage waveform would you expect to see across the Triac? Choose from Fig. 11-37.

11-16. Suppose that the voltage of Fig. 11-37A is applied to the circuit of Fig. 11-24. If R is adjusted so that the capacitor C charges to the Diac's breakover voltage 3.75 ms after the start of each positive alternation, what will the load voltage look like?

11-17. If waveform of Fig. 11-37A is the voltage applied e to the circuit of Fig. 11-24, and you view waveform Fig. 11-37C across the Triac, what waveform would you expect to view across the load? Choose from Fig. 11-37.

11-18. If the waveform of Fig. 11-37A is applied to the circuit of 11-24 and R is adjusted so that the capacitance C charges to the Diac's breakover voltage 4.9 ms after the start of each positive alternation, what are the voltage waveforms (a) across the load, and (b) across the Triac? Choose from Fig. 11-37.

11-19. Describe the voltage waveform you would expect to see at point x to ground in the circuit of Fig. 11-33 if $E=40$ V, $R_1=30$ kΩ, $C_1=2$ μF, and the voltage operated switch is a neon bulb whose ionization voltage is 70 V.

11-20. In the circuit of Fig. 11-33, describe the voltage waveform that you would expect to see (peak-to-peak amplitude and approximate frequency) at point x to ground if $E=40$ V, $R_1=30$ kΩ, $C_1=1$ μF, and the voltage operated switch is a Diac with a $V_{BO}=10$ V?

11-21. Suppose that in Fig. 11-27, the source $E=24$ V and the GCS has the characteristic shown in Fig. 11-30 and can safely conduct up to 10 A anode current. (a) What component values will you use? (b) What minimum load resistance can this circuit safely use?

11-22. Suppose that in the circuit of Fig. 11-27, the source voltage $E=30$ V and the GCS has the characteristic shown in Fig. 11-30 and can safely conduct up to 15 A anode current. (a) What R_a and R_b component values will you use? (b) What minimum load resistance can this circuit safely use?

11-23. The load current in the circuit of Fig. 11-28 is to be kept under 8 A. What value of R_F will you use if $E=50$ V and the for-

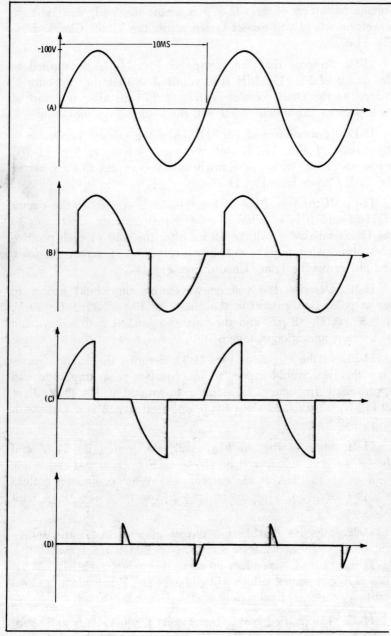

Fig. 11-37. Possible waveforms of

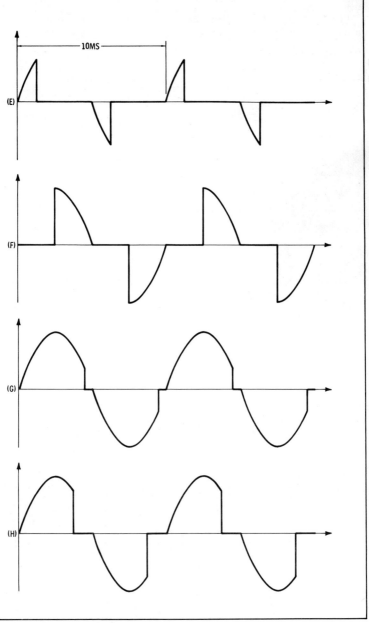

Diac-Triac phase control circuit.

ward breakover voltage of the four-layer diode is 4 V?

11-24. If $E = 100$ V and the Shockley diode $V_{FOB} = 4$ V in the circuit of Fig. 11-28, what value of R_F will you use to protect the load from currents over 16A?

12

General Characteristics of the Field-Effect Transistor

The *field-effect transistor* (FET) is a three-terminal device capable of providing amplification (gain) when used in the proper circuitry. Its three terminals are called the *source, drain,* and *gate*. In general, current in the source and drain leads is controlled with a voltage applied to the gate with respect to the source. The fact that the FET can be used as an amplifier makes it look similar to the junction transistor, the FET's source, drain, and gate leads being analogous to the transistor emitter, collector, and base leads respectively. The FET, however, has one contrasting characteristic, which is its very high input impedance, typically ranging from 10^6 to 10^{14} Ω, depending on type. For this reason, the FET is often compared to the vacuum tube instead of the transistor. And in fact, FET's are now used in applications that at one time only tubes were suitable for. The FET also has a low temperature coefficient, which means that its characteristics do not change very much with changing temperatures.

In the general category of FET's, there are two more specific types that have notable differences: (a) The *junction field-effect transistor* (JFET) and (b) the *insulated gate field-effect transistor* (IGFET). The construction and characteristics of both types are covered in this chapter.

394 GENERAL CHARACTERISTICS OF THE FIELD-EFFECT TRANSISTOR

Although the FET's development may be traced back to the 1920's, it is not until recently that it has become practical because of its previous high cost.

Each FET is made to operate in one of three different modes: mode A, mode B, and mode C, each of which is explained in this chapter. The operating mode affects the way in which the gate is biased as you will see.

12-1 CONCEPTS OF THE JUNCTION FIELD-EFFECT TRANSISTOR

In its simplest form, the JFET is a bar of n-type semiconductor with p-type materials diffused into two sides as shown in Fig. 12-1. The source and drain lead are connected to the bar ends while the gate is connected to the diffused p-type materials. With a positive voltage on the drain with respect to the source, electrons flow into the bar through the source lead, then through the bar, and finally out via the drain lead. Sometimes, as specified by the manufacturer, the source and drain leads can be interchanged. Since the current flows through a channel of n-type material in Fig. 12-1, this device

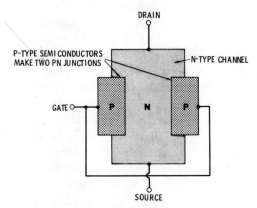

Fig. 12-1. Junction FET with n-type material sandwiched between p-type materials.

is called an *n-channel JFET*. If the channel (bar) is made of p-material with n-type materials diffused into its sides, the resulting JFET is called a *p-channel type*, which is operated with a negative drain-to-source voltage.

2-1 Concepts of the Junction Field-Effect Transistor

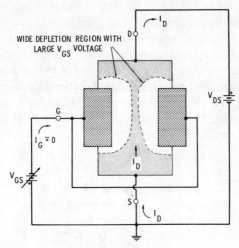

(A) Large V_{GS} gives I_D relatively narrow and high-resistance channel.

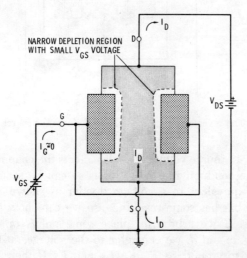

(B) Small V_{GS} gives I_D relatively wide and low-resistance channel.

Fig. 12-2. Resistance to drain current varies with V_{GS}.

As with any pn junction, if the p-type material is negative with respect to the n-material, the junction is reverse biased and a depletion region forms at the junction. Thus, if a negative gate-to-source voltage is applied to the n-channel JFET, a depletion region is formed as shown in Fig. 12-12. Note that more or less voltage $-V_{GS}$ causes the depletion region to spread more or less into the bar. Since the depletion region is depleted of its majority carriers, it is nonconductive and effectively narrows the channel through which drain I_D can flow. Thus, more negative gate-to-source voltages tend to make the channel narrower and this reduces I_D. Of course if $-V_{GS}$ is made large enough, the channel is completely closed off and $I_D \cong 0$ A. The $-V_{GS}$ voltage that causes $I_D \cong 0$ A is called the *gate-to-source cutoff voltage* and is often abbreviated $V_{GS(off)}$.

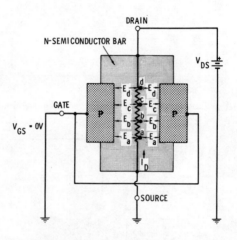

Fig. 12-3. The reverse voltage is smallest (E_a) near the source and largest (E_d) near the drain.

The drain-to-source voltage V_{DS} also affects the channel width and therefore the conductivity. Why this is so can be seen if we view the bar as equivalent to a voltage divider as shown in Fig. 12-3. That is, the bar has some resistance, so we can show it equivalent to a resistor. If we arbitrarily place some points on this voltage divider, a, b, c, and d, we can refer to the voltages at these points, with respect to ground, as E_a, E_b, E_c, and E_d. If there is resistance between each point and I_D flows, the voltages E_a through E_d are unequal, E_a being the smallest and E_d being the largest. Mathematically this relationship may be shown as

$$E_a < E_b < E_c < E_d$$

For the present, if we conveniently place the gate at ground potential ($V_{GS}=0$) the voltages E_a, E_b, etc., are the reverse voltages

2-1 Concepts of the Junction Field-Effect Transistor

across the pn junctions in the area of points a, b, etc., respectively. Thus, since E_d is the largest voltage, the greatest reverse bias exists at about point d. From this we may conclude that the depletion region spreads further into the n-type bar near the drain than

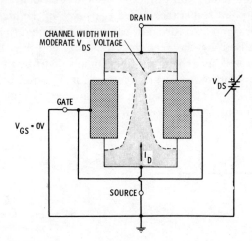

(A) *Conductive channel is wide with relatively low resistance for small V_{DS} and $V_{GS} = 0$ V.*

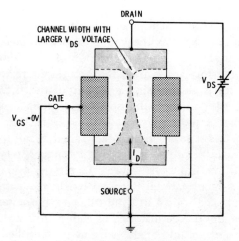

(B) *Conductive channel is narrow with relatively high resistance for larger V_{DS} and $V_{GS} = 0$ V.*

Fig. 12-4. Effect of V_{DS} on depletion regions.

near the source. The result is a narrower channel, especially near the drain, with larger V_{DS} values as shown in Fig. 12-4.

If we increase the voltage V_{DS} in the circuit of Fig. 12-4A, say from zero to larger positive values, we would expect the drain current I_D to increase too. And in fact, it does up to a point. Eventually V_{DS} reaches a point sometimes called the *pinch-off voltage* V_p, from which further increases do not cause proportional increases in I_D. This is caused by the fact that by increasing V_{DS} beyond V_p causes the channel near the drain to become so narrow its resistance increases in proportion to further increases in V_{DS} and this tends to hold I_D fairly constant. Typical drain characteristics of the JFET with the gate grounded are shown in Fig. 12-5.

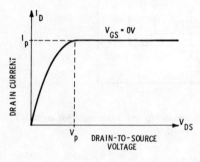

Fig. 12-5. Typical V-I curves for JFET with zero gate-to-source voltage.

The resistance seen looking into the gate is very high, typically 10^6 to 10^9 ohms, as long as the gate-to-source voltage is negative enough to keep the pn junctions reverse biased. Should the pn junctions become forward biased, the gate input resistance drops significantly. This is an undesirable feature of the JFET.

In practice it is difficult to diffuse p-type materials onto two sides of an n-type bar. A more practical JFET construction is shown in Fig. 12-6A. In this case a silicon base, called a *substrate*, is used onto which a layer of n-type material is formed. A p-type material is then diffused into the n-type layer. The n-type layer is the channel while the p-type material is connected to the gate lead. Current I_D flows via the source lead into the n-type material. It then proceeds under the p-type material, as shown in Fig. 12-6B, to the drain lead. The magnitude of this current is controlled by the potential across the gate and source leads. That is, if the gate is made several volts negative with respect to the source, a depletion region spreads well into the n-material all around under the p-ma-

2-1 CONCEPTS OF THE JUNCTION FIELD-EFFECT TRANSISTOR

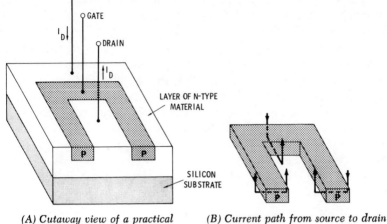

(A) Cutaway view of a practical JFET construction.

(B) Current path from source to drain is under the p-type material.

Fig. 12-6. JFET is formed on a chip of silicon substrate.

terial. This tends to reduce the cross sectional area (the space) under the p-material that will conduct current, consequently reducing it. On the other hand, if the gate is made less negative, the depletion region is reduced, which allows more current to flow. As with the previously discussed JFET structure in Fig. 12-1A, so long as the pn junction is kept at reverse bias, the resistance seen looking into the gate is very high.

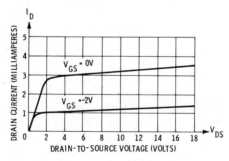

Fig. 12-7. Typical drain characteristics of JFET.

Example 12-1 The characteristics in Fig. 12-7 are for the JFET in Fig. 12-8. If the slide on the potentiometer R_1 is in the far right position as shown while the slide in potentiometer R_2 is progressively moved from left to right, what reading would you expect on the ammeter M_4 for each of the following readings on voltmeter M_3? (a) 1 V, (b) 2 V, (c) 4 V, (d) 16 V.

Answer 12-1 (a) 1.5 mA, (b) 2.7 mA, (c) 3 mA, (d) 3.5 mA.— With the slide of R_1 at the far right, the gate is zero volts with respect to the source. That is, $V_{GS} = 0$ V and voltmeter M_2 reads 0 V. Thus, the upper curve in Fig. 12-7 applies. Note that as V_{DS} is increased from zero to 3 V, the drain current I_D increases significantly to about 3 mA. But with further increases, $V_{DS} > 3$ V, the drain current I_D increases very little. The pinch-off voltage V_p, which is the V_{DS} voltage beyond which I_D ceases to increase significantly, is about 3 V and the pinch-off current I_p is about 3 mA in this case where $V_{GS} = 0$ V.

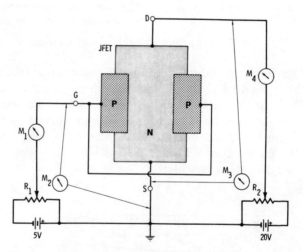

Fig. 12-8. JFET circuit with means of varying V_{GS} and V_{DS}.

Example 12-2 Referring to Figs. 12-7 and 12-8 again, if the slide on R_1 is moved to the left until voltmeter M_2 reads 2 V, (a) over what range can I_D be varied by adjustment of the slide on R_2? (b) What are the approximate pinch-off voltage and current values now? (c) What would you expect to read on ammeter M_1?

Answer 12-2 (a) 0 to 1.5 mA, (b) $V_p \cong 2$ V, $I_p \cong 1$ mA, (c) about zero.—(a) By varying the slide on R_2 from extreme left to extreme right, voltage V_{DS} is varied from zero to 20 V. We can predict the resulting current changes from the lower curve in Fig.

12-7 because R_1 was adjusted to make the gate -2 V with respect to the source ($V_{GS} = -2$ V). On this lower curve, I_D varies from zero to about 1.5 mA. (b) The lower curve levels off at a point whose coordinates are about 2 V and 1 mA. (c) Since this is an n-channel JFET, negative voltages on its gate reverse bias its pn junctions, causing insignificant or zero gate current.

Example 12-3 Still referring to Figs. 12-7 and 12-8, if R_2 is adjusted so that M_3 reads 10 V, what minimum and maximum current would you expect to read on M_4 if the slide on R_1 is moved from far right to far left?

Answer 12-3 About 3.2 mA to 0 mA.—With the slide on R_1 at the far right position, $V_{GS} = 0$ and with $V_{DS} = 10$ V, we can see in Fig. 12-7 that if we enter the V_{DS} axis at 10 V and follow vertically to

(A) Begin with lightly doped p-type substrate.

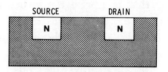

(B) N-type regions diffused into substrate

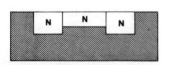

(C) Lightly doped n-type layer diffused into substrate to act as channel.

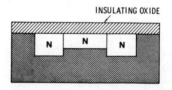

(D) Oxide layer formed over entire surface.

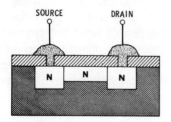

(E) Source and drain connected to n-type materials.

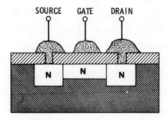

(F) Metallic layer over but not contacting channel becomes gate.

Fig. 12-9. Typical construction method of n-channel IGFET.

intersect the $V_{GS}=0$ V curve, we read about 3.2 mA to the left of the intersection. With the slide on R_1 at the far left, $V_{GS}=-5$ V. This is negative enough to cut off the JFET. Note on the characteristics that if V_{GS} is changed from 0 V to -2 V, the drain current I_D decreases from about 3.2 mA to 1.2 mA, a decrease of 2 mA. Thus, we can reason that if the gate is made even more negative by another 3 V (changing V_{GS} from -2 V to -5 V), the drain current will tend to decrease by more than 2 mA from $I_D=2$ mA, thus driving the drain into cutoff.

12-2 FUNDAMENTALS OF THE INSULATED GATE FIELD-EFFECT TRANSISTOR

As its name implies, the *insulated gate field-effect transistor* (IGFET) actually has its gate lead insulated from the semiconductors that comprise the structure of the FET. This IGFET is also known as the *metal oxide semiconductor field-effect transistor* (MOSFET). Its method of operation is similar to but yet different than that of the JFET.

Fig. 12-9 shows the steps taken to construct the typical IGFET. In Fig. 12-9A we start with a lightly doped p-type substrate. Two separate heavily doped n-type regions are then diffused into the substrate as shown in Fig. 12-9B. Next, a lightly doped n-type region is diffused between the heavily doped n-type regions as shown in Fig. 12-9C. It is this lightly doped region that eventually acts as the channel. Now an insulating oxide layer is formed over the entire surface as shown in Fig. 12-9D. Next, holes are cut or etched in the oxide, and metal, similar to solder, is evaporated into the holes to make contact with the heavily doped n materials as shown in Fig. 12-9E. Note that the source and drain leads are connected to these metallic contacts. Finally, the gate is connected to metal evaporated onto the oxide directly over the lightly doped n-type channel (Fig. 12-9F). Thus, the gate is insulated from the semiconductors by the oxide, which gives the gate its extremely high input impedance, typically 10^{10} to 10^{14} Ω. Since the channel in this case is made of n-type material, this is an n-channel IGFET and it requires a positive drain-to-source bias voltage. P-channel IGFET's are also available and they require negative drain bias voltages.

Since the n-type channel in Fig. 12-9 is lightly doped, it acts like a fairly large resistance. That is, with a V_{DS} applied and zero

12-2 The Insulated Gate Field-Effect Transistor

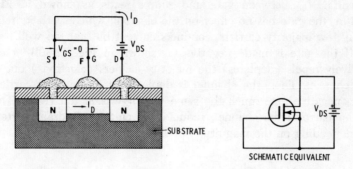

(A) With $V_{GS} = 0$, the lightly doped channel conducts moderate I_D.

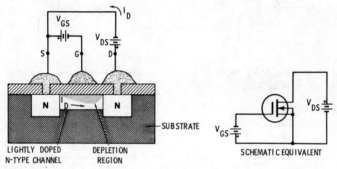

(B) With negative V_{GS} a depletion region spreads into the channel, reducing its conductivity and therefore I_D.

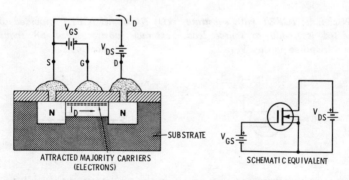

(C) With positive V_{GS}, more majority carriers are pulled into the channel, enhancing its conductivity and increasing I_D.

Fig. 12-10. Majority carriers may be pushed out of channel (depletion mode) or pulled into channel (enhancement mode), depending on gate polarity.

potential V_{GS} between gate and source leads as shown in Fig. 12-10A, the gate has no effect on the channel, which, with its relatively few majority carriers, conducts current but not too well.

If the gate is made negative, as shown in Fig. 12-10B, it effectively repels (depletes) the majority carriers (electrons) out of the channel. Thus, the channel is depleted of its majority carriers by a negative gate much the same way it was in the JFET. This action narrows the channel, reducing or cutting off drain current I_D, depending on the magnitude of the gate voltage.

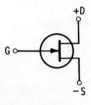

(A) N-channel JFET requires positive V_{DS}.

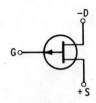

(B) P-channel JFET requires negative V_{DS}.

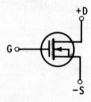

(C) N-channel IGFET with substrate connected internally to source lead requires positive V_{DS}.

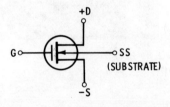

(D) N-channel IGFET provided with external substrate lead SS requires positive V_{DS}.

(E) P-channel IGFET with substrate connected internally to source lead requires negative V_{DS}.

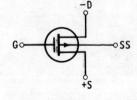

(F) P-channel IGFET provided with external substrate lead SS requires negative V_{DS}.

Fig. 12-11. Schematic symbols for JFET's and IGFET's.

12-2 THE INSULATED GATE FIELD-EFFECT TRANSISTOR

If a positive gate-to-source voltage is used as shown in Fig. 12-10C, it attracts electrons (majority carriers) into the channel, which enhances its conductivity. Thus, more or less positive gate voltages pull (induce) more or less electrons into the channel, which in turn cause more or less drain current respectively.

It is interesting to note that with this n-channel IGFET (Fig. 12-9) the drain current I_D is decreased by driving the gate with a more negative voltage. On the other hand, positive gate voltages actually increase I_D, which is quite a contrast from the way a JFET works. Remember that a positive gate is avoided with the JFET because it may forward bias the pn junction and thus greatly lower the input resistance of the gate. Since the gate is insulated from the semiconductor materials in the IGFET, the gate input resistance is unaffected by the polarity of the gate voltage within limits.

Symbols commonly used to represent JFET's and IGFET's are shown in Fig. 12-11. Note that if the gate arrow points inward (n-channel types), the drain is biased positively with respect to the source. On the other hand, if the arrow points outward (p-channel types), the drain is biased negatively.

The gate-to-source voltages V_{GS} that we can use with the IGFET's have a limited range. Too much V_{GS} may actually puncture the metal oxide insulation and thus ruin the IGFET. Because the gate resistance is so large, even static voltages, such as exists on your fingers, can puncture the oxide. To avoid damage, the manufacturers ship IGFET's with their leads taped together or with the leads shorted some other way such as with a shorting ring or wire. The leads must be kept shorted until the IGFET is soldered or otherwise connected into the circuit.

Another type of IGFET is shown in Fig. 12-12. Note that its construction is similar to the one just covered except that it does

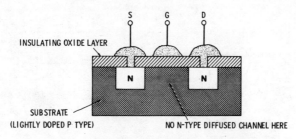

Fig. 12-12. Construction of induced-channel IGFET.

not have a diffused n-type channel. In this case only the substrate is between the source and drain. Therefore the substrate must act as the channel. Since it has very few charge carriers it conducts negligible I_D in absence of any gate voltage ($V_{GS}=0$). It takes positive gate-to-source voltages to increase I_D from zero. The reason is that a positive gate pulls electrons into the channel as shown in Fig. 12-13. This is similar to the action of a charged capacitor; the metallic gate area is the top plate, the insulating oxide is the dielectric, and the substrate is the bottom plate. With a positive

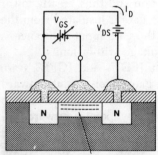

N-TYPE CHANNEL INDUCED BETWEEN SOURCE AND DRAIN BY POSITIVE GATE VOLTAGES

Fig. 12-13. Drain current I_D increases from about zero with positive gate voltage only, in induced-channel IGFET.

gate potential the metallic gate area charges positively and a corresponding negative charge is induced into the substrate. The negative charge is in the form of extra electrons which support current I_D through the channel. Thus,

$$I_{DS} \cong 0 \text{ with } V_{GS}=0.$$

But I_D increases with positive V_{GS} values. The symbols in Fig. 12-14 are sometimes used to represent induced-channel IGFET's.

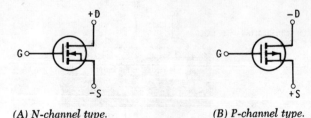

(A) N-channel type.   (B) P-channel type.

Fig. 12-14. Schematic symbols of induced-channel IGFET.

12-3 MODES OF OPERATION

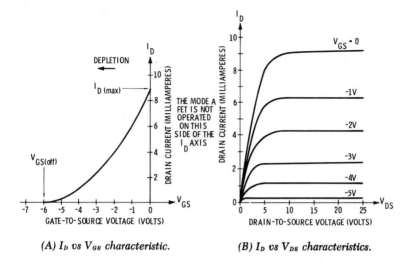

(A) I_D vs V_{GS} characteristic.

(B) I_D vs V_{DS} characteristics.

Fig. 12-15. Typical n-channel depletion mode (mode A) FET characteristics.

12-3 MODES OF OPERATION

FET's are manufactured and specified for one of three different modes of operation. The specified mode of an FET affects the way in which the gate lead is biased with respect to the source.

Depletion Mode or Mode A

The JFET is a mode-A type field-effect transistor. The I_D vs V_{GS} characteristics and drain characteristics of a typical n-channel mode-A FET are shown in Figs. 12-15A and 12-15B. Note in Fig. 12-15A that positive gate-to-source voltages V_{GS} are not used, to avoid forward biasing the gate-source junction. The gate is therefore biased between some negative cutoff voltage $V_{GS(off)}$ and 0 V. Since more or less negative gate voltage depletes the n-channel of more or less majority carriers, this mode-A FET is appropriately also called the *depletion mode FET*. Though mode-A types are mainly JFET's, a few IGFET's have been made.

Depletion-and-Enhancement Mode or Mode B

With units of this mode, whose characteristics are shown in Figs. 12-16A and 12-16B, the gate-to-source voltage V_{GS} can be varied in *polarity* as well as amplitude, within limits. When V_{GS} is negative, the channel is *depleted* of its majority carriers and

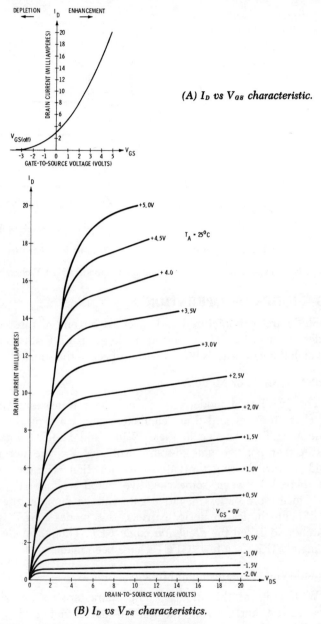

(A) I_D vs V_{GS} characteristic.

(B) I_D vs V_{DS} characteristics.

Fig. 12-16. Typical n-channel depletion-and-enhancement mode (mode B) FET characteristics.

12-3 MODES OF OPERATION

this tends to limit the magnitude of drain current I_D. When V_{GS} is positive, the majority carriers are induced into the channel and *enhances* the channel conductivity and the magnitude of I_D. Hence the term *depletion-and-enhancement mode* is descriptive of how I_D is varied. The IGFET constructed in Fig. 12-10 is a depletion-enhancement mode (mode B) type. Unlike the JFET, whose high input resistance is dependent on maintaining reverse biased pn junctions, the IGFET's extremely high input resistance is provided by the insulation between the gate and the channel.

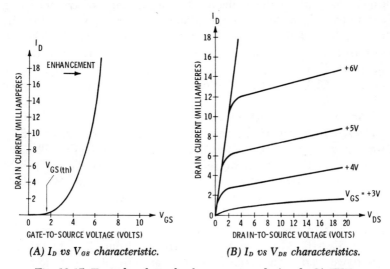

(A) I_D vs V_{GS} characteristic. (B) I_D vs V_{DS} characteristics.

Fig. 12-17. Typical n-channel enhancement mode (mode C) FET characteristic.

Enhancement Mode or Mode C

As shown in the characteristics in Figs. 12-17A and 12-17B, gate-to-source voltages more positive than $V_{GS(th)}$ induce majority carriers, more or less, depending on the magnitude of V_{GS}, into the channel, thus *enhancing* channel conductivity. Typically I_D is only a few microamperes at the threshold voltage $V_{GS(th)}$; therefore, for most practical purposes we may assume that $I_D = 0$ for V_{GS} values more negative than $V_{GS(th)}$. As with units made to operate mode B, the mode-C types are of IGFET construction (Fig. 12-12).

Example 12-4 In the circuit of Fig. 12-18, the JFET has the characteristics shown in Fig. 12-15. What current value would you

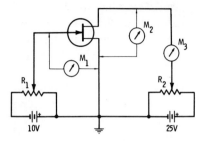

Fig. 12-18. JFET in circuit capable of varying V_{GS} from 0 to -10 V and V_{DS} from 0 to 25 V.

expect to read on the milliammeter M_3 in each of the following cases? (a) The slide on the potentiometer R_1 is adjusted so that M_1 reads 2 V, and R_2 is adjusted so that M_2 reads 15 V. (b) The adjustment on R_1 is left intact, M_1 reads 2 V, but R_2 is adjusted until M_2 reads 25 V. (c) The slide on R_1 is moved to the far right while M_2 reads 25 V. (d) The slide on R_1 is moved to the far left while M_2 reads 25 V.

Answer 12-4 (a) About 4.2 mA, (b) about 4.2 mA, (c) about 9.2 mA, (d) about 0.—In (a), since $V_{DS} = 15$ V and $V_{GS} = -2$ V, we can enter V_{DS} axis on the drain characteristics (Fig. 12-15B), vertically at 15 V. When we reach the -2-V curve we read about 4.2 mA to the left on the I_D axis. In (b), we enter the V_{DS} axis at 25 V and proceed vertically to the -2 V curve. Since the curve is horizontal in this area of the characteristics, we read 4.2 mA again to the left of the intersection. In (c), the slide to the far right makes $V_{GS} = 0$. After proceeding vertically from 25 V to the 0 V curve, we read about 9.2 mA on the I_D axis. In (d), the slide to the far left makes $V_{GS} = -10$ V. This gate voltage is beyond (more negative) than the gate cutoff voltage as shown in Fig. 12-15A. The JFET is effectively prevented from conducting with maybe a usually negligible leakage current.

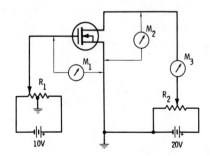

Fig. 12-19. IGFET in circuit capable of varying V_{GS} from -5 to $+5$ V and V_{DS} from 0 to 20 V.

12-3 MODES OF OPERATION

Example 12-5 The IGFET in Fig. 12-19, has the characteristics shown in Fig. 12-16. What current values would you expect to read on the milliammeter M_3 in each of the following cases? (a) The slide on R_1 is placed in the far left position while R_2 is adjusted so that M_2 reads 12 V. (b) The slide on R_1 is placed at center, directly above the ground point, while R_2 is adjusted so that M_2 reads 14 V. (c) The slide on R_1 is placed in the far right position while R_2 is adjusted so that M_2 reads 14 V.

Answer 12-5 (a) About 0, (b) about 3 mA, (c) about 20 mA.—Assuming that the ground on the tapped potentiometer R_1 is centered, the left end of R_1 is -5 V with respect to ground while the right end of R_1 is $+5$ V to ground. In (a), with the slide on R_1 in the far left position, $V_{GS} = -5$ V. Note in Fig. 12-16A that cutoff occurs with gate-to-source voltages more negative than -3 V; thus $I_D \cong 0$. In (b), the slide on R_1 is adjusted so that $V_{GS} = 0$. Since $V_{DS} = 14$ V, we enter the V_{DS} axis, in Fig. 12-16B, at 14 V and proceed vertically until we intersect the curve marked $V_{GS} = 0$. The left of the intersection we not that $I_D \cong 3$ mA. In (c), since the slide on R_1 is in the far right position, $V_{GS} = 5$ V. With $V_{DS} = 14$ V, we enter the V_{DS} axis at 14 V in Fig. 12-16B. Then proceeding vertically until we intersect the $+5$ V curve, we ready $I_D \cong 20$ mA to the left of the intersection.

Example 12-6 If the induced-channel IGFET in the circuit of Fig. 12-20 has the characteristics shown in Fig. 12-17, what happens

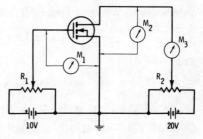

Fig. 12-20. IGFET in circuit capable of varying V_{GS} from 0 to 10 V and V_{DS} from 0 to 20 V.

to the reading on milliammeter M_3 when R_1 is positioned so that M_1 reads 6 V and the slide on R_2 is adjusted so that the reading on M_2 gradually increases from 6 V to 18 V?

Answer 12-6 The drain current I_D gradually increases from 12 mA to 14 mA.

12-4 PARAMETERS OF FET'S

Much as the parameters (constants) α, β, and r_e', used in Chaps. 7, 8, and 9, are useful in designing and predicting operating qualities of transistor circuits, so we likewise have FET parameters that are similarly useful. If we are to use FET's to amplify audio frequencies, there are two parameters of particular interest to us. These are: (1) the *forward transadmittance (transconductance)*, which is referred to with several symbols: y_{fs}, g_{fs}, g_m, and g_{21}, to name only the more common ones, and (2) the *output admittance*, which is also referred to with several symbols: y_{os}, g_{os}, $1/r_d$, and g_{22}.

Specifically, the equation for forward transadmitance is as follows:

$$y_{fs} = \frac{\Delta I_D}{\Delta V_{GS}} \bigg| \Delta V_{DS} = 0 \qquad (12\text{-}1)$$

where the vertical bar means "for the condition(s) . . ." Thus Eq. 12-1 says that the forward transadmittance is equal to the ratio of the change in drain current to the change in gate-to-source voltage while the change in drain-to-source voltage is zero; that is, the V_{DS} is held constant. We can approximate the y_{fs} of an FET from its drain characteristics. Take for example the JFET drain characteristics in Fig. 12-15B. We can hold $V_{DS} = 15$ V constantly and change V_{GS} say from -3 V to -2 V and note the change in I_D. That is, we follow the vertical line from 15 V, in Fig. 12-15B, to where it intersects curves -3 V and -2 V. To the left of these intersections we read about 2.3 mA and 4.2 mA. Thus,

$$\Delta I_D \cong 4.2 - 2.3 = 1.9 \text{ mA}$$

and

$$\Delta V_{GS} = 3 - 2 = 1 \text{ V}$$

while $\Delta V_{DS} = 0$; that is, $V_{DS} = 15$ V constantly. Now substituting these values into Eq. 12-1 we find that

$$y_{fs} \cong \frac{1.9 \text{ mA}}{1 \text{ V}} = 1.9 \text{ mmhos}$$

The *output admitance* is given by the following equation:

$$y_{os} = \frac{\Delta I_D}{\Delta V_{DS}} \bigg| \Delta V_{GS} = 0 \qquad (12\text{-}2)$$

12-4 PARAMETERS OF FET's

That is, this parameter is equal to the change in drain current over the change in drain-to-source voltage while the change in gate-to-source voltage is zero, i.e., V_{GS} is held constant. Again referring to the JFET whose drain characteristics are shown in Fig. 12-15B, we can hold V_{GS} constant at say -2 V. Now we pick two points on the -2 V curve, preferably close together, such as directly over 10 V and 15 V on the V_{DS} axis. In this case, due to the -2 V curve being almost perfectly horizontal in this area, we read $I_D \cong$ 4.2 mA to the left of both points. Thus, while V_{GS} was held constant at -2 V, a drain-to-source voltage change of $\Delta V_{DS} \cong 15-10 = 5$ V causes a drain current change of $\Delta I_D \cong 4.2 - 4.2 = 0$ mA. Substituting into Eq. 12-2 we find that in this case

$$y_{os} \cong \frac{0}{5 \text{ V}} = 0 \text{ mhos}$$

The output admittance is about zero in this case because the V_{GS} curves on the drain characteristics (Fig. 12-15B) are practically horizontal lines for drain-to-source voltages greater than 10 V. Drain characteristics with V_{GS} curves that slope up to the right, as in Figs. 12-16B and 12-17B, have values of y_{os} which are nonzero, as you will see.

Example 12-7 Referring to the IGFET whose drain characteristics are shown in Fig. 12-16B, find (a) the forward transadmittance y_{fs} by holding V_{DS} constant at 8 V while varying V_{GS} from $+2.5$ V to $+3$ V, and find (b) the output admittance y_{os} by holding V_{GS} constant at $+2.5$ V while varying V_{DS} from 8 V to 14 V.

Answer 12-7 (a) $y_{fs} \cong 3.6$ mmhos, (b) $y_{os} \cong 83.3$ μmhos.—In (a), we enter the V_{DS} axis in Fig. 12-16B vertically at 8 V and proceed vertically until we intersect the $+2.5$ V and $+3$ V curves. To the left of these intersections, on the I_D axis, we read 10 mA and about 11.8 mA respectively. Substituting these values into Eq. 12-1 we find that

$$y_{fs} \cong \frac{11.8 - 10}{3 - 2.5} = \frac{1.8 \text{ mA}}{0.5 \text{ V}} = 3.6 \text{ mmhos}$$

In (b), we note where vertical projections from 8 V and 14 V on the V_{DS} axis intersect the $+2.5$ V curve. To the left of these intersections we read 10 mA and about 10.5 mA respectively. These values substituted into Eq. 12-2 show us that

$$y_{os} \cong \frac{10.5-10}{14-8} = \frac{0.5 \text{ mA}}{6 \text{ V}} = 0.0833 \text{ mmho} = 83.3 \text{ μmhos}$$

Example 12-8 Suppose that we place an IGFET in a circuit like that of Fig. 12-20 and that initially M_1 reads 4 V, M_2 reads 10 V, and meter M_3 reads 2 mA. When we adjust R_1 so that M_1 reads 5 V, the readings on M_3 and M_2 increase and decrease respectively. But after adjusting R_2 so that M_2 reads 10 V again, we read 4 mA on meter M_3. Now we adjust R_2 until M_3 reads 2 mA again while M_1 is kept at 5 V. The reading on M_2 now is 6 V. With these observations and Eqs. 12-1 and 12-2, determine the approximate values of y_{fs} and y_{os}.

Answer 12-8 $y_{fs} = 2$ mmhos, $y_{os} = 0.5$ mmho.—The facts are listed in Table 12-1. The intial values are listed in row I. After

Table 12-1. IGFET Circuit Data

Adjustment	M_1 or V_{GS}	M_2 or V_{DS}	I_D
I	4 V	10 V	2 mA
II	5 V	10 V	4 mA
III	5 V	6 V	2 mA

the adjustments on R_3 and R_2 we obtained values in row II. And after readjustment of R_2, the values obtained are in row III. Note in rows I and II that V_{DS} is constant at 10 V but changes have occured in values V_{GS} and I_D. Thus, by Eq. 12-1

$$y_{fs} = \frac{4-2}{5-4} = \frac{2 \text{ mA}}{1 \text{ V}} = 2 \text{ mmhos}$$

Also note that V_{GS} is constant at 5 V in rows II and III but changes have occured to V_{DS} and I_D. So by Eq. 12-2,

$$y_{os} = \frac{4-2}{10-6} = \frac{2 \text{ mA}}{4 \text{ V}} = 0.5 \text{ mmho}$$

is the output admittance.

12-5 METHODS OF BIASING FET'S FOR USE AS AMPLIFIERS

Typical methods of bias on FET's of modes A, B, and C are shown in Fig. 12-21. Because of the large gate input resistance, the dc current in R_G is essentially zero in Fig. 12-21A and therefore the

12-5 METHODS OF BIASING FET's FOR USE AS AMPLIFIERS

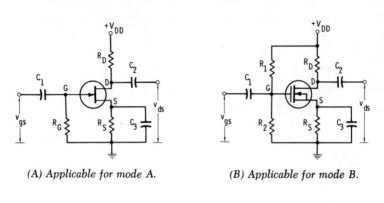

Fig. 12-21. Useful FET signal amplifier circuits.

voltage across R_G, which is the voltage at point G to ground, is about zero. The drain current I_D, for all practical purposes, is the current in R_S. Thus, I_D through R_G causes a positive voltage at point S with respect to ground. Since the gate G is 0 V while the source S is some positive voltage to ground, the gate is negative with respect to the source by the amount of voltage drop across R_S. A negative quiescent V_{GS} is necessary if we want to use an n-channel mode-A JFET as a signal amplifier. In this circuit (Fig. 12-12A), by Ohm's law

$$R_S = \frac{V_{GS}}{I_D} \qquad (12\text{-}3)$$

where
  V_{GS} is the absolute value of the required quiescent gate-to-source voltage,
  I_D is the required quiescent drain current.

The value of R_G is selected more or less arbitrarily. It can be very large—up to 2 MΩ or so—so as to keep the input resistance to the circuit very high.

Example 12-9 If the JFET with the characteristics in Fig. 12-15 is to be used in the circuit of Fig. 12-21A, what value of R_S will you use if the quiescent drain current is to be about 4 mA?

Answer 12-9 $R_S \cong 500$ Ω.—Projecting to the left of 4 mA from the I_D axis in Fig. 12-15A, we intersect the curve approximately over -2 V on the V_{GS} axis. Therefore by Eq. 12-3

$$R_S \cong \frac{2 \text{ V}}{4 \text{ mA}} = 0.5 \text{ k}\Omega$$

so that $R_S = 500$ Ω.

The circuit of Fig. 12-21B enables us to obtain very small bias voltages such as often required with mode-B type IGFET's. That is, because V_{GS} can be varied from negative to positive values with continuous control of I_D (see Fig. 12-16A), the required quiescent V_{GS} might be zero or perhaps a small positive voltage, depending on the specific characteristics of the unit. Since the resistance seen looking into the gate is extremely high, the voltage divider resistors R_1 and R_2 are in series to dc currents.

Example 12-10 Referring to the circuit of Fig. 12-21B, if $V_{DD} = 20$ V, $R_1 = 3.6$ MΩ, $R_2 = 400$ kΩ, $R_S = 300$ Ω, $R_D = 2$ kΩ, and the voltage at point S to ground is $+1.5$ V, what are the quiescent values of I_D, V_{GS}, and V_{DS}?

Answer 12-10 $I_D = 5$ mA, $V_{GS} = 0.5$ V, $V_{DS} = 8.5$ V. — The drain current I_D is the current through R_S. Since the voltage across, and the resistance of, R_S are known to be 1.5 V and 300 Ω, with Ohm's law we find that

$$I_D = \frac{1.5 \text{ V}}{300 \text{ }\Omega} = 5 \text{ mA}$$

The voltage at point G to ground, which is the voltage across R_2, can be found in series circuit fashion. That is, since R_1 and R_2 are in series with voltage V_{DD} applied,

$$V_G = V_{R_2} = \frac{V_{DD} R_2}{R_1 + R_2} = \frac{(20 \text{ V})(0.4 \text{ M}\Omega)}{3.6 \text{ M}\Omega + 0.4 \text{ M}\Omega} = 2 \text{ V}$$

12-6 SMALL-SIGNAL VOLTAGE GAIN IN FET AMPLIFIERS

The gate-to-source voltage is the difference in the gate-to-ground voltage V_G and the source-to-ground voltage V_S and in this case is

$$V_{GS} = V_G - V_S = 2 - 1.5 = 0.5 \text{ V}$$

Examine the circuit and note that the sum of the voltages across R_D and R_S plus the drain-to-source voltage V_{DS} must be equal to the source voltage V_{DD}. That is,

$$V_{DD} = R_D I_D + V_{DS} + R_S I_D$$

Rearranging terms, we can show that

$$V_{DS} = V_{DD} - R_D I_D - R_S I_D$$

Now substituting the knowns into this equation we get

$$V_{DS} = 20 \text{ V} - (2 \text{ k}\Omega)(5 \text{ mA}) - 1.5 \text{ V} = 8.5 \text{ V}$$

is the drain-to-source voltage.

In the circuit of Fig. 12-21C, the absence of resistance in the source lead tends to make the gate G quite positive with respect to the source S. A comparison can be made by using the gate voltage divider resistors given in the last problem; that is, $R_1 = 3.6$ MΩ and $R_2 = 400$ kΩ. With these resistors in the circuit of Fig. 12-21C, the gate G is 2 V with respect to ground whereas the source S is 0 V to ground. In this case

$$V_{GS} = V_G - V_S = 2 - 0 = 2 \text{ V}$$

Clearly then, the same divider gives us a more positive gate than it did in Fig. 12-21B.

Example 12-11 If you are to establish a quiescent $V_{GS} = 5$ V in the circuit of Fig. 12-21C, what ratio of R_1 to R_2 will you use if $V_{DD} = 30$ V?

Answer 12-11 $R_1/R_2 = 5/1$.—The voltage across R_2, call it V_2, is V_{GS}. Since in this case V_{GS} must be 5 V, the remaining 25 V is dropped across R_1. Thus, $R_1/R_2 = 25$ V/5 V $= 5/1$.

12-6 SMALL-SIGNAL VOLTAGE GAIN IN FET AMPLIFIERS

The voltage gain A_e of each circuit in Fig. 12-21 is defined as the ratio of signal output v_{ds} to the signal input v_{gs}. That is,

$$A_e = \frac{v_{ds}}{v_{gs}} \tag{12-4}$$

where

v_{ds} is the drain-to-source signal voltage,
v_{gs} is the gate-to-source signal voltage.

The term *small-signal voltage gain* applies if the peak-to-peak value of output voltage v_{ds} is less than about one-tenth the dc source voltage V_{DD}. You may assume that we are dealing with such small signals for the remainder of this section.

The amount of voltage gain A_e obtained is determined by the FET's forward transadmittance y_{fs} and output admittance y_{os}, and the circuit component values. We will see why this is true after analyzing a couple of previously given equations and some ac equivalent circuits.

On re-examination and rearrangement of Eq. 12-1, we can show that since

$$y_{fs} = \frac{\Delta I_D}{\Delta V_{GS}} \bigg| \Delta V_{DS} = 0$$

then

$$\Delta I_D = \Delta V_{GS} y_{fs} \bigg| \Delta V_{DS} = 0 \tag{12-1A}$$

which is to say that the *change* in drain current ΔI_D is equal to the product of the *change* in gate-to-source voltage ΔV_{GS} and the forward transadmittance y_{fs} under conditions where there is no change in the drain-to-source voltage V_{DS}, that is, $\Delta V_{DS} = 0$.

Actually the condition $\Delta V_{DS} = 0$ is not necessary in most cases if we plan to bias the FET for operation in its drain characteristics where the V_{GS} curves are nearly straight lines and parallel. That is, if we use the JFET whose drain characteristics are given in Fig. 12-15B, with V_{DS} voltages greater than about 10 V, *all* of the V_{GS} curves are nearly straight and parallel. This allows simplification of Eq. 12-1A to

$$\Delta I_D \cong \Delta V_{GS} y_{fs} \tag{12-1B}$$

for straight and parallel V_{GS} curves.

An input signal v_{gs} causes a varying (change in) V_{GS}. Actually the amplitudes of v_{gs} and ΔV_{GS} are proportional. Similarly, then, as

12-6 SMALL-SIGNAL VOLTAGE GAIN IN FET AMPLIFIERS

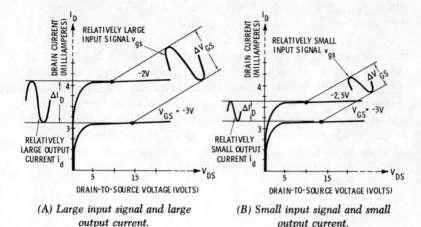

(A) Large input signal and large output current.

(B) Small input signal and small output current.

Fig. 12-22. Effect of input signal v_{gs} (ΔV_{GS}) on i_d and ΔI_D shown projected on typical JFET drain characteristics.

a result of the input signal v_{gs}, the dc drain current I_D has a signal component i_d superimposed on it, where the amplitudes of i_d and ΔI_D are proportional. Fig. 12-22 shows that a larger v_{gs} signal causes a larger ΔV_{GS} and ΔI_D. Fig. 12-22A shows an input signal v_{gs} causing $\Delta V_{GS} \cong 3-2 = 1$ V which in turn causes an output signal current i_d with a peak-to-peak amplitude of $\Delta I_D \cong 4.1 - 3.1 = 1$ mA. Fig. 12-22B shows an input signal v_{gs} causing $\Delta V_{GS} = 3 - 2.5 = 0.5$ V which in turn causes an output signal current i_d with a peak-to-peak amplitude of $\Delta I_D \cong 3.6 - 3.3 = 0.5$ mA. In equation form,

$$\frac{\Delta I_D}{\Delta V_{GS}} \cong \frac{i_d}{v_{gs}} \cong y_{fs}$$

which when rearranged gives us

$$\Delta I_D \cong \Delta V_{GS} y_{fs} \qquad (12\text{-}1\text{B})$$

and

$$i_d \cong v_{gs} y_{fs} \qquad (12\text{-}1\text{C})$$

This last equation enables us to determine the signal component i_d in the drain lead if the applied signal voltage v_{gs} and transfer admittance y_{fs} are known.

An ac equivalent of the circuit in Fig. 12-21A is shown in Fig. 12-23A. As the source v_{gs} sees it, R_G and $r_{in(gate)}$ are in parallel

420 General Characteristics of the Field-Effect Transistor

where $r_{in(gate)}$ is about 10^6 to 10^9 Ω, the JFET's gate input resistance. Thus, v_{gs} works into $r_{in(stage)} = R_G \| r_{in(gate)}$.

The signal output current i_d is essentially independent of all factors except v_{gs} and y_{fs}, as indicated by Eq. 12-1C. The output of the FET may therefore be considered a constant current source and independent of drain-to-source voltage and drain load resistance changes. The constant current source i_d works into parallel resistances whose admittances are y_{os} and y_L as shown in Fig. 12-23A. As previously mentioned, y_{os} is the FET's output admittance. Admittance y_L is the admittance of the load in the drain circuit. That is, in the case of the circuit in Fig. 12-12A,

$$y_L = \frac{1}{R_D} \qquad (12\text{-}5)$$

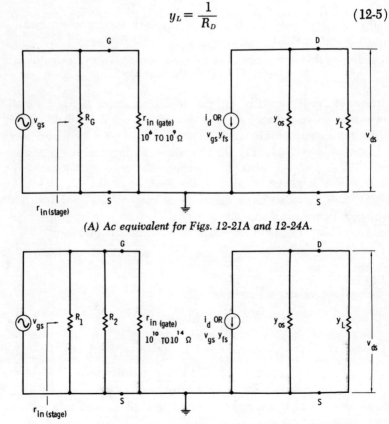

(A) Ac equivalent for Figs. 12-21A and 12-24A.

(B) Ac equivalent for Figs. 12-21B, 12-21C, 12-24B, and 12-24C.

Fig. 12-23. Ac equivalents of circuits in Figs. 12-21 and 12-24.

12-6 SMALL-SIGNAL VOLTAGE GAIN IN FET AMPLIFIERS

If the total load, as seen by the drain, is more complicated, as in Fig. 12-24A, then

$$y_L = \frac{1}{r_L} \qquad (12\text{-}6)$$

where

$$r_L = R_D \| R \qquad (12\text{-}7)$$

Resistance R_S is not included in the equivalents of Fig. 12-23 because it is bypassed with C_3. That is, signal sees the R_S and C_3 combination as nearly a short circuit provided that $X_{C3} \ll R_S$.

The circuit in Fig. 12-23B is an ac equivalent of the circuits in Figs. 12-21B and 12-21C and Figs. 12-24B and 12-24C. Note that the signal source v_{gs} sees the voltage divider resistors R_1 and R_2 in parallel. That is, part of the signal current from the source v_{gs} flows up through R_1 to ground through the source V_{DD} while some of the

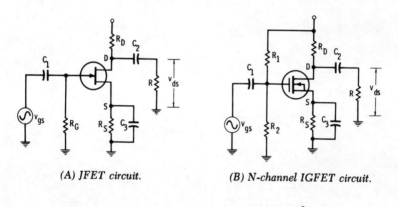

(A) JFET circuit.   *(B) N-channel IGFET circuit.*

(C) Induced-channel IGFET circuit.

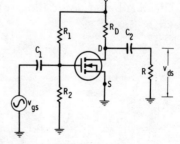

Fig. 12-24. Looking from drain and source, each FET sees ac load $r_L = R_D \| R$.

current flows down through R_2. In this case, then, $r_{in(stage)} = R_1 \| R_2 \| r_{in(gate)}$.

In both equivalents in Fig. 12-23, the constant current source i_d works into parallel admittances y_{os} and y_L. The total admittance that the current source i_d sees is the sum of the parallel admittances, or

$$y_t = y_{os} + y_L \qquad (12\text{-}8)$$

Of course if the admittance of the load y_L is much greater than the FET's output admittance y_{os}, which often is the case, then

$$y_t \cong y_L \qquad (12\text{-}8\text{A})$$

which is a simplification of Eq. 12-8.

The signal voltage across this total admittance y_t is the output voltage v_{ds}. This voltage is, by Ohm's law,*

$$v_{ds} = \frac{i_d}{y_t} \qquad (12\text{-}9)$$

Since by Eq. 12-1C, $i_d \cong y_{fs} v_{gs}$, then

$$v_{ds} \cong \frac{y_{fs} v_{gs}}{y_t} \qquad (12\text{-}10)$$

or in more detail, by Eq. 12-8

$$v_{ds} \cong \frac{y_{fs} v_{gs}}{y_{os} + y_L} \qquad (12\text{-}10\text{A})$$

or if $y_L \gg y_{os}$, then by Eq. 12-8A

$$v_{ds} \cong \frac{y_{fs} v_{gs}}{y_L} \qquad (12\text{-}10\text{B})$$

Dividing both sides of the last two equations by v_{gs} gives us gain equation

$$A_e = \frac{v_{ds}}{v_{gs}} \cong \frac{y_{fs}}{y_{os} + y_L} \qquad (12\text{-}11)$$

or

$$A_e = \frac{v_{ds}}{v_{gs}} \cong \frac{y_{fs}}{y_L} \bigg|_{y_L \gg y_{os}} \qquad (12\text{-}11\text{A})$$

---

* In ac circuits, impedance Z and admittance Y are recpirocals of each other; that is, $Z = 1/Y$. Therefore Ohm's law, $E = ZI$, can also be shown as $E = I/Y$.

12-6 SMALL-SIGNAL VOLTAGE GAIN IN FET AMPLIFIERS

Example 12-12 If in the circuit of Fig. 12-24A, $R_G = 100$ kΩ, $R_D = 40$ kΩ, $R = 160$ kΩ, and the JFET's $y_{fs} = 7000$ μmhos, and $y_{os} = 20$ μmhos, what are this circuit's (a) input resisance $r_{in(stage)}$ as the source v_{ds} sees it, and (b) the voltage gain A_e?

Answer 12-12 (a) About 100 kΩ, (b) $A_e \cong 137$.—(a) As shown in the equivalent circuit in Fig. 12-23A, R_G and $r_{in(gate)}$ are in parallel. But since $r_{in(gate)}$ for a JFET is typically between 10^6 to 10^9 Ω, it negligibly affects the value of $r_{in(stage)}$ in this case. That is, because $R_G \ll r_{in(gate)}$,

$$r_{in(stage)} = R_G \| r_{in(gate)} \cong R_G = 100 \text{ k}\Omega$$

In this case, by Eq. 12-7,

$$r_L = R_D \| R = 40 \text{ k}\Omega \| 160 \text{ k}\Omega = 32 \text{ k}\Omega$$

This ac load admittance is, by Eq. 12-6, is

$$y_L = \frac{1}{r_L} = \frac{1}{32 \text{ k}\Omega} \cong 31.1 \text{ }\mu\text{mhos}$$

Since y_{fs} and y_{os} are specified, we can substitute them and the above y_L into the gain equation (Eq. 12-11) as follows:

$$A_e \cong \frac{y_{fs}}{y_{os} + y_L} \cong \frac{7000}{20 + 31.1} \cong 137$$

Example 12-13 Suppose that we remove the load R from the circuit described in the previous example so as to make it like the circuit in Fig. 12-21A. What is the ratio v_{ds}/v_{gs} in this case?

Answer 12-13 $A_e = v_{ds}/v_{gs} \cong 156$.—In this case, as seen by the drain, r_L is R_D alone. Thus, $y_L = 1/r_L = 25$ μmhos. As in the last problem, we substitute the known parameters of the FET, along with the new y_L, into Eq. 12-11 and find that

$$A_e \cong \frac{7000}{20 + 25} \cong 156$$

Example 12-14 If in the circuit in Fig. 12-24B, $R_1 = 1$ MΩ, $R_2 = 100$ kΩ, $R_D = 20$ kΩ, $R = 5$ kΩ, and the IGFET's $y_{fs} = 5000$ μmhos, and $y_{os} = 2$ μmhos, what are this circuit's (a) input resistance $r_{in(stage)}$, and (b) the voltage gain A_e?

Answer 12-14 (a) $r_{in(stage)} \cong 91$ kΩ, $A_e \cong 20$.—In this case the source v_{gs} sees R_1, R_2, and $r_{in(gate)}$ in parallel, as shown in the equivalent circuit (Fig. 12-23B). Since $r_{in(gate)}$ has typically a value between 10^{10} to 10^{14} Ω, we may assume that it is effectively an open, compared to R_2, and therefore its effect on the input resistance may be disregarded. That is, as v_{gs} sees it,

$$r_{in(stage)} = R_1 \| R_2 \| r_{in(gate)} \cong R_1 \| R_2 = 1 \text{ M}\Omega \| 100 \text{ k}\Omega = 91 \text{ k}\Omega \cong 100 \text{ k}\Omega$$

The drain sees an ac resistance

$$r_L \cong R_D \| R = 20 \text{ k}\Omega \| 5 \text{ k}\Omega = 4 \text{ k}\Omega$$

This load admittance is

$$y_L = \frac{1}{4 \text{ k}\Omega} = 0.25 \text{ mmho} = 250 \text{ }\mu\text{mhos}$$

In this case, since $y_L \gg y_{os}$, that is, since 250 μmhos are much larger than 2 μmhos, the gain equation (Eq. 12-11A) is adequate, and therefore

$$A_e \cong \frac{y_{fs}}{y_L} = \frac{5000}{250} = 20$$

Example 12-15 Suppose that we remove the load R from the circuit described in the previous example so as to make the circuit like that of Fig. 12-21B, what is the ratio v_{ds}/v_{gs} in this case?

Answer 12-15 $A_e = v_{ds}/v_{gs} \cong 96.3 \cong 100$.—With R removed, $r_L = R_D = 20$ kΩ. Thus

$$y_L = \frac{1}{20 \text{ k}\Omega} = 50 \text{ }\mu\text{mhos}$$

If we are interested in accuracy, by Eq. 12-11,

$$A_e \cong \frac{y_{fs}}{y_{os} + y_L} = \frac{5000}{2 + 50} \cong 96.3$$

But since y_L is more than 10 times the value of y_{os}, we can use Eq. 12-11A,

$$A_e \cong \frac{y_{fs}}{y_L} = \frac{5000}{50} = 100$$

12-7 LARGE SIGNALS AND LOAD LINES

As with the amplifiers using junction transistors covered in previous chapters, sketching load lines for FET amplifiers helps us see how voltage and current values change when the input signal is

12-7 LARGE SIGNALS AND LOAD LINES

applied. More importantly, load lines also show us the limitations within which these voltages and currents can be varied without clipping the signal waveforms.

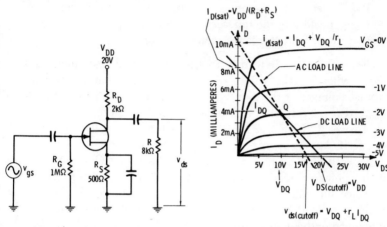

(A) JFET amplifier biased for mode A operation.

(B) Ac and dc load lines for JFET amplifier.

Fig. 12-25. Amplifier with load lines drawn on JFET's drain characteristics.

Load lines for the JFET in Fig. 12-25A are shown on its drain characteristics, Fig. 12-25B. As with previous amplifiers, both dc and ac load lines go through the operating point Q. The quiescent drain current I_{DQ} is read directly to the left of point Q. Similarly the quiescent drain-to-source voltage V_{DQ} is read directly below point Q. The equations used to find the end points of both load lines are similar to those used in the transistor circuits. As shown in Fig. 12-25B, the end points on the dc load line are

$$V_{DS(\text{cutoff})} = V_{DD} \qquad (12\text{-}12)$$

and

$$I_{D(\text{sat})} = \frac{V_{DD}}{R_D + R_S} \qquad (12\text{-}13)$$

The ac load line end-points are

$$v_{ds(\text{cutoff})} = V_{DQ} + r_L I_{DQ} \qquad (12\text{-}14)$$

and

$$i_{d(\text{sat})} = I_{DQ} + \frac{V_{DQ}}{r_L} \tag{12-15}$$

where
    V_{DQ} is specifically the quiescent drain-to-source voltage,
    I_{DQ} is specifically the quiescent drain current,
    r_L is the ac load resistance seen by the drain.

Example 12-16 Referring to the circuit and load lines in Fig. 12-25, what are the values of V_{DQ} and I_{DQ}? What are the dc-to-ground voltages at the gate V_{GG}, at the source V_{SG}, and at the drain V_{DG}? Also solve for the end points of both dc and ac load lines.

Answer 12-16 $V_{DQ} \cong 10$ V, $I_{DQ} \cong 4$ mA, $V_{GG} \cong 0$ V, $V_{SG} \cong 2$ V, $V_{DG} \cong 12$ V, $V_{DS(\text{cutoff})} = 20$ V, $I_{D(\text{sat})} = 8$ mA, $v_{ds(\text{cutoff})} \cong 16.4$ V, $i_{d(\text{sat})} \cong 10.25$ mA.—Directly below the operating Q and to the left of this point we read approximately 10 V and 4 mA respectively.

Since the gate has 10^6 to 10^9 Ω input resistance, the dc current in the gate lead is negligible. Thus, the voltage across R_G and the voltage from the gate to ground is about zero.

The voltage across R_S is the voltage from the source to ground, which must be 2 V in this case. That is, since the operating point Q is on the curve identified as $V_{GS} = -2$ V, the potential across gate and source must be -2 V. With the gate at about 0 V to ground, the source must be $+2$ V with respect to ground if the gate is to be negative with respect to the source by 2 V.

The drain-to-ground voltage is equal to the sum of the voltages across the drain and source leads and across R_S. We have just determined above that these voltages are 10 V and 2 V respectively.

By Eq. 12-12 $V_{DS(\text{cutoff})} = V_{DD} = 20$ V, and by Eq. 12-13

$$I_{D(\text{sat})} = \frac{V_{DD}}{R_D + R_S} = \frac{20 \text{ V}}{2.5 \text{ k}\Omega} = 8 \text{ mA}$$

On the ac load line, by Eq. 12-14,

$$v_{ds(\text{cutoff})} = V_{DQ} + r_L I_{DQ} \cong 10 \text{ V} + (1.6 \text{ k}\Omega)(4 \text{ mA}) = 16.4 \text{ V}$$

and by Eq. 12-15

$$i_{d(\text{sat})} = I_{DQ} + \frac{V_{DQ}}{r_L} \cong 4 \text{ mA} + \frac{10 \text{ V}}{1.6 \text{ k}\Omega} = 10.25 \text{ mA}$$

where $r_L = R_D \| R = 2 \text{ k}\Omega \| 8 \text{ k}\Omega = 1.6 \text{ k}\Omega$.

Example 12-17 In the circuit of Fig. 12-25, if the signal source v_{gs} is sinusoidal with a peak-to-peak value of 2 V, what will

12-7 LARGE SIGNALS AND LOAD LINES

the output signal v_{ds} look like? (b) What is the maximum peak-to-peak unclipped signal output voltage of this circuit?

Answer 12-17 (a) See Fig. 12-26A, (b) see Fig. 12-26B.—As shown in Fig. 12-26A, with the 2-V peak-to-peak input signal projected off the ac load line, this signal causes V_{GS} to vary from its quiescent value (-2 V) to -1 V (point x) on positive peaks and to -3 V (point y) on the negative peaks. Projecting down from

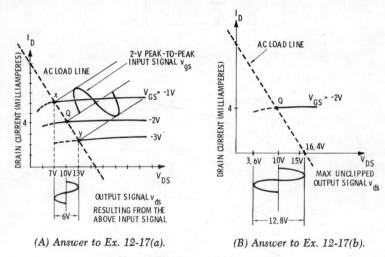

(A) Answer to Ex. 12-17(a).   (B) Answer to Ex. 12-17(b).

Fig. 12-26. Answers to Ex. 12-17.

points x and y we intersect the V_{SD} axis at about 7 V and 13 V, indicating that the peak-to-peak output is approximately $13 - 7 = 6$ V. (b) If an unclipped output signal is required, the amplifier must not be driven into cutoff or saturation. As shown in Fig. 12-26B, the cutoff point is located 6.4 V to the right of the 10-V quiescent drain-to-source voltage. Thus $v_{ds\text{(peak-to-peak)}} \cong 2(6.4 \text{ V}) = 12.8$ V.

Fig. 12-27A shows a mode-B type IGFET with drain characteristics shown in Fig. 12-27B. The end points of the load line are found with the same equations used with the last circuit (Fig. 12-25A). In this case, though, the gate G is not 0 V but instead some positive voltage with respect to ground, and therefore the voltage across R_S is not the gate-to-source bias voltage V_{GS}. Now the bias V_{GS} is the difference in the voltages at points G and S with respect to ground.

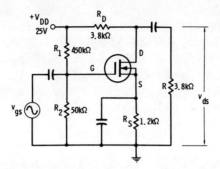

(A) IGFET amplifier biased for mode-B operation.

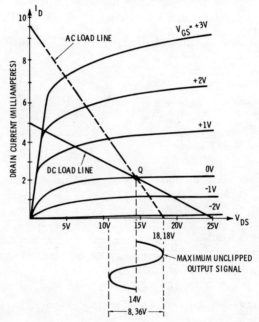

(B) Ac and dc load lines.

Fig. 12-27. Amplifier with load lines drawn on IGFET's drain characteristics.

Example 12-18 Referring to the circuit and its load lines in Fig. 12-27, what are the approximate values of V_{DQ} and I_{DQ}? What are the dc-to-ground voltages at points G, S, and D? Also solve for the end points of both dc and ac load lines, and determine the maximum unclipped peak-to-peak signal output voltage capability of this circuit.

12-7 LARGE SIGNALS AND LOAD LINES

Answer 12-18 $V_{DQ} \cong 14$ V, $I_{DQ} \cong 2.2$ mA, $V_{GG} = 2.5$ V, $V_{SG} \cong 2.6$ V, $V_{DG} \cong 16.6$ V, $V_{DS(\text{cutoff})} = 25$ V, $I_{D(\text{sat})} \cong 5$ mA, $V_{ds(\text{cutoff})} \cong 18.18$ V, $i_{d(\text{sat})} \cong 9.6$ mA, $v_{ds} \cong 8.36$ V peak to peak.—Directly below the operating point Q we estimate 14 V on the V_{DS} axis. To the left of point Q we read $I_{DQ} \cong 2.2$ mA on the I_D axis.

The voltage at the gate to ground is found by solving for the voltage across R_2. Since the gate current is zero, R_1 and R_2 are in series with V_{DD} and therefore

$$V_{GG} = V_2 = \frac{V_{DD}R_2}{R_1 + R_2} = \frac{25(50)}{450 + 50} = 2.5 \text{ V}$$

The voltage at point S is the voltage across R_S. Since the current through R_S is $I_{DQ} \cong 2.2$ mA, by Ohm's law, then,

$$V_{SG} = R_S I_{DQ} = (1.2 \text{ k}\Omega)(2.2 \text{ mA}) = 2.64 \text{ V} \cong 2.6 \text{ V}$$

As a check, we can now solve for V_{GS} by finding the difference between voltages at points G and S:

$$V_{GS} = V_{GG} - V_{GS} \cong 2.5 - 2.6 = -0.1 \text{ V}$$

This near-zero V_{GS} voltage fits reasonably well with the operating point Q shown in Fig. 12-27B; that is, point Q is on the $V_{GS} = 0$ V curve and the above $V_{GS} \cong 0.1$ V $\cong 0$ V. Of course the answer that $V_{GG} = 2.5$ V is all right too, because an assumption that point Q, in Fig. 12-27B, is precisely on the $V_{GS} = 0$ V curve leads to the conclusion that the potential difference between points G and S is zero, which is to say that both points must be at 2.5 V to ground.

The drain-to-gate voltage V_{DG} must be the sum of voltages V_{DQ} and V_{SG}:

$$V_{DG} = V_{DQ} + V_{SG} \cong 14 + 2.6 = 16.6 \text{ V}$$

By Eqs. 12-12 through 12-15 we find, in this case, that

$$V_{DS(\text{cutoff})} = V_{DD} = 25 \text{ V}$$

$$I_{D(\text{sat})} = \frac{25 \text{ V}}{3.8 \text{ k}\Omega + 1.2 \text{ k}\Omega} = 5 \text{ mA}$$

$$v_{ds(\text{cutoff})} = V_{DQ} + r_L I_{DQ} \cong 14 \text{ V} + (1.9 \text{ k}\Omega)(2.2 \text{ mA}) = 18.18 \text{ V}$$

$$i_{d(\text{sat})} = I_{DQ} + \frac{V_{DQ}}{r_L} \cong 2.2 \text{ mA} + \frac{14 \text{ V}}{1.9 \text{ k}\Omega} \cong 9.6 \text{ mA}$$

where $r_L = R_D \| R = 1.9$ kΩ.

As shown in Fig. 12-27B, the signal can swing operation from the 14 V quiescent value to cutoff on the ac load line (18.18 V). Thus, an unclipped signal can have as large as $18.18 - 14 = 4.18$ V peak value or a $2(4.18 \text{ V}) = 8.36$ V peak-to-peak value.

Example 12-19 Suppose that you have a circuit like that of Fig. 12-24C with values $V_{DD} = 20$ V, $R_1 = 400$ kΩ, $R_2 = 100$ kΩ, $R_D = 2.5$ kΩ, and $R = 10$ kΩ. The IGFET's drain characteristics are shown in Fig. 12-17. On these characteristics sketch the dc and ac load lines. What is the approximate maximum unclipped output signal (peak-to-peak) capability?

Answer 12-19 See Fig. 12-28.—In this circuit the voltage across R_2 is the gate-to-source bias voltage V_{GS}. Solving for this voltage in series circuit fashion we have

$$V_{GS} = \frac{V_{DD}R_2}{R_1 + R_2} = \frac{(20 \text{ V})(100)}{400 + 100} = 4 \text{ V}$$

Thus, the operating point Q is on the $V_{GS} = 4$ V curve as shown. The ac and dc load lines are calculated by Eqs. 12-12 through 12-15, as in Ex. 12-18. The results are shown in Fig. 12-28.

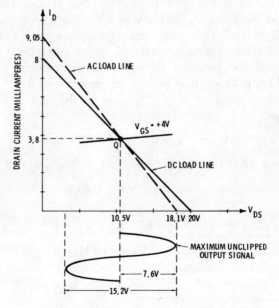

Fig. 12-28. Answer to Ex. 12-19.

12-8 SELECTING COMPONENTS FOR FET AMPLIFIERS

When building FET amplifiers we can easily select proper component values by working with the FET drain characteristics. For example, if we have a JFET with the drain characteristics in Fig. 12-15B and intend to use it in a circuit like that of Fig. 12-21A, we can start by arbitrarily choosing a safe drain saturation current $I_{D(sat)}$. We can assume that all current listed on the I_D scale of the drain characteristics are safe saturation currents. Thus, if $V_{DD} = 25$ V and we choose $I_{D(sat)} = 10$ mA, the dc load line appears as shown in Fig. 12-29. Now we select an operating point that allows large variations of I_D and V_{DS}, without driving the FET into cutoff or regions where the V_{GS} curves *are not* nearly parallel and horizontal.

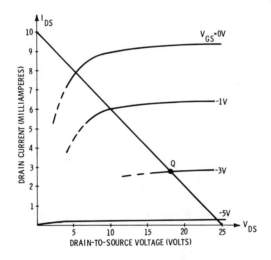

Fig. 12-29. Selecting operating point for JFET amplifier.

In this case then, we don't want variations in V_{GS}, caused by input signal, to go more negative than -5 V or more positive than about -1 V. That is, if input signal drives operation on the load line significantly above the -1 V curve, the instantaneous operating point intersects the $V_{GS} = 0$ V curve at a point where this curve is not parallel with the other V_{GS} curves crossing the load line. Such operation can cause excessive distortion. Of course, if the instantaneous operating point is driven below the -5 V curve, cutoff and sharp clipping of the output signal takes place. Therefore, since we should confine variations of V_{GS} between about -1 V and -5 V, we will select a quiescent V_{GS} of -3 V, which is halfway between -1 V and -5 V. To the left of the point where the $V_{GS} = -3$ V curve crosses the load line we estimate that $I_{DQ} \cong 2.5$ mA. Directly below this point we estimate $V_{DQ} \cong 18$ V.

Example 12-20 If the circuit in Fig. 12-21A has a JFET with the drain characteristics of Fig. 12-29 and is to have a load line and operating point Q as shown in Fig. 12-29 what values of R_S and R_D should be used?

Answer 12-20 $R_S \cong 1.2$ kΩ, $R_D \cong 1.3$ kΩ.—In this circuit, the voltage across R_S causes the potential difference between the gate and source leads (since R_G is very large) and therefore by Eq. 12-3

$$R_S \cong \frac{3 \text{ V}}{2.5 \text{ mA}} = 1.2 \text{ k}\Omega$$

Rearranging Eq. 12-13 we can show that

$$R_D + R_S = \frac{V_{DD}}{I_{D(sat)}} = \frac{25 \text{ V}}{10 \text{ mA}} = 2.5 \text{ k}\Omega$$

Thus since $R_D + R_S = 2.5$ kΩ and $R_S \cong 1.2$ kΩ, then $R_D \cong 2.5 - 1.2 = 1.3$ kΩ. The value of R_G is not critical. Usually up to several megohms can be used with no trouble.

Example 12-21 What is the voltage gain A_e of the circuit described in the last example if the JFET's $y_{fs} = 1.9$ mmhos and $y_{os} \cong 0$?

Answer 12-21 $A_e \cong 2.47$.—In this case since $R_D \cong 1.3$ kΩ, then, by Eq. 12-5,

$$y_L = \frac{1}{R_D} \cong \frac{1}{1.3 \text{ k}\Omega} \cong 0.77 \text{ mmho}$$

Since $y_{os} \cong 0$, then by Eq. 12-11A

$$A_e \cong \frac{y_{fs}}{y_L} \cong \frac{1.9 \text{ mmhos}}{0.77 \text{ mmho}} \cong 2.47$$

We could have substituted Eq. 12-5 into Eq. 12-11A as worked as follows:

$$A_e \cong \frac{y_{fs}}{y_L} \cong y_{fs} R_D \cong (1.9 \text{ mmhos})(1.3 \text{ k}\Omega) \cong 2.47$$

12-8 SELECTING COMPONENTS FOR FET AMPLIFIERS

Obviously the gain A_e of the circuit described in the last example is not very large. By examining the gain equations above, we can see that the gain A_e can be increased using a larger R_D, which is to say a smaller y_L. A larger R_D decreases the slope (steepness) of the load line. So if we select component values based on a load line whose slope is relatively small, the gain A_e will be relatively large.

Example 12-22 Suppose that you have a circuit as in Fig. 12-21A with a JFET whose drain characteristics are as shown in Fig. 12-15B. If $V_{DD} = 20$ V and you are to make $I_{D(\text{sat})} = 2$ mA, what approximate quiescent value of V_{GS}, I_{DQ}, and V_{DQ} do you plan on using? Approximately what values of R_S and R_D will you select? What is the gain A_e of the circuit if $y_{fs} = 1.9$ mmhos and $y_{os} \cong 0$?

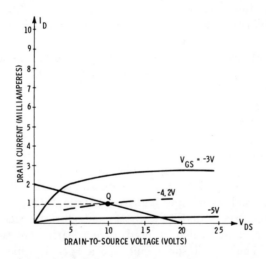

Fig. 12-30. Load line for JFET circuit of Ex. 12-22.

Answer 12-22 See Fig. 12-30, $V_{GS} \cong -4.2$ V, $I_{DQ} \cong 1$ mA, $V_{DQ} \cong 10$ V, $R_S \cong 4.2$ kΩ, $R_D \cong 5.8$ kΩ, $A_e \cong 11$.—In this case we center the operating point Q between the $V_{GS} = -5$ V and the -3 V curves. Roughly, then, the quiescent gate-to-source voltage is $V_{GS} \cong 4.2$ V, $I_{DQ} \cong 1$ mA to the left of point Q and $V_{DQ} \cong 10$ V below this point. Thus, by Eq. 12-3,

$$R_S \cong \frac{4.2 \text{ V}}{1 \text{ mA}} = 4.2 \text{ k}\Omega$$

and, by Eq. 12-13,

$$R_D + R_S = \frac{V_{DD}}{I_{D(\text{sat})}} = \frac{20 \text{ V}}{2 \text{ mA}} = 10 \text{ k}\Omega$$

and therefore

$$R_D = 10 \text{ k}\Omega - R_S \cong 10 - 4.2 = 5.8 \text{ k}\Omega$$

The voltage gain is, by Eq. 12-11A,

$$A_e \cong \frac{y_{fs}}{y_L} \cong \frac{1.9 \text{ mmhos}}{1/5.8 \text{ k}\Omega} \cong 11$$

12-9 THE EFFECTS OF TEMPERATURE CHANGES

In a previous chapter we learned that the dc collector current, in junction transistor circuits, tends to vary noticeably with temperature changes. Fortunately, however, temperature instability is not as big a problem with FET circuits. In most cases the drain current is affected by temperature changes and should be considered when building FET amplifiers that are required to operate in a variety of environments.

Increased temperatures tend to decrease the mobility of the charge carriers in the channel of the FET. Of course, fewer carriers means reduced drain current. On the other hand, the increased temperature decreases the depletion region, which tends to increase I_D. These opposing tendencies give the FET its relatively low temperature coefficient, which means that drain current changes are small with large changes in temperature.

With some types of FET's, the manufacturers specify a quiescent drain current, which, if used, gives nearly zero temperature coefficient. That is, in some units, at a specific quiescent drain current the decreased conductivity of the channel due to decreased charge carrier mobility at higher temperatures is cancelled by the increased conductivity of the channel, due to the reduced depletion region. This results in a very stable operating point.

Typically the temperature coefficient is not zero and it is good practice to use circuit designs that offer temperature stability. Actually the simple use of resistance in the source lead, as in Figs. 12-24A and 12-24B, provides temperature stability. That is, if I_D increases, due to a temperature change in these circuits, point S becomes more positive to ground. The voltage at point G, however,

12-9 THE EFFECTS OF TEMPERATURE CHANGES

remains constant, causing the gate G to become more negative with respect to the source S. Thus, as I_D increases, V_{GS} becomes more negative, which tends to decrease I_D. The more negative V_{GS} does not actually decrease I_D but it does prevent it from rising as much as it would if V_{GS} had remained constant. Of course if a temperature change causes I_D to decrease, point S becomes less positive and therefore point G becomes more positive with respect to point S. The result of a more positive V_{GS} is only a small decrease in I_D. Thus, if we are to build a circuit that is required to work under conditions of changing temperatures, we should plan to use resistance in the source lead and, as a rule of thumb, drop about 10% of the source voltage V_{DD} across it. If it is necessary to drop more than 10% of the V_{DD} voltage across R_S, as was the case in the last example, the temperature stability is better.

The JFET has a pn junction within it which is normally reverse biased via the gate and source leads. It is this reverse biased pn junction that gives the JFET its high input resistance $r_{in(gate)}$. Like the typical diode pn junction, its leakage increases, and therefore $r_{in(gate)}$ decreases as temperature increases. For the silicon JFET, $r_{in(gate)}$ decreases by a factor of 2 for every 60°C rise in temperature. In the IGFET, $r_{in(gate)}$ is unaffected by temperature changes.

Example 12-23 Suppose that we are to build a circuit like that of Fig. 12-21B with an IGFET whose drain characteristics are shown in Fig. 12-16. What approximate values of R_S, R_D, and ratio R_1/R_2 will you use if $I_{D(sat)} = 10$ mA and $V_{DD} = 20$ V? Plan on providing some temperature stability.

Answer 12-23 $R_S \cong 670$ Ω, $R_D \cong 1.33$ kΩ, $R_1/R_2 = 9$.—The load line in this case is drawn from 10 mA on the I_D axis to 20 V on the V_{DS} axis as shown in Fig. 12-31. Note that this load line crosses the $V_{GS} = +2.5$ V curve at a point where the curve is not horizontal or parallel with the other curves. Therefore to avoid distortion, we should avoid driving the gate with as much +2.5 V. We can restrict operation between the +2 V and the −2 V curves. A logical operating point then is on the $V_{GS} = 0$-V curve which is, voltage wise, halfway between the +2 V and −2 V curves. So if we plan on using $V_{GS} = 0$ V as the quiescent value, points G and S in the circuit must be at the same potential. Since temperature stability is desired, point S should be about 2 V to ground that is 10% of V_{DD}. Note

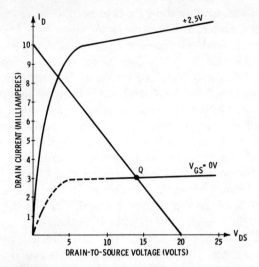

Fig. 12-31. Load line for IGFET circuit of Ex. 12-23.

that to the left of the operating point Q (Fig. 12-31) the drain current $I_D \cong 3$ mA and since the required drop across $R_S \cong 2$ V, we find by Ohm's law that

$$R_S \cong \frac{2 \text{ V}}{3 \text{ mA}} \cong 666 \; \Omega \cong 670 \; \Omega$$

Since, by Eq. 12-13,

$$10 \text{ mA} = \frac{V_{DD}}{R_D + R_S}$$

then

$$R_D = \frac{V_{DD}}{10 \text{ mA}} - R_S \cong 2000 \; \Omega - 670 \; \Omega = 1.33 \text{ k}\Omega \cong 1.3 \text{ k}\Omega$$

With point S at 2 V to ground, the voltage at point G (across R_2) must be 2 V, too, if we want to make $V_{GS} \cong 0$. Thus, with $V_{DD} =$ 20 V and 2 V across R_2, the remaining 18 V must be across R_1. Since R_1 and R_2 are effectively in series, we can show that

$$\frac{R_1}{R_2} = \frac{18 \text{ V}}{2 \text{ V}} = \frac{9}{1} = 9$$

Example 12-24 What is the voltage gain A_e of the circuit described in the last example if the FET's $y_{fs} \cong 2$ mmhos and $y_{os} \cong 30$ μmhos?

12-9 The Effects of Temperature Changes

Answer 12-24 $A_e \cong 2.56$.—The admittance of the load is

$$y_L = \frac{1}{r_L} = \frac{1}{R_D} \cong \frac{1}{1.33 \text{ k}\Omega} \cong 0.75 \text{ mmho}$$

Substituting this with the known values of y_{fs} and y_{os} into Eq. 12-11 we get

$$A_e \cong \frac{2000}{750 + 30} = 2.56$$

Example 12-25 Select a value of R_D and a ratio of R_1/R_2 for a circuit like that of Fig. 12-21C whose source voltage $V_{DD} = 20$ V and whose IGFET has characteristics shown in Fig. 12-17B. Limit the value of $I_{D(\text{sat})}$ to 2 mA. Also find the approximate gain A_e of the circuit if $y_{fs} \cong 1.5$ mmhos and y_{os} is negligible.

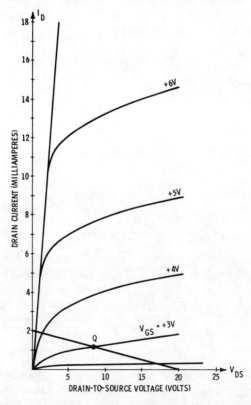

Fig. 12-32. Operating point for IGFET circuit of Ex. 12-25.

Answer 12-25 $R_D = 10$ kΩ, $R_1/R_2 \cong 5.66/1$, $A_e \cong 15$.—With no R_S in this case ($R_S = 0$), Eq. 12-13 simplifies to

$$I_{D(sat)} = \frac{V_{DD}}{R_D} \qquad (12\text{-}13\text{A})$$

Rearranging and substituting the knowns in this case we get

$$R_D = \frac{V_{DD}}{I_{D(sat)}} = \frac{20 \text{ V}}{2 \text{ mA}} = 10 \text{ k}\Omega$$

Drawing a load line from 2 mA on the I_D axis to 20 V on the V_{DS} axis as shown in Fig. 12-32, we note that the $+3$ V curve crosses the load line at about its center. The intersection of the $+3$-V curve and the load line therefore is a reasonably good operating point Q. In this circuit, we obtain a quiescent $V_{GS} = +3$ V by choosing resistors R_1 and R_2 such that there is a 3 V drop across R_2, which means that the remaining 17 V must be dropped across R_1. Therefore, in series circuit fashion

$$\frac{R_1}{R_2} = \frac{17 \text{ V}}{3 \text{ V}} \cong 5.66$$

Since in this case $r_L = 10$ kΩ, then

$$y_L = \frac{1}{r_L} = 0.1 \text{ mmho or } 100 \text{ } \mu\text{mhos}$$

With a negligible y_{os} we can substitute our knowns into Eq. 12-11A to find that

$$A_e \cong \frac{y_{fs}}{y_L} = \frac{1.5 \text{ mmhos}}{0.1 \text{ mmho}} = 15$$

12-10 A PRACTICAL CONSIDERATION

In Ex. 12-23 we found that the required ratio of $R_1/R_2 = 9$. In this case, then, with a circuit like that of Fig. 12-21B, we can use an infinite variety of R_1 and R_2 combinations. For example, each of the following has a ratio $R_1/R_2 = 9$:

(I) $R_1 = 90$ MΩ and $R_2 = 10$ MΩ (III) $R_1 = 180$ kΩ and $R_2 = 20$ kΩ
(II) $R_1 = 9$ MΩ and $R_2 = 1$ MΩ (IV) $R_1 = 36$ kΩ and $R_2 = 4$ kΩ

Since FET's are usually used where very high input resistance is required, it would be rather pointless to use relatively low values of R_1 and R_2 which would "drag down" the input resistance. This

12-10 A Practical Consideration

is more obvious after examining the ac equivalent circuit (Fig. 12-23B). In other words, since R_1 and R_2 are in parallel, $r_{in(stage)}$ is smaller than the smaller of these two resistors. Specifically, assuming that $r_{in(gate)}$ is extremely large the $r_{in(stage)}$, with each of the above combinations, may be shown as follows:

(I) $r_{in(stage)} \cong R_1 \| R_2 = 90 \text{ M}\Omega \| 10 \text{ M}\Omega = 9 \text{ M}\Omega$
(II) $r_{in(stage)} \cong 9 \text{ M}\Omega \| 1 \text{ M}\Omega = 900 \text{ k}\Omega$
(III) $r_{in(stage)} \cong 180 \text{ k}\Omega \| 20 \text{ k}\Omega = 18 \text{ k}\Omega$
(IV) $r_{in(stage)} \cong 36 \text{ k}\Omega \| 4 \text{ k}\Omega = 3.6 \text{ k}\Omega$

Combinations (I) and (II) do not excessively "drag down" the input resistance but combinations (III) and (IV) do. Thus, the latter two combinations of R_1 and R_2 in circuits like those of Figs. 12-24B and 12-24C are impractical. Input resistances of 18 kΩ or 3.6 kΩ can be obtained with a junction transistor circuit; use of an FET is not necessary.

There is a good argument for relatively small values of R_1 and R_2, however. Small values draw more current from the V_{DD} voltage source and therefore act like a bleeder resistance across this source. This improves the voltage regulation of the V_{DD} voltage. Also the amplitudes of the noise and transient voltages across R_1 and R_2 are reduced if smaller values of these resistances are used.*

A small modification can be made on FET amplifiers that use a voltage divider in the gate circuit. It allows us to use a relatively small $R_1 \| R_2$ combination but yet provides a large $r_{in(stage)}$. This modification is made by the use of an additional resistance R_3 as shown in Fig. 12-33. In these circuits, as long as a low leakage coupling capacitor C_1 is used, the dc current through and the voltage drop across R_3 is negligible. Thus, the voltage across R_2 is still equal to the gate-to-ground voltage. Typically an R_3 of about 1 MΩ is used. Note in the ac equivalent of Fig. 12-33, shown in Fig. 12-34, that the source v_{gs} sees an $r_{in(stage)}$ that is usually larger than R_3. Thus, the quality of a high input resistance is preserved.

Example 12-26 Suppose that in the circuit of Fig. 12-33A, $V_{DD} = 25$ V, $R_1 = 1.2$ MΩ, $R_2 = 50$ kΩ, $R_3 = 1$ MΩ, $R_D = 10$ kΩ, $R_S = 2$

---

*Even extremely small stray inductances, like the inductances of the leads in the circuit, can produce fairly large transient (spiked) voltages if the circuit resistance is extremely large. Such transients can puncture the insulating oxide of an IGFET or at least cause noise and static in the output of the amplifier.

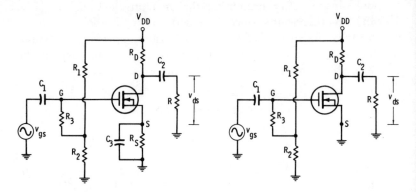

(A) With source resistance.   *(B) With grounded source.*

Fig. 12-33. Practical method of biasing gate.

kΩ, $R = 800$ kΩ, and the voltage at point D to ground is 15 V. Find the (a) resistance seen by $r_{in(stage)}$, (b) dc voltage at point G to ground, (c) dc voltage at point S to ground, (d) the quiescent gate-to-source voltage V_{GS}, and (e) the output signal voltage v_{ds} if $v_{gs} = 5$ mV (rms) and the IGFET's $y_{fs} = 2$ mmhos and $y_{os} \cong 0$.

Answer 12-26 (a) $r_{in(stage)} \cong 1.048$ MΩ, (b) 1 V, (c) 2 V, (d) -1 V, (e) 100 mV (rms).—(a) R_1 and R_2 are in parallel as shown in the equivalent circuit (Fig. 12-34) and their total resistance is 50 kΩ‖1200 kΩ = 48 kΩ. Thus, the signal source v_{gs} sees $R_3 + 48$ kΩ = 1.048 MΩ. This 1.048 MΩ is the $r_{in(stage)}$, assuming that $r_{in(gate)}$ is extremely large, which in fact it normally is. (b) Since essentially no dc current flows through R_3, the voltage across R_2 is the voltage at point G to ground. In series circuit fashion, then,

$$V_{R2} = \frac{V_{DD}R_2}{R_1 + R_2} = \frac{(25\text{ V})(50)}{1250} = 1\text{ V}$$

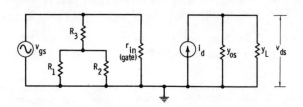

Fig. 12-34. Ac equivalent of circuits in Fig. 12-33.

(c) With 15 V at point D to ground, and $V_{DD} = 25$ V, the remaining 10 V must be across R_D. Thus, we can determine the quiescent current through R_D with Ohm's law:

$$I_{DQ} = \frac{10 \text{ V}}{10 \text{ k}\Omega} = 1 \text{ mA}$$

This 1 mA also flows in R_S. Therefore the voltage at S to ground is

$$V_{Rs} = R_S I_{DQ} = (2 \text{ k}\Omega)(1 \text{ mA}) = 2 \text{ V}$$

(d) Since the voltage at G is 1 V and the voltage at S is 2 V, point G is -1 V with respect to S. (e) We can first determine the voltage gain. Eq. 12-11A is adequate because $y_{os} \cong 0$. Thus,

$$A_e \cong \frac{y_{fs}}{y_L} \cong \frac{2 \text{ mmhos}}{1/10 \text{ k}\Omega} = 20$$

Now solving Eq. 12-4 for v_{ds} we can show that

$$v_{ds} = A_e v_{gs} \cong 20(5 \text{ mV}) = 100 \text{ mV (rms)}$$

REVIEW QUESTIONS

12-1. What is the range of input resistances of JFET's?

12-2. What words do the initials JFET represent?

12-3. What words do the initials IGFET represent?

12-4. What is another term for IGFET?

12-5. What is the range of input resistances of IGFET's?

12-6. What is the operating mode of a JFET?

12-7. Mode A is also called_____mode.

12-8. Mode B is also called_____mode.

12-9. Mode C is also called_____mode.

12-10. Why are IGFET's shipped with the leads shorted to each other?

12-11. What parameters do the symbols y_{fs} and y_{os} represent?

12-12. Briefly describe two ways that the parameters y_{fs} and y_{os} can be obtained for a certain FET if you have it and its drain characteristics.

12-13. Can the operating mode of an FET be changed by selective bias methods?

12-14. What do the terms I_{DQ} and V_{DQ} represent?

12-15. What do the terms v_{gs}, v_{ds}, and i_d represent?

12-16. What problems might arise if the gate voltage divider resistors, like R_1 and R_2 in Fig. 12-27A, are too large in resistance.

12-17. What does the term "low temperature coefficient" mean as applied to FET's?

PROBLEMS

12-1. The circuit shown in Fig. 12-19 is being used to obtain the drain characteristics of a certain FET. These characteristics are to be sketched on a graph such as that shown in Fig. 12-35. (a) Initially, the potentiometer R_1 is adjusted so that the voltmeter M_1 reads 0 V and is kept at that value. Then R_2 is adjusted by increments so that the readings on M_2 are increased. When M_2 reads 2 V, ammeter M_3 reads 2.4 mA; and when M_2 reads 10 V, ammeter M_3 reads 3 mA. Also when M_2 reads 18 V, ammeter M_3 reads 3.2 mA. (b) Now the slide on R_1 is moved to the right so that M_1 reads 2 V and it is kept at that value. The slide on R_2 is moved back to the far left and is adjusted by increments to the right. Now when M_2 reads 2 V, the ammeter M_3 reads 4.5 mA; and when M_2 reads 4 V, ammeter M_3 reads 6 mA. Also when M_2 reads 16 V, ammeter M_3 reads 7 mA. Sketch the V_{GS} curves obtained with this information.

12-2. Continuing with the test circuit and FET described in the previous problem, the potentiometer R_1 is adjusted so that M_1 reads 2 V while the slide is to the *left* of the ground point, which means that the gate is negative with respect to the source. (a) Now when M_2 reads 2 V, ammeter M_3 reads 0.5 mA; and when R_2 is adjusted so that M_2 reads 20 V, ammeter M_3 reads 0.8 mA. (b) Now the slide on R_1 is adjusted, still to the left of the ground point, so that M_1 reads 1 V. When R_2 is adjusted, causing M_2 to read 2 V and ammeter to read 1.5 mA; and then R_2 is readjusted, causing M_2 to read 20 V, ammeter M_3 reads 1.8 mA. Sketch the V_{GS} curves obtained with this information on a graph such as that shown in Fig. 12-35.

12-3. Referring back to the FET described in Prob. 12-1 above, what are its approximate pinch-off voltage V_p and pinch-off current I_p when $V_{GS} = 0$ V?

PROBLEMS

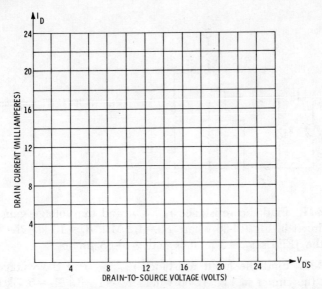

Fig. 12-35. FET drain characteristic graph.

12-4. Referring back to the FET described in Prob. 12-2, what are its approximate pinch-off voltage V_p and pinch-off current I_p when $V_{GS} = -1$ V?

12-5. If we use a drain-to-source voltage $V_{DS} = 16$ V and a gate-to-source voltage $V_{GS} = +1$ V on the FET described in Prob. 12-1, what drain current would you expect?

12-6. What is the approximate gate-to-source cutoff voltage of the FET described in Prob. 12-2?

12-7. If an FET with characteristics shown in Fig. 12-15 is to be used in a circuit as shown in Fig. 12-21A, what value of R_S will you use to establish a quiescent drain current $I_D = 4$ mA?

12-8. If an FET with the characteristics shown in Fig. 12-15 is to be used in a circuit as shown in Fig. 12-21A, what value of R_S will you use to establish a quiescent drain current $I_D = 2$ mA?

12-9. If a gate-to-source bias voltage of $+3$ V is required in the circuit of Fig. 12-33B, what ratio of R_1/R_2 will you use if $V_{DD} = 15$ V?

12-10. If the circuit in Fig. 12-33B has a $V_{DD} = 10$ V and the required $V_{GS} = 0.5$ V, what ratio R_1/R_2 should be used?

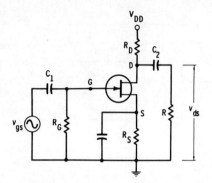

Fig. 12-36. JFET amplifier.

12-11. Find the resistance $r_{\text{in(stage)}}$ and the voltage gain A_e of the circuit in Fig. 12-36 where $R_G = 1.2$ MΩ, $R_D = 10$ kΩ, $R = 40$ kΩ, and the JFET's $y_{fs} = 3$ mmhos and $y_{os} = 35$ μmhos.

12-12. Find the input resistance $r_{\text{in(stage)}}$ and the voltage gain A_e of the circuit in Fig. 12-36 where $R_G = 1$ MΩ, $R_D = 5.4$ kΩ, $R = 4$ MΩ, and the JFET's $y_{fs} = 5$ mmhos and $y_{os} = 15$ μmhos.

12-13. Referring to the circuit described in Prob. 12-11 above, what is the output signal voltage v_{ds} if the input $v_{gs} = 10$ mV?

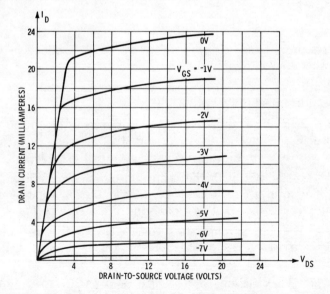

Fig. 12-37. FET drain characteristics (Probs. 12-15, 12-17, 12-21).

PROBLEMS

12-14. Referring to the circuit described in Prob. 12-12 above, what is the output voltage v_{ds} if the input $v_{gs} = 8$ mV?

12-15. Suppose that the JFET in Fig. 12-36 has the drain characteristics shown in Fig. 12-37, and the circuit has the following component values: $V_{DD} = 20$ V, $R_G = 2$ MΩ, $R_D = 1.2$ kΩ, $R_S = 1.3$ kΩ, and $R = 6$ kΩ. The voltage at point S is $+5$ V to ground. (a) Sketch dc and ac load lines on Fig. 12-37. (b) What is the maximum peak-to-peak unclipped output signal capability?

12-16. Suppose that the IGFET in Fig. 12-33B has the drain characteristics shown in Fig. 12-38 and the circuit has the following component values: $V_{DD} = 24$ V, $R_1 = 1.5$ MΩ, $R_2 = 100$ kΩ, $R_3 = 2$ MΩ, $R_D = 2$ kΩ, $R = 3$ kΩ. (a) Sketch dc and ac load lines. (b) What is the maximum peak-to-peak unclipped output signal capability?

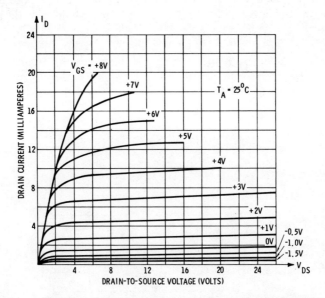

Fig. 12-38. FET drain characteristics (Probs. 12-16, 12-18, 12-22).

12-17. What is the mode of operation of the FET whose characteristics are shown in Fig. 12-37?

12-18. What is the mode of operation of the FET whose characteristics are shown in Fig. 12-38?

12-19. Referring to the circuit described in Prob. 12-15, what is the approximate output voltage v_{ds} if the input signal v_{gs} has a 2 V (peak-to-peak) amplitude? Solve graphically by projecting from the ac load line.

12-20. Referring to the circuit described in Prob. 12-16, what is the approximate output voltage v_{ds} if the input signal v_{gs} has a 2 V (peak-to-peak) amplitude? Solve graphically by projecting from the ac load line.

12-21. Select an operating point and component values R_S and R_D for a circuit like Fig. 12-36 whose $V_{DD} = 24$ V, load $R = 800$ kΩ, $I_{D(\text{sat})} = 20$ mA, and whose FET's characteristics are given in Fig. 12-37.

12-22. Select an operating point and component values R_S, R_D, and ratio of R_1/R_2 for a circuit like that in Fig. 12-33A whose $V_{DD} = 24$ V, load $R = 1.2$ MΩ, $I_{D(\text{sat})} = 12$ mA, and whose FET's characteristics are in Fig. 12-38. Plan to drop about 2 V dc across R_S.

13

Integrated Circuits

The term *integrated circuit,* or *IC,* is used to describe a group of very small electronic components that are constructed and permanently interconnected on or in a piece of semiconductor material.* The semiconductor serving as the base on which the integrated ciruits are made is called a *substrate*. Integrated circuits and their applications are often called *microcircuits* and *microelectronics* respectively.

For a number of reasons, applications of IC's are increasing rapidly and are expected to continue in this trend. The small size and weight of IC's are advantages in the areospace and computer industries where these qualities are premium. In addition to microminiaturization, IC's have greater reliability, better performance, and often lower cost than equivalent conventional circuits. Lower costs are possible because IC's are highly adaptable to automation and therefore to mass production. As the demand for certain types of IC's has increased, their unit cost has decreased. In fact, some types of conventional circuits, such as amplifiers and logic circuits, cannot compete in cost with integrated circuits that perform the same functions. Conventional circuits using individual (discrete) components are often called *discrete component circuits* when compared to IC's.

---

*The Institute of Electrical and Electronic Engineers defines an IC as "a combination of interconnected circuit elements inseparably associated on or within a continuous substrate."

13-1 GENERAL CONSTRUCTION OF IC'S

Though the technology of IC's is advanced, it is not new. It is an extension of the technology used in the manufacture of discrete components such as silicon diodes, junction transistors, and FET's. The methods used to make IC's are continually improving. It would be difficult to discuss all of the exact IC manufacturing techniques in use today. The producers of IC's try to keep their latest techniques secret. We can, however, look at some techniques which are basic and this will give us a good picture of the foundation on which the IC technology is built and growing.

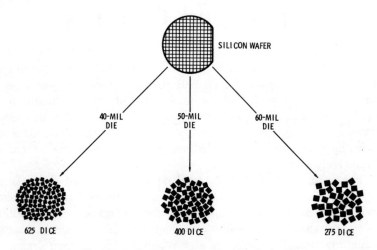

Fig. 13-1. Cutting wafer into individual IC's.

The making of IC's typically starts with a piece of circular silicon called a *wafer*. It is about 1 inch in diameter, 0.006 inch thick, and serves as the substrate mentioned earlier. Only a small portion of the wafer surface is needed for one IC. Therefore, many separate IC's are formed simultaneously on one wafer. Although the individual IC's are separate, they are all usually the same type. After the circuits are formed, the wafer is "cut" into smaller dice (chips) as shown in Fig. 13-1. Each chip contains one complete IC and typically contains 20 to 60 components and interconnections. After the IC's are checked, each one that works is encapsulated in a package about the size of a junction transistor, as shown in Fig. 13-2. Examples of actual IC's made on chips and housed in

13-1 GENERAL CONSTRUCTION OF IC's

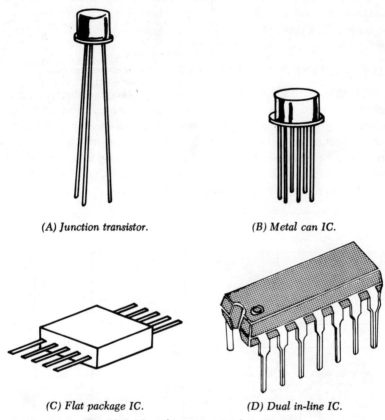

(A) *Junction transistor.* (B) *Metal can IC.*

(C) *Flat package IC.* (D) *Dual in-line IC.*

Fig. 13-2. IC packages compared to transistor.

such packages are shown in Fig. 13-3, where the components on the chip are shown contained within the area enclosed by the dotted line. Components external to an IC are usually shown outside the areas enclosed by the dotted lines.

Note the conspicuous absence of capacitors in the IC's in Fig. 13-3. Components like transistors, diodes, and resistors are made much more easily and economically than capacitors. Therefore, capacitors are avoided in IC designs. This restriction makes IC's look considerably different from discrete circuits that perform the same functions. The point is that, IC's are not simply smaller versions of equivalent discrete component circuits.

The number of usable IC's out of the total number started with is called the *yield* and is usually expressed as a percentage. It is common to have less than 25% yield when manufacturing the more complex IC's; that is, typically more than 75% of the IC's on a wafer do not measure up to standards and are scrapped.

13-2 DIFFUSED COMPONENTS

Starting with a lightly doped (high-resistance) substrate, semiconductor resistances can be made quite simply. By diffusing* either n or p type impurities, in segments, into the substrate, a desired amount of resistance can be formed as shown in Fig. 13-4. If a small resistance is required, a relatively large amount of impurity atoms are diffused into the substrate. Conversely, larger resistances require fewer impurities.

---
*Diffusion is defined in Sec. 2-5 of Chap. 2.

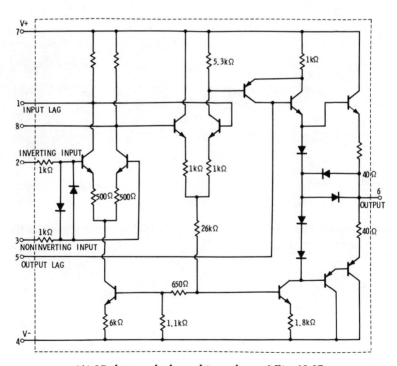

(A) IC that can be housed in package of Fig. 13-2B.

Fig. 13-3. Complex IC's like these are very small compared

13-2 DIFFUSED COMPONENTS

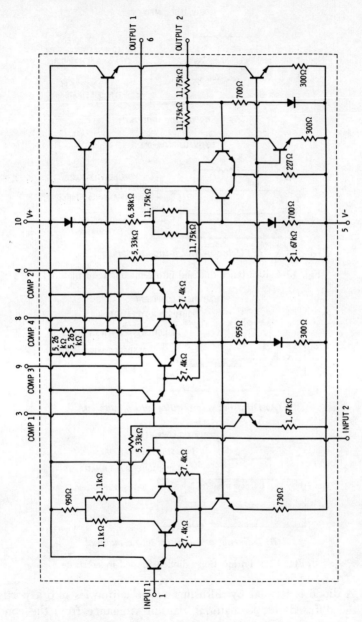

(B) IC that can be housed in package of Fig. 13-2C.

to equivalent circuits built with discrete components.

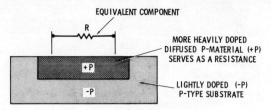

(A) Using p-type materials.

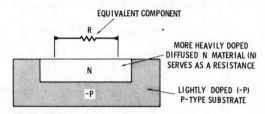

(B) Using p- and n-type materials.

Fig. 13-4. Resistance diffused into semiconductor material.

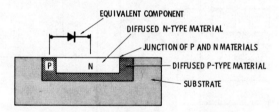

(A) Diffusing n impurities into p material.

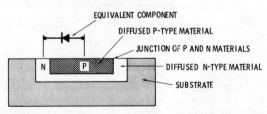

(B) Diffusing p impurities into n material.

Fig. 13-5. Pn junction (diode) formed in substrate.

A diode is formed by diffusing n-type impurities into a portion of the diffused p-type material, causing a change from the condition shown in Fig. 13-4A to the condition shown in Fig. 13-5A. That is, by adding enough n-type impurities into the p-type ma-

13-3 COMPONENT ISOLATION

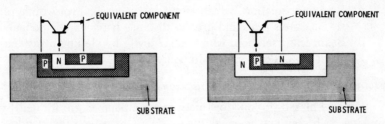

(A) Pnp transistor formed in substrate.   *(B) Npn transistor formed in substrate.*

Fig. 13-6. Diffusing third layer forms transistor.

terial, the n-type impurities can dominate and produce an n-type semiconductor.

Continuing in the same way, we could expect it possible to diffuse another doped semiconductor layer into the previously diffused layer, as shown in Fig. 13-6. Thus, we see how pnp or npn transistors can be formed.

Theoretically, many resistances, diodes, and transistors can be made simultaneously, along with interconnections, to form desired circuits on a single chip. Practically, however, due to the relatively small distances between components, the substrate provides shunt current paths between them. Thus, the component construction covered in this section, though basically simple, lacks practicability necessary step in studying the evolution of IC's.

13-3 COMPONENT ISOLATION

An IC made of many components on a single chip is called a *monolithic IC*. Since the components are on or in a common semiconductor substrate in monolithic IC's, steps are taken to ensure isolation between components. The methods of achieving isolation largely distinguishes classes of monolithic IC's from each other. Some of these methods are:

1. Use of reverse biased pn junctions between components.
2. Epitaxial isolation, which often also depends on the use of reverse biased pn junction between elements. The epitaxial process, discussed in more detail later, enables manufacturers to make IC's with characteristics previously unattainable.
3. Dielectric isolation, which is achieved by forming a thin layer of insulating oxide or dielectric material around each component to be isolated.

4. Resistive isolation, which requires the use of very high resistance substrates and greater distance between components.
5. Beam lead isolation, which is achieved by etching away material around each region to be isolated.

Isolating With Reverse Biased Junctions

Reverse biased pn junctions between components are formed by making each component on its own "island" of n-type semiconduc-

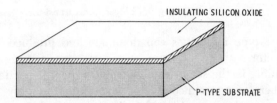

(A) Substrate with insulating oxide layer.

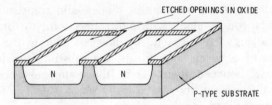

(B) Diffused n-type regions in substrate.

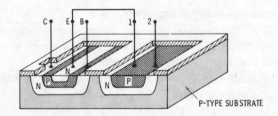

(C) Further diffusions form transistor and resistor.

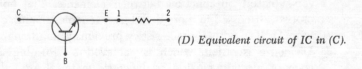

(D) Equivalent circuit of IC in (C).

Fig. 13-7. Forming reverse biased component on substrate.

13-3 Component Isolation

tor. This process may start with a piece of substrate on which a layer of insulating oxide has been formed as shown in Fig. 13-7A. Openings are etched in the oxide through which n-type impurities are then diffused into the n-type regions as shown in Fig. 13-7B. The desired components are then diffused into the n-type regions

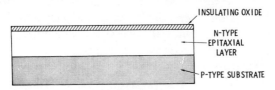

(A) *Epitaxial layer grown over substrate.*

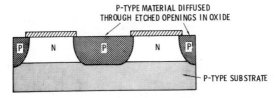

(B) *Islands of n-type material formed.*

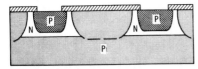

(C) *Reoxidizing and reetching new openings allows diffusion of p impurities into n materials.*

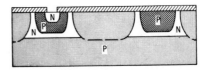

(D) *Reoxidizing entire surface and reetching allows diffusion of another n region which will be emitter.*

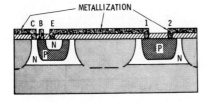

(E) *After reoxidation and etching of proper openings, vaporized metal makes circuit connections.*

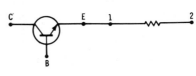

(F) *Equivalent circuit of the integrated circuit which is described in (E) of this figure.*

Fig. 13-8. Epitaxial isolation process.

as shown in Fig. 13-7C. Note in this example that the left component is an npn transistor. The right-hand component is a resistor. In the latter, the p-material alone conducts the circuit current and is doped more or less to provide the required amount of resistance. With leads connected, this IC is equivalent to the circuit in Fig. 13-7D. The leads interconnecting components are not wires as Fig. 13-7C might suggest. Actually, though not shown, an insulating oxide is formed over all diffused regions. This protects these regions from contamination with unwanted impurity atoms. Small "windows" are etched in the oxide through which metallic connections are vaporized. The interconnections are also vaporized on the oxide. This step in the manufacturing process is called *metallization*.

The Epitaxial Process

In the process discussed in the previous paragraph, in which there are three diffusions to make a transistor (once for the col-

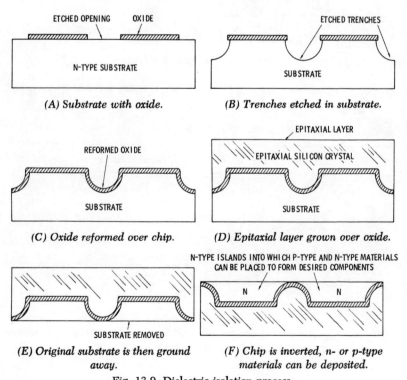

Fig. 13-9. Dielectric isolation process.

13-3 COMPONENT ISOLATION

lector, again for the base, and finally for the emitter), consistent transistor characteristics are difficult to control. The epitaxial process offers better control of characteristics. The definition of *epitaxy* is the growing of one material on another. In IC manufacturing this often means that an n-type epitaxial layer is grown, by deposition, on a p-type substrate. An insulating layer is then formed over the layer as shown in Fig. 13-8A. After openings are etched through the oxide, p-type impurities are diffused through the epitaxial layer and into the p-type substrate as shown in Fig. 13-8B. This forms "islands" of n-type semiconductors in which isolated components can be formed as shown in Fig. 13-8C. As in the previous method, these "islands" of n-type semiconductor form reverse biased pn junctions with the p-type substrate. That is, electrons cannot flow from one "island" to another within the IC structure without encountering a reverse biased pn (diode) junction on the way. Each diffusion is preceded by oxidation of the entire surface and etching of openings in the proper places. Fig. 13-8D shows that the right-hand component is covered with oxide while the left-hand component (transistor) has its emitter material diffused into it. The connections to the components are made through openings in the oxide while interconnections between components are vaporized on the oxide as shown in Fig. 13-8E. Thus, the IC in Fig. 13-8E becomes equivalent to the circuit in Fig. 13-8F.

Dielectric Isolation

This process can start with an n-type substrate on which an oxide is formed. Openings are etched in the oxide and trenches are etched in the substrate as shown in Figs. 13-9A and 13-9B. Oxide is reformed over the entire chip and then an epitaxial layer is grown over the oxide as shown in Figs. 13-9C and 13-9D. Next the original

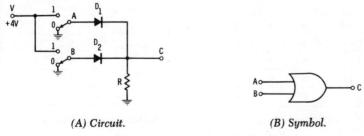

(A) Circuit.   (B) Symbol.

Fig. 13-10. Two-input diode OR gate.

13-4 THE NOR GATE WITH A DIODE OR GATE AND INVERTER

substrate is ground away. The chip can now be inverted and the required components can be formed in the well isolated "islands" of n-type materials. That is, each component is isolated by an insulating oxide layer (dielectric).

IC's are used extensively in digital computers and logic circuits in general. IC's are available to perform a variety of logic functions; a useful one is the NOR function. The circuit used to perform it is called a NOR gate. The term "NOR gate" is an abbreviation of the expression "NOT an OR gate." Before discussing exactly what the NOR gate does, a review of what the OR gate does is necessary (Sec. 4-7, Chap. 4).

Briefly, the OR gate in Fig. 13-10 has two inputs, A and B, though more inputs are commonly used. Note in this case that when the switch A is in the 0 position (down), the voltage at A is zero with respect to ground. When it is in the 1 position (up), the voltage at A is 4 V to ground. The same may be said for switch B. In this circuit, for simple notation, we can refer to 4 V as "1" and 0 V as "0." The 1 and 0 are standard logic symbols. If the 1 represents a more positive voltage than 0, as is the case here, the circuit is said to be working with *positive logic*. We can similarly refer to a +4 V output at C as a 1 output at C. Of course then, 0 V or nearly 0 V at C may be referred to as a 0 output.

The OR gate of Fig. 13-10 has a 4 V (1) output if *either* or *both* inputs are 4 V. Thus, a 1 input at A *or* at B gives us a 1 output at C. Only when both inputs are 0 is the output 0. In this circuit, assuming that the diodes are ideal, all combinations of input and resulting voltages can be tabulated on a *truth table* as shown

ROW	A	B	C
1	0V	0V	0V
2	4V	0V	4V
3	0V	4V	4V
4	4V	4V	4V

(A) Using voltages.

ROW	A	B	C
1	0	0	0
2	1	0	1
3	0	1	1
4	1	1	1

(B) Using positive-logic states.

Fig. 13-11. Truth tables for two-input OR gate.

13-4 THE NOR GATE WITH A DIODE OR GATE AND INVERTER

in Fig. 13-11A. Using logic symbols simplifies the truth table as shown in Fig. 13-11B. Row 1 indicates that if inputs A and B are both 0, the output C is 0. Rows 2, 3, and 4 indicate that if either or both inputs A and B are 1, the output C is 1.

There are a number of circuits that perform the NOR function. One such circuit is a combination or a diode OR gate and an *inverter*. Fig. 13-12A shows an inverter. The transistor, in this case, is used as a switch by being operated alternately in cutoff and saturation. That is, when the input switch is down in the 0 position, the base

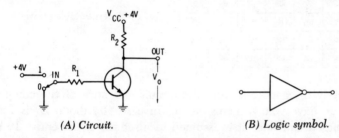

(A) Circuit.   *(B) Logic symbol.*

Fig. 13-12. Transistor inverter.

current is $I_B \cong 0$, causing the collector current $I_C \cong 0$. The transistor thus is cut off and acts like an open, looking into its collector and emitter leads. The 4 V of the dc source therefore appears across the collector-emitter terminals, which is the output voltage. When the input switch is up in the 1 position, base current I_B flows such as to saturate the transistor. This causes the output voltage to drop to a few tenths of a volt (approximately zero). Now we can see why this circuit is called an inverter. With 0 V *in*, it has 4 V *out*, whereas with 4 V *in*, it has about 0 V *out*. The inverter circuit tends to switch from saturation to cutoff more rapidly with smaller values of collector resistance R_2. However, the resistance of R_2 should not be so low as to cause excessive collector current during saturation. Thus, the saturation current $I_{C(\text{sat})}$ must be less than the manufacturer's maximum recommended collector current $I_{C(\text{max})}$:

$$I_{C(\text{sat})} \cong \frac{V_{CC}}{R_2} < I_{C(\text{max})} \qquad (13\text{-}1)$$

if $V_{CE} \cong 0$ during saturation. Use of small collector resistances also causes relatively large current to be drawn from the source V_{CC} when the transistor is "on" (conducting).

Example 13-1 If the source $V = 10$ V in the circuit of Fig. 13-10A, assuming that the diodes are ideal, what are the voltages across the diodes in each of the following cases? (a) Both inputs A and B are 0, (b) input A is 1 while B is 0, (c) input A is 0 while B is 1, (d) both inputs A and B are 1.

Answer 13-1 See Fig. 13-13.—In this case the logic symbol 1 represents 10 V, and 0 represents 0 V. In case (a) the source V is

CASE	A	B	C	V_{D1}	V_{D2}
a	0V	0V	0V	0V	0V
b	10V	0V	10V	0V	10V
c	0V	10V	10V	10V	0V
d	10V	10V	10V	0V	0V

Fig. 13-13. Truth table answer to Ex. 13-1.

completely disconnected from the circuit and none of the components can have voltage across them. In case (b) diode D_1 is forward biased and since it is assumed ideal, it places the entire 10-V source across the reverse biased D_2. In case (c), diode D_2 is forward biased placing 10 V in reverse bias across D_1. Both diodes are forward biased in case (d); thus, like a short circuit, they each have zero voltage across them.

Example 13-2 If we assume that the transistor in Fig. 13-12 is an ideal switch, that is, when saturated it acts like a short, when cut off it acts like an open, and when the base-emitter junction is forward biased, $V_{BE} = 0$ V, then what is the voltage across each resistor and the output voltage with each of the following input conditions: (a) the switch is on 0 position, (b) the switch is in 1 position?

Answer 13-2 See Fig. 13-14.—(a) With the switch in the 0 position there is no input voltage. Thus, the base current and voltage across R_1 must be zero. The transistor is cut off, resulting in the entire 4-V source across the collector and emitter terminals, and therefore there is no current in or voltage across R_2. (b) On the

V_{IN}	V_{R1}	V_{R2}	V_{OUT} OR V_{CE}
0V	0V	0V	4V
4V	4V	4V	0V

Fig. 13-14. Answer to Ex. 13-2.

13-4 THE NOR GATE WITH A DIODE OR GATE AND INVERTER 461

other hand, when the switch is in the 1 position, the base-emitter junction is forward biased. Since we assume that $V_{BE}=0$ V in this case, the 4 V input voltage appears across R_1. Also assuming that the base current is large enough to saturate the transistor, and that $V_{CE}=0$ V, the 4 V V_{CC} voltage appears across R_2.

Example 13-3 If the manufacturer's data sheets for the transistor in Fig. 13-12 specify that $I_{C(max)}=2$ mA, what minimum R_2 value would you recommend?

Answer 13-3 $R_2=2$ kΩ or more.—Since the collector current should never exceed 2 mA, we plan to keep $I_{C(sat)}$ at 2 mA or less. Solving Eq. 13-1 for R_2 we find that the minimum value of this collector resistor is

$$R_2 = \frac{V_{CC}}{I_{C(sat)}} = \frac{4 \text{ V}}{2 \text{ mA}} = 2 \text{ kΩ}$$

Example 13-4 Work Ex. 13-2 assuming that the transistor is not an ideal switch and that during saturation $V_{BE} \cong 0.7$ V and $V_{CE} \cong 0.8$ V.

V_{IN}	V_{R1}	V_{R2}	V_{OUT} OR V_{CE}
0V	0V	0V	4V
4V	3.3V	3.2V	0.8V

Fig. 13-15. Answer to Ex. 13-4.

Answer 13-4 See Fig. 13-15.—As in Ex. 13-2, when the switch is in the 0 position, no current flows in either the base or collector leads, neglecting leakage, causing zero volts across R_1 and R_2 and the entire 4 V of the source across the collector and emitter. However, when the switch is in the 1 position, the sum of the voltage across R_1 and across the base-emitter junction must be 4 V. That is,

$$V_{R1} + V_{BE} = 4 \text{ V}$$

Since $V_{BE} \cong 0.7$ V, then

$$V_{R1} = 4 - 0.7 = 3.3 \text{ V}$$

Assuming that I_B is large enough to saturate the transistor, the collector-to-emitter drop $V_{CE} \cong 0.8$ V. Since the sum of the drops across R_2 and the transistor output must equal to 4 V in this case, that is,

$$V_{R2}+V_{CE}=4 \text{ V}$$

then

$$V_{R2} \cong 4-0.8=3.2 \text{ V}$$

A combination of an OR gate and an inverter, as shown in Fig. 13-16A, is a NOR gate. Its truth table is in Fig. 13-16B. Note in this truth table that the inverter output, point D, is exactly opposite to its input, point C. That is, when point C is 0, point D is 1, and vice versa. Since point C is the output of the diode OR gate and point D is opposite to it, the function at D is said to be *not an* OR *function;* in short, it is called a NOR function.

Example 13-5 Assuming that the diodes are ideal and that the transistor is an ideal switch in Fig. 13-16A, what are the voltages across R_1, R_2, R_3 and at point D with each of the following conditions? (a) Inputs A and B are 0, (b) input A is 1 and B is 0, (c) input A is 0 and B is 1, (d) both inputs A and B are 1?

Answer 13-5 See Fig. 13-17.—In case (a), with no inputs, there are no voltages across R_3 or R_1. Thus, the base current is zero,

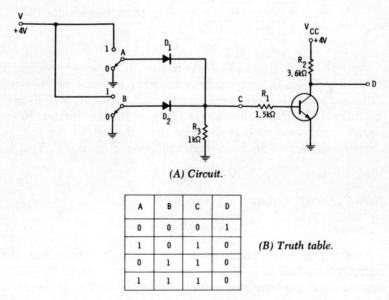

(A) *Circuit.*

A	B	C	D
0	0	0	1
1	0	1	0
0	1	1	0
1	1	1	0

(B) *Truth table.*

Fig. 13-16. NOR gate.

13-4 THE NOR GATE WITH A DIODE OR GATE AND INVERTER

CASE	A	B	C	V_{R_3}	V_{R_1}	V_{R_2}	D
a	0V	0V	0V	0V	0V	0V	4V
b	4V	0V	4V	4V	4V	4V	0V
c	0V	4V	4V	4V	4V	4V	0V
d	4V	4V	4V	4V	4V	4V	0V

Fig. 13-17. Answer to Ex. 13-5.

CASE	A	B	C	V_{R_3}	V_{R_1}	V_{R_2}	D
a	0V	0V	0V	0V	0V	0V	4V
b	4V	0V	3.4V	3.4V	2.7V	3.2V	0.8V
c	0V	4V	3.4V	3.4V	2.7V	3.2V	0.8V
d	4V	4V	3.4V	3.4V	2.7V	3.2V	0.8V

Fig. 13-18. Answer to Ex. 13-6.

which cuts off the transistor. The full V_{CC} source voltage appears across the collector and emitter and consequently there is no drop across R_2 either. In cases (b) through (d), either or both diodes are forward biased. Thus, the entire 4 V input voltage appears across R_3. Likewise, this 4 V appears across R_1 through the forward biased base-emitter junction. Therefore base current flows, saturating the transistor. Assuming that the transistor is an ideal switch, the saturated transistor or output acts like a short, resulting in the entire 4 V across R_2 and zero volts across the collector and emitter leads.

Example 13-6 Rework Ex. 13-5 where the components are not ideal. Assume that each diode drops 0.6 V when conducting, the forward biased base-emitter junction drops 0.7 V, and $V_{CE} = 0.8$ V when the transistor is saturated.

Answer 13-6 See Fig. 13-18.—As in the previous example, no input voltage causes a voltage at point D to ground only. Neglecting leakage, there is no voltage across any other component. When

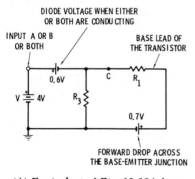

(A) *Equivalent of Fig. 13-16A from input V point of view.*

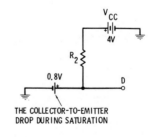

(B) *Equivalent circuit of transistor output.*

Fig. 13-19. Equivalent circuits of NOR gate.

4 V is applied to either or both inputs A or B, the equivalent circuit of this NOR gate, as seen by the 4 V input, may be shown as in Fig. 13-19A. Note that the diode's 0.6 V drop is in series with the 4 V input and that their sum is across R_3. The voltage across R_3 therefore is

$$V_{R3} = +4 - 0.6 = 3.4 \text{ V}$$

Since the left side of R_1 is at 3.4 V to ground and its right side is at 0.7 V, the difference between these potentials is across R_1; thus

$$V_{R1} = 3.4 - 0.7 = 2.7 \text{ V}$$

The inverter's output equivalent circuit is shown in Fig. 13-19B. The voltages at the top and bottom of R_2 are $+4$ V and $+0.8$ V respectively. Their difference is the voltage across R_2, which in this case is

$$V_{R2} = 4 - 0.8 = 3.2 \text{ V}$$

13-5 THE TRANSISTOR NOR GATE

The NOR function can also be performed with a *resistor transistor logic* (RTL) circuit as shown in Fig. 13-20A. In this circuit, with 0 V at both inputs A and B, both transistors are cut off. Thus, there is no current in or voltage across R_1, causing the full V_{cc} voltage to appear at the output C.

If a voltage V is applied to input A, while input B is kept at 0 V, transistor Q_1 saturates. The voltage across Q_1 then drops to zero or nearly zero. This, of course, causes the output voltage at C to drop to about zero too even though transistor Q_2 is still cut off.

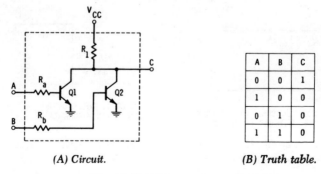

(A) *Circuit.*   (B) *Truth table.*

Fig. 13-20. RTL two-input NOR gate.

13-5 THE TRANSISTOR NOR GATE

With input A at 0 V and with voltage V at input B, transistor Q_1 is cut off and Q_2 is saturated. The saturated transistor, as before, "drags" point C to about ground potential.

Of course, with both transistors saturated, as would be the case if V volts are applied to both inputs, the output is about zero volts.

Thus, the output point C has about the V_{CC} voltage to ground only if both inputs are 0 V. With voltage V at either or both inputs, the output is nearly zero. The truth table of this NOR gate is shown in Fig. 13-20B. RTL circuits offer a lot of flexibility that is not available with diode logic circuits, and therefore they are popular in IC constructions. As you will see, an RTL integrated circuit can often be wired externally to perform a variety of functions, not only the NOR function.

A four-input NOR gate can easily be made by putting two two-input NOR gates in parallel as shown in Fig. 13-21. In this circuit, the

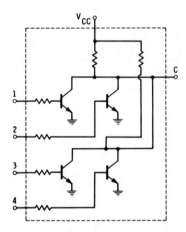

Fig. 13-21. Four-input NOR gate constructed with two two-input NOR gates.

output at C is 1 only when all four inputs are 0. A logic 1 at any one or more inputs causes a 0 to appear at the output.

Example 13-7 (a) In Fig. 13-20, what are the maximum possible values of I_B and I_C that can flow if $V_{CC} = 4$ V, $R_a = R_b = 1.5$ kΩ, $R_1 = 3.6$ kΩ, and each input is either 0 V or 4 V? (b) What minimum required value of β must each transistor have if saturation is definitely required to occur with a 4 V input?

Answer 13-7 (a) $I_{B(max)} \cong 2.66$ mA, $I_{C(max)} \cong 1.11$ mA, (b) $\beta \geq 0.415$.—(a) The maximum base current flows with a 4 V input. Assuming that the drop across the forward biased base-emitter junction is zero,

$$I_{B(max)} = \frac{4\text{ V}}{1.5\text{ k}\Omega} \cong 2.66\text{ mA}$$

When the transistor is saturated and $V_{CE} = 0$ V, there is 4 V across R_1 and thus

$$I_{C(max)} = \frac{4\text{ V}}{3.6\text{ k}\Omega} \cong 1.11\text{ mA}$$

(b) As covered in a previous chapter,

$$\beta \cong \frac{I_C}{I_B}$$

In this circuit then, we require a

$$\beta \cong 1.11\text{ mA}/2.66\text{ mA} \cong 0.415$$

Typically the β's of transistors are much larger than 0.415, which ensures that typically the transistors are well saturated with an $I_B \cong 2.66$ mA. This unusually large I_B should cause no difficulty because it flows across a forward biased base-emitter junction whose voltage drop is small. That is, the power dissipated in this junction, which is the product $V_{BE}I_B$, is fairly small, too.

13-6 THE DUAL TWO-INPUT GATE IC

Two RTL NOR gates, as shown in Fig. 13-22A, constructed in a single package is called a *dual two-input gate*. Transistors Q_1 and

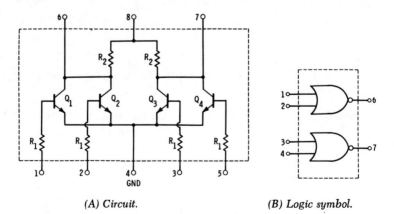

(A) Circuit. *(B) Logic symbol.*

Fig. 13-22. Dual two-input gate IC.

13-6 THE DUAL TWO-INPUT GATE IC

Q_2 and their associated resistors form one NOR gate. Of course, then, transistors Q_3 and Q_4 are parts of the other NOR gate. These NOR gates can be used singly and independently, or paralleled to form a four-input gate, or cross connected to form other useful circuits, as you will see. The numbered leads in Fig. 13-22 represent the pin numbers on the package that contains this circuit. That is, the terminals on packages shown in Figs. 13-2B and 13-2C are identified by numbers in the manufacturer's specification sheets to correspond with the numbered leads on the circuit diagram.

Example 13-8 What external connection(s) are necessary on a package containing the circuit of Fig. 13-22 to make it work as a four-input NOR gate?

Answer 13-8 Connect $+V_{CC}$ to pin 8, ground or common to pin 4, and connect pin 6 to pin 7.

This *dual two-input gate* can be wired externally, as shown in Fig. 13-23, to be an amplifier circuit. Note, that in this circuit, pins

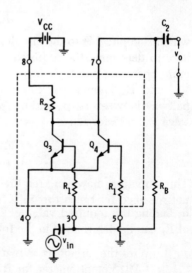

Fig. 13-23. Portion of dual two-input gate used as signal amplifier.

4 and 5 are grounded, essentially placing zero voltage across the base-emitter junction of Q_4. Therefore, Q_4 is cut off and acts like an open. It does not contribute to the circuit operation. Transistor Q_3, however, is active and obtains base bias via resistors R_B and R_1. Typically R_1 is very small compared to R_B and its effect on bias

may be neglected. Redrawn, this circuit may be shown as in Fig. 13-24. We can now see that this is an amplifier that uses base bias with collector feedback (review Sec. 8-3, Chap. 8, as needed).

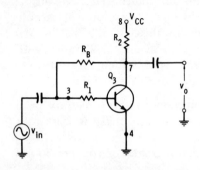

Fig. 13-24. Equivalent circuit of dual two-input gate of Fig. 13-23.

The general range of R_1 and R_2 values is specified by the manufacturer. It would seem logical therefore that Eq. 8-17,

$$I_C \cong \frac{V_{CC}}{R_C + R_B/\beta}$$

which is applicable to the circuit of Fig. 8-5, Chap. 8, could be used here to determine the required value of R_B necessary for the optimum operating point. Of course, $R_2 = R_C$ in this case. That is, with V_{CC} and R_2 known, we can select a quiescent I_C that is about halfway between $I_{C(\text{sat})}$ and 0. With the value of I_C known, we can solve Eq. 8-17 for R_B:

$$R_B \cong \left(\frac{V_{CC}}{I_C} - R_C\right)\beta$$

This equation, however, requires the factor β, which typically is not specified for transistors in IC's. Therefore an empirical[*] method of finding the required value of R_B is better in this case. The value of R_B can be determined in the following steps:

1. Wire the circuit as shown in Fig. 13-23 but use a 500 kΩ or 1 MΩ potentiometer for R_B instead of a fixed resistor. Adjust it for maximum resistance initially.
2. Place a dc voltmeter across the collector and emitter terminals (pins 4 and 7 in this case) and decrease the resistance of R_B until V_{CE} becomes half the V_{CC} source voltage.

---

[*]Reliance on observation of the circuit.

13-6 THE DUAL TWO-INPUT GATE IC

3. Disconnect R_B and measure its resistance with an ohmmeter or other suitable resistance measuring device.
4. Replace the potentiometer with a fixed resistor that is equal to, or nearly equal to, the resistance measured in step 3.

These preceding steps will place this circuit operation at very nearly the optimum operating point if the circuit works into a resistance much larger than R_2.

If the circuit is required to work into a load resistance that is not much larger than the collector resistance R_2, you can "close in" on the optimum operating point as follows:

5. After step 2 above, place a signal generator v_{in} at the input (pin 3, Fig. 13-23), and a scope at the output (pin 7) to ground. With the generator output frequency adjusted within the range of frequencies this circuit is expected to handle, increase its output amplitude from zero until the scope shows a little clipping of the amplifier output waveform. Then readjust R_B until the output is no longer clipped.
6. Now proceed to increase the signal generator output amplitude again until the signal on the scope begins to clip again. Again readjust R_B until the scope output is unclipped.
7. Repeat the above steps until increases in the generator output cause simultaneous clipping of both positive and negative alternations of the signal viewed on the scope or until adjustments on R_B cannot restore the slightly clipped portions of the amplifier output.

Example 13-9 Suppose that we connect a portion of a dual two-input gate as shown in Fig. 13-23 with $V_{CC} = 9$ V, R_B a 0 to 500 kΩ potentiometer, and a 5 kΩ load resistance R_L connected to the right of C_2. The manufacturer specifies that typically $R_1 \cong 1.5$ kΩ and $R_2 \cong 3.6$ kΩ. Then the resistance R_B is decreased until $V_{CE} = 4.5$ V.
(a) Is this circuit biased at the optimum operating point? Why?
(b) R_B is removed and measures 215 kΩ, what nearest two-significant-figure fixed resistance will you use for R_B if you decide to bias at this operating point?

Answer 13-9 (a) Not likely, because the resistance R_L is not much greater than R_2. Only if R_L is say 10 times R_2 or larger, then for most practical purposes we have the optimum operating point when $V_{CE} = V_{CC}/2$. (b) Use 220 kΩ.

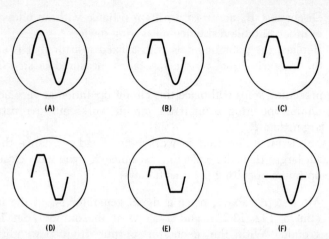

Fig. 13-25. Possible waveforms of dual two-input gates wired as amplifiers.

Example 13-10 Referring to the previous example, suppose that we replace the potentiometer as R_B and apply an input signal with a variable-amplitude signal generator. After adjusting the generator output and R_B a few times, we obtain waveforms shown in Fig. 13-25. Which one(s) represent operation at the optimum operating point?

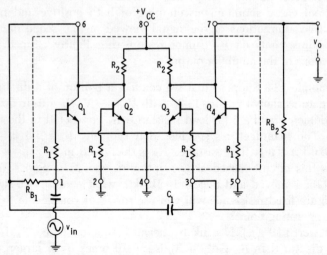

Fig. 13-26. Dual two-input gate (within the dotted area) wired as a two-stage amplifier.

13-6 THE DUAL TWO-INPUT GATE IC

Answer 13-10. Fig. 13-25D.—This figure shows slight and equal amounts of clipping on both alternations, thus bias is at or very near the optimum operating point. Review Sec. 8-14 in Chap. 8 if necessary. Fig. 13-25A is not correct because a stage not being driven hard enough to cause some clipping tells us little about the operating point. That is, if the input signal is small, the output is not clipped even though the operating point may be near cutoff or saturation. Figs. 13-25C and 13-25E are not correct because the clipping is excessive and not symmetrical. Nonsymmetry indicates that bias is not on the optimum operating point. With excessive clipping, it is difficult to tell whether the waveform is symmetrical or not.

A cascaded two-stage amplifier can also be constructed with a dual two-input gate IC. An example of this is shown in Fig. 13-26. Note that the inputs of transistors Q_2 and Q_4 are grounded, which means that both are cut off and therefore effectively disengaged from the circuit. The externally connected resistors R_{B1} and R_{B2} provide base bias here as in the single-stage amplifier of Fig. 13-23. The capacitor C_1 couples the signal from the collector of the first stage to the base of the second. This two-stage arrangement is more obvious in the equivalent circuit of Fig. 13-27. An amplified

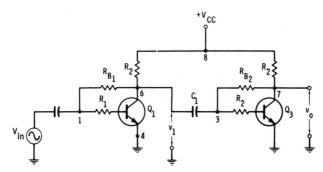

Fig. 13-27. Equivalent of circuit in Fig. 13-26.

version of the signal v_{in} appears at the collector (pin 6) of the first stage. This signal v_1 is the input of the second stage, which provides more amplification. The total voltage gain of this circuit is the product of the voltage gains of the individual stages. That is, if the gain of the first stage $A_{e1} = v_1/v_{in}$, the gain of the second stage $A_{e2} = v_o/v_1$, and the total gain is $A_{et} = v_o/v_{in}$, then

$$A_{et} = A_{e1} \times A_{e2} \qquad (13\text{-}2)$$

is the overall gain.

The values of R_{B1} and R_{B2} can be determined empirically as explained in a previous paragraph. Although transistor characteristics may vary considerably on separate IC's, even of the same type, the characteristics are fairly consistent for transistors of the same chip or *monolithic* IC.* Thus, if you find the required value of base bias resistance of the second stage in Fig. 13-26, the same value will very likely work with the first stage too. The fact that the first stage is handling a smaller signal makes the bias resistance in that stage less critical anyway.

Example 13-11 If the stage shown in Fig. 13-23 has a voltage gain of about 40, what approximate total voltage gain would you expect of two such stages cascaded as shown in Fig. 13-26?

Answer 13-11 Roughly 1600.—Assuming that the gains of both stages are about 40, their product, by Eq. 13-2, is 1600. Of course, this is a very rough estimate but does serve to show that the total gain of two stages is not merely twice the gain of one stage but instead is typically quite large.

Example 13-12 Suppose that in the circuit of Fig. 13-26, with a signal v_{in} applied and both base bias resistors equal to 220 kΩ, you obtain an output v_o that looks like Fig. 13-25B. In order to get optimum operation from this circuit, what changes will you make on (a) resistance R_{B1} and on (b) resistance R_{B2}? Answer either *increase, decrease,* or *no change*.

Answer 13-12 (a) No change, (b) decrease a little.—(a) The fact that the output of the second stage is not badly clipped indicates that the 220 kΩ base bias resistances provide operating points that are not too far from the optimum values. Since clipping occurs on the positive alternations, it indicates that the second stage is being driven into cutoff on portions of these alternations. The operating point is too low on this stage's load line. (b) A slightly smaller R_{B2} will move the operating point up the load line a little, thus lessening the chance of clipping due to cutoff. Since only a little distortion occurred in this second stage, which handles a

---

*The monolithic IC has all of its components formed on the same chip as opposed to the IC's that are constructed with more than one chip and sometimes include small discrete components.

13-7 THE ASTABLE MULTIVIBRATOR

relatively large signal, the same operating point is very likely all right in the first stage that handles a much smaller signal. That is, a small signal will not likely drive the first stage into cutoff or saturation even though the operating point is not optimum.

13-7 THE ASTABLE MULTIVIBRATOR

The astable multivibrator is an important circuit in digital computer systems. Its purpose is to provide "square wave" output voltages. Such output signal voltages are available without necessity of an input signal. Therefore the astable multivibrator is also called a *free-running* multivibrator. Often the astable circuit is used to keep circuit operations in digital systems in step with one another. Thus, it serves as master timer that synchronizes circuits in the digital system and is often referred to by still another term: a *clock*.

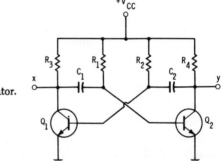

Fig. 13-28. Astable multivibrator.

As shown in Fig. 13-28, the astable multivibrator contains two common-emitter stages with the output (collector) of each capacitively coupled to the input (base) of the other. For reasons to be explained later, the transistors do not work as signal amplifiers but instead as switches. They are alternately driven into cutoff and saturation. When transistor Q_1 is cut off, Q_2 is saturated. After a time, Q_1 saturates and Q_2 cuts off. This process continually repeats itself, resulting in output signals shown in Fig. 13-29. The output v_x and v_y are signals obtained from points x and y respectively in Fig. 13-28. The frequency (repetitions per second) of the square wave is determined by the values of R_1, C_1, R_2, and C_2, as you will see.

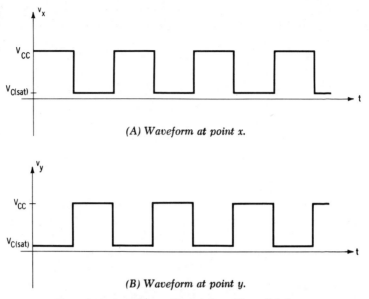

(A) Waveform at point x.

(B) Waveform at point y.

Fig. 13-29. Output waveforms of astable multivibrator.

We can analyze the astable multivibrator operation by starting with some specific component values such as $V_{CC} = 12$ V, $R_1 = R_2 = 50$ kΩ, $R_3 = R_4 = 4$ kΩ, $C_1 = C_2 = 0.1$ μF, and assuming that the transistors are ideal switches.

When V_{CC} is initially applied, both Q_1 and Q_2 tend to turn on because of the forward base bias provided to each transistor through R_1 and R_2. Also, initially both capacitors start to charge (plates connected to the collectors become positive) via the for-

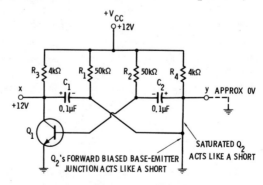

Fig. 13-30. Equivalent circuit of astable multivibrator when Q_2 is saturated.

13-7 THE ASTABLE MULTIVIBRATOR

ward biased base-emitter junctions of the transistors and the collector resistors R_3 and R_4. However, due to inherent differences in the components, one transistor will race toward saturation faster than the other. Suppose that Q_2 saturates before Q_1. This causes the voltage at point y to become zero, which means that this point is effectively grounded as shown in the equivalent circuit of Fig. 13-30. Note that this places C_2 and its charge across the base-emitter junction of Q_1 in reverse bias. This action is more obvious in Fig. 13-31A. The reverse bias into the base of Q_1 quickly cuts it off and therefore causes the voltage point x to rise suddenly to 12 V.

The charge on C_2 does not hold Q_1 cut off permanently. As shown in Fig. 13-31A, this capacitor discharges through R_2 and V_{cc} via ground. The time it takes C_2 to discharge is determined by the

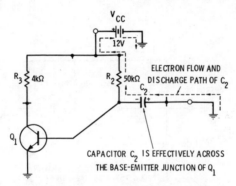

(A) *Charge on C_2 holds Q_1 in cutoff.*

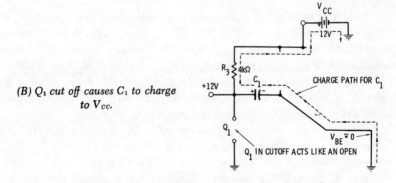

(B) *Q_1 cut off causes C_1 to charge to V_{cc}.*

Fig. 13-31. Equivalent circuits of astable multivibrator showing C_1 charge and C_2 discharge paths when Q_1 is cut off and Q_2 is saturated.

values of R_2 and C_2. Note in Fig. 13-31B that while Q_1 is cut off, C_1 charges to 12 V through R_3 and the forward biased base-emitter junction of Q_2. Because R_3, the resistance in the charge path of C_1, is much smaller than R_2, capacitor C_1 becomes fully charged before C_2 discharges.

Actually, the electron flow shown in Fig. 13-31A not only discharges C_2 but also tends to recharge this capacitor in the opposite polarity. That is, even after C_2 completely discharges, the electron flow continues, pushed by the source V_{CC}. This tends to recharge C_2, the plate connected to the base becoming positive. However, C_2 recharges in this way to only a few tenths of a volt because the base-emitter junction of Q_1 becomes forward biased, which quickly drives Q_1 out of cutoff and into saturation.

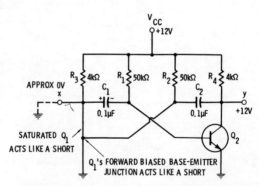

Fig. 13-32. Equivalent circuit of astable multivibrator when Q_1 is saturated.

When transistor Q_1 saturates it effectively places point x at ground potential as shown in Fig. 13-32. Thus, capacitor C_1, which was previously charged to 12 V in the polarity shown, is placed across the base-emitter junction of Q_2 as shown in Fig. 13-33A. This causes Q_2 to cut off suddenly and C_2 recharges as shown in Fig. 13-33B. In this way, C_2 becomes "rearmed" with a charge that will serve to cut off Q_1 again at the instant Q_2 saturates.

Transistor Q_2, of course, stays cut off until C_1 discharges and recharges to a few tenths of a volt in the opposite polarity. That is, when the base of Q_2 becomes a few tenths of a volt positive with respect to ground, this transistor turns on and saturates again.

In summary, when Q_2 saturates, it places C_2 across the base-emitter junction of Q_1. The charge on C_2 quickly cuts off Q_1. This

13-7 THE ASTABLE MULTIVIBRATOR

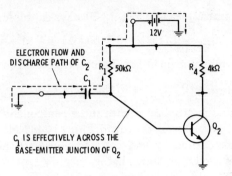

(A) Charge on C_1 holds Q_2 in cutoff.

(B) Q_2 in cutoff causes C_2 to charge to V_{CC}.

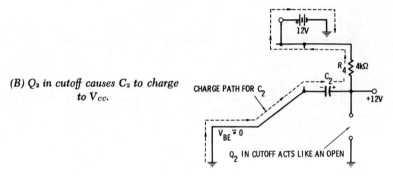

Fig. 13-33. Equivalent circuits of astable multivibrator showing C_2 charge and C_1 discharge paths when Q_2 is cut off and Q_1 is saturated.

transistor Q_1 stays cut off as long as it takes C_2 to discharge. When eventually the voltage across C_2 drops to about zero, Q_1 saturates, placing C_1 across the base-emitter junction of Q_2. The charge on C_1 quickly cuts off Q_2. With time, C_1 discharges and when its voltage drops to about zero, Q_2 saturates again, thus starting the whole cycle over again.

If generally T_{cutoff} is the time a transistor in the astable multivibrator is cut off, its value can be closely approximated with the equation

$$T_{\text{cutoff}} \cong 0.7\ RC \qquad (13\text{-}3)$$

where
  R is the base bias resistance of the transistor,
  C is the coupling capacitance to the base of the transistor.

For example, transistor Q_1 in Fig. 13-30 stays cut off for a time

$$T_{(cutoff)1} \cong 0.7R_2C_2 = 0.7(50 \text{ k}\Omega)(0.1 \text{ }\mu\text{F}) = 3.5 \text{ ms}$$

Similarly the transistor Q_2 in Fig. 13-32 stays cut off for a time

$$T_{(cutoff)2} \cong 0.7R_1C_1 = 0.7(50 \text{ k}\Omega)(0.1 \text{ }\mu\text{F}) = 3.5 \text{ ms}$$

Thus, an astable multivibrator with these component values has the output voltage waveforms shown in Fig. 13-34. Note that the time of one complete cycle (repetition) is equal to the sum of the cutoff times of each transistor. That is,

$$T \cong T_{(cutoff)1} + T_{(cutoff)2} \qquad (13\text{-}4)$$

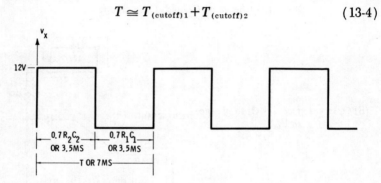

(A) *Voltage at point x.*

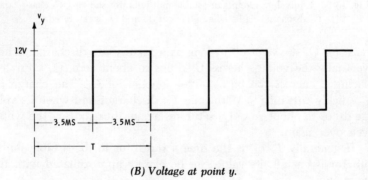

(B) *Voltage at point y.*

Fig. 13-34. Output voltages of astable multivibrator of Fig. 13-30.

In the circuit of Fig. 13-30 then, the time of one complete cycle, from either output x or y, is

$$T \cong 3.5 + 3.5 = 7 \text{ ms}$$

13-7 THE ASTABLE MULTIVIBRATOR

In general, if the base bias resistors R and the coupling capacitors C are the same for both transistors, the period of one cycle is

$$T \cong 1.4RC \qquad (13\text{-}5)$$

which, of course, is twice the right-hand side of Eq. 13-3.

The number of cycles that occur in one second is called the *frequency* f of the multivibrator and it is equal to the reciprocal of the period of one cycle. That is,

$$f = \frac{1}{T} \cong \frac{1}{1.4RC} \cong \frac{0.7}{RC} \qquad (13\text{-}6)$$

where the base bias resistances, and coupling capacitances, are equal.

In order to have each transistor alternate between cutoff and saturation, its ratio of base bias resistance R_B to the collector resistance R_C must be less than its β. If the circuit is to operate properly, it must satisfy the condition

$$\frac{R_B}{R_C} < \beta \qquad (13\text{-}7)$$

Example 13-13 If in Fig. 13-30, $R_1 = R_2 = 100$ kΩ, $R_3 = R_4 = 5$ kΩ, $C_1 = C_2 = 0.2$ μF, and $V_{CC} = 9$ V, what is the peak-to-peak amplitude and the frequency of the output voltage at either point x or point y?

Answer 13-13 $v_{o(p-p)} \cong 9$ V, $f \cong 35$ Hz.—Since the transistors alternate between cutoff and saturation, the output voltage varies from 9 V, the V_{CC} source voltage, to almost zero, the collector-to-ground voltage during saturation. Thus, the difference between the maximum and minimum values of the output voltage is about 9 V. The output frequency is

$$f \cong \frac{0.7}{RC} = \frac{0.7}{(100 \text{ k}\Omega)(0.2 \text{ μF})} = 35 \text{ Hz}$$

by Eq. 13-6.

Example 13-14 Referring to the circuit described in the previous example, what minimum β must each transistor have for proper operation, and what collector current must these transistors be capable of conducting?

Answer 13-14 $\beta_{\min} \cong 20$, $I_{C(\max)} = 1.8$ mA.—In this case, the base resistors R_B of both transistors are 100 kΩ and their collector resistors are 5 kΩ. Thus, by Eq. 13-7 the transistor β's must be greater than the ratio

$$\frac{R_B}{R_C} = \frac{100\ \text{k}\Omega}{5\ \text{k}\Omega} = 20$$

When each of these transistors is saturated, it conducts maximum current, which of course, is the saturation current. In this case

$$I_{C(\text{sat})} = \frac{V_{CC}}{R_C} = \frac{9\ \text{V}}{5\ \text{k}\Omega} = 1.8\ \text{mA}$$

is the saturation current.

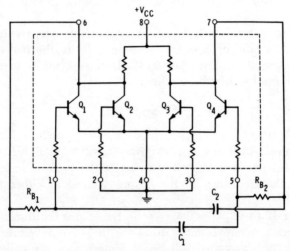

(A) Dual two-input gate externally wired to be astable multivibrator.

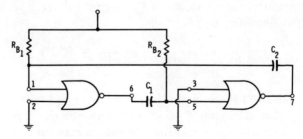

(B) Dual two-input gate shown with logic symbols and wired to be astable multivibrator.

Fig. 13-35. Astable multivibrator constructed with IC and discrete components.

13-7 THE ASTABLE MULTIVIBRATOR

An astable multivibrator can easily be made with a dual two-input gate IC. By grounding one input and capacitively coupling the output of each NOR gate to the remaining input of the other, and also by providing base bias resistors externally as shown in Fig. 13-35A, an astable multivibrator is formed. As in the circuit of Fig. 13-30, the frequency of the square wave, obtained off either collector (pins 6 or 7), is as before determined by the values of the coupling capacitors and base bias resistors. Therefore Eq. 13-6 is still applicable.

Note in Fig. 13-35A that the inputs of transistors Q_2 and Q_3 are grounded, which effectively disables them. If we remove Q_2 and Q_3 and rearrange the components, the circuit of Fig. 13-35 can be made to look very much like the circuit of Fig. 13-30 except for an additional resistance in series with each base. Typically these base input resistances, shown as R_1 in Fig. 13-26, are much smaller than the externally connected base bias resistances and therefore negligibly affect the circuit operation.

Example 13-15 If in Fig. 13-35A, we use a $V_{CC} = 12$ V, $R_{B1} = R_{B2} = 100$ kΩ, and $C_1 = C_2 = 0.02$ μF, (a) what is the period of each cycle and (b) what is the frequency of the square wave obtained from either pin 6 or pin 7?

Answer 13-15 (a) $T \cong 2.8$ ms, (b) $f \cong 357$ Hz.—By Eqs. 13-5 and 13-6 we find that the period is

$$T \cong 1.4RC = 1.4(100 \text{ k}\Omega)(0.02 \text{ }\mu\text{F}) = 2.8 \text{ ms}$$

and

$$f \cong \frac{1}{T} = \frac{1}{2.8 \text{ ms}} \cong 357 \text{ Hz}$$

is the frequency of the circuit.

Example 13-16 Suppose that you are required to build an astable multivibrator with an output frequency f of about 1000 Hz. Using a dual two-input gate and 0.1-μF coupling capacitors, what values of base bias resistances will you use?

Answer 13-16 $R_{B1} = R_{B2} \cong 7.15$ kΩ.—Again referring to Eq. 13-6, we can rearrange it to solve for R. That is, in this case

$$R \cong \frac{1}{1.4fC} = \frac{1}{1.4(1 \times 10^3)(0.1 \times 10^{-6})} \cong 7.15 \text{ k}\Omega$$

is the required base bias resistance.

13-8 THE OPERATIONAL AMPLIFIER

The *operational amplifier* is a dc amplifier, which means that, within limits, a change in average dc voltage at its input causes a proportional change in average dc at its output. This kind of action does not occur with amplifiers that are capacitively coupled. That is, the amplifiers in previous chapters and the amplifiers in Figs. 13-24, 13-26, and 13-27 in this chapter have either or both input and output coupling capacitors. These capacitors block dc; therefore, if the dc voltage at the input is gradually changed, it does not affect the dc level at the output. The operational amplifier, on the other hand, is frequently used to amplify dc voltages. For example, an operational amplifier with a gain of 10 will have a 20-mV dc output if the input dc is 2 mV. Similarly, with either 1 mV or 3 mV input dc voltages, the output dc voltages are either 10 mV or 30 mV respectively Because the output voltage is proportional to the input in the operational amplifier, it is usually classified with *linear devices* in the manufacturers' data sheets. The term "linear" is descriptive because if we plot a graph of output versus input voltages of an operational amplifier, the result is a straight-line function. Of course, the linear properties of the operational amplifier exist within limits that are specified by the manufacturer. That is, operational amplifiers are subject to saturation as were previous amplifiers.

Operational amplifiers are available constructed with discrete components, and as monolithic IC's. Compared to versions made with discrete components, in many applications, the IC operational amplifiers offer qualities such as increased reliability, small size, and lower cost.

The circuit of Fig. 13-3A is an example of an IC operational amplifier. The details of the inner workings of this circuit are not within the scope of this text. We will instead look at the IC opera-

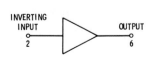

Fig. 13-36. Symbol for simple operational amplifier.

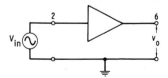

Fig. 13-37. Ratio v_o/v_{in} is called "open-loop voltage gain."

13-8 THE OPERATIONAL AMPLIFIER

tional amplifier as a "black box" and thus concern ourselves with its operation as seen looking into the external leads.

Being linear devices, operational amplifiers are very useful as ac signal as well as dc amplifiers. We will, in fact, start our analysis of operational amplifiers from the point of view of ac signals. The characteristics thus found are also applicable to dc signals, as you will see.

A common symbol for the operational amplifier is shown in Fig. 13-36. For the present, only two of its external leads are shown: the *inverting input* lead and the *output* lead. As the names of these leads imply, the output is an inversion of the input signal. This means that if a sine wave voltage is applied to the input, the output is also a sine wave but 180° out of phase with the input. This also means that if the input signal is dc and becomes more positive, the output becomes more negative, and vice versa. If zero dc-to-ground voltage is applied to the input (pin 2),[*] ideally the output (pin 6) is zero dc volts to ground too. Practically, though, it may take a few microvolts, or at most a few millivolts in less expensive units, applied to pin 2 to make the voltage at pin 6 zero. The value of dc voltage required at the input (pin 2) to cause zero dc at the output (pin 6) is called the *input offset voltage*. The manufacturers specify the largest possible input offset voltage with each type of operational amplifier they make. Often the input offset voltage is so small that it may be ignored for practical purposes. This means that we can assume that the output dc voltage is zero when the input dc voltage is zero.

The ratio of the output v_o to the input v_{in} of the circuit shown in Fig. 13-37 is called the *open-loop voltage gain* A_{EOL}. Typically A_{EOL} is very high, ranging from 15,000 into millions. Thus,

$$A_{EOL} \cong \frac{v_o}{v_{in}} \qquad (13\text{-}8)$$

and therefore

$$v_{in} \cong \frac{v_o}{A_{EOL}} \qquad (13\text{-}8A)$$

Since A_{EOL} is very large, it is apparent, in this last equation that v_{in} must be very small compared to v_o. In fact, the voltage at the

---

[*]The inverting input and the output terminals are not pins 2 and 6 on all types of operational amplifiers.

input (pin 2) is so small that this input is considered to be at ground potential for most practical purposes.

The input resistance of the operational amplifier, as seen by the source v_{in} in Fig. 13-37, is very large. For simplicity we will assume that it is infinite and therefore the current into the operational amplifier is practically zero.

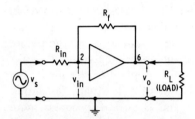 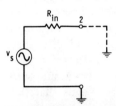

Fig. 13-38. Use of input R_{in} and feedback R_f resistances modifies the gain, viz, $A_e = R_f/R_{in}$.

Fig. 13-39. Equivalent of input of circuit of Fig. 13-38, with nearly all of v_s across R_{in}.

Use of input and feedback resistances, as in Fig. 13-38, enables us to modify the overall gain. That is, the ratio v_o/v_s, in Fig. 13-38, is determined mainly by the values of resistances R_{in} and R_f. How and why R_{in} and R_f affect the gain v_o/v_s may be shown as follows:

Since v_{in} at pin 2, Fig. 13-38, is very small and nearly at ground potential, the equivalent of the input circuit as seen by v_s can be shown as in Fig. 13-39. Thus, nearly all of the v_s voltage appears across R_{in}. By Ohm's law we can show that the signal current through R_{in} is

$$i \cong \frac{v_s}{R_{in}}$$

which, when rearranged, becomes

$$v_s \cong R_{in}i \qquad (13\text{-}9)$$

the generator voltage.

Similarly, looking to the left of pin 6, we see the circuit shown in Fig. 13-40. Note that because the input (pin 2) is nearly grounded, the output voltage v_o appears across R_f. The current through R_f thus is

$$i \cong \frac{v_o}{R_f}$$

13-8 THE OPERATIONAL AMPLIFIER

and therefore

$$v_o \cong R_f i \quad (13\text{-}10)$$

is the output voltage.

Since the current into the operational amplifier input lead (pin 2) is about zero, the current in R_{in} must nearly equal the current in R_f as shown in Fig. 13-41. Therefore, because the overall gain of the circuit in Fig. 13-38 is $A_e = v_o/v_s$, we can substitute Eqs. 13-9 and 13-10 into this gain equation and show that

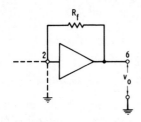

Fig. 13-40. Equivalent of output of circuit of Fig. 13-38.

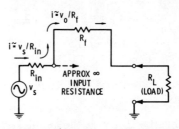

Fig. 13-41. Equivalent circuit showing that i_{Rin} and i_{Rf} are nearly equal.

$$A_e = \frac{v_o}{v_s} \cong \frac{R_f i}{R_{in} i} = \frac{R_f}{R_{in}}$$

Note that the currents i, being nearly equal, cancel, which leaves us with a very useful equation:

$$A_e \cong \frac{R_f}{R_{in}} \quad (13\text{-}11)$$

However, it is valid only if R_f is connected from the output to the inverting input.

Example 13-17 The manufacturer specifies that the maximum dc output voltage of its operational amplifier, in Fig. 13-42, is 15 V and that its open-loop voltage gain is $A_{EOL} \cong 300,000$. Find (a) the maximum input that we can use that will not exceed the output

Fig. 13-42. Simple operational amplifier dc circuit.

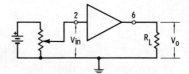

15 V capability, and (b) the dc output voltage if the input V_{in} is adjusted to $+20\ \mu V$.

Answer 13-17 (a) $V_{in(max)} \cong 50\ \mu V$, (b) $v_o \cong -6$ V.—Since A_{EOL} and the maximum output are known in this case, we can find the maximum input with Eq. 13-8A. That is, since

$$v_{in} \cong \frac{v_o}{A_{EOL}}$$

then

$$V_{in(max)} \cong \frac{V_{o(max)}}{A_{EOL}} \cong \frac{15\ V}{300{,}000} = 50\ \mu V$$

Similarly if given the input voltage V_{in}, we can solve for V_o in Eq. 13-8. Thus, since $A_{EOL} \cong v_o/v_{in}$, then $v_o \cong A_{EOL}v_{in}$, and, in this case,

$$V_o \cong A_{EOL}V_{in} \cong 300{,}000 \times 20\ \mu V = 6\ V$$

But since there is an inversion from input to output, the output is negative because the input is positive to ground. Therefore $V_o \cong -6$ V.

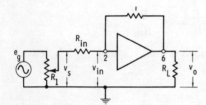

Fig. 13-43. Simple operational amplifier ac circuit.

Example 13-18 In the circuit of Fig. 13-43, if $R_{in} = 100$ kΩ, $R_f = 1$ MΩ, and the potentiometer R_1 is adjusted so that $v_s = 25$ mV, (a) what is the output voltage v_o? (b) If the signal generator e_g is replaced with a dc source E (negative to ground) and the potentiometer R_1 is adjusted so that $V_o = -40$ mV, what's the output V_o?

Answer 13-18 (a) $v_o \cong 250$ mV, (b) $V_o \cong 400$ mV.—(a) The overall gain of the stage can be determined first by Eq. 13-11. In this case,

$$A_e \cong \frac{R_f}{R_{in}} = \frac{1\ M\Omega}{100\ k\Omega} = 10$$

Thus, rearranging and plugging in our known v_s,

$$v_o = A_e v_s \cong 10(25\text{ mV}) = 250\text{ mV}$$

(b) Remember that the operational amplifier is a dc amplifier, which means that this gain of 10 is also applicable to dc. With a dc $V_S = -40$ mV, we find that the output dc is

$$V_o = A_e V_S \cong 10(40\text{ mV}) = 400\text{ mV}$$

This output is positive to ground because the input is negative and the amplifier inverts the signal.

13-9 PRACTICAL CONSIDERATIONS AND APPLICATIONS OF THE OPERATIONAL AMPLIFIER

As an aid to the circuit designer, manufacturers of operational amplifiers usually provide data as shown in Fig. 13-44. The table in Fig. 13-44A shows that, depending on the voltage gain desired, there is a small variety of component values that are connected

TYPICAL CHARACTERISTICS OF OP AMP TYPE --
+V = +15V, -V = -15V, T = 25° C, R_L = 2kΩ

VOLTAGE GAIN	R_{in}	R_f	R_1	R_2	R_3	C_1
1	10kΩ	10kΩ	5kΩ	390Ω	10kΩ	2nF
10	1kΩ	10kΩ	1kΩ	1kΩ	10kΩ	2nF
100	1kΩ	100kΩ	1kΩ	10kΩ	OPEN	2nF
1000	1kΩ	1MΩ	1kΩ	0Ω	OPEN	10pF
A_{EOL}	0Ω	OPEN	0Ω	OPEN	OPEN	0

(A) *Manufacturer's data.*

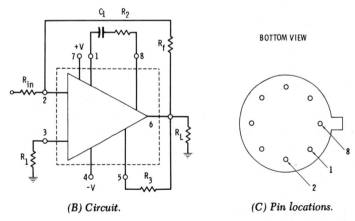

(B) *Circuit.*   (C) *Pin locations.*

Fig. 13-44. Typical data supplied by manufacturers of operational amplifiers.

externally. These components and the manner in which they are connected to the pins of the IC are shown in Fig. 13-44B. The pin locations on the case are shown in Fig. 13-44C. The components other than R_{in} and R_f do not directly determine the gain. They provide frequency compensation and prevent oscillations. An amplifier circuit oscillating can act much like an astable multivibrator, which, needless to say, is not proper operation for a linear circuit like an operational amplifier.

Note that the operational amplifier is powered with two dc source voltages: one positive to ground ($+V$); the other is negative to ground ($-V$). These sources must be well regulated. A power supply circuit that often is suitable is shown in Fig. 13-45.

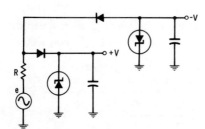

Fig. 13-45. Power supply capable of providing plus and minus voltages with respect to ground.

Example 13-19 The operational amplifiers in Fig. 13-46 are to be used to obtain each of the voltage gain factors listed in Fig. 13-44A, though not necessarily in the order shown. Using the data in Fig. 13-44A, what values of R_{in} and R_f will you use in each circuit, Figs. 13-46A through 13-46D? Note that since the other components are already specified, you must first find the gain that requires these specified components, in each circuit.

Answer 13-19 (a) $R_{in} = 1$ kΩ, $R_f = 1$ MΩ, (b) $R_{in} = 0$, R_f is open, (c) $R_{in} = 1$ kΩ, $R_f = 10$ kΩ, (d) $R_{in} = 1$ kΩ, $R_f = 100$ kΩ.—(a) Since $R_2 = 0$, R_3 is an open, and $C_1 = 10$ pF in Fig. 13-46A, the voltage gain must be 1000 (see data), requiring a ratio $R_f/R_{in} = 1000$. (b) In Fig. 13-46B since R_1 and C_1 are zero, and since R_2 and R_3 are open, which are the values listed in the bottom row of data, the voltage gain is A_{EOL}, the open-loop gain. (c) The components used in Fig. 13-46C are those listed in the fourth row from the bottom in Fig. 13-44A; thus the amplifier voltage gain must be 10 and the ratio $R_f/R_{in} = 10$. (d) The components in Fig. 13-46D are in the third row from the bottom of the data in Fig. 13-44A, and therefore the voltage gain is 100, requiring a ratio $R_f/R_{in} = 100$.

13-10 THE OPERATIONAL AMPLIFIER AS A MULTIPLIER

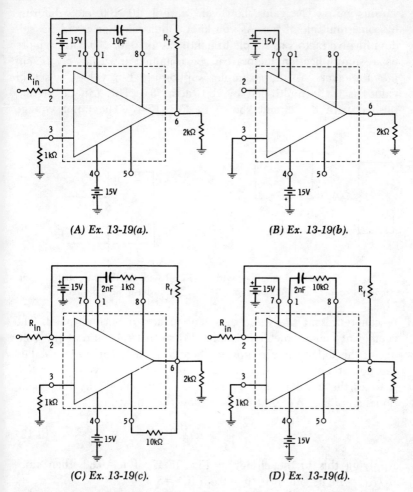

Fig. 13-46. Operational amplifier IC's connected for different voltage gains.

13-10 THE OPERATIONAL AMPLIFIER AS A MULTIPLIER

The operational amplifier is the very heart of the analog computer in which it is often required to multiply (amplify) a signal voltage. In the analog computer, voltages are used to represent (be analogous to) quantities that are to be multiplied. In the last section, you may have noticed that the data given on the table in Fig. 13-44 enabled us to build operational amplifier stages with various gains. The gains, however, were 1, 10, 100, etc.—differing by some multiple of ten. As you know, multiplication problems seldom involve factors as simple to handle as 1, 10, or 100. Other, more usual, multiplication factors can be obtained by using *coefficient potentiometers*. As an example, suppose that a value or set of values is to be multiplied by the factor 8.3. This can be accomplished with the circuit shown in Fig. 13-47. The input voltage

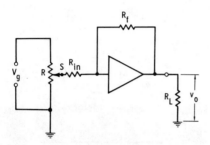
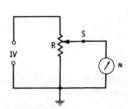

Fig. 13-47. Operational amplifier with input coefficient potentiometer.

Fig. 13-48. Coefficient potentiometer adjusted for desired μ.

v_g can represent the value or set of values to be multiplied. The coefficient potentiometer R is adjusted so that the ratio of the voltage at the slider S to the input voltage v_g is 0.83. This ratio is called the potentiometer μ (Greek letter mu). Thus, the voltage at S is μ times the input v_g, and is multiplied (amplified) by the operational amplifier. We can therefore show that the gain is

$$\frac{v_o}{v_g} \cong \mu \frac{R_f}{R_{in}} \qquad (13\text{-}12)$$

Applying this to the circuit of Fig. 13-47, if $\mu = 0.83$, then since $R_f/R_{in} = 10$, its gain is $v_o/v_g = 0.83(10) = 8.3$.

The potentionmeter R is adjusted for a $\mu = 0.83$, or any other required fraction, by placing a convenient voltage across it, such as 1 V as shown in Fig. 13-48, and adjusting the slide until 0.83 V or whatever other fraction μ you need is read on the voltmeter M.

Of course, in the circuit of Fig. 13-47, if v_g is positive to ground, then v_o is negative, and vice versa. If this inversion is unacceptable. the solution is simple: Connect another operational amplifier into the system as shown in Fig. 13-49. If no gain, only phase inversion, is required of the second stage, its ratio R_f/R_{in} is made equal to 1. The second stage can also serve to provide gain if we need it. For

13-11 OPERATIONAL AMPLIFIER AS A SUMMING AMPLIFIER

example, if we need a multiplying factor of 56, we can use a couple of operational amplifiers as in Fig. 13-49 such that the potentiometer $\mu = 0.56$ and the ratios R_f/R_{in} of both stages are 10.

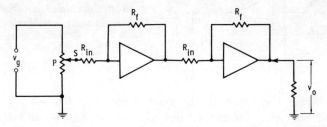

Fig. 13-49. Two cascaded operational amplifiers provide noninverted output v_o.

Example 13-20 In the circuit of Fig. 13-49, 1 V is applied across the potentiometer R and the slide is adjusted until the voltage at S is 0.43 V. The first stage ratio is $R_f/R_{in} = 10$, the second stage $R_f/R_{in} = 100$. What is the output of the second stage v_o when the input to the potentiometer v_g is a 3-mV rms sine wave?

Answer 13-20 $v_o = 1.29$ V.—The coefficient potentiometer μ is

$$\mu = \frac{0.43 \text{ V}}{1 \text{ V}} = 0.43$$

We can determine the gain from v_g to the output of the first stage with Eq. 13-12:

$$\frac{v_{o1}}{v_g} = 0.43 \frac{R_f}{R_{in}} = 0.43(10) = 4.3$$

The second stage adds more gain by the factor 100, its ratio R_f/R_{in}. This gives us a total gain of

$$\frac{v_o}{v_g} = 4.3(100) = 430$$

Therefore with $v_g = 3$ mV, we can solve the above equation for v_o and show that $v_o = 430 v_g = 430(3 \text{ mV}) = 1.29$ V rms.

13-11 THE OPERATIONAL AMPLIFIER AS A SUMMING AMPLIFIER

A *summing amplifier* has more than one input and its output is the sum, or a multiple of the sum, of the individual inputs. The operational amplifier used as a summing amplifier is shown in Fig.

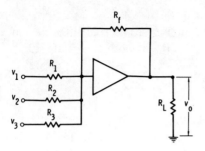

Fig. 13-50. Operational amplifier used as summing amplifier.

13-50. Since the input to the operational amplifier is nearly at ground potential, the right sides of resistors R_1, R_2, and R_3 are effectively grounded. Therefore the input voltages v_1, v_2, and v_3 are across resistors R_1, R_2, and R_3 respectively. The currents in these individual resistors are found by Ohm's law:

The current in R_1 is

$$i_1 \cong \frac{v_1}{R_1} \qquad (13\text{-}13\text{A})$$

The current in R_2 is

$$i_2 \cong \frac{v_2}{R_2} \qquad (13\text{-}13\text{B})$$

And the current in R_3 is

$$i_3 \cong \frac{v_3}{R_3} \qquad (13\text{-}13\text{C})$$

Since the input resistance to the operational amplifier is very large, the algebraic sum of the individual input currents is equal to the current i in R_f. As before, the current i in R_f is approximately the output voltage v_o divided by R_f (Eq. 13-10). Thus, by Kirchhoff's law,

$$i_1 + i_2 + i_3 \cong i$$

Substituting Eqs. 13-13A through 13-13C and Eq. 13-10 into the above, we get

$$\frac{v_1}{R_1} + \frac{v_2}{R_2} + \frac{v_3}{R_3} \cong \frac{v_o}{R_f}$$

If we use equal input resistors, that is, $R_1 = R_2 = R_3 = R$, then the above equation modifies to

13-11 Operational Amplifier as a Summing Amplifier

$$\frac{v_1}{R} + \frac{v_2}{R} + \frac{v_3}{R} \cong \frac{v_o}{R_f}$$

or

$$\frac{1}{R}(v_1 + v_2 + v_3) \cong \frac{v_o}{R_f}$$

Rearranging the above we can show that

$$v_o \cong \frac{R_f}{R}(v_1 + v_2 + v_3)$$

Since the signal is inverted with a single operational stage, the previous equation is more correctly shown as

$$v_o \cong -\frac{R_f}{R}(v_1 + v_2 + v_3) \quad (13\text{-}14)$$

If we want a summing amplifier with no gain, we can use a ratio $R_f/R = 1$, which simplifies the above equation to

$$v_o \cong -(v_1 + v_2 + v_3) \quad (13\text{-}15)$$

so that the output voltage is the negative sum of the input voltages.

Example 12-21 Suppose that in the circuit of Fig. 13-50, we apply input signal voltages that are in phase with each other: $v_1 = 2.5$ mV, $v_2 = 3.5$ mV, and $v_3 = 1.5$ mV. If the input resistors are each 100 kΩ and the feedback resistor $R_f = 1$ MΩ, what output signal voltage v_o would you expect?

Answer 12-21 $v_o \cong 75$ mV (out of phase with the input voltages).—Eq. 13-14 is applicable here:

$$v_o \cong -\frac{1 \text{ M}\Omega}{100 \text{ k}\Omega}(2.5 \text{ mV} + 3.5 \text{ mV} + 1.5 \text{ mV})$$

$$\cong -10(7.5 \text{ mV}) = -75 \text{ mV}$$

is the output voltage.

Example 12-22 Suppose we apply the following dc voltages to the inputs of the circuit in Fig. 13-50: $V_1 = 2.75$ mV, $V_2 = -3.25$ mV, $V_3 = 4.5$ mV. What is the output voltage if all resistors are equal to 100 kΩ?

Answer 12-22 $v_o = -4$ mV.—Since all the resistors are equal, Eq. 13-15 is applicable. Thus,

$$V_o \cong -(2.75 \text{ mV} - 3.25 \text{ mV} + 4.5 \text{ mV}) \cong -4 \text{ mV}$$

is the dc output voltage.

13-12 MISCELLANEOUS INFORMATION ABOUT IC's

IC's made for use in logic circuits and digital computers are usually referred to as *logic IC's* or *digital IC's* as opposed to *linear* types of IC's such as the operational amplifier. These digital IC's are available in a variety of designs. The types of designs are usually classified under general abbreviated headings such as: ECL, TTL, DTL, RTL, MOS, and a few others. The meanings of some popular ones are as follows:

Emitter Coupled Logic

The abbreviation ECL means Emitter Coupled Logic. The ECL integrated circuits are so named because their outputs are obtained from the emitters of transistors built on the chip. Examples of ECL IC's are shown in Fig. 13-51.

The ECL IC's are designed to provide output logic levels 1 and 0, as are other types of logic circuits, but without ever driving the output transistors into saturation. A transistor not in saturation may be driven into cutoff faster than one that is saturated. Thus, the ECL IC offers extremely high speed operation, which means that changes in input voltage cause almost instantaneous changes in the output voltage levels.

Since the emitter follower configuration inherently has fairly good power output capability, the ECL IC is capable of driving several inputs of other logic circuits. This characteristic is referred to as high *fan-out* capability.

Transistor Transistor Logic

The abbreviation TTL means Transistor Transistor Logic. The TTL IC's are identified by the inputs that have multiple-emitter transistors as shown in Fig. 13-52. The multiple emitters are independent of each other and a current in either one or more of them causes a current in the single collector lead. The TTL IC's offer a large amount of logic in a small amount of space and they are among the fastest of saturating-type logic circuits. That is, most types of IC's obtain logic levels 1 or 0 in the output by driving an output transistor into cutoff or saturation. As mentioned previously, saturated transistors respond more slowly than unsaturated ones.

13-12 MISCELLANEOUS INFORMATION ABOUT IC's

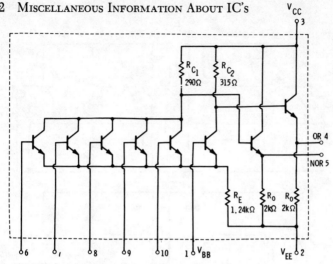

(A) Multi-input gate that can be used as either OR or NOR gate.

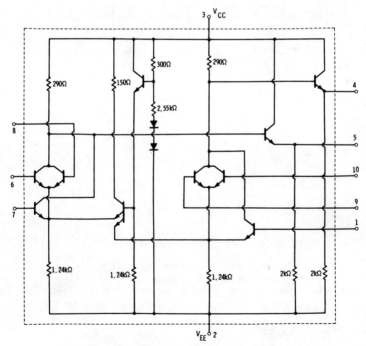

(B) Dual three-input that gate can be used as NOR gate.

Fig. 13-51. ECL type IC's.

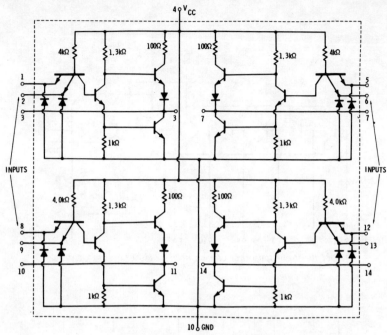

(A) Quad two-input NAND gate.

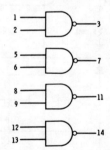

(B) Symbols for quad two-input NAND gate.

Fig. 13-52. TTL type IC functioning as four separate two-input NAND gates.

The TTL IC's are saturating type but respond relatively fast to changes in their inputs and approach speeds of the ECL ICs.

Diode Transistor Logic

The abbreviation DTL usually means Diode Transistor Logic. The DTL IC's typically have diodes in series with their input

13-12 Miscellaneous Information About IC's

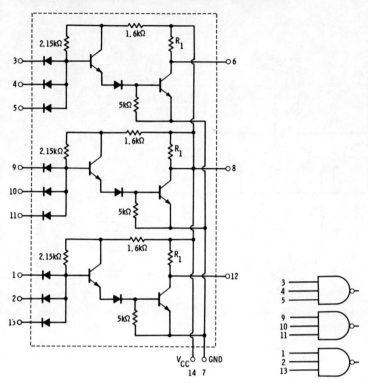

(A) Schematic diagram of triple three-input gate.

(B) Logic symbol of (A) if positive logic is used.

Fig. 13-53. DTL type IC.

leads as shown in Fig. 13-53. This type of IC is known for its moderate speed, good economy, and flexibility. The flexibility is available, in part, because its logic function can be altered by tying two or more of its output leads together. This kind of modification is not possible with most other types of IC's.

Some manufacturers use the letters DTL to mean Digital Transistor Logic. In their data sheets you'll find the abbreviation DTL used to refer to IC's that contain no diodes. Instead, these IC's contain transistors and resistors only.

Resistor Transistor Logic

The abbreviation RTL means Resistor Transistor Logic. Examples of RTL IC's are shown in Fig. 13-54. They are easily recog-

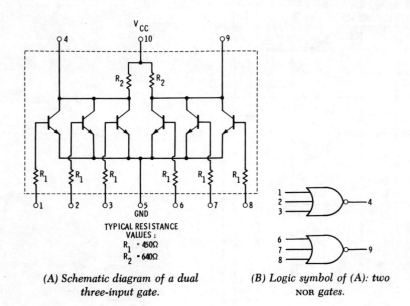

(A) Schematic diagram of a dual three-input gate.

(B) Logic symbol of (A): two NOR gates.

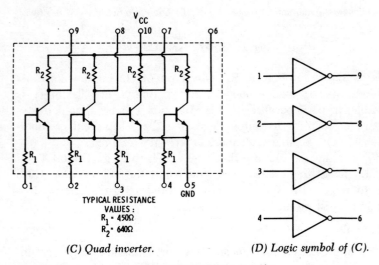

(C) Quad inverter.

(D) Logic symbol of (C).

Fig. 13-54. RTL type IC's.

13-12 Miscellaneous Information About IC's

nized by their inputs. Each input is connected to the base of a transistor through a series resistor.

RTL IC's have advantages of low cost, good availability, and flexibility. Their flexibility is apparent if we refer back to the dual two-input gate IC in Fig. 13-22, which is an RTL type. Note in the symbols of Fig. 13-22 that this RTL IC can be used as two separate NOR gates. Also it was connected as a single amplifier stage or two cascaded amplifier stages as shown in Figs. 13-23 and 13-26 respectively. In addition we used it as an astable multivibrator in a connection shown in Fig. 13-35.

Compared to other logic functions, RTL IC's are able to withstand more abuse in the form of improper connections and voltages to the external pins, which makes them very suitable to experimental work and for familiarizing less experienced technicians and engineers with IC's. Also comparatively, RTL IC's have slower speeds, more noise, and require more dc supply power.

Metal Oxide Silicon Logic

The abreviation MOS means Metal Oxide Semiconductor as it does when referring to FET's. In fact, MOS IC's use FET's as the active devices shown in Fig. 13-55. Note that this particular IC has three n-channel and three p-channel IGFET's. With various external connections, this IC can be used to perform as three separate inverters, as a three-input NOR gate, or as a three-input NAND gate.

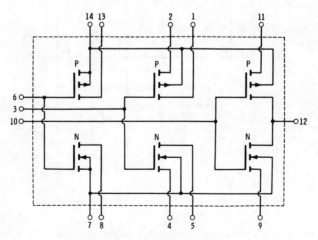

Fig. 13-55. MOS type IC.

Advantages of MOS IC's are their good temperature stability, low power dissipation, and very high input resistance. The very high input resistance, which is characteristic of FET devices, gives the MOS IC very high fan-out capability. This means, that if necessary, many MOS IC's can be connected to the output of a single MOS IC with no loading difficulties. Typically, one MOS IC can drive as many as fifty inputs of other MOS IC's. Comparatively, in systems using ICs other than the MOS type, each IC is only able to drive a few inputs of other IC's.

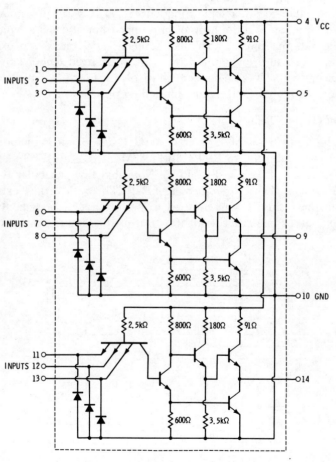

Fig. 13-56. Triple three-input NAND gate with positive logic, or NOR gate with negative logic.

13-12 Miscellaneous Information About IC's

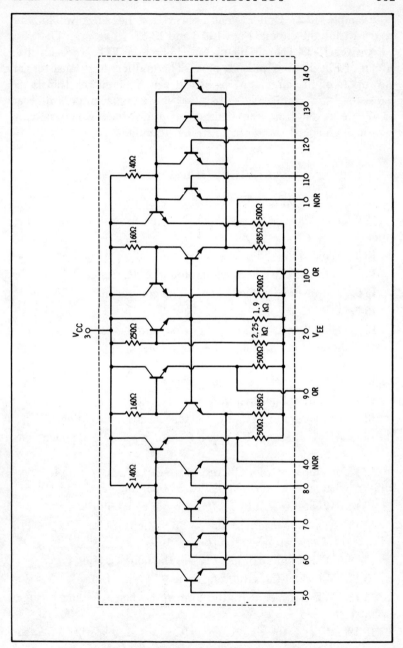

Fig. 13-57. Dual four-input NOR/OR gate.

Example 13-23 Under what abbreviated headings would you expect to find the circuits Figs. 13-56 and 13-57?

Answer 13-23 The circuit of Fig. 13-56 is a TTL type, and the circuit of Fig. 13-57 is an ECL type.—The multiemitter transistors at the inputs of the circuit in Fig. 13-56 classify it in the transistor-transistor logic family. Note that the NOR and OR outputs in Fig. 13-57 are taken from the emitters of transistors, which puts it into the high-speed emitter coupled logic family.

REVIEW QUESTIONS

13-1. What is an integrated circuit?

13-2. How does a monolithic IC differ from a nonmonolithic type?

13-3. What is the difference between a wafer and a chip?

13-4. Why are capacitors seldom found in IC designs?

13-5. Name three methods of isolating components in IC's.

13-6. Define the term *epitaxy*.

13-7. How does an OR gate differ from a NOR gate?

13-8. Based on your answer to the previous problem, what do you think is the difference between an AND gate and a NAND gate?

13-9. What does the term *yield* mean?

13-10. What is a multivibrator?

13-11. What are two other terms that mean "multivibrator?"

13-12. How does an IC differ from a discrete component circuit?

13-13. Compared to RC (resistance-capacitance) coupled amplifiers, what are the characteristics of operational amplifiers?

13-14. What is meant by an IC's fan-out capability?

13-15. What do each of the abbreviations ECL, TTL, DTL, RTL, and MOS mean when referring to IC's?

13-16. Which family of logic IC's is the nonsaturating type?

13-17. Which of the saturating logic IC's is the fastest?

13-18. What is meant by the term *fast* when referring to logic circuits?

13-19. What is the difference in IC's that are classified as *linear* from those classified as *digital*?

PROBLEMS

13-1. Referring to Fig. 13-16A, if source $V = 9$ V, source $V_{CC} = 9$ V, switch A is in the up (1) position, and switch B is in the down (0) position, assuming that the diodes are ideal and the transistor is an ideal switch, what are the voltages (a) across diode D_1, (b) across diode D_2, (c) at point C to ground, (d) across R_1, (e) across R_2, and (d) at point D to ground?

13-2. Referring to Fig. 13-16A, if source $V = 12$ V, source $V_{CC} = 9$ V, switch A is in the up (1) position, and switch B is the down (0) position, assuming that the diodes are ideal and that the transistor is an ideal switch, what are the voltages (a) across diode D_1, (b) across diode D_2, (c) at point C to ground, (d) across R_1, (e) across R_o, and (d) at point D to ground?

13-3. Suppose that in the circuit of Fig. 13-16A, the diodes drop 0.7 V when conducting, the transistor forward biased base-emitter junction drops 0.8 V, and that $V_{CE(\text{sat})} = 0.9$ V. If both switches A and B are down (0) position, what are the voltages (a) across diode D_1, (b) across diode D_2, (c) at point C to ground, (d) across R_1, (e) across R_2, and (d) at point D to ground? Assume that source $V = 9$ V and $V_{CC} = 12$ V.

13-4. Suppose that in the circuit of Fig. 13-16A, the diodes drop 0.7 V when conducting, the transistor forward biased base-emitter junction drops 0.8 V, and that $V_{CE(\text{sat})} = 0.9$ V. If source $V = 5$ V, source $V_{CC} = 9$ V, and both switches A and B are in the up (1) position, what are the voltages (a) across D_1, (b) across D_2, (c) at point C to ground, (d) across R_1, (e) across R_2, and (d) at point D to ground?

13-5. Referring back to Prob. 13-1, what is the collector saturation current?

13-6. Referring back to Prob. 13-2, what is the collector saturation current?

13-7. Referring back to Prob. 13-3, what maximum collector current must the transistor be capable of?

13-8. Referring back to Prob. 13-4, what maximum collector current must the transistor be capable of?

13-9. If in Fig. 13-20, $V_{CC} = 20$ V, $R_1 = 5$ kΩ, and the transistor β's are 50, what is the largest possible series resistance that we can use with each transistor that will cause saturation when the input voltage is +20 V?

13-10. If in Fig. 13-20, $V_{CC} = 12$ V, $R_1 = 6$ kΩ, and the transistor $β$'s are 80, what is the largest possible series resistance that we can use with each transistor that will cause saturation? Assume 12 V is logic 1 and 0 V is logic 0.

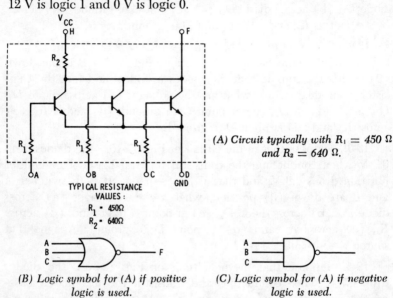

(A) Circuit typically with $R_1 = 450$ Ω and $R_2 = 640$ Ω.

(B) Logic symbol for (A) if positive logic is used.

(C) Logic symbol for (A) if negative logic is used.

Fig. 13-58. Three-input gate IC.

A	B	C	F
0V	0V	0V	
0V	0V	4V	
0V	4V	0V	
0V	4V	4V	
4V	0V	0V	
4V	0V	4V	
4V	4V	0V	
4V	4V	4V	

A	B	C	F

A	B	C	F

(A) For possible combinations of input voltages.

(B) Truth table for positive logic.

(C) Truth table for negative logic.

Fig. 13-59. Truth tables for three-input gate IC of Fig. 13-58.

PROBLEMS

13-11. If in Fig. 13-58, $V_{CC} = 12$ V and the inputs a, b, and c are either 0 V or 4 V to ground, what is the output voltage at f for each of the input combinations shown in the truth table in Fig. 13-59A? Place your answers in column f.

13-12. Referring to the previous problem and Fig. 13-58A, use logic symbols 0 and 1 instead of 0 V and 4 V respectively (positive logic). Fill in the truth table of Fig. 13-59B.

13-13. Still referring to Prob. 13-11 and Fig. 13-58A, if we use negative logic, that is, refer to 0 V as logic 1 and 4 V as logic 0, what does the truth table look like? Fill in Fig. 13-59C using symbols 1 and 0 only.

13-14. About what maximum current would you expect drawn from a 9-V V_{CC} source (through R_2), in the circuit of Fig. 13-58A?

13-15. If $V_{CC} = 9$ V and the inputs are either 0 V or 9 V to ground in Fig. 13-58A, what minimum β must the transistors have if they are expected to saturate properly?

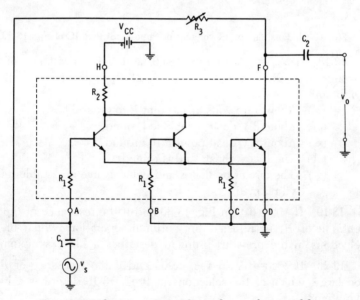

Fig. 13-60. Three-input gate IC wired to work as amplifier.

13-16. By varying the potentiometer R_3 and the output amplitude of the signal source v_s in Fig. 13-60, a variety of output waveforms, as viewed with a scope, are obtained. Which of the waveforms shown in Fig. 13-61 indicates nearly an optimum operating point? Assume that v_s is sinusoidal.

13-17. Suppose that accidentally the right side of the potentiometer R_3, in Fig. 13-60, is connected to point H instead of point F. After adjustment of R_3 an unclipped output signal is obtained. Will the circuit perform as well this way? Why?

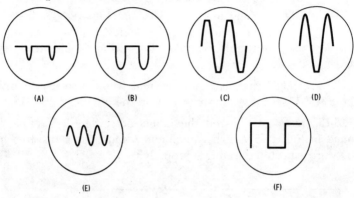

Fig. 13-61. Possible output voltages of three-input gate IC of Fig. 13-60.

13-18. If $V_{CC} = 6$ V in Fig. 13-60, and if the resistance of the potentiometer R_3 is too small, which of the following is a likely symptom?

(a) About zero volts dc at point F to ground.
(b) About 3 V dc at point F to ground.
(c) About 6 V dc at point F to ground.
(d) The IC overheats and is ruined.
(e) The circuit oscillates and behaves like an astable multivibrator.

13-19. If $V_{CC} = 6$ V in Fig. 13-60 and if the resistance of the potentiometer R_3 is adjusted for optimum operation, which of the selections in the previous problem describes a likely symptom?

13-20. If $V_{CC} = 6$ V in Fig. 13-60 and if the resistance of R_3 is too large, which of the selections in Prob. 13-18 describes a likely symptom?

13-21. If the voltage gain of the circuit shown in Fig. 13-60 is about 200, what gain would you expect from two such stages connected in cascade?

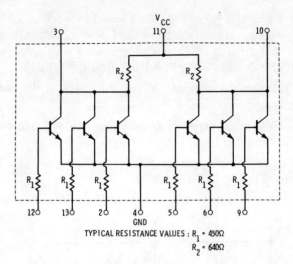

Fig. 13-62. Dual three-input gate.

13-22. Using 100 kΩ base bias resistors and 0.2 μF coupling capacitors, show how you would make an astable multivibrator with the dual three-input gate shown in Fig. 13-62.

13-23. What is the output frequency of the astable multivibrator built with the components described in the previous problem?

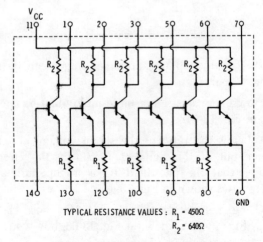

Fig. 13-63. Six-input IC.

13-24. Using the IC in Fig. 13-62 wired to work as a clock and 100 kΩ base bias resistors, which of the following values of coupling capacitors will provide an output square wave with a frequency of about 360 Hz? 3.6 μF, 360 pF, 0.1 μF, 0.02 μF, 20 μF.

13-25. How many separate inverters can you make out of the IC in Fig. 13-63?

13-26. How many separate free-running multivibrators can you make out of the IC in Fig. 13-63?

13-27. How many separate two-input NOR gates can be constructed out of the IC in Fig. 13-64?

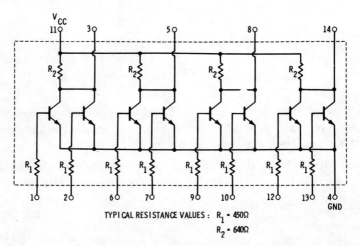

Fig. 13-64. Eight-input IC.

13-28. How many separate amplifier stages can be made with the IC in Fig. 13-64?

13-29. How many separate astable multivibrators can you make with the IC in Fig. 13-64?

13-30. What is the ratio of the signal voltage at pin 6 to the signal voltage out of the signal generator in the circuit of Fig. 13-65. Assume that the packaged IC is an operational amplifier with characteristics shown in Fig. 13-44.

13-31. Using the operational amplifier IC whose characteristics are shown in Fig. 13-44A, show, in Fig. 13-66, how you would wire it to obtain a voltage gain of 75.

PROBLEMS

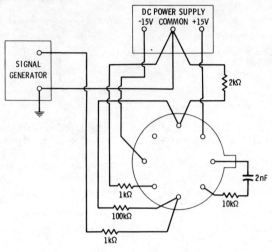

Fig. 13-65. Wiring of operational amplifier circuit.

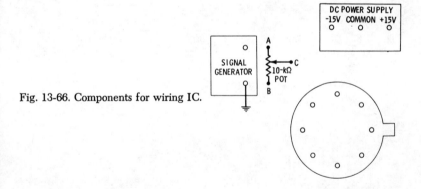

Fig. 13-66. Components for wiring IC.

13-32. The package in the circuit of Fig. 13-67 is an operational amplifier with characteristics shown in Fig. 13-44A. What is the voltage to ground at pin 6?

13-33. Referring to Fig. 13-67 and the previous problem, what is the voltage at pin 6 to ground if the 2.5 mV dc source is replaced with a short to ground.

13-34. An operational amplifier with characteristics shown in Fig. 13-44A is used in the circuit of Fig. 13-68. The 10 kΩ potentiometer is adjusted so that the voltage at B is 30% of the voltage at A. What is the output voltage (at pin 6) to ground?

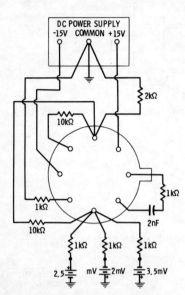
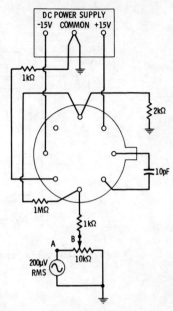

Fig. 13-67. Determining voltage gain and output of operational amplifier.

Fig. 13-68. Using coefficient potentiometer with operational amplifier.

Answer the remaining questions with one of the following: ECL, TTL, DTL, RTL, and MOS.

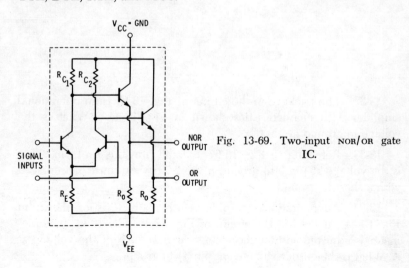

Fig. 13-69. Two-input NOR/OR gate IC.

PROBLEMS

13-35. In which family of IC's does the circuit of Fig. 13-69 belong?

13-36. In which family of IC's does the circuit of Fig. 13-64 belong?

13-37. In which family of IC's does the circuit of Fig. 13-70 belong?

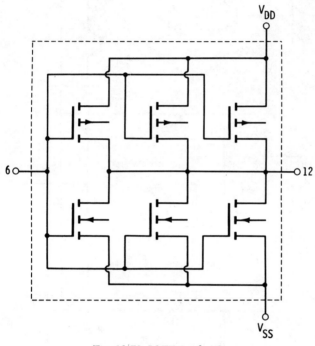

Fig. 13-70. IGFET-only IC.

13-38. In which family of IC's does the circuit of Fig. 13-71 belong?

13-39. In which family of IC's does the circuit of Fig. 13-72 belong?

13-40. In which family of IC's does the circuit of Fig. 13-73 belong?

13-41. In which family of IC's does the circuit of Fig. 13-74 belong?

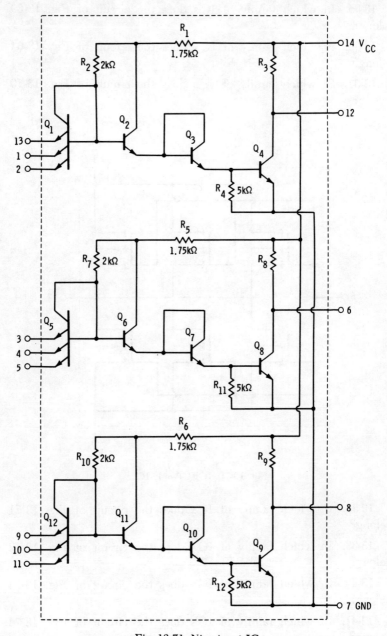

Fig. 13-71. Nine-input IC.

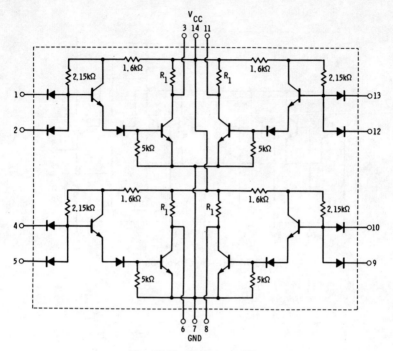

Fig. 13-72. Eight-input IC.

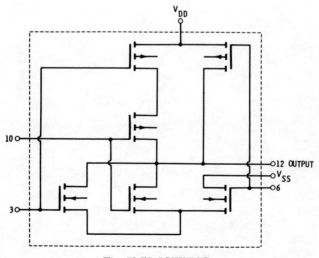

Fig. 13-73. IGFET IC.

514 INTEGRATED CIRCUITS

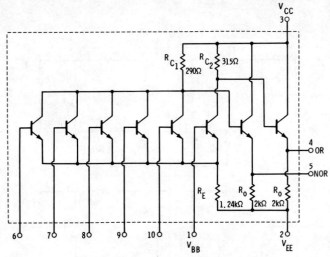

Fig. 13-74. Five-input NOR/OR gate IC.

14

Miscellaneous Solid-State Components

14-1 THE TUNNEL DIODE

The tunnel diode is a two-terminal solid-state device that is usually used for its negative-resistance characteristics. A device has *negative resistance* if a portion of its volt-ampere (*V-I*) characteristics has a negative (decreasing from left to right) slope. For example, note in the typical *V-I* characteristics of the tunnel diode in Fig. 14-1A, that if the forward bias voltage V_F increases from about

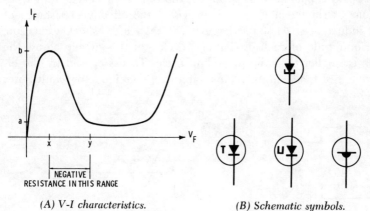

(A) *V-I characteristics.*   (B) *Schematic symbols.*

Fig. 14-1. Tunnel diode data.

x to y, the forward current I_F actually decreases from about b to a. On the other hand, if V_F decreases in the range between x and y, the forward current I_F increases in the domain between a and b. This action is referred to as a "negative-resistance" characteristic because current changes in a direction opposite to what one would expect in a resistance under the same change in voltage. Graphically we can compare the tunnel diode negative resistance V-I curve to the V-I curve of an ordinary (positive) resistance shown in Fig. 14-2. Note that the slope of the curve is positive (increasing from left to right) in Fig. 14-2.

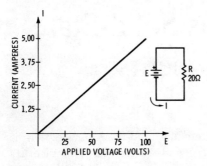

Fig. 14-2. V-I characteristics of 20-Ω resistance.

A useful application of a tunnel diode negative resistance is to use it to cancel a circuit positive resistance effectively. For example, if a dc source is applied momentarily across a tank circuit (inductance and capacitance in parallel), the tank is "shocked" into oscillations as shown in Fig. 14-3, provided the effective resistance of the inductance is not too large. Note that, after shock excitation, the amplitude of each oscillation (cycle) of the voltage across the tank is smaller than the preceding cycle. This is caused by the effective resistance in this tank circuit. (The inductor contributes

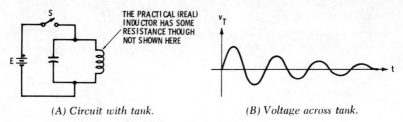

(A) Circuit with tank.   (B) Voltage across tank.

Fig. 14-3. Real inductor and capacitor is parallel produce damped wave after S is momentarily closed.

14-1 THE TUNNEL DIODE

most of the resistance.) This resistance dissipates power as long as the oscillations persist. The oscillations "die out" after all of the power initially delivered to the tank, by the dc source, is consumed. Theoretically, then, if we could have a tank with no resistance, that is with only pure capacitance in parallel, oscillations in it would continue forever after an initial shock as shown in Fig. 14-4. In

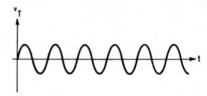

Fig. 14-4. Waveform across shock-excited circuit of pure capacitance and inductance in parallel.

other words, with theoretically no resistance in the circuit, nothing will dissipate the power initially delivered to the tank. This initial power transfers from the magnetic field of the inductor to the electric field of the capacitor and back again from the capacitor to the inductor, repeating forever.

Although a tank made of pure inductance and capacitance is an impossibility, we can simulate one by using a negative-resistance device, like a tunnel diode, along with a practical tank circuit. The negative resistance of the tunnel diode can be used to cancel the effective resistance of the tank, causing it to behave like a tank with no resistance. Fig. 14-5A shows a practical tank and its equivalent circuit. Fig. 14-5B shows an equivalent of a practical tank shunted with a negative resistance device along with the equivalent of this combination. The negative resistance cancels the effect of R_{eq}, resulting in an equivalent circuit containing pure L and C in parallel. Thus, a tunnel diode can cause continuous oscillations in a practical tank circuit.

Since the tunnel diode requires some forward bias voltage in order to act as a negative resistance, a dc bias source is used along with it as is shown in Fig. 14-6. The potentiometer R is adjusted so that the tunnel diode forward voltage is about centered between x and y voltages in Fig. 14-1. Typically, this is only a few tenths of a volt. Capacitor C_1 puts the tunnel diode across the tank and it prevents the diode dc bias voltage from being shorted through the inductor. Its reactance is chosen to be negligible at the frequency of oscillations in the tank. Neglecting stray reactances, we can show that the tank oscillating frequency is

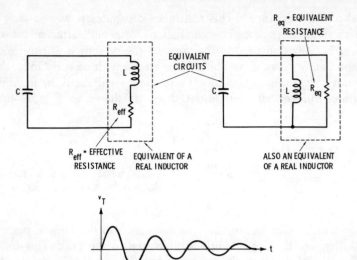

(A) Real inductance can be considered as inductance in series with resistance R_{eff} or in parallel with resistance R_{eq}.

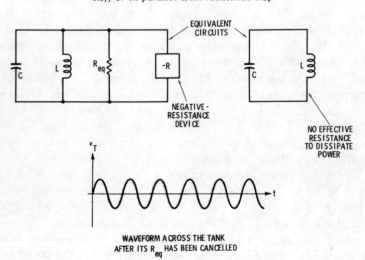

(B) Negative resistance across tank tends to cancel effects of equivalent resistance R_{eq}.

Fig. 14-5. Use of negative resistance to cancel inductor resistance in tank circuit.

$$f \cong \frac{1}{2\pi \sqrt{LC_2}}$$

where L and C_2 are the tank inductance and capacitance respectively. Thus, the frequency of oscillations across the tank in Fig. 14-6 can be chosen or varied by selecting or varying the values of L or C_2.

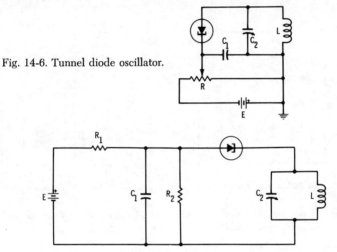

Fig. 14-6. Tunnel diode oscillator.

Fig. 14-7. Tunnel diode oscillator with variable capacitor C_2 which allows output frequency changes.

After adjusting the potentiometer R, in the circuit of Fig. 14-6, so that oscillations start in the tank, the potentiometer can be replaced with a less expensive voltage divider as shown in Fig. 14-7. Resistance R_1 is chosen to be equal to the potentiometer resistance left of the slider. Similarly then, R_2 is made equal to the resistance right of the slider in potentiometer R. The reactance of the capacitor C_1 must be negligible at the oscillating frequency. Capacitor C_1 serves to place the tunnel diode across the tank circuit as signal sees it.

14-2 THE VARACTOR

The varactor is a solid-state capacitor whose capacitance can be varied by changing a dc reverse bias voltage across it.* The varactor

---

*The varactor is generically known as the varactor diode, silicon capacitor, and voltage-variable capacitor. It is also known under trade names as Capsil, Epicap, Paramp Diode, Semicap, Varicap, and Voltacap.

is constructed much like an ordinary diode and contains a pn junction. Like the ordinary diode, a reverse bias voltage depletes the pn junction of its majority carriers (Sec. 2-3). That is, as shown in Fig. 14-8, the holes in the p material are pulled toward the negative

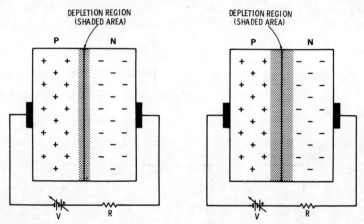

(A) *Depletion region forms at reverse biased pn junction.*

(B) *Larger voltage V causes wider depletion region.*

Fig. 14-8. "Insulating" depletion region separates the p and n materials, causing varactor junction to have capacitance.

terminal of the source V, whereas the electrons are pulled toward the positive terminal. The region near the junction (shaded area) is thus depleted of current carriers and acts much like an insulator. This depleted region becomes wider with larger reverse voltages and, conversely, it becomes narrower with smaller voltages.

Since a reverse bias causes a depleted region to appear between the p and n materials, it also causes the junction to have capacitance. The p material behaves as one plate, the n material as the other plate, and the depleted region acts as the dielectric. Since capacitance varies inversely with the thickness of the dielectric, the pn junction capacitance varies in a somewhat inverse relationship with the reverse bias voltage. That is, with larger reverse voltages, the depletion region widens, causing a decrease in the varactor capacitance. On the other hand, of course, a smaller reverse voltage increases the capacitance. This capacitance versus voltage relationship is not linear, which means that if the reverse voltage is

14-2 THE VARACTOR

doubled, the capacitance does not necessarily decrease to half its previous value. Some symbols for the varactor that are in use are shown in Fig. 14-9.

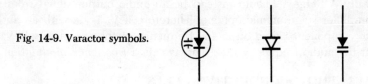

Fig. 14-9. Varactor symbols.

The applications of varactors are numerous and discussion of them is beyond the scope of this book. We can, however, look at one, such as in the circuit of Fig. 14-10, which is a tunnel diode oscillator whose output frequency can be varied by the voltage V_v. The reactances of capacitors C_1 and C_2 are negligible and this effectively places the tunnel diode and the varactor across the

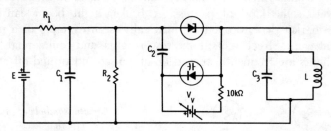

Fig. 14-10. Tunnel diode oscillator frequency is varied by varying voltage V_v.

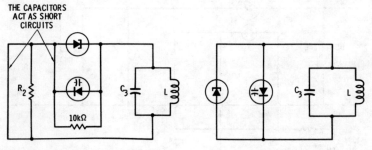

(A) *Capacitive reactances negligible at oscillator frequency.*

(B) *Tunnel diode and varactor effectively across tank.*

Fig. 14-11. Ac equivalent circuits of tunnel diode oscillator in Fig. 14-10.

tank as shown in Fig. 14-11. The varactor capacitance thus is part of the tank total capacitance. If voltage V_r is varied, the tank total capacitance varies, causing the output frequency to vary, too.

The voltage V_r, in Fig. 14-10, is not necessarily varied manually. Voltage V_r could instead be an audio output signal from an amplifier stage or microphone. Therefore if V_r is dc with an audio ac component riding on it, the frequency of the oscillator will vary at the audio frequency rate. This is called *frequency modulation*.

14-3 PHOTOCONDUCTIVE CELLS

A *photoconductive cell* or *photocell*, is a two-terminal resistive device. Its resistance changes if the intensity of light shining on it changes. In absence of light, the cell resistance is maximum. This resistance typically drops to less than a tenth of its maximum when the cell is exposed to light.

Applications of photocells are numerous. They can be used where changes in light intensity can be used, or must be used, to activate some kind of process. A broken light beam causing a supermarket door to open automatically is an example of a photocell at work. Street lights, radio tower lights, and commercial building lights are frequently turned on at sunset and turned off at sun-

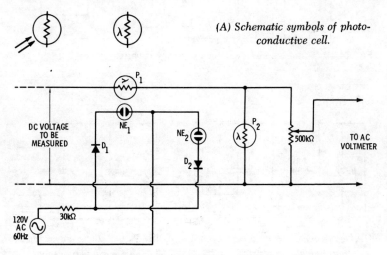

(A) Schematic symbols of photoconductive cell.

(B) Circuit enables dc to be measured with ac voltmeter.

Fig. 14-12. Application of photoconductive cell.

14-3 Photoconductive Cells

rise by electronic equipment that contains photoconductive cells. Symbols used to represent the photoconductive cell are shown in Fig. 14-12A.

A simple but interesting application of photoconductive cells is shown in Fig. 14-12B. This circuit allows us to use a peak-to-peak reading ac voltmeter to determine the value of a dc voltage with fairly good accuracy. This could be useful for measuring small dc voltages, in the millivolt range.

The neon lamps NE_1 and NE_2 are placed physically close to photoconductive cells P_1 and P_2 respectively. Thus, if NE_1 ionizes, its light will cause the resistance of P_1 to decrease significantly. Similarly when NE_2 ionizes, the resistance of P_2 decreases. Diode D_1 allows NE_1 to ionize on positive alternations only, whereas D_2 allows NE_2 to ionize on the negative alternations. This causes cell P_1 to have low resistance and cell P_2 to have high resistance on the positive alternations. Conversely, the resistance of P_1 is high and the resistance of P_2 is low on the negative alternations. This causes a chopped (somewhat square looking) voltage waveform across the 500-kΩ potentiometer. This "square" wave has a peak-to-peak value approximately equal to the dc input magnitude. This circuit therefore converts a pure dc to a dc voltage with an ac component on it. The ac reading voltmeter can respond to this ac component.

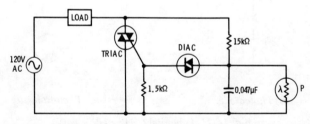

Fig. 14-13. Light beam on photoconductive cell turns load power off.

Another simple application of the photoconductive cell is shown in Fig. 14-13. In this circuit, the presence or absence of light on the cell turns off power or turns on power to the load respectively. That is, in the absence of light the cell resistance is high. This enables the capacitor C to charge to the Diac breakover voltage early in each alternation. When the Diac breaks over, it discharges the capacitor mainly through the gate lead of the Triac, which turns it on and provides a current path through the load. With light on the cell, its resistance is low. This prevents build up of voltage across the capacitor *C*, and the Diac never triggers the Triac into conduction. The load current remains essentially zero.

14-4 PHOTOVOLTAIC (SOLAR) CELLS

When exposed to light, the photovoltaic cell generates a voltage across its two terminals. Thus, it is capable of converting solar energy into electrical energy. By connecting many such cells in various series and parallel arrangements, a variety of output voltages and current capabilities can be obtained. Thus, a bank of photovoltaic cells can be used to convert solar energy from the sun to electrical energy which can be used to operate electronic equipment. In fact, this has been done in manmade satellites and other

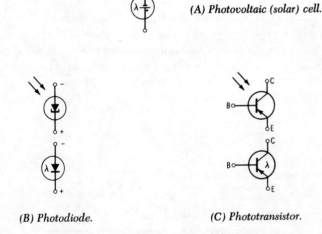

(A) Photovoltaic (solar) cell.

(B) Photodiode.

(C) Phototransistor.

Fig. 14-14. Other light-activated devices.

space vehicles. These solar cells are also used in portable lightmeters for photography, soundtrack pickup in movie projectors, punched tape and card readers in computer hardware, and they have many other applications. A symbol used for the photovoltaic cell is shown in Fig. 14-14A.

14-5 PHOTODIODES AND PHOTOTRANSISTORS

The photodiode is another light-sensitive two-terminal device. Its conductivity changes with changes in the light intensity to

14-5 Photodiodes and Phototransistors

which it is exposed. Its symbols and a simple application are shown in Figs. 14-14B and 14-15 respectively. As shown in Fig. 14-14B, the terminal voltage, whose polarity is shown, increases with an increase in light polarity.

Without light, the photodiode in Fig. 14-15 conducts very little base current I_B and the transistor is practically cut off. Therefore the output voltage $V_o \cong 0$. When the photodiode is exposed to light, the current I_B through it increases, which in turn increases the collector current I_C and the output voltage V_o. For example, a light beam on the photodiode can be broken by moving parts on a conveyor. Thus, each time the beam is broken, the output V_o changes. This output V_o then can be used to drive a counter or to actuate the next phase of the automated industrial process.

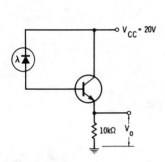

Fig. 14-15. Variations of light on photodiode cause variations in output voltage V_o.

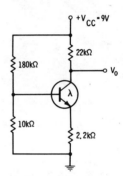

Fig. 14-16. Input light variations cause variations in output voltage V_o.

The phototransistor, whose symbol is shown in Fig. 14-14C, is like an ordinary junction transistor in that it has three terminals: a base, an emitter, and a collector. And like the ordinary junction transistor, the collector current in the phototransistor varies with base current variations. Unlike the ordinary transistor, however, the phototransistor collector current can also be varied directly with changes in light intensity.

The phototransistor can be used as in the circuit of Fig. 14-16. Variations of light on the phototransistor cause collector current variations and consequently the output V_o varies, too. Thus, this circuit (Fig. 14-16) and the photodiode circuit (Fig. 14-15) convert input light variations into output voltage variations.

14-6 THE LIGHT-EMITTING DIODE

The light-emitting diode (LED), sometimes called a *laser diode*, contains a pn junction that emits light when passing a forward bias current. The amount of light output is nearly a linear function of the forward current I_F as shown in Fig. 14-17A. A schematic symbol in use and typical packages for the LED are shown in Figs. 14-17B and 14-17C.

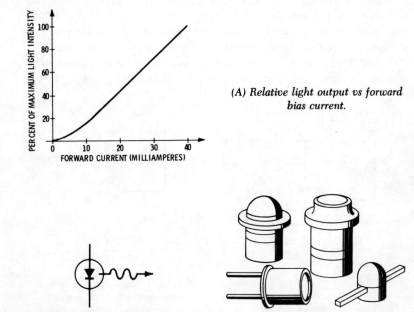

(A) *Relative light output vs forward bias current.*

(B) *Schematic symbol sometimes used to represent LED.*

(C) *Typical packages sometimes used for LED.*

Fig. 14-17. Light-emitting diode.

The LED can be used in the collector lead of a transistor or the drain lead of an FET as shown in the amplifiers in Fig. 14-18. A signal applied to the base or the gate causes the light emitted from the LED to vary at the signal rate. Thus, the light output is said to be modulated by the signal. As shown, the LED's light output can be used to operate light-sensitive devices like the photodiode or phototransistor. In this way a signal can be coupled from one stage to another via a light beam, which allows total electrical isolation between stages.

14-7 THE LIGHT-ACTIVATED THYRISTOR

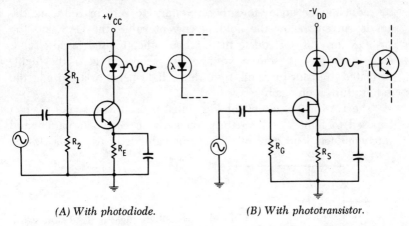

(A) With photodiode.   (B) With phototransistor.

Fig. 14-18. Coupling signal from one stage to another with LED.

14-7 THE LIGHT-ACTIVATED THYRISTOR

In the thyristor family, silicon controlled rectifiers that can be turned on with a light beam are available. Some types have two terminals; others have three. The two-terminal type (Fig. 14-19A) is called a *light-activated reverse blocking diode thyristor* or a *light-activated switch* (*LAS*). It turns on if a beam of light is applied to it. The three-terminal type (Fig. 14-19B) is called a *light-activated reverse blocking triode thyristor* or a *light-activated SCR* (*LASCR*).

(A) Light-activated switch (LAS).   (B) Light-activated SCR (LASCR).

Fig. 14-19. Symbols for light-activated thyristors.

Like an ordinary SCR, the LASCR has three terminals: a cathode, anode, and gate. The LASCR is turned on with either a positive pulse of voltage at the gate or by exposure to a light beam.

Since the LAS and the LASCR are thyristors, they are on-or-off devices. That is, once they are conducting, removal of the light beam or gate voltage will not turn them back off. As with the SCR, by reducing the anode-to-cathode voltage to zero or reducing the anode current below the holding current value, the LAS and the LASCR turn off. Of course then, when either is used in ac circuit, each time the ac source voltage goes through zero the light-activated thyristor turns off, requiring a light beam or a pulse at the gate to turn it on again.

The LAS in the circuit of Fig. 14-20 is used as a switch. When exposed to light, the LAS turns on early in *each* alternation and thus conducts load current whose rms value is nearly equal to the

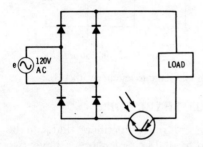

Fig. 14-20. Light on LAS admits load current; no light causes current to cease.

rms current that would flow if the source *e* were placed directly across the load. If the light is removed, the LAS turns off when the source voltage *e* passes through zero and therefore load current ceases. The bridge rectifier allows the LAS to conduct on every alternation. That is, it can conduct only when the anode is positive with respect to the cathode, whereas it cannot be made to conduct normally when the anode is negative.

The circuit in Fig. 14-21 is an interesting application of the LASCR. The potentiometer R_1 is adjusted so that the UJT triggers the LASCR late in each alternation. Thus, in absence of light on

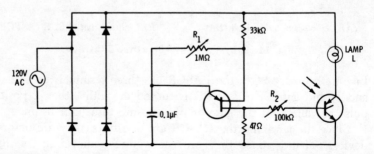

Fig. 14-21. Light turns on lamp with filament protected against surge currents.

the LASCR, a current with a waveform shown in Fig. 14-22A flows in the lamp. The rms value of this current is relatively small and the lamp filament gets hot but not hot enough to cause it to glow appreciably. When a light beam is applied to the LASCR, it turns on at the beginning of each alternation, resulting in a lamp current waveform as shown in Fig. 14-22B. This current has a relatively large rms value and causes the lamp to glow brightly.

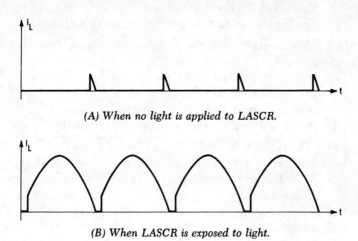

(A) When no light is applied to LASCR.

(B) When LASCR is exposed to light.

Fig. 14-22. Lamp current waveforms for LASCR lamp circuit of Fig. 14-21.

This circuit minimizes thermal stress on the lamp filament. That is, by keeping the filament hot always, the filament resistance is kept considerably larger than its cold resistance. This prevents a large surge current at the instant the LASCR turns on, which lengthens the life of the filament.

PROBLEMS

14-1. Suppose that after building the tunnel diode oscillator of Fig. 14-6, it fails to start oscillating when the source voltage E is connected. Which of the following is likely the problem?
 (a) The tunnel diode is definitely defective.
 (b) The source voltage E is likely defective.
 (c) The potentiometer R needs adjusting.
 (d) The ground is disconnected.

14-2. Which of the following statements *is not* true concerning the tunnel diode oscillator circuit of Fig. 14-7?
  (a) The frequency of the sine wave output can be varied by varying the tank capacitance C.
  (b) The tunnel diode is biased to operate in the center of its negative-resistance characteristics.
  (c) From the point of view of the oscillating signal, the tunnel diode is shunted across the tank.
  (d) The purpose of R_2 is to prevent excessive oscillations.

14-3. Which of the following statements *is not* true concerning the varactor?
  (a) It is normally used with forward bias dc voltages.
  (b) Its capacitance varies as the dc voltage varies.
  (c) It is also known as a silicon capacitor.
  (d) Its capacitance is larger with smaller reverse bias voltages.

14-4. The varactor in circuit Fig. 14-10:
  (a) Supplies dc bias to the tunnel diode.
  (b) Provides negative resistance to the circuit to keep the tank oscillating.
  (c) Varies the frequency of the output if the voltage V_v is varied.
  (d) Regulates the voltage across the tank in case voltage V_v is varied.

14-5. Which of the following statements is incorrect referring to the circuit of Fig. 14-12B? (More than one selection may be incorrect.)
  (a) The ac voltmeter measures the value of the sine wave output voltage.
  (b) Neon lamps NE_1 and NE_2 ionize and deionize alternately 60 times a second.
  (c) When NE_1 is ionized, NE_2 is deionized, and vice versa.
  (d) Both photocells, P_1 and P_2, change from high resistance to low resistance for an instant, then both change from low resistance to high resistance the next instant.

14-6. The main function of the circuit in Fig. 14-12B is to:
  (a) Provide blinking neon lamps.
  (b) Test photocells to see if they are working properly.
  (c) Rectify the 120-V source voltage.
  (d) Enable us to measure dc voltages with an ac voltmeter.

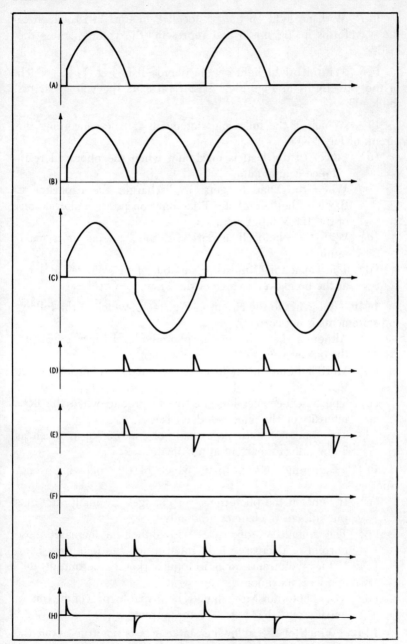

Fig. 14-23. Possible waveforms of photocell and LAS circuits.

14-7. With no light on the photocell P in Fig. 14-13, which of the waveforms in Fig. 14-23 best represents the voltage across the load?

14-8. With a bright light on the photocell in Fig. 14-13, which of the waveforms in Fig. 14-23 best represents the voltage across the load?

14-9. Which of the following statements is true concerning the circuit of Fig. 14-13?
  (a) Power to the load is maximum when the photocell resistance is minimum.
  (b) When the Diac triggers, it discharges the capacitor C through the gate of the Triac only on positive alternations of the 120 V input voltage.
  (c) With the photocell in darkness, the load power is maximum.
  (d) The capacitor C must be capable of dc voltages as large as the breakover voltage of the Triac.

14-10. Referring to the circuit in Fig. 14-15, which of the following statements is not correct?
  (a) Although the light on the photodiode changes gradually, the output voltage V_o changes abruptly.
  (b) More light on the photocell causes increased output voltage V_o.
  (c) The base becomes less positive to ground when the light intensity on the photocell decreases.
  (d) The transistor must be capable of conducting as much as 2 mA collector current approximately.

14-11. Referring to Fig. 14-16, which of the following is not true?
  (a) If light on the phototransistor changes gradually, the output voltage V_o changes gradually.
  (b) The voltage V_o decreases with either an increased base current or decreased light intensity on the phototransistor.
  (c) If the phototransistor is in total darkness, the output voltage V_o is maximum.
  (d) The phototransistor should be capable of conducting a maximum of 0.37 mA collector current.

14-12. Which of the following statements is not true about the light-emitting diode?

Problems

(a) Its light output intensity is almost linearly related to the magnitude of the forward current through it.
(b) Variations in the forward bias current through the LED cause variations in its light output.
(c) Larger forward bias currents through the LED cause decreased light output intensities.
(d) LED's can be used to provide modulated light outputs.

14-13. Suppose that you are required to use a quiescent current of 10 mA in the LED in the circuit of Fig. 14-18A. What circuit resistance values will provide this if $V_{CC} = 20$ V?

(a) $R_1 = 180$ kΩ, $R_2 = 20$ kΩ, $R_E = 2$ kΩ.
(b) $R_1 = 18$ kΩ, $R_2 = 2$ kΩ, $R_E = 200$ Ω.
(c) $R_1 = 200$ kΩ, $R_2 = 20$ kΩ, $R_E = 200$ Ω.
(d) $R_1 = 20$ kΩ, $R_2 = 2$ kΩ, $R_E = 2$ kΩ.

14-14. Which of the following is not true concerning Fig. 14-18B?

(a) Changing the value of R_G will have little effect on the average light output of the LED.
(b) Changing the value of R_s will have little or no effect on the average light output of the LED.
(c) The output light intensity of the LED increases on the positive alternations of the input signal and decreases on negative alternations.
(d) An input signal v_s causes a modulated light output.

14-15. Which of the following statements is true concerning the LAS and the LASCR?

(a) Their anode currents can be gradually varied by varying the light intensity on them.
(b) The LAS can be used to replace the photocell whereas the LASCR can be used to replace the phototransistor in all applications.
(c) Both devices tend to turn off with more light exposure.
(d) Both devices normally have two states: conducting or nonconducting.

14-16. In absence of light on the LAS in Fig. 14-20, which of the waveforms in Fig. 14-23 best represents the load voltage?

14-17. With light on the LAS in Fig. 14-20, which of the waveforms in Fig. 14-23 best represents the load voltage?

14-18. What is the main function of the circuit in Fig. 14-21?
  (a) It is a lamp dimmer circuit with light feedback from the lamp to the LASCR for overload protection.
  (b) It provides a small average current through the lamp filament, thus warming it prior to turn-on of the LASCR with a light beam.
  (c) The lamp indicates overload on the bridge rectifier.
  (d) It is a self-regulating lamp control; more light from the lamp turns off the LASCR whereas less light turns the LASCR on, thus providing constant current through the lamp.

Appendix

Derivations of Constants and Equations

A-1 THE HALF-WAVE RECTIFIER

If the diode is ideal in a half-wave rectifier circuit as in Fig. 4-11A (see Fig. A-1), the average dc voltage E_{dc} across the load R can be found with the general equation

$$E_{dc} = \frac{1}{T} \int_{t_1}^{t_2} e_s(t)\, dt$$

where

T is the period of the source e_s,

t_1 and t_2 are the instants of time within which the voltage appears across R.

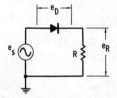

Fig. A-1. Half-wave rectifier circuit.

Specifically, if e_s is sinusoidal, the time of each cycle $T = 2\pi$ radians and the voltage across R appears during the first half of each cycle; that is, within times 0 to π radians. Also, the independent variable here is $2\pi ft = \omega t$ rather than simply t. Therefore

$$E_{dc} = \frac{1}{2\pi} \int_0^\pi E_m \sin \omega t \, d(\omega t)$$

where E_m is the peak value of e_s. Factoring out the constant E_m and proceeding with the integration, we can show that

$$E_{dc} = \frac{E_m}{2\pi} \int_0^\pi \sin \omega t \, d(\omega t) = \frac{E_m}{2\pi} \Big[-\cos \omega t \Big]_0^\pi$$

$$= \frac{E_m}{2\pi} \Big[-\cos \pi - (-\cos 0) \Big] = \frac{E_m}{2\pi} \Big[1 + 1 \Big]$$

so that

$$E_{dc} = \frac{E_m}{\pi} \cong 0.318 \, E_m \qquad (4\text{-}1A)$$

A-2 THE BRIDGE RECTIFIER (FULL-WAVE)

When dealing with a full-wave bridge rectifier as was shown in Fig. 4-22A (see Fig. A-2A) voltage appears across the load R on both alternations of e_s as was shown in Fig. 4-22C (see Fig.

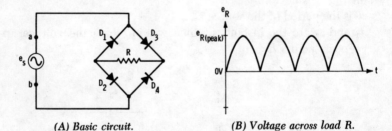

(A) Basic circuit. (B) Voltage across load R.

Fig. A-2. Full-wave bridge rectifier.

A-2B). The dc average of one alternation is equal to the dc average of any number of similar alternations. Thus, for simplicity, we can integrate the first alternation (0 to π radians) to find E_{dc} across the load R. If e_s is sinusoidal and the diodes are ideal,

$$E_{dc} = \frac{1}{T} \int_{t_1}^{t_2} e_s(t) \, dt = \frac{1}{\pi} \int_0^\pi E_m \sin \omega t \, d(\omega t)$$

A-3 THE DYNAMIC EMITTER-TO-BASE RESISTANCE

$$E_{dc} = \frac{E_m}{\pi} \int_0^\pi \sin \omega t \, d(\omega t) = \frac{E_m}{\pi} \Big[-\cos \omega t \Big]_0^\pi$$

$$= \frac{E_m}{\pi} \Big[-\cos \pi - (-\cos 0) \Big] = \frac{2 E_m}{\pi} \qquad (4\text{-}4A)$$

We can use $T = \pi$ in this case because the voltage waveform across the load R repeats itself every π radians.

A-3 THE DYNAMIC EMITTER-TO-BASE RESISTANCE

The dynamic resistance r_e' seen looking into the emitter of a forward biased base-emitter junction can be found with the equation

$$r_e' = \frac{kT/q}{I_E}$$

where

k is Boltzmann's constant (1.38×10^{-16} erg/K),
T is the absolute temperature in kelvins,
q is the charge of electron $= 1.602 \times 10^{-19}$ coulomb,
I_E is the emitter current in amperes.

Here the absolute temperature T is expressed as so many kelvins, which is 273° plus the Celsius temperature. Therefore, a room temperature of 25°C can be expressed as 298 K. At room temperature, the dynamic resistance r_e' is given as

$$r_e' = \frac{kT/q}{I_E} \cong \frac{26 \text{ mV}}{I_E}$$

which may be expressed, to keep r_e' in convenient numbers, as

$$r_e' \cong \frac{25 \text{ mV}}{I_E} \qquad (7\text{-}3)$$

Typically, this dynamic resistance is in the range

$$\frac{25 \text{ mV}}{I_E} \leq r_e' \leq \frac{50 \text{ mV}}{I_E}$$

A-4 THE END POINTS OF AN AC LOAD LINE

In general form a linear equation can be written as follows:

$$y = mx + b \qquad (A\text{-}1)$$

where
- y is the value of the vertical component,
- x is the value of the horizontal component,
- b is the y-intercept of the line.

Now we can introduce the *slope* of the equation:

$$m = \frac{y_2 - y_1}{x_2 - x_1} = \frac{\Delta y}{\Delta x}$$

where (x_1, y_1) and (x_2, y_2) are the coordinates of any two points on the line given by $y = mx + b$.

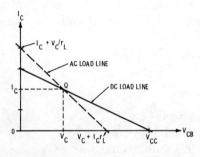

Fig. A-3. Finding end points of ac load line.

The ac load line in Fig. 7-32 (see Fig. 4-3) is a linear function with which we can use i_C to represent values on the vertical axis, and v_C to represent values on the horizontal axis. Let $i_{C(\text{sat})}$ be the point where the ac load line crosses the I_C axis, which is the y-intercept. Thus, Eq. A-1 above can be shown equivalent to

$$i_C = \frac{\Delta i_C}{\Delta v_C} v_C + i_{C(\text{sat})} \qquad (A\text{-}2)$$

where the slope m is such that

$$\left| \frac{\Delta i_C}{\Delta v_C} \right| = \frac{1}{r_L}$$

and $\Delta i_C / \Delta v_C$ is negative in Eq. A-2 above because the ac load line has a negative slope; that is, it slopes down from left to right.

Thus, Eq. A-2 above can be shown and rearranged as follows:

$$i_C = -\frac{v_C}{r_L} + i_{C(\text{sat})} \qquad (A\text{-}3)$$

or

A-5 DERIVATIONS OF EQS. 8-10 AND 8-12

$$i_{C(\text{sat})} = i_C + \frac{v_C}{r_L} \quad (A-4)$$

Now if we know the coordinates of some points on the load line, say the operating point, whose coordinates in general are I_C and V_C, we can substitute them into Eq. A-4 above and get

$$i_{C(\text{sat})} = I_C + \frac{V_C}{r_L} \quad (A-5)$$

At the x-intercept, i.e., the point where the ac load line crosses the horizontal (V_{CB}) axis, the coordinates are

$$i_C = 0, \qquad v_C = v_{C(\text{cutoff})}$$

Substituting these and Eq. A-5 into Eq. A-3, we get

$$0 = -\frac{v_{C(\text{cutoff})}}{r_L} + I_C + \frac{V_C}{r_L}$$

Rearranging the above, to solve for $v_{C(\text{cutoff})}$, we can show that

$$\frac{v_{C(\text{cutoff})}}{r_L} = I_C + \frac{V_C}{r_L}$$

Thus,

$$V_{C(\text{cutoff})} = r_L I_C + V_C = V_C + r_L I_C \quad (A-6)$$

Eqs. A-5 and A-6 are very useful when sketching the ac load line of a transistor amplifier on its collector characteristics.

A-5 DERIVATIONS OF EQS. 8-10 AND 8-12

Referring to the circuit in Fig. 8-4 (see Fig. A-4) which uses emitter feedback, Eq. 8-10, which can be used to determine the circuit quiescent collector current, is found as follows: Writing a Kirchhoff's voltage equation for the loop containing components V_{CC}, R_B and R_E, we get

$$V_{CC} = R_B I_B + R_E I_E + V_{BE}$$

Since $I_E = I_C + I_B$ and $V_{BE} \cong 0$, then

$$V_{CC} \cong R_B I_B + R_E(I_C + I_B) = R_B I_B + R_E I_C + R_E I_B = (R_B + R_E) I_B + R_E I_C \quad (A-7)$$

By Eq. 8-6 $I_C = \beta I_B + I_{CEO}$. Therefore, by rearranging terms, we can show that

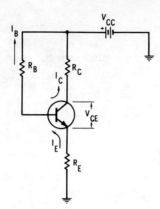

Fig. A-4. Emitter feedback circuit.

$$I_B = \frac{I_C - I_{CEO}}{\beta}$$

Substituting this last equation in Eq. A-7 we get

$$V_{CC} \cong (R_B + R_E)\left(\frac{I_C - I_{CEO}}{\beta}\right) + R_E I_C = \frac{R_B I_C}{\beta}$$

$$- \frac{R_B I_{CEO}}{\beta} + \frac{R_E I_C}{\beta} - \frac{R_E I_{CEO}}{\beta} + R_E I_C$$

or

$$V_{CC} \cong \left(\frac{R_B}{\beta} + \frac{R_E}{\beta} + R_E\right) I_C - I_{CEO}\left(\frac{R_B + R_E}{\beta}\right)$$

Solving the above for I_C, we have

$$I_C \cong \frac{V_{CC} + I_{CEO}\left(\frac{R_B + R_E}{\beta}\right)}{(R_B + R_E + \beta R_E)/\beta} = \frac{V_{CC}}{(R_E + R_B + \beta R_E)/\beta} + \frac{I_{CEO}(R_B + R_E)}{R_B + R_E + \beta R_E}$$

If $\beta \gg 1$, then $1 \ll \beta$ and hence $R_E \ll \beta R_E$ and R_E may be dropped from the denominators. Now dividing $R_B + \beta R_E$ by β in the denominator of the first term,

$$I_C \cong \frac{V_{CC}}{R_E + R_B/\beta} + \left(\frac{R_B + R_E}{R_B + \beta R_E}\right) I_{CEO} \qquad (8\text{-}10)$$

Now by substituting Eq. 8-6 into the left side of Eq. 8-10, shown above,

A-6 DERIVATIONS OF EQS. 8-16 AND 8-18

$$\beta I_B + I_{CEO} \cong \frac{V_{CC}}{R_E + R_B/\beta} + \left(\frac{R_B + R_E}{R_B + \beta R_E}\right) I_{CEO}$$

Solving the above for I_B we get

$$I_B \cong \frac{V_{CC}}{\beta R_E + R_B} + \frac{R_B + R_E}{\beta(R_B + \beta R_E)} I_{CEO} - \frac{I_{CEO}}{\beta}$$

Factoring out I_{CEO},

$$I_B \cong \frac{V_{CC}}{\beta R_E + R_B} + \left(\frac{R_B + R_E}{\beta(R_B + \beta R_E)} - \frac{1}{\beta}\right) I_{CEO}$$

$$\cong \frac{V_{CC}}{\beta R_E + R_B} + \left(\frac{R_B + R_E}{\beta(R_B + \beta R_E)} - \frac{R_B + \beta R_E}{\beta(R_B + \beta R_E)}\right) I_{CEO}$$

$$\cong \frac{V_{CC}}{\beta R_E + R_B} + \left(\frac{(1-\beta)R_E}{\beta(R_B + \beta R_E)}\right) I_{CEO}$$

If $\beta \gg 1$, the above simplifies to

$$I_B \cong \frac{V_{CC}}{\beta R_E + R_B} - \left(\frac{R_E}{R_B + \beta R_E}\right) I_{CEO} \qquad (8\text{-}12)$$

A-6 DERIVATIONS OF EQS. 8-16 AND 8-18

Referring to Fig. 8-5 (see Fig. A-5) and writing a Kirchhoff's voltage equation for the loop containing components V_{CC}, R_C, R_B, and the forward biased base-emitter junction, we get

$$V_{CC} = R_C I_C + R_B I_B + V_{BE}$$

Assuming that $V_{BE} \cong 0$ and by solving for I_B we can show that

$$V_{CC} \cong R_C I_C + R_B I_B$$

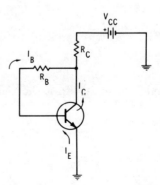

Fig. A-5. Collector feedback circuit.

and that

$$I_B \cong \frac{V_{CC} - R_C I_C}{R_B} \quad (A\text{-}8)$$

Now by solving Eq. 8-6 for I_B we find that since $I_C = \beta I_B + I_{CEO}$, then

$$I_B = \frac{I_C - I_{CEO}}{\beta} \quad (A\text{-}9)$$

Now setting the right sides of Eq. A-8 and A-9 equal to each other:

$$\frac{V_{CC} - R_C I_C}{R_B} \cong \frac{I_C - I_{CEO}}{\beta}$$

or

$$\beta V_{CC} - \beta R_C I_C \cong R_B I_C - R_B I_{CEO}$$

so that

$$R_B I_C + \beta R_C I_C \cong \beta V_{CC} + R_B I_{CEO}$$

Now solving for I_C

$$(R_B + \beta R_C) I_C \cong \beta V_{CC} + R_B I_{CEO}$$

$$I_C \cong \frac{\beta V_{CC} + R_B I_{CEO}}{R_B + \beta R_C} = \frac{\beta V_{CC}}{R_B + \beta R_C} + \left(\frac{R_B}{R_B + \beta R_C}\right) I_{CEO}$$

Using β to divide the numerator and denominator of the first term in the expression at the right side of the equals sign above, we finally get

$$I_C \cong \frac{V_{CC}}{R_C + R_B/\beta} + \left(\frac{R_B}{R_B + \beta R_C}\right) I_{CEO} \quad (8\text{-}16)$$

If we start with the Kirchhoff's voltage equation again,

$$V_{CC} \cong R_C I_C + R_B I_B$$

and substitute into it the right side of Eq. 8-6, i.e., $I_C = \beta I_B + I_{CEO}$, we get

$$V_{CC} \cong R_C(\beta I_B + I_{CEO}) + R_B I_B$$

or

$$V_{CC} \cong R_C \beta I_B + R_C I_{CEO} + R_B I_B$$

A-7 DERIVATIONS OF EQS. 8-23 AND 8-26

or

$$V_{CC} \cong R_C \beta I_B + R_B I_B + R_C I_{CEO} = (\beta R_C + R_B)I_B + R_C I_{CEO}$$

Solving this last equation for I_B we get

$$I_B \cong \frac{V_{CC} - R_C I_{CEO}}{\beta R_C + R_B}$$

or

$$I_B \cong \frac{V_{CC}}{\beta R_C + R_B} - \left(\frac{R_C}{\beta R_C + R_B}\right) I_{CEO} \quad (8\text{-}18)$$

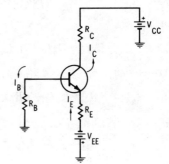

Fig. A-6. Emitter feedback circuit with two voltage sources.

A-7 DERIVATIONS OF EQS. 8-23 AND 8-26

Referring to Fig. 8-6 (see Fig. A-6) and writing a Kirchhoff's voltage equation for the loop continuing components V_{EE}, R_E, R_B, and the forward biased base emitter junction, we get

$$V_{EE} = R_B I_B + R_E I_E + V_{BE} \quad (\text{A-}10)$$

Since $I_E = I_C + I_B$ and assuming that V_{BE} is negligible, Eq. A-10 above can be rewritten as

$$V_{EE} \cong R_B I_B + R_E(I_C + I_B) = R_B I_B + R_E I_C + R_E I_B$$

By rearranging and factoring out I_B we can show that

$$V_{EE} \cong (R_B + R_E)I_B + R_E I_C \quad (\text{A-}11)$$

Now since $I_C = \beta I_B + I_{CEO}$ (Eq. 8-6), then

$$I_B = \frac{I_C - I_{CEO}}{\beta} \quad (\text{A-}12)$$

Solving Eq. A-11 above for I_B we get

$$I_B \cong \frac{V_{EE} - R_E I_C}{R_B + R_E} \quad \text{(A-13)}$$

Setting the right sides of Eqs. A-12 and A-13 equal to each other and solving for I_C:

$$\frac{I_C - I_{CEO}}{\beta} \cong \frac{V_{EE} - R_E I_C}{R_B + R_E}$$

or

$$(I_C - I_{CEO})(R_B + R_E) \cong \beta V_{EE} - \beta R_E I_C$$

or

$$R_B I_C + R_E I_C - R_B I_{CEO} - R_E I_{CEO} \cong \beta V_{EE} - \beta R_E I_C$$

Transposing $\beta R_E I_C$ and $R_B I_{CEO}$, and $R_E I_{CEO}$,

$$R_B I_C + R_E I_C + \beta R_E I_C \cong \beta V_{EE} + R_B I_{CEO} + R_E I_{CEO}$$

or

$$(R_B + R_E + \beta R_E) I_C \cong \beta V_{EE} + (R_B + R_E) I_{CEO} \quad \text{(A-14)}$$

If $\beta \gg 1$ then $R_E + \beta R_E \cong \beta R_E$. Substituting the latter into Eq. A-14 above and solving for I_C we can show that

$$(R_B + \beta R_E) I_C \cong \beta V_{EE} + (R_B + R_E) I_{CEO}$$

so that

$$I_C \cong \frac{\beta V_{EE} + (R_B + R_E) I_{CEO}}{R_B + \beta R_E}$$

$$\cong \frac{\beta V_{EE}}{R_B + \beta R_E} + \left(\frac{R_B + R_E}{R_B + \beta R_E}\right) I_{CEO}$$

Dividing the numerator and denominator of the first term on the right above by β results in

$$I_C \cong \frac{V_{EE}}{R_E + R_B/\beta} + \left(\frac{R_B + R_E}{R_B + \beta R_E}\right) I_{CEO} \quad \text{(8-23)}$$

Now if we solve Eq. A-11 for I_C we can set it equal to the right side of Eq. 8-6; that is, by Eq. A-11,

$$V_{EE} \cong (R_B + R_E) I_B + R_E I_C$$

A-8 DERIVATIONS OF EQS. 8-34 AND 8-36

then

$$I_C \cong \frac{V_{EE} - (R_B + R_E)I_B}{R_E}$$

Therefore

$$\beta I_B + I_{CEO} \cong \frac{V_{EE} - (R_B + R_E)I_B}{R_E}$$

or

$$\beta R_E I_B + R_E I_{CEO} \cong V_{EE} - (R_B + R_E)I_B$$

Rearranging and factoring gives

$$\beta R_E I_B + (R_B + R_E)I_B \cong V_{EE} - R_E I_{CEO}$$

and,

$$(\beta R_E + R_E + R_B)I_B \cong V_{EE} - R_E I_{CEO}$$

Now isolating I_B shows that

$$I_B \cong \frac{V_{EE} - R_E I_{CEO}}{\beta R_E + R_E + R_B}$$

If $\beta \gg 1$, the above simplifies to

$$I_B \cong \frac{V_{EE} - R_E I_{CEO}}{\beta R_E + R_B} = \frac{V_{EE}}{\beta R_E + R_B} - \left(\frac{R_E}{\beta R_E + R_B}\right) I_{CEO} \quad (8\text{-}26)$$

A-8 DERIVATIONS OF EQS. 8-34 AND 8-36

The base bias circuitry, R_B and V_B, shown in Fig. A-7B is a Thevenin's equivalent of the base bias circuitry in Fig. A-7A. That is, Figs. A-7A and A-7B show equivalents, where, by Thevenin's theorem,

$$R_B = \frac{R_1 R_2}{R_1 + R_2} \quad (A\text{-}15)$$

and

$$V_B = \frac{V_{CC} R_2}{R_1 + R_2} \quad (A\text{-}16)$$

Writing a Kirchhoff's voltage equation for the loop containing components V_B, R_B, R_E, and the forward biased base-emitter junction, in Fig. A-7B, we get

$$V_B = R_B I_B + R_E I_E + V_{BE}$$

Assuming that $V_{BE} \cong 0$, the above equation simplifies to

$$V_B \cong R_B I_B + R_E I_E \qquad (A\text{-}17)$$

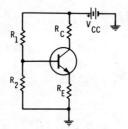
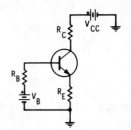

(A) *With voltage divider bias.* (B) *With voltage supply bias.*

Fig. A-7. Circuits having equivalent base bias circuitry.

Now since $I_E = I_C + I_B$ we can substitute the right-hand side of this equation into Eq. A-17 along with Eqs. A-15 and A-16. The result is

$$\frac{V_{CC} R_2}{R_1 + R_2} \cong \left(\frac{R_1 R_2}{R_1 + R_2} \right) I_B + R_E (I_C + I_B) \qquad (A\text{-}18)$$

Working to isolate I_B we can show

$$\frac{V_{CC} R_2}{R_1 + R_2} - R_E I_C \cong \left(\frac{R_1 R_2}{R_1 + R_2} + R_E \right) I_B$$

or

$$\frac{V_{CC} R_2 - R_E I_C (R_1 + R_2)}{R_1 + R_2} \cong \left(\frac{R_1 R_2 + R_E (R_1 + R_2)}{R_1 + R_2} \right) I_B$$

or

$$I_B \cong \frac{V_{CC} R_2 - R_E I_C (R_1 + R_2)}{R_1 R_2 + R_E (R_1 + R_2)}$$

Setting the right-hand side of the above equation equal to the right-hand side of Eq. 8-6,

$$I_B = \frac{I_C - I_{CEO}}{\beta}$$

we get

A-8 Derivations of Eqs. 8-34 and 8-36

$$\frac{V_{CC}R_2 - R_E I_C(R_1+R_2)}{R_1R_2 + R_E(R_1+R_2)} \cong \frac{I_C - I_{CEO}}{\beta}$$

or

$$\beta V_{CC}R_2 - \beta R_E I_C(R_1+R_2) \cong (I_C - I_{CEO})[R_1R_2 + R_E(R_1+R_2)]$$
$$\cong R_1R_2 I_C + R_E I_C(R_1+R_2) - R_1R_2 I_{CEO}$$
$$- R_E I_{CEO}(R_1+R_2)$$

which rearranged becomes

$$\beta V_{CC}R_2 + R_1R_2 I_{CEO} + R_E I_{CEO}(R_1+R_2) \cong R_1R_2 I_C + \beta R_E I_C(R_1+R_2)$$
$$+ R_E I_C(R_1+R_2)$$

Now if $R_1R_2 \gg R_E(R_1+R_2)$ and $30 \leq \beta \leq \infty$, which typically is the case, we simplify and solve for I_C as follows:

$$\beta V_{CC}R_2 + R_1R_2 I_{CEO} \cong R_1R_2 I_C + \beta R_E I_C(R_1+R_2)$$

$$I_C \cong \frac{\beta V_{CC}R_2 + R_1R_2 I_{CEO}}{R_1R_2 + \beta R_E(R_1+R_2)}$$

A relatively large β makes $R_1R_2 \ll \beta R_E(R_1+R_2)$ which allows us to simplify the previous equation further:

$$I_C \cong \frac{\beta V_{CC}R_2 + R_1R_2 I_{CEO}}{\beta R_E(R_1+R_2)}$$

or

$$I_C \cong \frac{\beta V_{CC}R_2}{\beta R_E(R_1+R_2)} + \left[\frac{R_1R_2}{\beta R_E(R_1+R_2)}\right] I_{CEO} \qquad (8\text{-}34)$$

Now we can substitute the right-hand side of $I_C = \beta I_B + I_{CEO}$ into the previous equation (Eq. 8-34) and solve for I_B:

$$\beta I_B + I_{CEO} \cong \frac{V_{CC}R_2}{R_E(R_1+R_2)} + \left[\frac{R_1R_2}{\beta R_E(R_1+R_2)}\right] I_{CEO}$$

or

$$I_B \cong \frac{V_{CC}R_2}{\beta R_E(R_1+R_2)} + \left[\frac{R_1R_2}{\beta R_E(R_1+R_2)}\right]\frac{I_{CEO}}{\beta} - \frac{I_{CEO}}{\beta}$$

Since typically $R_1R_2 \ll \beta R_E(R_1+R_2)$,

$$\left[\frac{R_1R_2}{\beta R_E(R_1+R_2)}\right] \ll 1$$

the above simplifies to

$$I_B \cong \frac{V_{CC}R_2}{\beta R_E(R_1+R_2)} - \frac{I_{CEO}}{\beta} \qquad (8\text{-}36)$$

A-9 DERIVATION OF EQ. 11-8

From the expression

$$\frac{V_{s(\max)}R_4}{R_3+R_{B1B2}+R_4} < V_{GT}$$

which was derived in Chap. 11, we can rearrange to get

$$V_{s(\max)}R_4 < V_{GT}(R_3+R_{B1B2}+R_4)$$

By the distributive principle,

$$V_{s(\max)}R_4 < V_{GT}R_3 + V_{GT}R_{B1B2} + V_{GT}R_4$$

Collecting terms with the factor R_4,

$$V_{s(\max)}R_4 - V_{GT}R_4 < V_{GT}R_3 + V_{GT}R_{B1B2}$$

Factoring

$$R_4(V_{s(\max)} - V_{GT}) < V_{GT}(R_3+R_{B1B2})$$

So finally

$$R_4 < \frac{V_{GT}(R_3+R_{B1B2})}{V_{s(\max)} - V_{GT}} \qquad (11\text{-}8)$$

A-10 SELECTING BYPASS AND COUPLING CAPACITORS

When one is selecting a coupling or a bypass capacitor, its reactance is chosen to be about one-tenth of the resistance seen from the point of view of the capacitor. This equivalent resistance R_{eq} is viewed after the circuit voltage sources are replaced by their equivalent resistances. In solving for the capacitance needed to provide this relatively low reactance, the lowest intended operating frequency f is used. For example, in Fig. A-8, two common-base amplifiers are coupled by capacitor C. Capacitor C sees the equivalent circuits shown in Fig. A-9. Therefore, as shown in Fig. A-9B, capacitor C sees an equivalent resistance of

$$R_{eq} = R_{C1} + R_{E2} \| r_e'$$

Thus, the reactance of C must be

A-9 DERIVATION OF EQ. 11-8

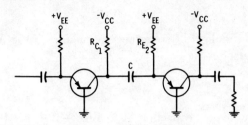

Fig. A-8. Common-base amplifier with capacitor coupling.

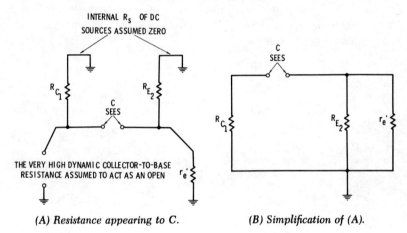

(A) Resistance appearing to C.   *(B) Simplification of (A).*

Fig. A-9. Equivalent circuits of Fig. A-8 from point of view of coupling capacitor C.

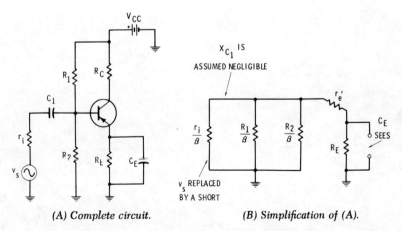

(A) Complete circuit.   *(B) Simplification of (A).*

Fig. A-10. Resistance appearing to bypass capacitor.

$$X_C = 0.1 R_{eq} \tag{A-19}$$

Finally, at the lowest operating frequency f, the value of the coupling capacitor is

$$C = \frac{1}{2\pi f X_C} \tag{A-20}$$

Similarly, bypass capacitor C_E in Fig. A-10A sees the equivalent circuit shown in Fig. A-10B. Note that r_i, R_1, and R_2 are in parallel and that r_e' is in series with them. Also R_E is in parallel with the series-parallel combination. Thus, as C_E sees it, in this case the equivalent resistance is

$$R_{eq} = R_E \| \frac{(r_e' + r_i\|R_1\|R_2)}{\beta}$$

From this point the value of C_E is found by using Eqs. A-19 and A-20 given previously.

Answers to Review Questions

Chapter 1

1-1. The core contains the atomic nucleus and all electrons except the valence electrons. In the case of silicon the core contains 14 protons of the nucleus and 10 electrons. The remaining four electrons are in the valence band.

1-2. A neutral atom is one that has a total number of electrons equal to the total number of protons in its nucleus.

1-3. A valence electron is one in the outer shell (valence band) of the atom. A free electron is one that has broken away from its parent atom. That is, when a valence electron is torn away from the valence band of an atom, it becomes free to drift through the material.

1-4. The parent atom is an atom to which a free electron belonged before it was torn away.

1-5. When a free electron breaks away from a parent atom, the place where the electron was is called a "hole." That is, a hole may be thought of as an absence of an electron.

1-6. When an electron breaks out of the valence band of the parent atom, a hole appears. The free electron and hole produced by this break are called an "electron-hole pair."

1-7. If we increase the temperature of silicon or germanium crystal, we will cause an increase in the number of electron-hole pairs. The electrons produced in this way are free electrons. More electron-hole pairs and therefore more free electrons are produced in pure germanium crystal than in pure silicon crystal at any given temperature. This characteristic of germanium is often undesirable, as you will see later. Also a pentavalent impurity added to the crystal during its fabrication adds free electrons.

1-8. When the electron of an electron-hole pair recombines with the hole, we have recombination.

1-9. Silicon or germanium is doped when impurities are added to it to increase the number of free electrons or holes.

1-10. Doping with trivalent impurity produces more holes in a semiconductor.

1-11. With a voltage applied, the holes tend to drift toward the negative side, whereas the electrons drift toward the positive side; free electrons and holes drift in opposite directions.

1-12. No, pure silicon produces relatively few electron-hole pairs at room temperature.

1-13. Doping will increase the conductivity of silicon. Increases in temperature will too, but for most practical purposes this is only a negligible amount.

1-14. Pentavalent materials have five valence electrons and trivalent materials have three.

1-15. Arsenic is a pentavalent material and gallium is trivalent.

1-16. Donor impurity atoms provide free electrons and are pentavalent. Acceptor impurity atoms are trivalent and provide holes when used in semiconductors.

1-17. An n-type semiconductor is a material which has been made into a semiconductor by the addition of a donor-type impurity, usually a pentavalent impurity.

1-18. None. The free electrons in n-type semiconductors are supplied by donor atoms which have no charge and five valence electrons.

1-19. A p-type semiconductor is a material which has been made into a semiconductor by the addition of an acceptor-type impurity, usually a trivalent impurity.

1-20. None. The trivalent (acceptor) impurity atoms added to silicon or germanium to make p-type material have no charge.

1-21. Electrons are majority carriers and holes are the minority carriers in an n-type semiconductor.

1-22. Holes are the majority carriers and electrons are the minority carriers in a p-type semiconductor.

1-23. The free electrons and holes are free to drift. All atoms are stationary in crystals.

1-24. Recombination.

Chapter 2

2-1. (a) Away from, (b) toward.

2-2. Majority.

2-3. Larger.

2-4. Zener.

2-5. Temperature increases, increased reverse bias voltage, and even increases in the amount of light (light energy) applied to a semiconductor may increase the number of electron-hole pairs. The minority carriers thus increased support an increased leakage current.

2-6. The PIV of a diode is its peak inverse voltage rating. When a manufacturer specifies the PIV voltage, voltages greater than it should not be applied in reverse bias to the diode.

2-7. Yes, because the forward biased diode conducts current relatively well and thus current is allowed to flow through the lamp.

2-8. Since a silicon diode is used, we could expect about 0.7 V across it and thus the remaining 11.3 V is across the lamp. If a germanium diode were used instead, about 0.3 V would be dropped across it and the remaining 11.7 V would be across the lamp.

2-9. Four methods are: (1) grown method, (2) alloy method, (3) diffusion method, and (4) epitaxial method.

Chapter 3

3-1. An actual diode may be assumed to be *ideal*, have *simplified V-I characteristics* as in Fig. 3-3A, or to have *approximated V-I characteristics* as in Fig. 3-4B.

3-2. The ideal diode acts like a short circuit when forward biased and like an open circuit when reverse biased.

3-3. The simplified diode characteristics assume that the diode has a constant 0.7 V or 0.3 V drop, depending on whether it is silicon or germanium type, when forward biased.

3-4. The approximated *V-I* characteristics are shown as a straight line drawn through the actual *V-I* curve of the diode. This straight line is not vertical but instead leans to the right slightly (has a positive slope), indicating that actually the forward voltage V_F increases slightly with increases in forward current I_F.

3-5. With large applied voltages and fairly large circuit resistances, diodes can usually be assumed to be ideal.

3-6. With small applied voltages and fairly large circuit resistances, simplified diode characteristics may be necessary, depending on the accuracy required.

3-7. If the applied voltage is large and the circuit resistance is small, approximated diode characteristics may be necessary.

3-8. A *load line* on the actual V-I diode characteristics must be drawn to determine V_F and I_F if the source voltage and circuit resistance are small.

Chapter 9

9-1. One or unity.

9-2. The common-collector amplifier can be used to work a high-internal-resistance signal source to a low-resistance load without excessive attenuation of signal voltage amplitude. Moderate power amplification of signal is also a characteristic of the common-collector amplifier.

9-3. The emitter follower.

9-4. A signal source is loaded down when the load placed on it has a much smaller resistance than its internal resistance.

9-5. Yes, but not as large as with common-emitter amplifiers in general.

9-6. In general, it is approximately beta times larger than the ac load resistance in the emitter circuit.

Chapter 11

11-1. The term *intrinsic standoff ratio* η applies to unijunction transistors (UJT's). Here η times the source voltage V_{BB} plus the voltage drop across the UJT forward biased pn junction gives the voltage value that, when applied to the emitter with respect to base 1, triggers the UJT.

11-2. The peak voltage V_p is the minimum required voltage on the emitter with respect to base 1 that will trigger the UJT.

11-3. The terms *valley voltage* and *valley current* apply to UJT's. After turn-on, a UJT's emitter to base 1 voltage V_E quickly drops to a minimum value at which the UJT turns off again. This minimum voltage is called the "valley voltage." The emitter current that exists at the instant the valley voltage is reached, is called the "valley current."

11-4. A simple relaxation oscillator can be made with a dc voltage source, a resistor, a capacitor, and a voltage operated switch. The voltage operated switch can be a UJT, a Diac, a Shockley diode, or a neon bulb which is placed in shunt with the capacitor. This combination is then placed in series with the dc source and resistor. The dc source voltage must be larger than the trigger voltage of the voltage operated switch. The value of source voltage, capacitance, and resistance affect the oscillator frequency.

11-5. The Triac can be triggered to conduct current in either direction, whereas the SCR can normally be triggered to conduct only in one direction.

11-6. Both the Triac and SCR turn off if the currents in them are reduced to values below the holding currents I_H.

11-7. No, Triacs do not have much time between alternations in which recombinations of majority charge carriers and turn-off can occur. If the voltage is not passing through zero at the same instant that current is, as would be the case in inductive circuits, the Triac may not reliably turn off.

11-8. Shunt capacitance protects a Triac from rapidly rising voltages (large de/dt), at instances of turn-off. This provides better turn-off reliability.

11-9. Mainly the Diac is a two-terminal device whereas the Triac has three terminals.

11-10. The Diac can be triggered into conduction in either direction but the four-layer (Shockley diode) is intended for unilateral conduction. Also the voltage required to trigger the four-layer diode is typically smaller than the trigger voltage of the Diac.

Chapter 12

12-1. 10^6 to 10^9 Ω; that is, 1,000,000 Ω to 1,000,000,000 Ω.

12-2. *J*unction *F*ield-*E*ffect *T*ransistor.

12-3. *I*nsulated *G*ate *F*ield-*E*ffect *T*ransistor.

12-4. MOSFET or Metal Oxide Semiconductor Field-Effect Transistor.

12-5. 10^{10} to 10^{14} Ω; that is, 10,000,000,000 Ω to 100,000,000,000,000 Ω.

12-6. JFET's operate in depletion mode, which is also called "mode A."

12-7. Depletion.

12-8. Depletion-enhancement.

12-9. Enhancement.

12-10. Their extremely high input resistance and relatively low voltage ratings make them vulnerable to damage from transient voltages.

12-11. Symbol y_{fs} represents an FET's *forward transadmittance*, also known as the *forward transconductance*, and is referred to sometimes with symbols: y_{fs}, g_{fs}, and g_{21}. The symbol y_{os} represents an FET *output admittance*, which is also known by other symbols: y_{os}, g_{os}, $1/r_d$ and g_{22}.

12-12. The y_{fs}, if not specified in the manufacturer's literature, can be determined from the FET drain characteristics or by measurements while the FET is in a test circuit. The latter two methods apply the equation

$$y_{fs} = \frac{\Delta I_D}{\Delta V_{GS}} \bigg|_{\Delta V_{DS}=0}$$

The parameter y_{os}, if not specified, can similarly be determined from the drain characteristics or by measurements while the FET is in a test circuit; that is, by using the equation

$$y_{os} = \frac{\Delta I_D}{\Delta V_{DS}} \bigg|_{\Delta V_{GS}=0}$$

12-13. No, the operating mode cannot be changed. Actually the operating mode dictates the type of bias that must be used.

12-14. The symbol I_{DQ} represents the quiescent drain current, which is the average dc drain current. Similarly, V_{DQ} is the quiesent drain-to-source voltage.

12-15. Symbol v_{gs} is the gate-to-source signal voltage, v_{ds} is the drain-to-source signal voltage, and i_d is the signal drain current.

12-16. Excessively large divider resistors in the gate bias circuitry make the gate-to-ground voltage subject to poor voltage regulation, noise, and transients. The transients, not only a cause of noise, can ruin IGFET's.

12-17. Referring to FET's, the term *low temperature coefficient* means that the quiescent drain current I_{DQ} changes very little with fairly large changes in temperature.

Chapter 13

13-1. It is a circuit or group of very small electronic components that are constructed and interconnected permanently on a piece of semiconductor.

13-2. A monolithic IC is one that is made on a single piece of semiconductor.

13-3. A wafer is a larger piece of semiconductor from which chips are cut after the forming processes of the IC components and interconnections are finished.

13-4. Compared to other components, capacitors cannot be economically made on IC's.

13-5. Use of reverse biased pn junctions between components, epitaxial isolation, and dielectric isolation.

13-6. Epitaxy means growing one layer of semiconductor material in a substrate.

13-7. A NOR gate is an OR gate followed by an inverter. Thus, the OR gate output is always opposite to a NOR gate output for any given combination of inputs.

13-8. A NAND gate is an AND gate followed by an inverter, and therefore the outputs of these two types of gates are always opposite for any given combination of inputs.

13-9. The yield is the percentage of usable (ones that work properly) IC's out of the total number started with.

13-10. A multivibrator (astable) is a circuit capable of providing a continuous chain of square waves.

13-11. The astable multivibrator is also called a "free-running multivibrator" or a "clock."

13-12. Discrete component circuits are those made with separate components such as transistors, resistors, capacitors, etc., as opposed to IC's that have components inseparably connected on a continuous substrate.

13-13. Mainly, operational amplifiers are able to respond to gradual changes of dc signals applied to the input, whereas *RC*-coupled stages are not. In the latter, coupling capacitors block the dc component of input signals and thus prevent their amplification.

13-14. A large fan-out capability of an IC is its ability to drive the inputs of several other IC's and not be overloaded.

13-15. ECL means Emitter-Coupled Logic, TTL means Transistor Transistor Logic, DTL means Diode Transistor Logic, RTL means Resistor Transistor Logic, and MOS means Metal Oxide Silicon.

13-16. ECL-type IC's are nonsaturating and are capable of fast switching.

13-17. TTL-type IC's have switching speeds that approach speeds of ECL types and are among the fastest of the saturating IC logic circuits.

13-18. The term "fast" means that an IC is able to change from an output logic 1 to 0 and from 0 to 1 in a short time.

13-19. Linear IC's such as the operational amplifier IC are capable of amplifying signals, such as sine waves, without distortion. Digital IC's are mainly intended to change abruptly with abrupt changes in their inputs. It is possible to use linear IC's to perform digital functions, and vice versa. Operational amplifiers can be wired to be astable multivibrators, and NOR gates can be wired to amplify signals, to mention only two of many possibilities.

Answers to Problems

Chapter 3

3-1. $E_2 = 0.7$ V.—The silicon diode is forward biased, and its forward voltage $V_F = 0.7$ V is in parallel with R_2.

3-3. $E_k \cong 40$ V.—Since the voltage drop across the diode is quite small compared to the voltage drop we expect to find across R_3, which is in series with it, we may say for practical purposes, R_2 and R_3 are in parallel. Their total resistance is $R_{2,3} = 800$ Ω. The voltage across this resistance $R_{2,3}$ is

$$E_{2,3} \cong \frac{ER_{2,3}}{R_1 + R_{2,3}} = 40 \text{ V}$$

Thus, the voltage across R_2 and at point k to ground is 40 V. Since the voltage across the diode and R_3 is 40 V and 0.7 V is across the diode, $E_3 \cong 39.3$ V.

3-5. Voltage across R is about 23.3 V.—E_a forward biases D_1. The voltage across R is the source E_a voltage minus the drop across the diode.

3-7. About 2 kΩ or more.—Since the diode is forward biased, most of the source 80 V is across the resistor R. If the current is to be limited to 40 mA, by Ohm's law $R \cong 80$ V/40 mA = 2 kΩ.

3-9. $E_o = 11.3$ V.—In this case the diode is forward biased and E_o is the source E minus the diode forward drop. Another point of view is to note that the source E, the diode, and resistance R are in series. With the diode dropping 0.7 V, the remaining 11.3 V must be across R, and E_o is the voltage across R.

3-11. $E_o = 0.3$ V.—The diode is forward biased and E_o is the voltage across the diode.

3-13. The voltage E_R across the resistor will increase.—The heat will produce many more minority carriers, thus increasing the reverse bias leakage; see Sec. 3-5.

3-15. $E_2 \cong 12$ V.—The diode is reverse biased in this case, thus acting like an open. Therefore, as the source E sees it, R_1 and R_2 are in series.

Chapter 4

4-1. e_o is zero on all positive alternations and sinusoidal on negative alternations with a peak of -8 V.—The diode, assumed ideal, acts like a short on all positive alternations. On negative alternations the diode acts like an open. Thus, the source e_s sees R_1 and R_2 in series. The peak voltage across R_2 therefore is

$$\frac{(-40 \text{ V})(2 \text{ k}\Omega)}{8 \text{ k}\Omega + 2 \text{ k}\Omega} = -8 \text{ V}$$

See Fig. B-1.

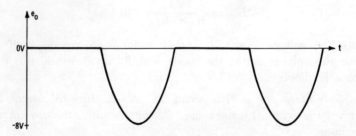

Fig. B-1. Waveform e_o.

4-3. $i_{D(\text{peak})} = 5$ mA.—On positive alternations, when the diode is forward biased, the entire source voltage e_s is across R_1. The peak current through R_1 is the peak current through the diode, which in this case is

$$i_{D(\text{peak})} = \frac{e_{s(\text{peak})}}{R_1} = \frac{40 \text{ V}}{8 \text{ k}\Omega} = 5 \text{ mA}$$

4-5. The positive alternations of e_D are sinusoidal with a peak of 6 V. Voltage e_D is zero on the negative alternations.—On positive alternations, e_D is the voltage across R_2. Since e_s sees R_1 and R_2 in series,

$$e_{D(\text{peak})} = \frac{e_{s(\text{peak})}R_2}{R_1+R_2} = \frac{(20\text{ V})(3\text{ k}\Omega)}{7\text{ k}\Omega + 3\text{k}\Omega} = 6\text{ V}$$

See Fig. B-2.

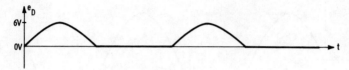

Fig. B-2. Waveform e_D.

4-7. PIV = 6 V.—The waveform found in Prob. 4-5 shows that there is 6 V peak across the diode at an instant on each cycle.

4-9. e_D appears to be a sine wave with peaks of 20 V. However, its positive alternations are clipped (sheared) off. Each positive alternation starts at -5 V, is clipped at 0 V, and has negative peaks of -25 V. See Fig. B-3.

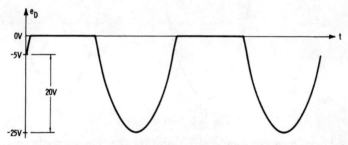

Fig. B-3. Clipped waveform e_D.

4-11. e_D appears sinusoidal. However, each cycle starts at -20 V, has its positive peak at -8 V, and its negative peak at -32 V. See Fig. B-4.

4-13. e_{xy} appears sinusoidal except that each positive alternation is clipped (sheared) off at 15 V. The negative peaks are normal with peaks of -20 V. See Fig. B-5.

4-15. $i_{D(\text{peak})} = 1$ mA.—As determined in the previous problem, the peak voltage across R is 5 V, which occurs when the diode is forward biased and conducting. The peak diode current therefore is the peak current through R, or, in this case,

$$i_{D(\text{peak})} = \frac{5\text{ V}}{5\text{ k}\Omega} = 1\text{ mA}$$

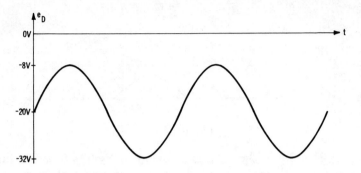

Fig. B-4. Unclipped waveform e_D.

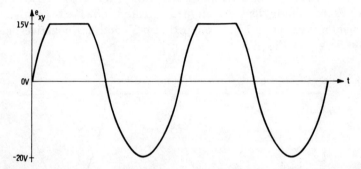

Fig. B-5. Waveform e_{xy}.

4-17. e_{ab} is a constant -10 V during positive alternations of e_s. For portions of negative alternations of e_s, e_{ab} dips negatively to -50 V peak. Therefore, e_{ab} appears as a constant -10 V with one negative-going alternation peaking at -50 V each cycle. See Fig. B-6.

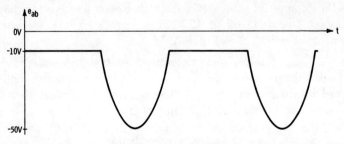

Fig. B-6. Waveform e_{ab}.

4-19. PIV = 40 V.—The maximum voltage across the diode appears during negative peaks of the input cycle e_s. The 10 V dc source bucks $e_{s(peak)} = -50$ V, leaving a net 40 V across the diode.

4-21. e_R is a series of positive and negative pulses (humps). The positive ones have peaks of 15 V and the negative ones have peaks of -20 V. See Fig. B-7.

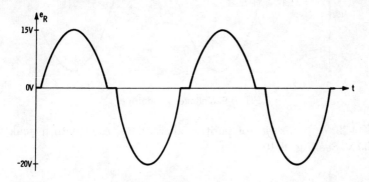

Fig. B-7. Waveform e_R.

4-23. e_D is zero for the first quarter cycle of e_s because the capacitor is charging then. Thereafter, e_D appears sinusoidal but with a -12 V dc component. That is, e_D varies from 0 V to -24 V peak-to-peak. See Fig. B-8.

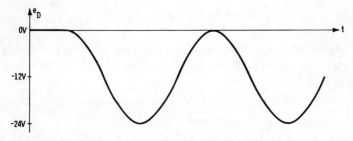

Fig. B-8. Waveform e_D with dc component.

4-25. 12 V dc, top plate positive with respect to the bottom plate.

4-27. On the peak of the first positive alternation, C is charging, thus clipping e_{xy} at 5 V. Thereafter e_{xy} varies sinusoidally, with positive peaks reaching $+5$ V and negative peaks reaching -25 V. See Fig. B-9.

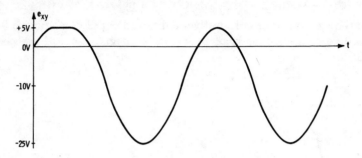

Fig. B-9. Voltage waveform e_{xy}.

4-29. e_o is a series of positive alternations, each with a peak of 100 V. See Fig. B-10.

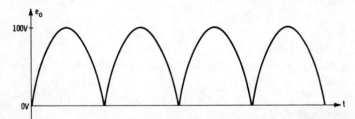

Fig. B-10. Voltage waveform e_o.

4-31. Minimum $R = 36 \, \Omega$.—Assuming that the 500 mA rating is the maximum peak current capability, we can find the minimum R with Ohm's law:

$$R_{\min} = \frac{e_{s(\text{peak})}}{500 \text{ mA}} = \frac{18 \text{ V}}{0.5 \text{ A}} = 36 \, \Omega$$

4-33. PIV = 24 V.—During negative peaks of e_s the diode has the sum of e_s and e_o across it.

4-35. Minimum $R \cong 21.7 \, \Omega$.—Since the secondary end to center tap has a peak of 17 V, we can determine its average dc voltage with Eq. 4-4B. That is, $E_{\text{dc}} \cong 0.637 \times 17 \cong 10.85$ V.

Since the maximum rated dc current is 500 mA, Ohm's law shows that

$$R_{min} \cong \frac{10.85 \text{ V}}{500 \text{ mA}} \cong 21.7 \text{ }\Omega$$

4-37. Minimum $R \cong 100 \text{ }\Omega$.—Since $e_R \cong 100$ V dc, as determined in the previous problem, with a rated dc current of 1 A, the smallest load resistance R that can be used safely is

$$R_{min} \cong \frac{100 \text{ V}}{1 \text{ A}} = 100 \text{ }\Omega$$

4-39. $V_{dc} \cong 12.74$ V.—As in the previous problem, we use Eq. 4-1:

$$E_{dc} \cong \frac{0.637 \times 40}{2} \cong 12.74 \text{ V}$$

4-41. $V_{dc} \cong 80$ V.—As in the previous problem, the capacitor C charges to the peak of e_s and practically stays there.

4-43. $V_{dc} \cong 13.6$ V.—As in the previous problem, Eq. 4-4B applies. In this case the transformer is 4:1 step-down, which means that the entire secondary voltage is one-fourth of 120 V rms, or 30 V rms, or about 42.5 V (peak). The secondary voltage to the center tap therefore is 21.24 V (peak). So in this case, $E_{dc} \cong 0.637 \times 21.24 \cong 13.6$ V.

4-45. $V_{dc} \cong 21.2$ V.—As in the previous problem, with S closed, the capacitor charges to the peak voltage across either secondary lead and the center tap, which was found to be about 21.24 V in Prob. 4-43 above.

4-47. $V_{dc} \cong 24$ V.—In this case, each capacitor charges to 12 V and their sum is across the load R.

4-49. $V_{dc} \cong 72$ V.—In this case the peak input to the assembly is about 113 V. Thus, with Eq. 4-6B we find that $E_{dc} \cong 0.637(113 \text{ V}) \cong 72$ V.

4-51. $V_{dc} \cong 113$ V.—With the switch S closed, the capacitor C charges to about the peak of the input, which is about 113 V in this case.

4-53. Still assuming ideal diodes, $V_o = 0$ V, $E_{D1} = 12$ V, $E_{D2} = E_{D3} = 0$ V, and $E_R = 18$ V.—The 12 V at A to ground forward biases diodes D_2 and D_3 but reverse biases D_1.

4-55. Assuming that the diodes are ideal, $V_o = 0$ V, $E_{D1} = E_{D2} = E_{D3} = 0$ V, and $E_R = 20$ V.—With all three points A, B, and C at ground potential, all three diodes are forward biased by the 20-V source through R.

4-57. With ideal diodes, $V_o = 12$ V, $E_{D1} = E_{D2} = E_{D3} = 0$ V, and $E_R = 32$ V.

Chapter 5

5-1. $r_z \cong 182 \, \Omega$.—For,

$$r_z \cong \frac{20.5 \text{ V} - 20 \text{ V}}{3 \text{ mA} - 0.25 \text{ mA}} = \frac{0.5 \text{ V}}{2.75 \text{ mA}} = 182 \, \Omega$$

5-3. It looks like the right-hand side of Fig. 5-8A where $r_z = 182 \, \Omega$ and $V_z = 20$ V.

5-5. $I_{Z(\min)} \cong 2$ mA, $I_{Z(\max)} \cong 5$ mA, $V_o \cong 20$ V.—The circuit currents are

$$I_{Z(\min)} \cong \frac{30 \text{ V} - 20 \text{ V}}{5 \text{ k}\Omega} = 2 \text{ mA}$$

$$I_{Z(\max)} \cong \frac{45 \text{ V} - 20 \text{ V}}{5 \text{ k}\Omega} = 5 \text{ mA}$$

5-7. $I_{Z(\min)} \cong 1.94$ mA, $I_{Z(\max)} \cong 4.85$ mA, $V_{o(\max)} \cong 20.88$ V, $V_{o(\min)} \cong 20.35$ V.—In this case,

$$I_{Z(\min)} \cong \frac{30 \text{ V} - 20 \text{ V}}{5000 \, \Omega + 182 \, \Omega} \cong 1.94 \text{ mA}$$

$$I_{Z(\max)} \cong \frac{45 \text{ V} - 20 \text{ V}}{5000 \, \Omega + 182} \cong 4.85 \text{ mA}$$

$$V_{o(\min)} \cong 20 \text{ V} + (1.94 \text{ mA})(182 \, \Omega) \cong 20.35 \text{ V}$$

$$V_{o(\max)} \cong 20 \text{ V} + (4.85 \text{ mA})(182 \, \Omega) \cong 20.88 \text{ V}$$

5-9. About 0.5 W.

5-11. 860 mW.—The zener must be derated by
$(7 \text{ mW}/°\text{C})(45°\text{C} - 25°\text{C}) = (7 \text{ mW}/°\text{C})(20°\text{C}) = 140$ mW
and will be capable of dissipating $1 \text{ W} - 0.14 \text{ W} = 0.86$ W.

5-13. $R_{s(\max)} = 1.25$ kΩ.—Assuming an ideal zener, Eq. 5-1 is applicable:

$$R_{s(\max)} = \frac{E_{\min} - V_Z}{I_{L(\max)}} = \frac{90 \text{ V} - 80 \text{ V}}{8 \text{ mA}} = 1.25 \text{ k}\Omega$$

CHAPTER 5

where $I_{L(max)} = 80 \text{ V}/10 \text{ k}\Omega = 8 \text{ mA}$.

5-15. $R_{s(max)} \cong 1.11 \text{ k}\Omega$.—In this case

$$R_{s(max)} = \frac{E_{min} - V_Z}{I_{L(max)} + I_{Z(min)}} = \frac{90 \text{ V} - 80 \text{ V}}{8 \text{ mA} + 1 \text{ mA}} \cong 1.11 \text{ k}\Omega$$

5-17. $R_{s(max)} = 4 \text{ k}\Omega$.—In this case the load current is zero. Thus

$$R_{s(max)} = \frac{16 \text{ V} - 12 \text{ V}}{0 + 1 \text{ mA}} = \frac{4 \text{ V}}{1 \text{ mA}} = 4 \text{ k}\Omega$$

5-19. $E_{o(min)} = 12.05 \text{ V}$, $E_{o(max)} = 12.1 \text{ V}$.—With load current zero, the current in R_s is also the current in the zener. When the input $E = 16 \text{ V}$, then

$$I_{Z(min)} = \frac{E - V_Z}{R_s} = \frac{16 \text{ V} - 12 \text{ V}}{4 \text{ k}\Omega} = 1 \text{ mA}$$

This causes a minimum output voltage

$$E_{o(min)} = V_Z + r_Z I_{Z(min)} = 12 \text{ V} + 0.05 \text{ V} = 12.05 \text{ V}$$

When the input $E = 20 \text{ V}$, then

$$I_{Z(max)} = \frac{20 \text{ V} - 12 \text{ V}}{4 \text{ k}\Omega} = 2 \text{ mA}$$

which causes a maximum output voltage

$$E_{o(max)} = 12 \text{ V} + 50(2 \text{ mA}) = 12.1 \text{ V}$$

5-21. $R_s = 200 \text{ }\Omega$, $P_{Z(max)} = 360 \text{ mW}$.—First we find the minimum and maximum load currents:

$$I_{L(max)} = \frac{12 \text{ V}}{300 \text{ }\Omega} = 40 \text{ mA}, \qquad I_{L(min)} = \frac{12 \text{ V}}{1.2 \text{ k}\Omega} = 10 \text{ mA}$$

Thus

$$R_s = \frac{E_{max} - V_Z}{I_{L(max)}} = \frac{20 \text{ V} - 12 \text{ V}}{40 \text{ mA}} = \frac{8 \text{ V}}{40 \text{ mA}} = 200 \text{ }\Omega$$

And

$$P_{Z(max)} = V_Z \left(\frac{E_{max} - V_Z}{R_s} - I_{L(min)} \right) = 12 \text{ V} \left(40 \text{ mA} - 10 \text{ mA} \right)$$

$$= 360 \text{ mW}$$

5-23. $R_s = 190 \text{ }\Omega$, $P_{Z(max)} = 384 \text{ mW}$.—As in Prob. 5-21,

$$I_{L(\max)} = 40 \text{ mA}, \quad \text{and} \quad I_{L(\min)} = 10 \text{ mA}$$

In this case we include the minimum zener current and get

$$R_s = \frac{20 \text{ V} - 12 \text{ V}}{40 \text{ mA} + 2 \text{ mA}} = 190 \text{ }\Omega$$

With a slightly smaller R_s, compared with that of Prob. 5-21, we can expect a slightly larger maximum power dissipation in the zener:

$$P_{Z(\max)} = V_Z \left(\frac{E_{\max} - V_Z}{R_s} - I_{L(\min)} \right) = (12 \text{ V})(42 \text{ mA} - 10 \text{ mA})$$

$$= 384 \text{ mW}$$

5-25. $R_s = 4 \text{ k}\Omega$, $P_{Z(\max)} = 1 \text{ W}$.—In this case

$$I_{L(\max)} = \frac{100 \text{ V}}{25 \text{ k}\Omega} = 4 \text{ mA}, \quad I_{L(\min)} = 0$$

Thus

$$R_s = \frac{120 \text{ V} - 100 \text{ V}}{4 \text{ mA} + 1 \text{ mA}} = \frac{20 \text{ V}}{5 \text{ mA}} = 4 \text{ k}\Omega$$

And therefore

$$P_{Z(\max)} = 100 \text{ V} \left(\frac{140 \text{ V} - 100 \text{ V}}{4 \text{ k}\Omega} - 0 \right) = 100 \text{ V} \left(\frac{40 \text{ V}}{4 \text{ k}\Omega} \right)$$

$$= 1000 \text{ mW} = 1 \text{ W}$$

5-27. $E_{o(\min)} = 100.12 \text{ V}$, $E_{o(\max)} = 101.2 \text{ V}$.—The minimum output voltage $E_{o(\min)}$ occurs when the zener current is minimum. A minimum zener current flows when the power supply voltage E is minimum and the load current is maximum; that is, when $E = 120 \text{ V}$ and $I_L = 4 \text{ mA}$, the current through R_s is

$$I_s \cong \frac{120 \text{ V} - 100 \text{ V}}{R_s} = 5 \text{ mA}$$

Thus, the zener current is

$$I_Z = I_s - I_L \cong 5 - 4 = 1 \text{ mA}$$

and therefore the minimum output voltage is

$$E_{o(\min)} = V_Z + r_Z I_{Z(\min)} \cong 100 \text{ V} + (120 \text{ }\Omega)(1 \text{ mA}) = 100.12 \text{ V}$$

On the other hand, the maximum output $E_{o(\max)}$ occurs when the zener current is maximum and a maximum zener current flows when the power supply voltage is maximum but the load current minimum. In this case then, when $E = 140$ V and $I_L = 0$, the current in R_s is

$$I_s \cong \frac{140 \text{ V} - 100 \text{ V}}{R_s} = 10 \text{ mA}$$

Therefore $I_Z \cong 10$ mA and

$$E_o = 100 \text{ V} + (120 \text{ }\Omega)(10 \text{ mA}) = 101.2 \text{ V}$$

5-29. $E_x = -120$ V, $E_y = 100$ V, the PIV of D_1 and D_2 is about 220 V.—During positive peaks, D_1 acts like a short and the sum of the voltages across D_3 and C_2 appear across the reverse biased D_2. During negative peaks, D_2 acts like a short and the sum of the voltages across D_4 and C_1 appears across the reverse biased D_1. The PIV of each rectifier diode therefore is simply the sum of the absolute voltage values at x and y.

Chapter 6

6-1. $I_E \cong 2.4$ mA, $V_{CB} \cong 6$ V, $P_C \cong 14.4$ mW.—From Eq. 6-1

$$I_E \cong \frac{V_{EE}}{R_E} = \frac{24 \text{ V}}{10 \text{ k}\Omega} = 2.4 \text{ mA}$$

$$V_{CB} = V_{CC} - R_C I_C \cong 24 \text{ V} - (7.5 \text{ k}\Omega)(2.4 \text{ mA}) = 6 \text{ V}$$

$$P_C \cong V_{CB} I_C \cong (6 \text{ V})(2.4 \text{ mA}) = 14.4 \text{ mW}$$

6-3. $h_{FB} = 0.98$.—Since $h_{FB} = \alpha_{\text{dc}}$ and $h_{FE} = \beta_{\text{dc}}$, then by Eq. 6-6

$$h_{FB} = \alpha_{\text{dc}} = \frac{\beta_{\text{dc}}}{\beta_{\text{dc}} + 1} = \frac{50}{51} = 0.98$$

6-5. $I_B \cong 48$ μA.—As we found in Prob. 6-1, $I_E \cong I_C \cong 2.4$ and since

$$\beta_{\text{dc}} \cong \frac{I_C}{I_B}$$

then in this case, with $\beta_{\text{ds}} = 50$ from Prob. 6-3,

$$I_B \cong \frac{I_C}{\beta_{\text{dc}}} \cong \frac{2.4 \text{ mA}}{50} = 48 \text{ }\mu\text{A}$$

6-7. $R_E \cong 6$ kΩ.—With the source $V_{CC}=18$ V and $V_{CB}=8$ V, the remaining 10 V must be dropped across R_C. Thus since $R_C I_C \cong R_C I_E \cong 10$ V, then

$$I_E \cong \frac{10 \text{ V}}{R_C} = \frac{10 \text{ V}}{5 \text{ k}\Omega} = 2 \text{ mA}$$

and therefore

$$R_E = \frac{V_{EE}}{I_E} \cong \frac{12 \text{ V}}{2 \text{ mA}} = 6 \text{ k}\Omega$$

6-9. $R_E \cong 9$ kΩ.—With $V_{CC}=22$ V and $V_{CB}=12$ V, the remaining 10 V is across R_C. Thus $R_C I_C \cong R_C I_E \cong 10$ V and

$$I_E \cong \frac{10 \text{ V}}{R_C} = \frac{10 \text{ V}}{5 \text{ k}\Omega} = 2 \text{ mA}$$

and therefore

$$R_E \cong \frac{V_{EE}}{I_E} = \frac{18 \text{ V}}{2 \text{ mA}} = 9 \text{ k}\Omega$$

6-11. $P_C = 24$ mW.—Here $P_C = V_{CB} I_C = (12 \text{ V})(2 \text{ mA}) = 24$ mW.

6-13. Npn.

6-15. $V_{CB} \cong 4$ V, $P_C \cong 8$ mW.—From Eq. 6-1, $I_E \cong V_{EE}/R_E = 20$ V/10 kΩ = 2 mA. Since $I_C \cong I_E$, then $I_C \cong 2$ mA and $V_{CB} = V_{CC} - I_C R_C \cong 20$ V $-$ (2 mA)(8 kΩ) = 20 V $-$ 16 V = 4 V. Also, $P_C = V_{CB} I_C \cong (4 \text{ V})(2 \text{ mA}) = 8$ mW.

Chapter 7

7-1. $I_{C(\text{sat})} \cong 4$ mA, $R_E \cong 7.5$ kΩ.—With Eq. 7-1 we find that

$$I_{C(\text{sat})} \cong \frac{V_{CC}}{R_C} \cong \frac{40 \text{ V}}{10 \text{ k}\Omega} = 4 \text{ mA}$$

Thus, 4 mA of emitter current causes the collector to saturate. The emitter bias resistance that admits this current is found by Ohm's law:

$$R_E \cong \frac{V_{EE}}{I_{E(\text{sat})}} \cong \frac{30 \text{ V}}{4 \text{ mA}} = 7.5 \text{ k}\Omega$$

7-3. $R_E \cong 60$ kΩ.—With $V_{CC} = 40$ V and the collector-to-ground voltage at 35 V, the remaining 5 V must be dropped across R_C.

Chapter 7

If we want 5 V across R_C, the current through it, which is approximately the value of the emitter current too, is

$$I_C \cong I_E \cong \frac{5\text{ V}}{R_C} = \frac{5\text{ V}}{10\text{ k}\Omega} = 0.5\text{ mA}$$

The emitter bias resistance that will admit this current is

$$R_E \cong \frac{V_{EE}}{I_E} \cong \frac{30\text{ V}}{0.5\text{ mA}} = 60\text{ k}\Omega$$

7-5. The current readings on meters M_1 and M_2 will change negligibly, whereas the voltage reading on meter M_3 will increase.

7-7. $I_C \cong 1.5$ mA, $V_C \cong 25$ V.—With $V_{EE} = -30$ V,

$$I_E \cong \frac{V_{EE}}{R_E} \cong \frac{30\text{ V}}{20\text{ k}\Omega} = 1.5\text{ mA}$$

The collector-to-ground voltage *is* the collector-to-base voltage in this circuit, and is equal to the source voltage V_{CC} minus the drop across R_C.

$$V_C = V_{CB} = V_{CC} - R_C I_C \cong 40\text{ V} - (10\text{ k}\Omega)(1.5\text{ mA}) = 25\text{ V}$$

7-9. $r_L = 20$ kΩ.—As the signal in the collector sees it, resistances R_C and R_L are in parallel. Thus, by Eq. 7-2,

$$r_L = \frac{R_C R_L}{R_C + R_L} = R_C \| R_L = \frac{(30\text{ k}\Omega)(60\text{ k}\Omega)}{30\text{ k}\Omega + 60\text{ k}\Omega} = 20\text{ k}\Omega$$

7-11. $r_{\text{in}} \cong 50$ Ω.—First we find the quiescent collector and emitter currents:

$$I_C \cong I_E \cong \frac{V_{EE}}{R_E} = \frac{30\text{ V}}{60\text{ k}\Omega} = 0.5\text{ mA}$$

Next we determine the approximate dynamic ac resistance seen looking into the emitter, by using Eq. 7-3:

$$r_e' \cong \frac{25\text{ mV}}{I_E} \cong \frac{25\text{ mV}}{0.5\text{ mA}} = 50\text{ }\Omega$$

The signal source v_s sees R_E and r_e' in parallel. Therefore

$$r_{\text{in}} \cong \frac{R_E r_e'}{R_E + r_e'} \cong \frac{(60\text{ k}\Omega)(50\text{ }\Omega)}{60\text{ k}\Omega + 50\text{ }\Omega} \cong 50\text{ }\Omega$$

7-13. $i_{e\text{(peak)}} \cong i_{c\text{(peak)}} \cong 0.1$ μA.—From Prob. 7-11, we have $r_e' = 50$ Ω. Since the source v_s is across r_e', in this case, the signal current through this dynamic resistance is

$$i_e \cong \frac{v_s}{r_e'} \cong \frac{5\ \mu\text{V}}{50\ \Omega} = 0.1\ \mu\text{A (peak)}$$

which is approximately the signal collector current i_c too.

7-15. $A_e \cong 400$, $v_{cb\text{(peak)}} \cong 2$ mV.—The simplest way to determine the voltage gain A_e is with Eq. 7-9. That is,

$$A_e \cong \frac{r_L}{r_e'} \cong \frac{R_L \| R_C}{r_e'} = \frac{60\ \text{k}\Omega \| 40\ \text{k}\Omega}{50\ \Omega} = \frac{24\ \text{k}\Omega}{50\ \Omega} = 480$$

Thus, by Eq. 7-6A,

$$A_e = \frac{v_{cb}}{v_{eb}}$$

and since in this case $v_s \cong v_{eb}$, then by rearranging Eq. 7-6A we can show that

$$v_{cb} = A_e(v_{eb}) \cong A_e(v_s)$$

And therefore,

$$v_{cb\text{(peak)}} \cong 400 \times 5\ \mu\text{V} = 2\ \text{mV}$$

Another approach is to find the output voltage v_{cb} first with Ohm's law, Eq. 7-8. That is, since

$$v_{cb} = r_L i_c \tag{7-8}$$

then

$$v_{cb\text{(peak)}} = (20\ \text{k}\Omega)(0.1\ \mu\text{A}) = 2\ \text{mV}$$

Now by Eq. 7-6A

$$A_e = \frac{v_{cb}}{v_{eb}} \cong \frac{2\ \text{mV}}{5\ \mu\text{V}} = 400$$

7-17. $I_C \cong 0.5$ mA, $V_C \cong 25$ V.—The dc quiescent collector and emitter currents are calculated in the answer to Prob. 7-11. That is, as found in Prob. 7-11, $I_C \cong I_E \cong 0.5$ mA. The collector voltage V_C is the collector-to-base voltage V_{CB}, which can be determined by subtracting the voltage drop across R_C from the source V_{CC}:

$$V_C = V_{CB} = V_{CC} - R_C I_C \cong 40\ \text{V} - (30\ \text{k}\Omega)(0.5\ \text{mA}) = 25\ \text{V}$$

7-19. (a) $A_i = \alpha \cong 1$, (b) $A_e \cong 75$, (c) $A_p \cong 75$, (d) $I_C \cong 1.25$ mA, (e) $V_{CB} \cong 5.25$ V, (f) $P_C \cong 6.56$ mW.—(a) The current gain is the ratio of output current i_c to the input current i_e: see

Eqs. 7-10 and 7-10A. This ratio is the alpha (α) of the transistor, which is approximately equal to 1. (b) We can determine the voltage gain A_e with Eq. 7-9 but first we find

$$r_L = R_L = R_C = 3\text{ k}\Omega \| 3\text{ k}\Omega = 1.5\text{ k}\Omega$$

and

$$I_E \cong \frac{V_{EE}}{R_E} = \frac{12\text{ V}}{9.6\text{ k}\Omega} = 1.25\text{ mA}$$

Thus

$$r_e' \cong \frac{25\text{ mV}}{I_E} = \frac{25\text{ mV}}{1.25\text{ mA}} = 20\text{ }\Omega$$

So finally,

$$A_e \cong \frac{r_L}{r_e'} \cong \frac{1500\text{ }\Omega}{20\text{ }\Omega} = 75$$

(c) The power gain is the product of the voltage and current gains, Eq. 7-12. Therefore,

$$A_p = A_e A_i \cong 75 \times 1 = 75$$

(d) Of course $I_C \cong I_E$, which was determined to be 1.25 mA above. (e) The collector-to-base voltage is

$$V_{CB} = V_{CC} - R_C I_C \cong 9\text{ V} - (3\text{ k}\Omega)(1.25\text{ mA}) = 5.25\text{ V}$$

(f) $P_C = V_{CB} I_C \cong (5.25\text{ V})(1.25\text{ mA}) \cong 6.56\text{ mW}$

7-21. $r_s = 1.5$ kΩ.—From a no load to a load on this source, the output voltage charges from 6 mV to 4.8 mV, which is a change in voltage $\Delta e_o = 1.2$ mV. With no load the signal generator output current is $i_o = 0$. With a load

$$i_o = \frac{4.8\text{ mV}}{6\text{ k}\Omega} = 0.8\text{ }\mu\text{A}$$

Thus, the current change is $\Delta i_o = 0.8 - 0 = 0.8$ μA. Now substituting into Eq. 7-13 we find that

$$r_s = \frac{\Delta e_o}{\Delta i_o} = \frac{1.2\text{ mV}}{0.8\text{ }\mu\text{A}} = 1.5\text{ k}\Omega$$

7-23. The voltage across the load is $e_o = 2.4$ mV.—In this case the 6-mV source sees a total resistance of $r_s + R_L = 1.5 + 1 = 2.5$ kΩ. The load current therefore must be

$$i_o = \frac{6 \text{ mV}}{2.5 \text{ k}\Omega} = 2.4 \ \mu\text{A}$$

This current flows through the 1-kΩ load resistance, which determines the output voltage

$$e_o = (1 \text{ k}\Omega)(2.4 \ \mu\text{A}) = 2.4 \text{ mV}$$

An alternate solution is the following:

$$e_o = \frac{v_s R_L}{r_s + R_L} = \frac{(6 \text{ mV})(1 \text{ k}\Omega)}{1.5 \text{ k}\Omega + 1 \text{ k}\Omega} = 2.4 \text{ mV}$$

7-25. (a) $r_e' \cong 50 \ \Omega$, (b) $i_{e(\text{peak})} \cong 20 \ \mu\text{A}$.—The emitter current is

$$I_E \cong \frac{V_{EE}}{R_E} = \frac{30 \text{ V}}{60 \text{ k}\Omega} = 0.5 \text{ mA}$$

Thus

$$r_e' \cong \frac{25 \text{ mV}}{I_E} = \frac{25 \text{ mV}}{0.5 \text{ mA}} = 50 \ \Omega$$

As seen to the right of the capacitor C_1,

$$r_\text{in} = R_E \| r_e' \cong 60 \text{ k}\Omega \| 50 \ \Omega \cong 50 \ \Omega$$

Thus, the source v_s sees r_s and r_in in series or a total resistance

$$r_s + r_\text{in} \cong 100 + 50 \cong 150 \ \Omega$$

And by Ohm's law

$$i_{e(\text{peak})} \cong \frac{3 \text{ mV}}{150 \ \Omega} = 20 \ \mu\text{A}$$

7-27. (a) $v_{cb(\text{peak})} \cong 240 \text{ mV}$, (b) $v_{cb}/v_{eb} \cong 240$, (c) $v_{cb}/v_s \cong 80$.
—(a) Since $i_{e(\text{peak})} \cong 20 \ \mu\text{A}$ from Prob. 7-25, we know that $i_{c(\text{peak})} \cong 20 \ \mu\text{A}$ too because $\alpha \cong 1$. This collector signal current flows through the ac load r_L, causing an output voltage

$$v_{cb(\text{peak})} = r_L i_{c(\text{peak})} \cong (12 \text{ k}\Omega)(20 \ \mu\text{A}) = 240 \text{ mV}$$

(b) By Eq. 7-9 and taking the value of r_e' from Prob. 7-25,

$$A_e = \frac{v_{cb}}{v_{eb}} \cong \frac{r_L}{r_e'} \cong \frac{12 \text{ k}\Omega}{50 \ \Omega} = 240$$

(c) Since r_s is not much greater than r_e', we use Eq. 7-14 to find

$$\frac{v_{cb}}{v_s} \cong \frac{r_L}{r_s + r_e'} \cong \frac{12 \text{ k}\Omega}{150 \text{ }\Omega} = 80$$

where $r_s + r_e' \cong 150 \text{ }\Omega$ as in Prob. 7-25.

7-29. (a) $v_{cb}/v_{eb} \cong 120$, (b) $v_{cb}/v_s \cong 2.5$, (c) $v_{cb(\text{peak})} \cong 1$ V. —(a) In this case,

$$\frac{v_{cb}}{v_{eb}} \cong \frac{r_L}{r_e'} \cong \frac{5 \text{ k}\Omega}{41.6 \text{ }\Omega} = 120$$

where

$$r_L = R_C \| R_L = 10 \text{ k}\Omega \| 10 \text{ k}\Omega = 5 \text{ k}\Omega$$

and

$$r_e' \cong \frac{25 \text{ mV}}{I_E} \cong \frac{25 \text{ mV}}{0.6 \text{ mA}} = 41.6 \text{ }\Omega$$

where

$$I_E \cong \frac{V_{EE}}{R_E} = \frac{12 \text{ V}}{20 \text{ k}\Omega} = 0.6 \text{ mA}$$

(b) Since r_s is much greater than r_e', the overall voltage gain is

$$\frac{v_{cb}}{v_s} \cong \frac{r_L}{r_s} \cong \frac{5 \text{ k}\Omega}{2 \text{ k}\Omega} = 2.5$$

(c) In this case, then,

$$i_{c(\text{peak})} \cong i_{e(\text{peak})} \cong \frac{v_{s(\text{peak})}}{r_s} = \frac{400 \text{ mV}}{2 \text{ k}\Omega} = 200 \text{ }\mu\text{A}$$

Thus, by Ohm's law, the output voltage is

$$v_{cb(\text{peak})} = r_L i_{c(\text{peak})} \cong (5 \text{ k}\Omega)(200 \text{ }\mu\text{A}) = 1 \text{ V}$$

Or we can show that since $v_{cb}/v_{eb} \cong 2.5$, then $v_{cb} \cong 2.5 v_s$, or

$$v_{cb(\text{peak})} \cong 2.5 v_{s(\text{peak})} = 2.5 \times 0.4 \text{ V} = 1 \text{ V}$$

7-31. The upper and lower end-points of the dc load line are 1.2 mA and 12 V respectively. The upper and lower end-points of the ac load line are 1.8 mA and 9 V respectively. The coordinates of the operating point, which are the quiescent collector current and voltage, are 0.6 mA and 6 V respectively.—Calculations for the quiescent values are:

$$I_C \cong I_E \cong \frac{V_{EE}}{R_E} = \frac{12 \text{ V}}{20 \text{ k}\Omega} = 0.6 \text{ mA}$$

$$V_C = V_{CC} - R_C I_C \cong 12 \text{ V} - (10 \text{ k}\Omega)(0.6 \text{mA})$$

$$\cong 6 \text{ V}$$

The end points of the dc load line are:

$$I_{C(\text{sat})} = \frac{V_{CC}}{R_C} = \frac{12 \text{ V}}{10 \text{ k}\Omega} = 1.2 \text{ mA}$$

$$V_{C(\text{cutoff})} = V_{CC} = 12 \text{ V}$$

Calculations for the end points of the ac load line are:

$$i_{C(\text{sat})} = I_C + \frac{V_C}{r_L} \cong 0.6 \text{ mA} + \frac{6 \text{ V}}{5 \text{ k}\Omega} = 1.8 \text{ mA}$$

$$v_{C(\text{cutoff})} = V_C + r_L I_C \cong 6 \text{ V} + (5 \text{ k}\Omega)(0.6 \text{ mA}) = 9 \text{ V}$$

7-33. The circuit is capable of about a 3 V peak or a 6 V peak-to-peak unclipped output voltage. Attempts to get even a little more than this out causes clipping due to saturation as can be viewed by inspection of this circuit ac load line and operating point.

7-35. (a) $R_E \cong 15$ kΩ, (b) maximum unclipped $v_{cb(\text{peak})} \cong 4$ V.— The quiescent collector current I_C required for an optimum operating point, which is approximately equal to the emitter bias current I_E, is determined with Eq. 7-21. Thus

$$I_C \cong I_E \cong \frac{V_{CC}}{R_C + r_L} \cong \frac{12 \text{ V}}{10 \text{ k}\Omega + 5 \text{ k}\Omega} = 0.8 \text{ mA}$$

where $r_L = R_C \| R_L = 5$ kΩ. The emitter bias resistance, therefore, must be

$$R_E \cong \frac{V_{EE}}{I_E} \cong \frac{12 \text{ V}}{0.8 \text{ mA}} = 15 \text{ k}\Omega$$

At the optimum operating point the collector-to-ground voltage V_C can be determined easily with Eq. 7-18. That is,

$$V_C = r_L I_C \cong (5 \text{ k}\Omega)(0.8 \text{ mA}) = 4 \text{ V}$$

This collector voltage is centered on the ac load line, allowing the signal to swing the collector voltage by a maximum of 4 V, to either saturation or cutoff, without clipping.

7-37. The output signal voltage v_{cb} has its positive alternations clipped at 8 V, and has sinusoidal unclipped negative alternations with a peak of 14 V.

CHAPTER 8   577

7-39. (a) $R_E \cong 5.33$ kΩ, (b) $v_{cb(peak)} \cong 14$ V and is an unclipped sine wave, (c) both positive and negative alternations of v_{cb} are clipped at 15 V.

Chapter 8: Part I

8-1. $I_C \cong 1.2$ mA, $V_{CE} \cong 6.24$ V, $P_C \cong 7.5$ mW.—Assuming that $V_{BE} \cong 0$ and since $I_{CEO} \cong 0$, by Eq. 8-8

$$I_B \cong \frac{V_{CC}}{R_B} \cong \frac{12 \text{ V}}{800 \text{ k}\Omega} = 15 \text{ }\mu\text{A}$$

Thus, using Eq. 8-5,

$$I_C \cong \beta I_B \cong 80(15 \text{ }\mu\text{A}) = 1.2 \text{ mA}$$

And, with Eq. 8-9,

$$V_{CE} = V_{CC} - R_C I_C \cong 12 \text{ V} - (4.8 \text{ k}\Omega)(1.2 \text{ mA}) = 6.24 \text{ V}$$

And therefore, from Eq. 6-10,

$$P_C \cong V_{CE} I_C \cong (6.24 \text{ V})(1.2 \text{ mA}) \cong 7.5 \text{ mW}$$

8-3. $I_{CEO} \cong 104$ μA, $I_{CBO} \cong 1.3$ μA.—With switch S open, the current in the 4.8-kΩ resistor is leakage I_{CEO}. Since this current causes 0.5-V drop across the 4.8-kΩ resistor, its value is found by Ohm's law:

$$I_{CEO} = \frac{0.5 \text{ V}}{4.8 \text{ k}\Omega} \cong 104 \text{ }\mu\text{A}$$

Therefore, since by Eq. 8-2, $I_{CEO} \cong \beta I_{CBO}$, then

$$I_{CBO} \cong \frac{I_{CEO}}{\beta} \cong \frac{104 \text{ }\mu\text{A}}{80} = 1.3 \text{ }\mu\text{A}$$

8-5. (a) $I_C \cong 1.35$ mA, $V_{CE} \cong 5.52$ V, $P_C \cong 7.45$ mW, (b) $I_C \cong 2.4$ mA, $V_{CE} \cong 0.5$ V, $P_C \cong 1.2$ mW.—(a) The answer to Prob. 8-1, which refers to the same circuit, shows that $I_B \cong 15$ μA. In this case, since we have a leakage current of 150 μA at 25°C, at this temperature,

$$I_C = \beta I_B + I_{CEO} \cong 80(15 \text{ }\mu\text{A}) + 150 \text{ }\mu\text{A} = 1.35 \text{ mA}$$

by Eq. 8-6. And

$$V_{CE} \cong 12 \text{ V} - (4.8 \text{ k}\Omega)(1.35 \text{ mA}) = 5.52 \text{ V}$$

And therefore

$$P_C \cong (5.52 \text{ V})(1.35 \text{ mA}) \cong 7.45 \text{ mW}$$

(b) Since the transistor is a germanium type, the leakage doubles every 10°C. In this case, then, with a temperature change from 25°C to 55°C, the leakage doubles three times. Therefore, at 55°C,

$$I_{CEO} \cong 150 \ \mu A (2)^3 = (0.15 \text{ mA})(8) = 1.2 \text{ mA}$$

Therefore, now,

$$I_C = \beta I_B + I_{CEO} \cong 1.2 + 1.2 = 2.4 \text{ mA}$$

And

$$V_{CE} = V_{CC} - R_C I_C \cong 12 \text{ V} - (4.8 \text{ k}\Omega)(2.4 \text{ mA}) \cong 0.5 \text{ V}$$

And therefore

$$P_C \cong V_{CE} I_C \cong (0.5 \text{ V})(2.4 \text{ mA}) = 1.2 \text{ mW}$$

8-7. $R_B \cong 770 \text{ k}\Omega.$—At the optimum operating point, by Eq. 7-21,

$$I_C = \frac{V_{CC}}{R_C + r_L} = \frac{12 \text{ V}}{4.8 \text{ k}\Omega + 4.8 \text{ k}\Omega} = 1.25 \text{ mA}$$

Therefore we need an

$$I_B \cong \frac{I_C}{\beta} = \frac{1.25 \text{ mA}}{80} \cong 15.6 \ \mu A$$

and a base bias resistance

$$R_B = \frac{V_{CC}}{I_B} \cong \frac{12 \text{ V}}{15.6 \ \mu A} \cong 770 \text{ k}\Omega$$

8-9. (a) $I_C \cong 1 \text{ mA}$, $V_{CE} \cong 12 \text{ V}$, (b) $I_{C(\text{sat})} \cong 3 \text{ mA}$, $V_{CE(\text{cutoff})} = 18 \text{ V}.$—(a) Since $I_{CEO} \cong 0$, by Eq. 8-11,

$$I_C = \frac{V_{CC}}{R_E + R_B/\beta} \cong \frac{18 \text{ V}}{1 \text{ k}\Omega + 1.7 \text{ M}\Omega/100} = 1 \text{ mA}$$

Therefore, by Eq. 8-14,

$$V_{CE} = V_{CC} - I_C(R_C + R_E) \cong 18 \text{ V} - (1 \text{ mA})(6 \text{ k}\Omega) = 12 \text{ V}$$

(b) With these quiescent I_C and V_{CE} values known, we now find from Eq. 8-15

CHAPTER 8

$$I_{C(\text{sat})} \cong \frac{V_{CC}}{R_C + R_E} = \frac{18 \text{ V}}{5 \text{ k}\Omega + 1 \text{ k}\Omega} = 3 \text{ mA}$$

At cutoff $V_{CE} = V_{CC} = 18$ V.

8-11. $R_B \cong 1.1$ MΩ.—As determined in Prob. 8-9, this circuit has $I_{C(\text{sat})} \cong 3$ mA. The required quiescent collector current is half of this or $I_C \cong 1.5$ mA. Solving Eq. 8-11 for R_B we get

$$R_B \cong \beta \left(\frac{V_{CC}}{I_C} - R_E \right) \cong 100 \left(\frac{18 \text{ V}}{1.5 \text{ mA}} - 1 \text{ k}\Omega \right) = 100(11 \text{ k}\Omega) = 1.1 \text{ M}\Omega$$

8-13. (a) $I_C \cong 1$ mA, $V_{CE} \cong 15$ V, (b) $I_{C(\text{sat})} = 4$ mA, $V_{CE(\text{cut-off})} = 20$ V.—Since leakage $I_{CEO} \cong 0$, by Eq. 8-16

$$I_C \cong \frac{V_{CC}}{R_C + R_B/\beta} = \frac{20 \text{ V}}{5 \text{ k}\Omega + 900 \text{ k}\Omega/60} = 1 \text{ mA}$$

Thus, by Eq. 8-20,

$$V_{CE} = V_{CC} - R_C I_C \cong 20 \text{ V} - (5 \text{ k}\Omega)(1 \text{ mA}) = 15 \text{ V}$$

And, from Eq. 8-21,

$$I_{C(\text{sat})} = \frac{V_{CC}}{R_C} = \frac{20 \text{ V}}{5 \text{ k}\Omega} = 4 \text{ mA}$$

And finally,

$$V_{CE(\text{cutoff})} = V_{CC} = 20 \text{ V}$$

8-15. $R_B \cong 300$ kΩ.—As determined in Prob. 8-13, $I_{C(\text{sat})} = 4$ mA. Half of this is the collector current at the optimum operating value. Thus, since $I_C = 2$ mA, we rearrange Eq. 8-16 and solve for R_B, that is,

$$R_B \cong \beta \left(\frac{V_{CC}}{I_C} - R_C \right) = 60(10 \text{ k}\Omega - 5 \text{ k}\Omega) = 300 \text{ k}\Omega$$

8-17. (a) $I_C \cong 1$ mA, $V_{CE} \cong 7$ V, (b) $I_{C(\text{sat})} \cong 1.47$ mA, $V_{CE(\text{cut-off})} = 22$ V.—(a) In this circuit, assuming that leakage has negligible effect, Eq. 8-25 gives

$$I_C \cong \frac{V_{EE}}{R_E} = \frac{10 \text{ V}}{10 \text{ k}\Omega} = 1 \text{ mA}$$

Also, by rearranging Eq. 8-30

$$V_{CE} \cong V_{CC} - R_C I_C \cong 12 \text{ V} - (5 \text{ k}\Omega)(1 \text{ mA}) = 7 \text{ V}$$

(b) With two dc voltage sources, Eq. 8-31 gives

$$I_{C(\text{sat})} \cong \frac{V_{CC}+V_{EE}}{R_C+R_E} = \frac{22 \text{ V}}{15 \text{ k}\Omega} \cong 1.47 \text{ mA}$$

And, finally, $V_{CE(\text{cutoff})} = V_{CC} + V_{EE} = 22$ V.

8-19. $R_E \cong 8.33$ kΩ.—With a collector-to-emitter voltage $V_{CE} = 6$ V, and a source $V_{CC} = 12$ V, the difference of 6 V is across R_C. The current in R_C therefore is

$$I_C = \frac{6 \text{ V}}{5 \text{ k}\Omega} = 1.2 \text{ mA}$$

Now solving Eq. 8-25 for R_E and substituting knowns into it, we get

$$R_E \cong \frac{V_{EE}}{I_C} = \frac{10 \text{ V}}{1.2 \text{ mA}} \cong 8.33 \text{ k}\Omega$$

8-21. (a) $I_C \cong 1.2$ mA, $V_{CE} \cong 18$ V, (b) $I_{C(\text{sat})} = 3$ mA, $V_{CE(\text{cutoff})} = 30$ V.—(a) Assuming that the effect of leakage is negligible, we may use Eq. 8-34:

$$I_C \cong \frac{R_2}{R_1+R_2}\left(\frac{V_{CC}}{R_E}\right) = \frac{40 \text{ k}\Omega}{250 \text{ k}\Omega}\left(\frac{30 \text{ V}}{4 \text{ k}\Omega}\right) = 1.2 \text{ mA}$$

Thus, by Eq. 8-39,

$$V_{CE} \cong V_{CC} - (R_C+R_E)I_C \cong 30 \text{ V} - (10 \text{ k}\Omega)(1.2 \text{ mA}) = 18 \text{ V}$$

(b) Now with Eq. 8-40

$$I_{C(\text{sat})} = \frac{V_{CC}}{R_C+R_E} = \frac{30 \text{ V}}{10 \text{ k}\Omega} = 3 \text{ mA}$$

And $V_{CE(\text{cutoff})} = V_{CC} = 30$ V.

8-23. $R_E \cong 4.8$ kΩ.—Solving Eq. 8-36 for R_E and then substituting the knowns into it, we get

$$R_E \cong \frac{R_2}{R_1+R_2}\left(\frac{V_{CC}}{I_C}\right) \cong \frac{40 \text{ k}\Omega}{250 \text{ k}\Omega}\left(\frac{30 \text{ V}}{1 \text{ mA}}\right) = 4.8 \text{ k}\Omega$$

since $I_C = \beta I_B + I_{CEO}$.

8-25. (a) $I_C \cong 1.189$ mA, (b) $I_C \cong 1.755$ mA.—In this case, since leakage I_{CEO} is not negligible, Eq. 8-23 is applicable. Therefore

$$I_C \cong \frac{V_{CC}}{R_E + R_B/\beta} + \left(\frac{R_B + R_E}{R_B + \beta R_E}\right) I_{CEO}$$

$$\cong 1 \text{ mA} + 0.945(0.2 \text{ mA}) \cong 1.189 \text{ mA}$$

At 37°C, an increase of 12°C, the leakage in a silicon transistor doubles twice. That is, at this higher temperature,

$$I_{CEO} \cong (0.2 \text{ mA})(2)^3 = 0.8 \text{ mA}$$

Thus, now by Eq. 8-23,

$$I_C \cong 1 + (0.945)0.8 \cong 1.755 \text{ mA}$$

8-27. (a) $I_C \cong 1.12$ mA, (b) $I_C \cong 1.48$ mA.—(a) For the circuit at 25°C, Eq. 8-16 gives

$$I_C \cong \frac{V_{CC}}{R_C + R_B/\beta} + \left(\frac{R_B}{R_B + \beta R_C}\right) I_{CEO}$$

$$\cong \frac{20 \text{ V}}{5 \text{ k}\Omega + 900 \text{ k}\Omega/60} + \left(\frac{900 \text{ k}\Omega}{1200 \text{ k}\Omega}\right)(0.16 \text{ mA})$$

$$\cong 1 + 0.12 = 1.12 \text{ mA}$$

(b) At 45°C, a change of 20°C, the leakage in the germanium transistor doubles twice. Thus

$$I_{CEO} \cong (0.16 \text{ mA})(2)^2 = 0.64 \text{ mA}$$

So at this higher temperature,

$$I_C \cong 1 + (0.75)0.64 = 1 + 0.48 = 1.48 \text{ mA}$$

8-29. (a) $I_C \cong 0.98$ mA, (b) $I_C \cong 1.07$ mA.—In this type of circuit, if leakage is 0.5 mA at 30°C, Eq. 8-23 gives

$$I_C \cong \frac{V_{EE}}{R_E + R_B/\beta} + \left(\frac{R_B + R_E}{R_B + \beta R_E}\right) I_{CEO}$$

$$\cong \frac{10}{10 \text{ k}\Omega + 40 \text{ k}\Omega/80} + \left(\frac{40 \text{ k}\Omega + 10 \text{ k}\Omega}{40 \text{ k}\Omega + 80(10 \text{ k}\Omega)}\right)(0.5 \text{ mA})$$

$$\cong 0.95 + 0.03 = 0.98 \text{ mA}$$

(b) At 50°C, at 20°C increase, the leakage doubles twice and becomes

$$I_{CEO} \cong 0.5(2)^2 = 2 \text{ mA}$$

Thus by Eq. 8-23 $I_C \cong 0.95 + 0.12 = 1.07$ mA.

8-31. (a) $I_C \cong 0.95$ mA, (b) $I_C \cong 0.975$ mA, (c) about 2.64%.—
(a) We can note the effect changes in β have on the collector current with Eq. 8-23. However, because the leakage $I_{CEO} \cong 0$, the second term on the right side drops out. Thus, when $\beta = 80$,

$$I_C \cong \frac{V_{EE}}{R_E + R_B/\beta} = \frac{10 \text{ V}}{10 \text{ k}\Omega + 40 \text{ k}\Omega/80} \cong 0.95 \text{ mA}$$

When $\beta = 160$, a 100% increase,

$$I_C \cong \frac{10 \text{ V}}{10 \text{ k}\Omega + 40 \text{ k}\Omega/160} \cong 0.975 \text{ mA}$$

The percent of increase is

$$\frac{0.975 - 0.95}{0.95} \times 100 \cong 2.64\%$$

8-33. $I_C \cong 2.4$ mA.—Assuming that $I_{CEO} \cong 0$, and since $I_B \cong 15$ μA, according to the answer to Prob. 8-1,

$$I_C \cong \beta I_B \cong 160(15 \text{ μA}) = 2.4 \text{ mA}$$

8-35. (a) $I_C \cong 1.284$ mA, $V_{CE} \cong 17.16$ V, (b) The I_C has increased a little and the V_{CE} has decreased a little. The changes in I_C and V_{CE}, however, are much smaller than the change in the transistor β.—(a) In this type of circuit, if leakage is not negligible, Eq. 8-34 applies:

$$I_C \cong \left(\frac{R_2}{R_1 + R_2}\right)\frac{V_{CC}}{R_E} + \left[\frac{R_1 R_2}{\beta R_E(R_1 + R_2)}\right] I_{CEO}$$

Therefore

$$I_C \cong \frac{40 \text{ k}\Omega}{250 \text{ k}\Omega}\left(\frac{30 \text{ V}}{4 \text{ k}\Omega}\right) + \left[\frac{(210 \text{ k}\Omega)(40 \text{ k}\Omega)}{100(4 \text{ k}\Omega)(250 \text{ k}\Omega)}\right](1 \text{ mA})$$

$$\cong 1.2 + (8.4 \times 10^{-2})1 = 1.284 \text{ mA}$$

And, by Eq. 8-39,

$$V_{CE} = V_{CC} - (R_C + R_E)I_C$$
$$\cong 30 \text{ V} - (10 \text{ k}\Omega)(1.284 \text{ mA}) = 17.16 \text{ V}$$

Chapter 8: Part II

8-37. (a) $I_E \cong 2$ mA, $r_{\text{in(stage)}} \cong 1.25$ kΩ, (b) $I_E \cong 1$ mA, $r_{\text{in(stage)}} \cong 2.5$ kΩ, (c) $I_E \cong 0.5$ mA, $r_{\text{in(stage)}} \cong 5$ kΩ, (d) $I_E \cong 0.25$ mA, $r_{\text{in(stage)}} \cong 10$ kΩ.—(a) When $R = 0$, $R_B = 1$ MΩ. Thus

$$I_B \cong \frac{V_{CC}}{R_B} = \frac{20 \text{ V}}{1 \text{ M}\Omega} = 20 \ \mu\text{A}$$

Since $\beta = 100$,

$$I_E \cong I_C \cong \beta I_B \cong 100(20 \ \mu\text{A}) = 2 \text{ mA}$$

The dynamic resistance looking into the base therefore is

$$r_b' \cong \frac{25 \text{ mV}}{I_B} \cong \frac{25 \text{ mV}}{20 \ \mu\text{A}} = 1.25 \text{ k}\Omega$$

The source v_s sees r_b' and the base bias resistance R_B in parallel; thus,

$$r_{\text{in(stage)}} \cong r_b' \| R_B \cong 1.25 \text{ k}\Omega$$

(b) When $R = 1 \text{ M}\Omega$, $R_B = 2 \text{ M}\Omega$. Thus

$$I_B \cong \frac{20 \text{ V}}{2 \text{ M}\Omega} = 10 \mu\text{A}$$

Since $\beta = 100$,

$$I_E \cong I_C \cong \beta I_B \cong 100(10 \ \mu\text{A}) = 1 \text{ mA}$$

In this case, then,

$$r_b' \cong \frac{25 \text{ mV}}{10 \ \mu\text{A}} = 2.5 \text{ k}\Omega$$

The source v_s now sees this r_b' in parallel with 2 MΩ base bias resistance R_B. Therefore,

$$r_{\text{in(stage)}} \cong 2.5 \text{ k}\Omega \| 2 \text{ M}\Omega \cong 2.5 \text{ k}\Omega$$

(c) When $R = 3 \text{ M}\Omega$, $R_B = 4 \text{ M}\Omega$. Therefore,

$$I_B \cong \frac{20 \text{ V}}{4 \text{ M}\Omega} = 5 \ \mu\text{A}$$

and

$$I_E \cong I_C \cong 100(5 \ \mu\text{A}) = 0.5 \text{ mA}$$

and therefore

$$r_b' \cong \frac{25 \text{ mV}}{5 \ \mu\text{A}} = 5 \text{ k}\Omega$$

Now the source v_s sees this r_b' in parallel with a 4 MΩ base bias resistance, or

$$r_{in(stage)} \cong 5 \text{ k}\Omega \| 4 \text{ M}\Omega \cong 5 \text{ k}\Omega$$

(d) When $R = 7$ MΩ, $R_B \cong 8$ MΩ. Therefore,

$$I_B \cong \frac{20 \text{ V}}{8 \text{ M}\Omega} = 2.5 \text{ }\mu\text{A}$$

which means that

$$I_E \cong I_C \cong 100(2.5 \text{ }\mu\text{A}) = 0.25 \text{ mA}$$

and that

$$r_b' \cong \frac{25 \text{ mV}}{2.5 \text{ }\mu\text{A}} = 10 \text{ k}\Omega$$

The source v_s sees

$$r_{in(stage)} \cong 10 \text{ k}\Omega \| 8 \text{ M}\Omega \cong 10 \text{ k}\Omega$$

8-39. (a) $A_e \cong 240$, (b) $A_i \cong 80$, (c) $A_p \cong 19{,}200$ (d) $v_{ce(peak)} \cong 0.48$ V.—The base current is

$$I_B \cong \frac{V_{CC}}{R_B} \cong \frac{18 \text{ V}}{1.44 \text{ M}\Omega} \cong 12.5 \text{ }\mu\text{A}$$

Since $\beta \cong 80$,

$$I_E \cong I_C \cong \beta I_B \cong 80(12.5 \text{ }\mu\text{A}) = 1 \text{ mA}$$

then

$$r_e' \cong \frac{25 \text{ mV}}{I_E} \cong \frac{25 \text{ mV}}{1 \text{ mA}} = 25 \text{ }\Omega$$

and

$$A_e \cong \frac{r_L}{r_e'} \cong \frac{6000}{25} \cong 240$$

where $r_L \cong 7.5 \text{ k}\Omega \| 30 \text{ k}\Omega = 6 \text{ k}\Omega$.

(b) The current gain is $A_i \cong \beta \cong 80$. Thus

$$A_p = A_e A_i \cong 240(80) = 19{,}200$$

In general, the signal output is $v_{ce} = A_e v_s$. In this case, then,

$$v_{ce(peak)} \cong 240(2 \text{ mV}) = 480 \text{ mV} = 0.48 \text{ V}$$

8-41. (a) $A_e \cong 256$, (b) $v_{ce(peak)} \cong 0.256$ V. (a) Assuming that the leakage is negligible,

CHAPTER 8

$$I_C \cong \frac{V_{CC}}{R_E + R_B/\beta} \cong \frac{18 \text{ V}}{0.5 \text{ k}\Omega + 8.5 \text{ k}\Omega} = 2 \text{ mA}$$

Thus

$$r_e' \cong \frac{25 \text{ mV}}{2 \text{ mA}} = 12.5 \text{ }\Omega$$

As seen by the collector,

$$r_L \cong R_C \| R_L = 4 \text{ k}\Omega \| 16 \text{ k}\Omega = 3.2 \text{ k}\Omega$$

With R_E bypassed,

$$A_e \cong \frac{r_L}{r_e'} \cong \frac{3200 \text{ }\Omega}{12.5 \text{ }\Omega} = 256$$

(b) Therefore $v_{ce(\text{peak})} = A_e v_{s(\text{peak})} \cong 256(1 \text{ mV}) = 0.256$ V.

8-43. (a) $A_e \cong 6.4$, (b) $v_{o(\text{peak})} \cong 1.28$ V.—(a) When R_E is not bypassed, by Eq. 8-51,

$$A_e \cong \frac{r_L}{R_E} \cong \frac{3200}{500} = 6.4$$

(b) Thus $v_{o(\text{peak})} \cong 6.4(200 \text{ mV}) = 1.28$ V.

8-45. $A_e \cong 10$, $v_{o(\text{peak})} \cong 4$ V.—In this circuit,

$$r_L \cong R_C \| R_L = 25 \text{ k}\Omega \| 100 \text{ k}\Omega = 20 \text{ k}\Omega$$

R_1 or 2 kΩ is not bypassed in the emitter lead. Therefore,

$$A_e \cong \frac{20 \text{ k}\Omega}{2 \text{ k}\Omega} = 10$$

which means that $v_{o(\text{peak})} \cong 10 v_{s(\text{peak})} = 10(400 \text{ mV}) = 4$ V.

8-47. $A_e \cong 192$, $v_{o(\text{peak})} \cong 384$ mV.—With this type of circuit

$$I_C \cong \frac{V_{CC}}{R_C + R_B/\beta} \cong \frac{12 \text{ V}}{6 \text{ k}\Omega + 720 \text{ k}\Omega/120} = 1 \text{ mA}$$

Thus

$$r_e' \cong \frac{25 \text{ mV}}{I_E} \cong \frac{25 \text{ mV}}{1 \text{ mA}} = 25 \text{ }\Omega$$

and as seen by the collector

$$r_L \cong R_C \| R_L = 6 \text{ k}\Omega \| 24 \text{ k}\Omega = 4.8 \text{ k}\Omega$$

Therefore the voltage gain is

$$A_e \cong \frac{r_L}{r_e'} \cong \frac{4.8 \text{ k}\Omega}{25 \text{ }\Omega} = 192$$

and consequently $v_{o(\text{peak})} \cong 192(2 \text{ mV}) = 384 \text{ mV}$.

8-49. (a) $A_e \cong 160$, (b) $A_p \cong 16{,}600$, (c) $r_{\text{in(stage)}} \cong 2.22 \text{ k}\Omega$, (d) $v_{o(\text{peak})} \cong 0.16$.—(a) Assuming that leakage is negligible, by Eq. 8-25,

$$I_E \cong \frac{V_{EE}}{R_E} \cong \frac{10 \text{ V}}{10 \text{ k}\Omega} = 1 \text{ mA}$$

Therefore

$$r_e' \cong \frac{25 \text{ mV}}{1 \text{ mA}} = 25 \text{ }\Omega$$

Also

$$r_L \cong R_C \| R_L \cong 5 \text{ k}\Omega \| 20 \text{ k}\Omega \cong 4 \text{ k}\Omega$$

Thus, since R_E is bypassed,

$$A_e \cong \frac{r_L}{r_e'} \cong \frac{4 \text{ k}\Omega}{25 \text{ }\Omega} = 160$$

and (b)

$$A_p = A_e \beta \cong 160(100) = 16{,}000$$

(c) As seen by the source v_s

$$r_{\text{in(stage)}} \cong R_B \| \beta r_e' \cong 20 \text{ k}\Omega \| \beta 25 \text{ }\Omega \cong 2.22 \text{ k}\Omega$$

(d) In this case, $v_{o(\text{peak})} \cong 160(1 \text{ mV}) = 160 \text{ mV}$.

8-51. (a) $r_{\text{in(stage)}} \cong 750 \text{ }\Omega$, (b) $v_{be(\text{peak})} \cong 7.2 \text{ mV}$, (c) $v_{o(\text{peak})} \cong 1.84 \text{ V}$.—(a) Since $r_e' \cong 12.5 \text{ }\Omega$, as was determined in Prob. 8-41, then

$$r_{\text{in(stage)}} \cong R_B \| \beta r_e' \cong 510 \text{ k}\Omega \| 60(12.5 \text{ }\Omega) \cong 750 \text{ }\Omega$$

(b) In general, since the emitter is at ac ground and since

$$v_{be} \cong \frac{v_g \, r_{\text{in(stage)}}}{r_i + r_{\text{in(stage)}}}$$

then in this case,

$$v_{be(\text{peak})} \cong \frac{(200 \text{ mV})(750 \text{ }\Omega)}{20 \text{ k}\Omega + 750 \text{ }\Omega} \cong 7.2 \text{ mV}$$

CHAPTER 8

(c) And since $v_{ce}/v_{be} \cong 256$,

$$v_{ce(\text{peak})} = v_{o(\text{peak})} \cong 256(7.2 \text{ mV}) = 1.84 \text{ V}$$

8-53. (a) $r_{\text{in(stage)}} \cong 28.4$ kΩ, (b) $v_{\text{in(peak)}} \cong 118$ mV, (c) $v_{o(\text{peak})} \cong 755$ mV.—(a) With R_E not bypassed,

$$r_{\text{in(stage)}} \cong R_B \| \beta R_E \cong 510 \text{ k}\Omega \| 60(0.5 \text{ k}\Omega) \cong 28.4 \text{ k}\Omega$$

(b) Therefore the signal voltage at the base with respect to ground is

$$v_{\text{in(peak)}} = \frac{v_s r_{\text{in(stage)}}}{r_i + r_{\text{in(stage)}}} \cong \frac{(200 \text{ mV})(28.4 \text{ k}\Omega)}{20 \text{ k}\Omega + 28.4 \text{ k}\Omega} \cong 118 \text{ mV}$$

(c) In this case $A_e \cong 6.4$ (see Prob. 8-43) and therefore

$$v_{o(\text{peak})} \cong 6.4(118 \text{ mV}) \cong 755 \text{ mV}$$

8-55. (a) $r_{\text{in(stage)}} \cong 232$ kΩ, (b) $v_{\text{in(peak)}} \cong 184$ mV, (c) $v_{o(\text{peak})} \cong 1.84$ V.—The ac resistance seen looking into the base of the transistor is about β times the unbypassed resistance in the emitter lead, assuming that the reactance of the bypass capacitor is negligible. Therefore, as seen to the right of the base coupling capacitor,

$$r_{\text{in(stage)}} \cong R_B \| \beta R_1 = 6.6 \text{ M}\Omega \| 120(2 \text{ k}\Omega) \cong 232 \text{ k}\Omega$$

The signal voltage from base to ground, therefore, is

$$v_{\text{in(peak)}} \cong \frac{(200 \text{ mV})(232 \text{ k}\Omega)}{20 \text{ k}\Omega + 232 \text{ k}\Omega} \cong 184 \text{ mV}$$

Since, by Eq. 8-51,

$$A_e \cong \frac{r_L}{R_E}$$

where R_E is the unbypassed portion of the resistance in the emitter, in this case,

$$A_e \cong \frac{r_L}{R_1} \cong \frac{20 \text{ k}\Omega}{2 \text{ k}\Omega} = 10$$

Therefore the signal output voltage is

$$v_{o(\text{peak})} \cong 10(184 \text{ mV}) = 1.84 \text{ V}$$

8-57. $v_{o(\text{peak})} \cong 2.96$ V.—Assuming that $I_{CEO} \cong 0$,

$$I_c \cong \frac{V_{cc}}{R_c + R_B/\beta} = \frac{12\text{ V}}{6\text{ k}\Omega + 720\text{ k}\Omega/120} = 1\text{ mA}$$

Thus

$$r_e' \cong \frac{25\text{ mV}}{1\text{ mA}} = 25\text{ }\Omega$$

and therefore

$$A_e \cong \frac{r_L}{r_e'} \cong \frac{4.8\text{ k}\Omega}{25\text{ }\Omega} = 192$$

As shown in the equivalent circuit in Fig. 8-25C, the signal source sees R_B/A_e in parallel with $\beta r_e'$. That is,

$$r_{in(stage)} \cong (R_B/A_e)\|\beta r_e' \cong 720\text{ k}\Omega/192\|120(25\text{ }\Omega) \cong 1.67\text{ k}\Omega$$

At the base with respect to ground,

$$v_{in(peak)} \cong \frac{(200\text{ mV})(1.67\text{ k}\Omega)}{20\text{ k}\Omega + 1.67\text{ k}\Omega} \cong 15.4\text{ mV}$$

Therefore at the collector to ground

$$v_{o(peak)} \cong 192(15.4\text{ mV}) \cong 2.96\text{ V}$$

8-59. $v_{o(peak)} \cong 2.92$ V.—Neglecting leakage,

$$I_E \cong \frac{V_{EE}}{R_E} \cong \frac{10\text{ V}}{10\text{ k}\Omega} = 1\text{ mA}$$

Thus

$$r_e' \cong \frac{25\text{ mV}}{1\text{ mA}} = 25\text{ }\Omega$$

and therefore

$$A_e \cong \frac{r_L}{r_e'} \cong \frac{4\text{ k}\Omega}{25\text{ }\Omega} = 160$$

Since R_E is bypassed,

$$r_{in(stage)} \cong R_B\|\beta r_e' \cong 20\text{ k}\Omega\|90(25\text{ }\Omega) \cong 2\text{ k}\Omega$$

so that the signal at the base

$$v_{in(stage)} \cong \frac{(200\text{ mV})(2\text{ k}\Omega)}{20\text{ k}\Omega + 2\text{ k}\Omega} \cong 18.2\text{ mV}$$

CHAPTER 8

The resulting signal output is

$$v_{o(\text{peak})} \cong 160(18.2 \text{ mV}) \cong 2.92 \text{ V}$$

8-61. $v_{o(\text{peak})} \cong 2.76$ V.—Since R_E is bypassed and I_{CEO} is assumed negligible, by Eq. 8-35

$$I_C \cong I_E \cong \frac{20 \text{ k}\Omega}{200 \text{ k}\Omega}\left(\frac{20 \text{ V}}{2 \text{ k}\Omega}\right) = 1 \text{ mA}$$

Therefore

$$r_e{}' \cong \frac{25 \text{ mV}}{1 \text{ mA}} = 25 \text{ }\Omega$$

and since R_E is bypassed,

$$A_e \cong \frac{r_L}{r_e{}'} \cong \frac{6.4 \text{ k}\Omega}{25 \text{ }\Omega} = 256$$

and

$$r_{\text{in(stage)}} \cong R_1 \| R_2 \| \beta r_e{}'$$
$$\cong 180 \text{ k}\Omega \| 20 \text{ k}\Omega \| 100(25 \text{ }\Omega) \cong 2.2 \text{ k}\Omega$$

Thus

$$v_{\text{in(stage)}} \cong \frac{(500 \text{ mV})(2.2 \text{ k}\Omega)}{100 \text{ k}\Omega + 2.2 \text{ k}\Omega} \cong 10.8 \text{ mV}$$

And finally

$$v_{o(\text{peak})} \cong 256(10.8 \text{ mV}) \cong 2.76 \text{ V}$$

8-63. $v_{o(\text{peak})} \cong 227$ mV.—Without R_E bypassed, the voltage gain is, by Eq. 8-51,

$$A_e \cong \frac{r_L}{R_E} \cong \frac{6.4 \text{ k}\Omega}{2 \text{ k}\Omega} \cong 3.2$$

The ac resistance seen looking into the base is about β times R_E; therefore, as seen by the signal generator,

$$r_{\text{in(stage)}} \cong 180 \text{ k}\Omega \| 20 \text{ k}\Omega \| 100(2 \text{ k}\Omega) \cong 16.5 \text{ k}\Omega$$

Therefore at the base with respect to ground,

$$v_{\text{in(peak)}} \cong \frac{(500 \text{ mV})(16.5 \text{ k}\Omega)}{116.5} \cong 71 \text{ mV}$$

and the signal output is

$$v_{o(\text{peak})} \cong 3.2(71 \text{ mV}) \cong 227 \text{ mV}$$

8-65. (a) $I_{C(\text{sat})} \cong 2.4$ mA, $V_{C(\text{cutoff})} \cong 18$ V, (b) $I_C \cong 1$ mA, $V_{CE} \cong 10.5$ V, (c) $i_{C(\text{sat})} \cong 2.75$ mA, $v_{CE(\text{cutoff})} \cong 16.5$ V, (d) $v_{ce(\text{peak})} \cong 6$ V.—(a) When the transistor acts as a short,

$$I_{C(\text{sat})} \cong \frac{V_{CC}}{R_C} = \frac{18 \text{ V}}{7.5 \text{ k}\Omega} = 2.4 \text{ mA}$$

and when it acts as an open,

$$V_{CE(\text{cutoff})} = V_{CC} = 18 \text{ V}$$

(b) As shown in Prob. 8-39, $I_C \cong 1$ mA. Therefore the quiescent V_C is

$$V_C = V_{CE} = V_{CC} - R_C I_C \cong 18 \text{ V} - (7.5 \text{ k}\Omega)(1 \text{ mA}) = 10.5 \text{ V}$$

(c) With the quiescent values known, we find the ac load line endpoints:

$$i_{C(\text{sat})} = I_C + \frac{V_C}{r_L} \cong 1 \text{ mA} + \frac{10.5 \text{ V}}{6 \text{ k}\Omega} \cong 2.75 \text{ mA}$$

and

$$v_{CE(\text{cutoff})} = V_C + r_L I_C \cong 10.5 \text{ V} + (6 \text{ k}\Omega)(1 \text{ mA}) = 16.5 \text{ V}$$

(d) The maximum unclipped ouput signal thus is

$$v_{CE(\text{cutoff})} \cong 6 \text{ V}$$

8-67. (a) $I_C \cong 1.33$ mA, $V_{CE} \cong 8$ V, (b) $R_B \cong 1.08$ MΩ, (c) $v_{ce(\text{peak})} \cong 8$ V.—(a) At the optimum operating point, by Eq. 7-21,

$$I_C = \frac{V_{CC}}{R_C + r_L} = \frac{18 \text{ V}}{7.5 \text{ k}\Omega + 6 \text{ k}\Omega} \cong 1.33 \text{ mA}$$

and

$$V_{CE} = V_C = V_{CC} - R_C I_C \cong 18 \text{ V} - (7.5 \text{ k}\Omega)(1.33 \text{ mA}) \cong 8 \text{ V}$$

(b) Using $\beta = 80$ and neglecting leakage,

$$I_B \cong \frac{I_C}{\beta} \cong \frac{1.33 \text{ mA}}{80} \cong 16.6 \text{ }\mu\text{A}$$

which means that we must use

$$R_B \cong \frac{V_{CC}}{I_B} \cong \frac{18 \text{ V}}{16.6 \text{ }\mu\text{A}} \cong 1.08 \text{ M}\Omega$$

(c) At the optimum operating point, the peak value of the maximum unclipped output signal capability is about equal to the quiescent V_{CE}. Thus, in this case,

$$v_{ce\text{(peak)}} \cong 8 \text{ V}$$

8-69. (a) $I_{C(\text{sat})} \cong 4.5$ mA, $V_{CE(\text{cutoff})} = 18$ V, (b) $I_C \cong 3.6$ mA, $V_{CE} \cong 3.6$ V, (c) $i_{C(\text{sat})} \cong 4.66$ mA, $v_{CE(\text{cutoff})} \cong 15.8$ V, (d) $v_{o(\text{peak})} \cong 3.6$ V.—(a) If the transistor acts as a short, the full V_{CC} voltage is dropped across R_C and R_E. Thus

$$I_{C(\text{sat})} \cong \frac{V_{CC}}{R_C + R_E} = \frac{18 \text{ V}}{4 \text{ k}\Omega} = 4.5 \text{ mA}$$

If the transistor is cut off (acts as an open), then

$$V_{CE(\text{cutoff})} = V_{CC} = 18 \text{ V}$$

(b) Also,

$$I_C \cong \frac{18 \text{ V}}{1 \text{ k}\Omega + 240 \text{ k}\Omega/60} = 3.6 \text{ mA}$$

Thus

$$V_{CE} = V_C \cong V_{CC} - (R_C + R_E)I_C \cong 18 \text{ V} - (4 \text{ k}\Omega)(3.6 \text{ mA}) = 3.6 \text{ V}$$

(c) Since R_E is not bypassed, the ac load resistance as seen by the transistor is

$$r_L' = r_L + R_E = 2.4 + 1 = 3.4 \text{ k}\Omega$$

Now the end points of the ac load line are

$$i_{C(\text{sat})} = I_C + \frac{V_C}{r_L'} \cong 3.6 \text{ mA} + \frac{3.6 \text{ V}}{3.4 \text{ k}\Omega} \cong 4.66 \text{ mA}$$

and

$$v_{CE(\text{cutoff})} = V_C + r_L' I_C \cong 3.6 \text{ V} + (3.4 \text{ k}\Omega)(3.6 \text{ mA})$$
$$\cong 3.6 + 12.2 = 15.8 \text{ V}$$

(d) The maximum unclipped signal output in this case has a peak value that is the distance from the operating point voltage $V_{CE} = 3.6$ V and saturation, where the instantaneous $v_{ce} = 0$, Thus $v_{o(\text{peak})} \cong 3.6$ V.

8-71. (a) $I_C \cong 2.8$ mA, $V_{CE} \cong 6.8$ V, (b) $R_B \cong 325$ kΩ, $v_{o(\text{peak})} \cong 6.8$ V.—(a) With this type of circuit, the optimum operating point is obtained with Eq. 8-57:

$$I_C = \frac{V_{CC}}{R_C + R_E + r_L} = \frac{18\text{ V}}{3\text{ k}\Omega + 1\text{ k}\Omega + 2.4\text{ k}\Omega} \cong 2.8\text{ mA}$$

Therefore at the operating point,

$$V_{CE} \cong 18\text{ V} - (4\text{ k}\Omega)(2.8\text{ mA}) = 18 - 11.2 = 6.8\text{ V}$$

(b) We can, of course, solve Eq. 8-11 for R_B but we can also determine R_B as follows: Since we established $I_C \cong 2.8$ mA, we will have about

$$R_E I_C \cong (1\text{ k}\Omega)(2.8\text{ mA}) = 2.8\text{ V}$$

dropped across R_E. Thus, neglecting V_{BE}, the voltage across R_B is about $18 - 2.8 = 15.2$ V. And since

$$I_B \cong I_C/\beta \cong \frac{2.8\text{ mA}}{60} \cong 46.6\text{ }\mu\text{A},$$

then, by Ohm's law,

$$R_B \cong \frac{15.2\text{ V}}{46.6\text{ }\mu\text{A}} \cong 325\text{ k}\Omega$$

In this case, the operating point is about 6.8 V from cutoff and from saturation. Thus, the peak of the maximum unclipped output is $v_{o(\text{peak})} \cong 6.8$ V.

8-73. (a) $I_{C(\text{sat})} \cong 1.47$ mA, $V_{CE(\text{cutoff})} \cong 22$ V, (b) $I_C \cong 1$ mA, $V_{CE} \cong 7$ V, (c) $i_{C(\text{sat})} \cong 2.75$ mA, $v_{ce(\text{cutoff})} \cong 11$ V, (d) $v_{o(\text{peak})} \cong 4$ V.—(a) In this case,

$$I_{C(\text{sat})} \cong \frac{V_{CC} + V_{EE}}{R_C + R_E} = \frac{12\text{ V} + 10\text{ V}}{5\text{ k}\Omega + 10\text{ k}\Omega} \cong 1.47\text{ mA}$$

and $V_{CE(\text{cutoff})} = V_{CC} + V_{EE} = 22$ V.
(b) At the operating point,

$$I_C \cong \frac{V_{EE}}{R_E} \cong \frac{10\text{ V}}{10\text{ k}\Omega} = 1\text{ mA}$$

and

$$V_C = V_{CE} \cong V_{CC} - R_C I_C \cong 12\text{ V} - (5\text{ k}\Omega)(1\text{ mA}) = 7\text{ V}$$

(c) The end points of the ac load line therefore are

CHAPTER 9   593

$$i_{C(\text{sat})} = I_C + \frac{V_C}{r_L} \cong 1 \text{ mA} + \frac{7 \text{ V}}{4 \text{ k}\Omega} = 2.75 \text{ mA}$$

and

$$v_{CE(\text{cutoff})} = V_C + r_L I_C \cong 7 + 4 = 11 \text{ V}$$

Thus, the maximum peak unclipped output signal is $v_{o(\text{peak})} \cong 4$ V.

Chapter 9

9-1. (a) $r_{\text{in(base)}} \cong 200$ kΩ, (b) $r_{\text{in(stage)}} \cong 18.2$ kΩ.—With the bias method in this circuit

$$I_E \cong \frac{V_{EE}}{R_E} = \frac{10 \text{ V}}{5 \text{ k}\Omega} = 2 \text{ mA}$$

so that

$$r_e' \cong \frac{25 \text{ mV}}{2 \text{ mA}} = 12.5 \text{ }\Omega$$

As shown in the equivalent circuit (Fig. 9-4B), the ac resistance into the stage is R_B in parallel with $\beta r_e' + \beta r_L$. Thus, in this case, looking into the base,

$$r_{\text{in(base)}} \cong \beta r_e' + \beta r_L \cong 80 \ (12.5 \text{ }\Omega) + 80 \ (2.5 \text{ k}\Omega) \cong 200 \text{ k}\Omega$$

(b) As seen to the right of C_1

$$r_{\text{in(stage)}} \cong 20 \text{ k}\Omega \| 200 \text{ k}\Omega \cong 18.2 \text{ k}\Omega$$

9-3. $v_o \cong 200$ mV.—The amplification of signal voltage from base to emitter in this circuit is

$$A_e \cong \frac{r_L}{r_e' + r_L} \cong \frac{2.5 \text{ k}\Omega}{12.5 \text{ }\Omega + 2.5 \text{ k}\Omega} \cong 1$$

With no internal resistance r_s, the entire 200 mV of the signal source is applied to the base. That is, $v_{\text{in(base)}} \cong 200$ mV. With a voltage gain of about 1, therefore, $v_o \cong 1(200 \text{ mV}) = 200$ mV.

9-5. $v_o \cong 47$ mV.—When $R = 0$, $r_{\text{in(stage)}} \cong 18.2$ kΩ as was determined in Prob. 9-1. Since $r_s \cong 60$ kΩ,

$$v_{\text{in(base)}} \cong \frac{(200 \text{ mV})(18.2 \text{ k}\Omega)}{60 \text{ k}\Omega + 18.2 \text{ k}\Omega} \cong 47 \text{ mV}$$

Since $A_e \cong 1$, then $v_o \cong 1(47 \text{ mV}) = 47 \text{ mV}$.

9-7. Maximum unclipped $v_{o(\text{peak})} \cong 8$ V.—By Eq. 9-1, $V_{CE} = V_{CC} = 12$ V. Since $I_E \cong V_{EE}/R_E = 10 \text{ V}/5 \text{ k}\Omega = 2 \text{ mA}$, in this circuit, $I_E \cong 2$ mA.

Also, $r_L = R_E \| (R + 5 \text{ k}\Omega) = 5 \text{ k}\Omega \| 20 \text{ k}\Omega = 4 \text{ k}\Omega$. Thus, the end points on the ac load line are, by Eqs. 9-12 and 9-13,

$$i_{E(\text{sat})} = I_E + \frac{V_{CE}}{r_L} \cong 2 \text{ mA} + \frac{12 \text{ V}}{4 \text{ k}\Omega} = 5 \text{ mA}$$

and

$$v_{CE(\text{cutoff})} = V_{CE} + r_L I_E \cong 12 \text{ V} + (4 \text{ k}\Omega)(2 \text{ mA}) = 12 + 8 = 20 \text{ V}$$

Thus, the maximum unclipped signal output is $v_{o(\text{peak})} \cong 8$ V.

9-9. (a) $r_{\text{in(stage)}} \cong 18.6 \text{ k}\Omega$, (b) $r_{\text{in(base)}} \cong 120 \text{ k}\Omega$.—(b) In this circuit,

$$I_E \cong I_C \cong \frac{R_2}{R_1 + R_2}\left(\frac{V_{CC}}{R_E}\right) \cong \frac{40 \text{ k}\Omega}{90 \text{ k}\Omega}\left(\frac{9 \text{ V}}{2 \text{ k}\Omega}\right) = 2 \text{ mA}$$

Thus

$$r_e' \cong \frac{25 \text{ mV}}{2 \text{ mA}} = 12.5 \text{ }\Omega$$

With $R = 0$, $r_L = 2 \text{ k}\Omega \| 3 \text{ k}\Omega = 1.2 \text{ k}\Omega$. The ac resistance seen looking into the base is

$$r_{\text{in(base)}} \cong \beta r_e' + \beta r_L \cong 100(12.5 \text{ }\Omega) + 100(1.2 \text{ k}\Omega) \cong 120 \text{ k}\Omega$$

(a) Looking to the right of C_1

$$r_{\text{in(stage)}} \cong 50 \text{ k}\Omega \| 40 \text{ k}\Omega \| 120 \text{ k}\Omega = 18.6 \text{ k}\Omega$$

9-11. $v_o \cong 19.2$ mV.—With $R = 0$, then $r_{\text{in(stage)}} \cong 18.6 \text{ k}\Omega$ (see Prob. 9-9). Therefore,

$$v_{\text{in(base)}} \cong \frac{(40 \text{ mV}) r_{\text{in(stage)}}}{r_s + r_{\text{in(stage)}}} \cong \frac{(40 \text{ mV})(18.6 \text{ k}\Omega)}{20 \text{ k}\Omega + 18.6 \text{ k}\Omega} \cong 19.2 \text{ mV}$$

This circuit has a voltage gain

$$A_e \cong \frac{r_L}{r_e' + r_L} \cong \frac{1200 \text{ }\Omega}{12.5 \text{ }\Omega + 1200 \text{ }\Omega} \cong 0.99$$

and therefore its output voltage is $v_o \cong 0.99(19.2 \text{ mV}) \cong 19.2 \text{ mV}$.

CHAPTER 10   595

9-13. Maximum unclipped $v_{o(peak)} \cong 2.4$ V.—Since $I_E \cong 2$ mA (see Prob. 9-9) and $V_{CE} = V_{CC} - R_E I_E \cong 5$ V, the end points on the ac load line are

$$i_{C(sat)} = I_E + \frac{V_{CE}}{r_L} \cong 2 \text{ mA} + \frac{5 \text{ V}}{1.2 \text{ k}\Omega} = 6.16 \text{ mA}$$

and

$$v_{CE(cutoff)} = V_{CE} + r_L I_E \cong 5 \text{ V} + (1.2 \text{ k}\Omega)(2 \text{ mA}) = 5 + 2.4 = 7.4 \text{ V}$$

Thus $v_{o(peak)} \cong 7.4 - 5 = 2.4$ V.

9-15. (a) $V_E \cong 0$ V, (b) $V_E \cong 10$ V if $I_{CEO} \cong 0$.—(a) When this type of circuit is working properly, you can expect to have almost the entire source voltage V_{EE} across R_E, which in this case is 10 V. Thus, the algebraic sum of the voltage across R_E and V_{EE} is the voltage V_E at the emitter. That is, $V_E \cong -10 + 10 = 0$ V. (b) With R_B open, $I_B = 0$, assuming that there is no leakage through the coupling capacitor. If leakage $I_{CEO} = 0$, then the collector current is

$$I_E \cong I_C \cong \beta I_B = \beta(0) = 0$$

Thus, with no current through R_E, there is no voltage drop across it. The voltage at the emitter-to-ground will be the source voltage $V_{EE} = +10$ V.

9-17. $P_C \cong 24$ mW.—Using the values from Prob. 9-8,

$$P_C = V_{CE} I_C \cong (12 \text{ V})(2 \text{ mA}) = 24 \text{ mW}$$

Chapter 10

10-1. (a) $V_{AK} = 40$ V, (b) voltage across the load resistance $V_R = 0$.—Assuming that the V_{FOM} of the SCR is greater than 40 V, the SCR does not turn on because its gate lead is open.

10-3. (a) Neglecting V_{GK}, the largest possible $i_{FG} \cong 20$ mA, (b) $I_A \cong 4$ A, (c) $V_{F(on)} \cong 1$ V, (d) load voltage $V_R \cong 39$ V.—(a) Neglecting the drop across the anode-cathode junction, and thus assuming that nearly the full source E is across R_g, the largest possible forward gate current is

$$i_{FG} \cong \frac{E_{max}}{R_g} = \frac{40 \text{ V}}{2 \text{ k}\Omega} = 20 \text{ mA}$$

(b, c, d) Since $I_{GT} \ll 20$ mA, the SCR will very likely turn on. And since typically $V_{F(\text{on})} \cong 1$ V, the remaining 39 V is across the load resistance and therefore

$$I_A \cong \frac{V_L}{R_L} \cong \frac{39\text{ V}}{10\text{ }\Omega} = 3.9 \text{ A} \cong 4 \text{ A}$$

10-5. The SCR triggers when $e \cong 150$ V on positive alternations, which is at about time t_2.—Assuming that the gate-to-cathode voltage $V_{GK} \cong 0$, the full source e appears across R_g during the positive alternations. When the current through R_g reaches 30 mA, the SCR triggers. Since $R_g = 5$ kΩ, the voltage required to cause 30 mA current through it is, by Ohm's law, $e \cong R_g I_{GT} \cong (5 \text{ k}\Omega)(30 \text{ mA}) = 150$ V.

10-7. The SCR does not turn on.—With $I_{GT} = 30$ mA and $R_g = 10$ kΩ, it would take about

$$e \cong (10 \text{ k}\Omega)(30 \text{ mA}) = 300 \text{ V}$$

of applied voltage to turn on the SCR. Since $e_{(\text{peak})} \cong 170$ V, the SCR does not turn on.

10-9. $I_f \cong 833$ mA (rms).—With $R_g \cong 0$, the SCR triggers early in each positive alternation of the input voltage e. Thus, practically the entire positive alternations appear across the load; thus, the load voltage appears to be a half-wave rectified waveform. If all alternations were across the load, its rms voltage would be about

$$e_{\text{peak}}/\sqrt{2} \cong 170 \text{ V}/1.414 \cong 120 \text{ V}$$

and the load current would be about

$$I = \frac{P}{e} = \frac{200 \text{ W}}{120 \text{ V}} = 1.67 \text{ A}$$

But since only half of the alternations are across the load, the rms load current is

$$\frac{1.67 \text{ A}}{2} = 0.833 \text{ A}$$

which is the SCR current.

10-11. (a) $V_{AK} = 30$ V, $E_R = 0$, (b) $V_{AK} = 60$ V, $E_R = 0$, (c) $V_{AK} = 90$ V, $E_R = 0$, (d) $V_{AK} \cong 1$ V, $E_R \cong 119$ V, (e) $V_{AK} \cong 1$ V, $E_R \cong 149$ V.—Since the switch S is open and since E is gradually increased, the SCR turns on only when the anode-to-cathode voltage exceeds the $V_{ROM(rep)}$ value, which is 100 V in this case. (E_R is the voltage across the load resistance.)

10-13. (a) $V_{AK} = 30$ V, $E_R = 0$, (b) $V_{AK} = 60$ V, $E_R = 0$, (c) $V_{AK} = 90$ V, $E_R = 0$, (d) $V_{AK} = 120$ V and $E_R = 0$ V if the SCR is ruined, or $V_{AK} \cong 100$ V and $E_R \cong 20$ V, if the SCR is not ruined. (e) $V_{AK} = 150$ V and $E_R = 0$ V if the SCR is ruined or $V_{AK} \cong 100$ V and $E_R \cong 50$ V if the SCR is not ruined.—In cases (a) through (c), the applied voltage E does not exceed the SCR reverse breakover voltage $V_{ROM(rep)}$, so the SCR acts like an open. In cases (d) and (e), however, the source E does exceed the 100-V $V_{ROM(rep)}$. As shown in Fig. 10-6, the SCR goes into avalanche conduction. The voltage across the SCR will remain at about the avalanche voltage *if* the unit is not destroyed. Typically the load resistance is relatively small and the SCR is "burned" open and ruined.

10-15. Low.—In this case the load waveform indicates that the SCR is triggering fairly early in each positive alternation. Thus, C is charging quickly to the V_{GT} voltage of the SCR, which means that R's value is low.

10-17. Over nearly the entire 180° of the positive alternations, which is within times t_o and t_6 in Fig. 10-29.

10-19. When the capacitor left plate is negative, diode D_2 keeps the capacitor voltage off the SCR gate-cathode junction.

10-21. Fig. 10-34D.—Assuming that switch S is open, the lamp at full brilliance is caused by the SCR being triggered early in each positive alternation. This means that capacitor C charges to the neon bulb ionization voltage early in *each* alternation. That is, C discharges through R_3 and the gate-cathode junction on positive alternations, and it discharges through R_3 and diode D on the negative alternations.

10-23. Fig. 10-34E.—Diode D prevents gate current during the negative alternations.

10-25. Fig. 10-35A.—The portion of the sine wave input that is missing in waveform G, Fig. 10-35 (the voltage across the SCR) must be across the lamp.

10-27. Fig. 10-35E.—Only the positive current spikes can flow across the gate-cathode junction.

10-29. $i_{FG(peak)} \cong 42.5$ mA.—In this case, $e_C \cong V_i = 60$ V. So by Eq. 10-5,

$$i_{FG(peak)} \cong \frac{60 \text{ V} - 40 \text{ V}}{470 \text{ }\Omega} = 42.5 \text{ mA}$$

10-31. $I_f \cong 250$ mA.—In this case, if e is applied directly to the lamp,

$$I \cong \frac{60 \text{ W}}{120 \text{ V}} = 0.5 \text{ A}$$

Thus, if the SCR turns on early on each positive alternation, the lamp current, which is the SCR current, is $I_f \cong 500$ mA/2 = 250 mA ignoring changes in lamp resistance.

10-33. Either the bulb NE or R_3 is defective, and acts like an open.—The waveform of the source voltage e appearing across C indicated that there is no discharge path through the neon bulb.

10-35. (a) Fig. 10-37G, (b) Fig. 10-37H.—With minimum R the SCR turns on early in each alternation, resulting in most of the full-wave rectified voltage across the load.

10-37. Fig. 10-37H.—With minimum R, the capacitor charges to the neon bulb ionization voltage early in each alternation.

10-39. $I_f \cong 11$ A.—Since full-wave rectification takes place,

$$I_f \cong \frac{110 \text{ V}}{10 \text{ }\Omega} = 11 \text{ A}$$

10-41. $i_{FG(peak)} \cong 115$ mA.—In this case, since $e_C \cong V_1 = 78$ V, then, by Eq. 10-5,

$$i_{FG(peak)} \cong \frac{78 \text{ V} - 32 \text{ V}}{400 \text{ }\Omega} \cong 115 \text{ mA}$$

10-43. Minimum required $V_{FOM(rep)}$ and $V_{ROM(rep)}$ are 156 V.—Assuming that the source e is sinusoidal,

$$V_{FOM(rep)} = V_{ROM(rep)} \geq 110\sqrt{2} \cong 156 \text{ V}$$

Chapter 11

11-1. $V_p \cong 16.7$ V.—Since in general, by Eq. 11-1,

$$V_p = V_F + \eta V_{BB}$$

Then assuming that $V_F \cong 0.7$ V, in this case,

$$V_p \cong 0.7 \text{ V} + 0.8(20 \text{ V}) = 0.7 + 16 = 16.7 \text{ V}$$

11-3. $V_{E(p\text{-}p)} \cong 11$ V.—The emitter peak voltage is

$$V_p \cong 0.7 \text{ V} + 0.82(15 \text{ V}) = 13 \text{ V}$$

Since the valley voltage is 2 V in this case, the peak-to-peak value of the sawtooth voltage at the emitter is

$$V_{E(p\text{-}p)} = V_p - V_v \cong 13 - 2 = 11 \text{ V}$$

11-5. $f \cong 19.2$ Hz.—Since $\eta = 0.63$, the period of each sawtooth is

$$T \cong RC = (52 \text{ k}\Omega)(1 \text{ }\mu\text{F}) = 52 \text{ ms}$$

Therefore, the frequency is

$$f = \frac{1}{T} \cong \frac{1}{52 \text{ ms}} = 19.2 \text{ Hz}$$

11-7. $V_{B1B2} \cong 39.5$ V.—Since in general,

$$V_{B1B2} \cong \frac{V_{s(\text{max})} R_{B1B2}}{R_t}$$

then, in this case,

$$V_{B1B2} \cong \frac{(170 \text{ V})(10 \text{ k}\Omega)}{43 \text{ k}\Omega} = 39.5 \text{ V}$$

11-9. Maximum $R_4 \cong 112$ Ω.—Since the peak of the rectified voltage is about 170 V, and since R_3, R_{B1B2}, and R_4 are, for practical purposes, in series, R_4 has the maximum voltage drop when R_{B1B2} is minimum. Thus, we use the minimum R_{B1B2} in Eq. 11-8, and in this case,

$$R_4 < \frac{V_{GT}(R_3 + R_{B1B2})}{V_{s(\text{max})} - V_{GT}} \cong \frac{0.5(33 \text{ k}\Omega + 5 \text{ k}\Omega)}{170 \text{ V} - 0.5 \text{ V}} \cong 112 \text{ }\Omega$$

11-11. Maximum $V_{R5} \cong 305$ mV.—Before the JT and the SCR turn on, resistances R_1, R_2, R_4, R_{B1B2}, and R_5 can, for practical purposes, be assumed to be in series. The largest drop occurs across the B_1 and B_2 terminals of the UJT when R_{B1B2} is minimum. Thus, in series circuit fashion,

$$V_{R5(\text{max})} \cong \frac{V_{s(\text{peak})} R_5}{R_1 + R_2 + R_4 + R_{B1B2} + R_5}$$

$$\cong \frac{(170 \text{ V})(80 \text{ }\Omega)}{10 \text{ }\Omega + 39 \text{ k}\Omega + 390 \text{ }\Omega + 5 \text{ k}\Omega + 80 \text{ }\Omega} \cong 305 \text{ mV}$$

11-13. $I_{max} \cong 4$ A (rms).—The Triac must be capable of conducting the effective current that would flow through it if it were to act as a dead short. In this case, then,

$$I_{max} \cong \frac{40 \text{ V}}{10 \text{ }\Omega} = 4 \text{ A (rms)}$$

11-15. Fig. 11-37E.—The portions of the applied voltage that do not appear across the load must appear across the Triac.

11-17. Fig. 11-37F.—The portions of the applied voltage that do not appear across the Triac must appear across the load.

11-19. The voltage at x to ground is 40 V dc.—The neon bulb never ionizes because the applied source voltage is less than the ionization voltage.

11-21. (a) $R_1 \cong 10$ kΩ, $R_2 = 10$ kΩ, $C_1 = C_2 = 1$ μF, $R_L \cong 30$ Ω, $R_b \cong 12$ Ω, (b) $R_{L(min)} = 2.4$ Ω.—Plotting a load line from 24 V on the forward gate voltage axis as near as possible to the P_{GM} curve, we intersect the forward gate current axis at about 0.8 A. Thus

$$R_a \cong \frac{24 \text{ V}}{0.8 \text{ A}} = 30 \text{ }\Omega$$

Similarly we sketch a load line from -24 V on the reverse gate voltage axis, but in this case we are limited by the -2-A maximum reverse gate current. Therefore,

$$R_b \cong \frac{24 \text{ V}}{2 \text{ A}} \cong 12 \text{ }\Omega$$

The minimum safe load resistance is found with Ohm's law. Use the applied voltage and the maximum safe anode current:

$$R_{L(min)} = \frac{24 \text{ V}}{10 \text{ A}} = 2.4 \text{ }\Omega$$

11-23. $R_F \cong 0.5$ Ω.—As shown in Ex. 11-10, the approximate value of the feedback resistor R_F is determined by dividing the Shockley diode forward breakover voltage by the maximum load current:

$$R_F \cong \frac{4 \text{ V}}{8 \text{ A}} = 0.5 \text{ }\Omega$$

Chapter 12

12-1. See Fig. B-11.

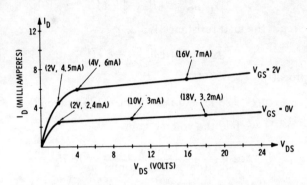

Fig. B-11. Answer to Prob. 12-1.

12-3. $V_p \cong 2$ V, $I_p \cong 2.5$ mA.—Note that in Fig. B-11, the curve for $V_{GS} = 0$ V levels off at about the point with coordinates (2 V, 2.5 mA).

12-5. $I_D \cong 5$ mA.—The characteristics of this FET are shown in Fig. B-11. We can estimate a curve representing $V_{GS} = 1$ V to be centered between the curves for $V_{GS} = 2$ V and $V_{GS} = 0$ V. Projecting vertically from 16 V on the V_{DS} axis, we intersect the estimated curve to the right of 5 mA on the I_D axis.

12-7. $R_S \cong 500$ Ω.—The curve in Fig. 12-15A shows that for $I_D = 4$ mA, the required gate-to-source bias is about -2 V. This means that the source-to-ground voltage must be about $+2$ V because the gate-to-ground voltage is about zero in this type of circuit. Thus, by Ohm's law,

$$R_s \cong \frac{2 \text{ V}}{4 \text{ mA}} = 500 \text{ }\Omega$$

12-9. $R_1/R_2 = 4$.—Since the desired V_{GS} is 3 V and the source is grounded, we need 3 V across R_2. This leaves the remaining 12 V across R_1. Since the current in R_3 is negligible because the input to the IGFET and the capacitor have extremely high dc resistance, resistors R_1 and R_2 may be assumed to be in series. Thus, the following ratio is valid:

$$R_1/R_2 = 12 \text{ V}/3 \text{ V} = 4/1 = 4$$

12-11. $r_{\text{in(stage)}} \cong 1.2$ MΩ, $A_e \cong 18.75$.—For practical purposes, the input resistance is R_G alone. That is,

$$r_{\text{in(stage)}} \cong R_G = 1.2 \text{ M}\Omega$$

In this circuit, the load resistance is

$$r_L = 10 \text{ k}\Omega \| 40 \text{ k}\Omega = 8 \text{ k}\Omega$$

Its reciprocal

$$y_L = 1/8 \text{ k}\Omega = 0.125 \text{ mmho} = 125 \text{ }\mu\text{mhos}$$

With Eq. 12-11 we find the voltage gain

$$A_e \cong \frac{3 \text{ mmhos}}{35 \text{ }\mu\text{mhos} + 125 \text{ }\mu\text{mhos}} = 18.75$$

12-13. $v_{ds} \cong 187.5$ mV.—The output v_{ds} is simply the input v_{gs} times the voltage gain A_e:

$$v_{ds} \cong 18.75(10 \text{ mV}) = 187.5 \text{ mV}$$

12-15. (a) On the dc load line $I_{D(\text{sat})} = 8$ mA and $V_{DS(\text{cutoff})} = V_{DD} = 20$ V; on the ac load line $i_{d(\text{sat})} \cong 15$ mA and $v_{ds(\text{cutoff})} \cong 15$ V, (b) The maximum unclipped output has a peak of about 4 V or a peak-to-peak of about 8 V.—The dc load line has the end points

$$I_{D(\text{sat})} = \frac{V_{DD}}{R_D + R_S} = \frac{20 \text{ V}}{1.2 \text{ k}\Omega + 1.3 \text{ k}\Omega} = 8 \text{ mA}$$

and

$$V_{DS(\text{cutoff})} = V_{DD} = 20 \text{ V}$$

The dc load line, found above, intersects the curve representing $V_{GS} = -5$ V at a point where approximately $I_{DQ} = 4$ mA and $V_{DQ} = 11$ V. We are interested in the -5 V curve because the problem specifies that the source is 5 V to ground, which means that the gate is -5 V with respect to the source in this type of circuit. Now the end points of the ac load line are found by Eqs. 12-15 and 12-14 as follows:

$$i_{d(\text{sat})} = I_{DQ} + \frac{V_{DQ}}{r_L} \cong 4 \text{ mA} + \frac{11 \text{ V}}{1 \text{ mA}} = 15 \text{ mA}$$

$$v_{ds(\text{cutoff})} = V_{DQ} + r_L I_{DQ} \cong 11 \text{ V} + (1 \text{ k}\Omega)(4 \text{ mA}) = 15 \text{ V}$$

Chapter 12

Since at the operating point $V_{DQ} \cong 11$ V and since $v_{ds(\text{cutoff})} = 15$ V, the drain-to-source voltage can swing 4 V from the 11 V operating point before exceeding cutoff.

12-17. Depletion mode (mode A).—Since it takes a negative gate-to-source voltage to cut off the FET and since all V_{GS} curves represent negative voltages, this FET is a depletion mode type. Of course, the circuit in Fig. 12-36 shows a JFET. Most JFET's are depletion mode types.

12-19. See Fig. B-13; $v_{ds} \cong 5$ V (peak to peak).

12-21. See Fig. B-13; $R_D \cong 630$ Ω, $R_s \cong 570$ Ω.—Note that the dc load line, drawn through 20 mA on the I_D axis and through 24 V on the V_{DS} axis has an operating point at the $V_{GS} \cong -4$ V curve. A gate-to-source bias of -4 V is chosen because it lies about halfway between the -1 V and the -7 V curves. This gives us plenty of room to swing the drain-to-source voltage while avoiding cutoff and saturation. This load line represents a total dc resistance of

$$R_D + R_S = \frac{V_{DD}}{I_{D(\text{sat})}} = \frac{24 \text{ V}}{20 \text{ mA}} = 1.2 \text{ k}\Omega$$

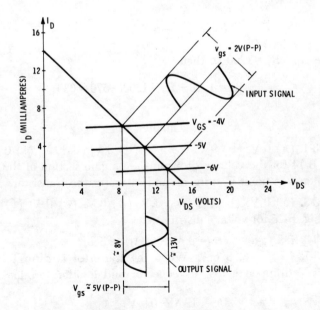

Fig. B-12. Answer to Prob. 12-9.

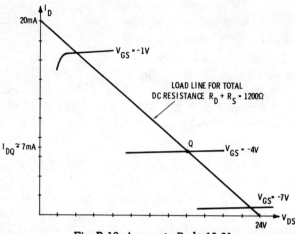

Fig. B-13. Answer to Prob. 12-21.

To the left of the operating point Q, the drain current $I_{DQ} \cong 7\text{mA}$. This means that in order to establish a $V_{GS} \cong -4$ V, the source voltage must be 4 V to ground. The source resistance that will provide this bias is found with Ohm's law:

$$R_s \cong \frac{4 \text{ V}}{7 \text{ mA}} \cong 570 \text{ }\Omega$$

Since $R_D + R_s = 1200$ Ω, then

$$R_D = 1200 - R_s = 1200 - 570 = 630$$

Chapter 13

13-1. (a) 0 V, (b) 9 V, (c) 9 V, (d) 9 V, (e) 9 V, (f) 0 V.—See Fig. B-14 for the voltage drops in the OR gate section of the circuit. A voltage at point C turns on and saturates the transistor.

13-3. (a) 0 V, (b) 0 V, (c) 0 V, (d) 0 V, (e) 0 V, (f) 12 V.— See Fig. B-15 for voltage distribution.

13-5. $I_{C(\text{sat})} = V_{CC}/R_2 = 9 \text{ V}/3.6 \text{ k}\Omega = 2.5 \text{ mA}$

13-7. $I_{C(\text{sat})} = 3.08$ mA.—When the transistor is saturated, there is 0.9 V dropped across its collector and emitter terminals. Thus

$$I_{C(\text{sat})} = \frac{12 \text{ V} - 0.9 \text{ V}}{3.6 \text{ k}\Omega} = 3.08 \text{ mA}$$

CHAPTER 13

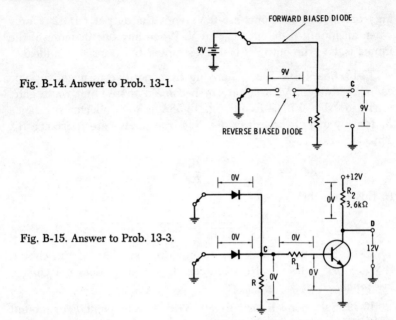

Fig. B-14. Answer to Prob. 13-1.

Fig. B-15. Answer to Prob. 13-3.

13-9. $R_a = R_b < 250$ kΩ.—Since, by Eq. 13-7, an inverter requires a ratio $R_B/R_C < \beta$, then $R_B < \beta R_C$, and in this case R_C is R_1 and R_B is R_a or R_b. Therefore, $R_B < 50(5 \text{ k}\Omega) = 250$ kΩ.

Table B-1. Truth Table of Three-Input NOR Gate

A	B	C	F
0	0	0	1
0	0	1	0
0	1	0	0
0	1	1	0
1	0	0	0
1	0	1	0
1	1	0	0
1	1	1	0

Table B-2. Truth Table of Three-Input NAND Gate

A	B	C	F
1	1	1	0
1	1	0	1
1	0	1	1
1	0	0	1
0	1	1	1
0	1	0	1
0	0	1	1
0	0	0	1

13-11. The circuit which is in Fig. 13-58A is a three-input NOR gate with positive logic. All of the transistors are off (nonconducting) only when all inputs are 0 V. Thus the output f is 12 V only when all inputs, a, b, and c are 0 V. When any one or more of the inputs is 4 V, the output f is 0 V, assuming the transistor is ideal.

13-13. See Table B-2.—Changing from positive to negative logic causes a three-input NOR gate to become a three-input NAND gate.

13-15. $\beta > 0.704$.—R_2 in Fig. 13-58A is the collector resistance R_C for any one of the transistors. R_1 is the series base resistance R_B. Thus, by Eq. 13-7,

$$\frac{R_B}{R_C} = \frac{R_1}{R_2} < \beta$$

In this case, then,

$$\frac{R_1}{R_2} = \frac{450\ \Omega}{640\ \Omega} \cong 0.704 < \beta$$

13-17. This will reduce the temperature stability of the circuit. It will work well only if the ambient temperature does not change too much.

13-19. Selection (b) is correct.—With 3 V at point F to ground and a 6-V dc source voltage, the remaining 3 V must be across R_2 in this circuit. This means that the operating point is centered on the dc load line, If the load resistance to the right of C_2 is large compared to R_2, the ac load line has about the same slope as the dc load line, which means that an operating point that is centered on the dc line is centered on the ac load line as well.

13-21. About 40,000.—Two stages in cascade each with a voltage gain of 200, have a total voltage gain of $200 \times 200 = 40{,}000$.

13-23. $f \cong 35.7$ Hz.—By Eq. 13-6,

$$f \cong \frac{1}{1.4RC} = \frac{1}{1.4 \times 1 \times 10^5 \times 2 \times 10^{-7}} \cong 35.7\ \text{Hz}$$

13-25. Six.—There are six transistors on this IC, each with its own load resistance R_2, input terminal, and output terminal.

13-27. Four two-input NOR gates.

13-29. Two.—Either the inputs 1, 6, 9, and 12 have to be grounded, thus disengaging their respective transistors, or inputs 2, 7, 10, and 13 have to be grounded, disengaging their respective transistors. This leaves only four active transistors that can be used in this IC. Since each astable circuit requires two transistors, a total of two astable multivibrators can be made with this IC. Actually, there are a number of ways this IC can be connected to be an astable circuit. Either input terminals 1 or 2, or 6 or 7, or 9 or 10, etc., can be grounded, giving a total of 16 ways in which this IC can be wired as an astable multivibrator.

13-31. See Fig. B-16.—The IC is wired exactly as in Fig. 13-65 except that in this case a coefficient potentiometer is used across the input terminals (point C and ground). This potentiometer is adjusted so that the voltage at C to ground is 0.75 of the voltage at point A to ground.

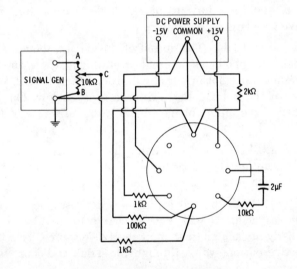

Fig. B-16. Answer to Prob. 13-31.

13-33. +55 mV.—The circuit gain is 10 as found in the previous problem. In this case, however, one of the input voltages is zero. The total input now is $0 - 2$ mV $- 3.5$ mV $= -5.5$ mV. Phase inversion and a gain of 10 gives us an output of 55 mV.

13-35. ECL.—This is an emitter-coupled-logic IC because the outputs are taken off the emitters of two transistors (far right).

13-37. MOS.—The presence of FET's indicates that this is a metal-oxide-silicon IC.

13-39. DTL.—The use of diodes *and* transistors in the IC indicate that it is in the diode-transistor-logic family.

13-41. ECL.—Like the circuit in Fig. 13-69, the outputs are taken off the emitters.

CHAPTER 14

14-1. (c).—Since the tunnel diode has negative-resistance characteristics within a very limited range of forward bias voltages, as shown in Fig. 14-1A, the most logical cause of failure of the circuit to oscillate is an improper bias adjustment with the potentiometer.

14-3. (a).—With *forward bias* voltages, the varactor conducts current as would any forward biased diode. A capacitor, of course, must have very large dc resistance.

14-5. (a) and (d).—The output of this circuit, though ac, is not sinusoidal. The photocells alternate from high to low and back to high resistance but not in phase with each other; that is, when the resistance of P_1 is high, the resistance of P_2 is low. On the next alternation of the input voltage (120 V), the resistance of P_1 is low and of P_2 is high.

14-7. Fig. 14-23C.—With no light, the resistance of the photocell P is high, causing the capacitor C to charge quickly to the Diac breakover voltage early in each alternation.

14-9. (c).—With the cell in darkness, its resistance is relatively large and capacitor C will tend to charge to the Diac breakover voltage early in each alternation of the 120-V source. This increases the effective current value through the Triac and load, which in turn increases the load power.

14-11. (b).—A decrease in light intensity decreases the collector current, which in turn will increase the output voltage V_o in this circuit.

14-13. (b).—Referring back to Chap. 8 and Eq. 8-35, which is for a transistor circuit whose base is biased with a voltage divider, we can show that

$$I_C \cong \frac{R_2}{R_1+R_2}\left(\frac{V_{CC}}{R_E}\right) = \frac{2 \text{ k}\Omega}{20 \text{ k}\Omega}\left(\frac{20 \text{ V}}{0.2 \text{ k}\Omega}\right) = 10 \text{ mA}$$

14-15. (d).—Both devices are in the thyristor family, which means that they have essentially two states: on or off. The LAS and the LASCR differ mainly in that the LASCR can be turned on with a current pulse in the gate lead, whereas the LAS cannot because it has no gate lead.

14-17. Fig. 14-23C, assuming that the light is intense enough.—With a weaker light, waveform D is possible.

Index

A

Absolute value, 169
Ac
 beta, 185
 load lines, common-base circuit, 165-171
 loads and load lines, common-collector amplifier, 280-289
Acceptor atom, 18
Admittance, output, FET, 412-413
Alloy method, pn junction, 27-28
Alpha, dc, 125
AND gate, diode, 80-84
Anode characteristics, SCR, 296-299
Approximated diode characteristics, 34-39
Arsenic atom, 17
Assemblies, rectifier, 75-79
Astable multivibrator, IC, 473-481
Atom
 acceptor, 18
 arsenic, 17
 core, 13
 donor, 17
 gallium, 18
 impurity, 16
 neutral, 12
 parent, 13
 silicon and germanium, 11-13
Atomic shells, 11-12
Avalanche conduction, 296
Average gate power, 303

B

Back-to-back zener diodes, 112-113
Band, valence, 12-13
Base
 bias
  voltage divider and emitter feedback, 205-211
  with emitter feedback, see Emitter feedback
 -emitter junction, forward bias on, 122-123
 lead, 120
Beam lead isolation, 454

Beta
 ac, 185
 dc, 125
Biasing methods, common-collector amplifier, 275-278
Bias voltage, diode, 47
Bidirectional
 diode thyristor, see Diac
 triode thyristor, see Triac
Bipolar semiconductor, 119
Bond, covalent, 13
Breakdown voltage, 26
Breakover voltage, Triac, 362
Bridge rectifier
 average dc voltage, 535-536
 diode, 66-71

C

Capacitor
 bypass, selecting, 548-550
 coupling, selecting, 548-550
Carrier
 charge, 16
 majority, 17
 minority, 17
Cascade connection, 182
Charge carrier, 16
Chip, semiconductor, 28
Clock, 473
Coefficient potentiometer, 489
Collector
 characteristics
  common-base circuit, 134-139
  common-emitter circuit, 211-215
 feedback
  for temperature stability, 196-205
  gain in common-emitter amplifier, 242-246
 lead, 120
 saturation current, 132
 -to-emitter leakage, 184
Common-base circuit
 ac load lines, 165-171
 collector
  characteristics, 134-139
  saturation current, 132

INDEX

Common-base circuit (Cont'd)
  current
    gain, 153
    transfer ratios, 144
  dynamic resistance, emitter-base 148-150
  high-resistance source vs gain, 153-157
  load lines, 136-139
  operating point
    affected by leakage, 174-176
    optimum, 171-174
  power gain, 153
  signal and load lines, 157-165
  V-I characteristics, 148-150
  voltage gain, 151
Common-collector amplifier
  ac loads and load lines, 280-289
  determining gains of, 278-280
  methods of biasing, 275-278
Common-emitter circuit
  ac load lines, 251-262
  collector
    characteristics, 211-215
    current vs base current, 183-190
    -emitter leakage, 184
    feedback, 196-205
  dc load lines, 215-222
  emitter feedback, 190-196, 200-205, 205-211
  gain
    current, 233
    power, 233
    using collector feedback, 242-246
    using emitter feedback, 236-242
    voltage, 232
    with other amplifiers, 246-251
  optimum operating point, 262-267
  rule of thumb design, 267-268
  V-I characteristics, 227-231
Complimentary pair, 294
Components
  diffused, IC, 450-453
  FET amplifiers, 431-434
  isolation, IC, 453-458
Conduction in silicon and germanium, 15-16
Conductor, unilateral, 21

Constant-current source, 143
Core, atom, 13
Covalent bond, 13
Crystals, silicon and germanium, 13-15
Current
  gain
    common-base circuit, 153
    common-collector circuit, 280
    common-emitter circuit, 233
    definition, 153
  holding, SCR, 294
  latching, SCR, 296
  leakage, 24
  transfer ratio, 144
  transistor, relations, 124-127
  valley, 348
Cutoff, definition, 132-134

D

Dc
  alpha, 125
  and ac, superposition of, 47-51
  beta, 125
  gate trigger
    current, 299
    voltage, 299
Depletion
  -and-enchancement mode, 407-409
  mode, 407
  region, 395-396
Derating
  curves, 98
  factor, zener diode, 98-99
Diac, 370-374
Dielectric isolation, 457-458
Diffused components, IC, 450-453
Diffusion method, pn junction, 28
Digital
  IC, 494
  transistor logic, 497
Diode
  AND gate, 80-84
  bias voltage, 47
  characteristics
    approximated, 34-39
    load lines on, 39
    operating point, 39
    simplified, 33-34

Diode (Cont'd)
  forward biased, 22-23
  ideal, 31-33
  operating point, 39
  OR gate, 84-86
  reverse
    biased, 23-25
    bias resistance, 39-44
  transistor logic, 496-497
  Shockley, 376-386
  tunnel, 515-519
  $V\text{-}I$ characteristics, 285-26
  zener
    power and derating, 97-99
    practical considerations, 107-110
    rectifier, 113-114
    $V\text{-}I$ characteristics, 94-97
    voltage
      limiter, 110-113
      regulator, 99-107
Discrete component circuits, 447
Donor atom, 17
Doping silicon and germanium, 16-19
Doubler, voltage, 71-75
Drain, FET, 393
DTL
  see Digital transistor logic
  see Diode transistor logic
Dual two-input gate IC, 466-473
Dynamic
  emitter resistance, 140, 148-150, 537
  resistance, base-emitter junction, 227-231

E

ECL, see Emitter coupled logic
Electron
  free, 13
  -hole pair, 14
  thermally excited, 13
  valence, 13
Emitter
  coupled logic, 494
  feedback
    and voltage-divider base bias, 205-211
    common-emitter circuit, 190-196
    gain in common-emitter circuit, 236-242

Emitter (Cont'd)
  with two supplies, 200-205
  follower, see Common-collector circuit
  lead, 120
  -to-base 2 leakage, 348
End point of ac load line, 537-539
Enhancement mode, FET, 409
Epitaxial
  method, pn junction, 28-29
  process isolation, 456-457

F

Fan-out, 494
Filter components, 56
Forward
  biased diode, 22-23
  bias on base-emitter junction, 122-123
  breakover voltage, SCR, 296
  transadmittance, FET, 412
Four-layer diode, see Shockley diode
Free
  electron, 13
  -running multivibrator, see Astable multivibrator
Frequency
  modulation, 522
  multivibrator, 479
  UJT relaxation oscillator, 352
Full-wave rectifier
  average dc voltage, 535-536
  diode, 59-66

G

Gain
  common-collector amplifier, 278-280
  common-emitter amplifier
    general, 232-236
    using collector feedback, 242-246
    using emitter feedback, 236-242
    with other amplifiers, 246-251
  open-loop voltage, 483
Gate
  controlled switch, 374-386
  FET, 393
  lead, SCR, 293
  -to-source cutoff voltage, JFET, 396
  trigger

INDEX

Gate (Cont'd)
    characteristics, SCR, 299-309
    current, dc, 299
    voltage, dc, 299
GCS, *see* Gate controlled switch
Grounded-base connection, *see* Common-base circuit
Grown method, pn junction, 26-27

H

Half-wave rectifier
    average dc voltage, 535-536
    diode, 51-59
High-resistance source, common-base circuit, 153-157
Holding current
    SCR, 294
    zener diode, 103
Hole
    definition, 14
    flow, 15-16
Hot spots, 305-306

I

IC, *see* Integrated circuit
Ideal
    collector characteristics, 213
    diode, 31-33
IGFET, *see* Insulated gate field-effect transistor
Impurity atoms, 16
Input offset voltage, 483
Insulated gate field-effect transistor
    components, selecting, 431-434
    fundamentals of, 402-407
    large signals and load lines, 424-430
    methods of biasing, 414-417
    modes of operation
        mode A, 407
        mode B, 407-409
        mode C, 409
    parameters of, 412-414
    practical considerations, 438-441
    small-signal voltage gain, 417-424
    temperature changes, 433-438
Integrated circuit
    astable multivibrator, 473-481
    component isolation
        beam lead, 454

Integrated Circuit (Cont'd)
    dielectric, 457-458
    epitaxial process, 456-457
    resistive, 454
    reverse biased junction, 454-457
    construction, 448-449
    definition, 447
    diffused components, 450-453
    dual two-input gate, 466-473
    logic used in
        ECL, 494
        DTL, 496-497
        MOS, 499-500
        RTL, 497-499
        TTL, 494, 496
    NOR gate
        diode, 458-464
        transistor, 464-466
    operational amplifier
        basic description, 482-487
        considerations and applications, 487-488
        multiplier, 489-491
        summing amplifier, 491-493
Intrinsic stand-off ratio, 345
Inverting input, 483
Isolation, IC component
    beam lead, 454
    dielectric, 457-458
    epitaxial process, 456-457
    resistive, 454
    reverse biased junction, 454-457

J

JFET, *see* Junction Field-Effect Transistor
Junction
    transistor, 119-127
    npn and pnp, 119-122
    pn, 21-22
Junction field-effect transistor
    components, selecting, 431-434
    concepts of, 394-402
    depletion mode operation, 407
    large signals and load lines, 424-430
    methods of biasing, 414-417
    parameters of, 412-414
    practical consideration, 438-441
    small-signal voltage gain, 417-424
    temperature changes, 433-438

L

Large signals and load lines, FET, 424-430
LAS, *see* Light-activated switch
LASCR, *see* Light-activated SCR
Laser diode, *see* Light-emitting diode
Latching current, SCR, 296
Leakage current, 24
LED, *see* Light-emitting diode
Light
  -activated
    SCR, 527-529
    switch, 527-529
  -emitting diode, 526
Lightly doped base, 120
Linear
  device, 482
  equation
    form of, 537-538
    slope of, 538
Loaded down, definition, 153-154
Load lines
  ac
    common
      -base circuit, 165-171
      -collector circuit, 280-289
      -emitter circuit, 251-262
  common-base circuit, 136-139
  diode characteristics, 39
  end points of, 537-539
Logic
  circuits, 79-86
  IC, 494
  positive, 458

M

Majority carrier, 17
Melt, semiconductor, 26
Metallization, 56
Metal oxide semiconductor
  field-effect transistor, *see* Insulated gate field-effect transistor
  logic, 499-500
Microcircuit, 447
Microelectronics, 447
Minority carrier, 17
Modes of operation, FET
  mode A, 407

Modes of operation, FET (Cont'd)
  mode B, 407-409
  mode C, 409
Monolithic IC, 453, 472
MOSFET, *see* Metal oxide semiconductor field-effect transistor
MOS logic, *see* Metal oxide semiconductor logic
Multiplier, operational amplifier, 489-491
Multivibrator, astable, IC, 473-481

N

N-channel, JFET, 394
Negative resistance, 515
Neutral atom, 12
Nor gate
  diode IC, 458-464
  transistor RTL, 464-466
Npn junctions, 119-122
N-type semiconductor, 17

O

Open
  -circuit voltage, 154
  -loop voltage gain, 483
Operating point
  common-base circuit
    affected by leakage, 174-176
    definition, 138
    optimum, 171-174
  common-emitter circuit, optimum, 262-267
  diode, 39
Operational amplifier, IC
  basic description, 482-487
  considerations and applications, 487-488
  multiplier, 489-491
  summing amplifier, 491-493
Or gate, diode, 84-86
Oscillation, period of, 351
Output admittance, FET, 412-413

P

Pair
  complementary, 294
  electron-hole, 14

Index

Parameters
 FET, 412-414
 transistor, 125
Parent atom, 13
P-channel JFET, 394
Peak
 inverse voltage, 26
 voltage, UJT, 346
Pentavalent material, 16
Period of oscillation, 351
Photocell, 522-524
Photoconductive cell, see Photocell
Photodiode, 524-525
Phototransistor, 524-525
Photovoltaic cell, 524
Pinch-off voltage, JFET, 398
Pn junction
 manufacturing methods
  alloy method, 27-28
  diffusion method, 28
  epitaxial method, 28-29
  grown method, 26-27
  rate grown method, 27
 nature of, 21-22
Pnp junctions, 119-122
Positive logic, 458
Power gain
 common
  -base circuit, 153
  -collector circuit, 280
  -emitter circuit, 233
 definition, 153
P-type semiconductor, 19

Q
Quiescent point, 138

R
Rate grown method, pn junction, 27
Recombination, 16
Rectifier
 assemblies, 75-79
 bridge, diode, 66-71
 full-wave, 59-66
 half-wave, 51-59
 zener diode, 113-114
Relaxation oscillator, 349
Resistance, diode reverse bias, 39-44
Resistive isolation, 454
Resistor transistor logic, 497-499

Reverse
 biased
  diode, 23-25
  junction isolation, 454-457
 bias resistance, diode, 39-44
 blocking triode thyristor, see Silicon controlled rectifier
Ripple, 54
RTL, see Resistor transistor logic
Rule of thumb design, common-emitter amplifier, 267-268

S
Saturation, definition, 131-134
SCR, see Silicon controlled rectifier
Seed, crystal, 27
Semiconductor
 bipolar, 119
 definition, 15
 n-type, 17
 p-type, 19
Shells, atomic, 11-12
Shockley diode, 376-386
Signal and load lines, common-base circuit, 157-165
Silicon
 and germanium
  atoms, 11-13
  conduction in, 15-16
  crystals, 13-15
  doping, 16-19
 capacitor, see Varactor
 controlled rectifier
  anode characteristics, 296-299
  applications, 309-330
  gate trigger characteristics, 299-309
  general characteristics, 292-296
  letter symbols, 330-332
Simplified diode characteristics, 33-34
Slope, of linear equation, 538
Small-signal voltage gain, FET, 417-424
Solar cell, see Photovoltaic cell
Source, FET, 393
Stability
 collector feedback for, 196-205
 emitter feedback
  base biased, 190-196

Stability (Cont'd)
    voltage-divider bias, 205-211
    with two supplies, for
        temperature, 200-205
Steady-state point, *see* Quiescent point
Substrate
    epitaxial method, 28
    IC, 447
    JFET, 398
Summing amplifier, operational
    amplifier, 491-493
Superposition
    of dc and ac, 47-51
    of signals, common-base amplifier,
        139-147
Switching time, UJT, 352

T

Table, truth, 458-459
Temperature changes, FET, 433-438
Thermally excited electron, 13
Three-layer silicon bidirectional
    trigger diode, 373
Thyristor, 292
Transadmittance, forward, FET, 412
Transconductance, FET, 412
Transistor
    junction, 119-127
    transistor logic, **494-496**
Triac, 361-370
Trivalent material, 16
Truth table, 458-459
TTL, *see* Transistor transistor logic
Tunnel diode, 515-519
Turnoff thyristor, *see* Gate controlled
    switch
Turns ratio, 60

U

UJT, *see* Unijunction transistor
Unijunction transistor
    applications, 353-361
    basic description, 344-353
Unilateral conductor, 21

V

Valence
    band, 12-13
    electrons, 13
Valley
    current, 348

Valley (Cont'd)
    voltage, 348
Varactor, 519-522
V-I characteristics
    common
        -base circuit, 148-150
        -emitter circuit, 227-231
    diode, 25-26
Voltage
    breakdown, 26
    breakover, triac, 362
    doubler, 71-75
    gain
        common
            -base circuit, 151
            -collector circuit, 279-280
            -emitter circuit, 232
        definition, 150
        small-signal, FET, **417-424**
    input offset, 483
    limiter, zener diode, 110-113
    open-circuit, 154
    peak
        inverse, 26, 346
        UJT, 346
    valley, 348
    -variable capacitor, *see* Varactor
    zener breakdown, 26

W

Wafer
    IC, 448
    semiconductor, 28

Y

Yield, 450

Z

Zener
    breakdown voltage, 26
    diode
        holding current, 103
        power and derating, 97-99
        practical considerations, **107-110**
        rectifier, 113-114
        V-I characteristics, 94-97
        voltage
            limiter, 110-113
            regulator, 99-107
    effect, 26
    region, 94
    voltage, 26, 94